T0254214

Applied Engineering Statistics

Applied Engineering Statistics

Second Edition

R. Russell Rhinehart
Robert M. Bethea

CRC Press is an imprint of the
Taylor & Francis Group, an **informa** business

Second edition published 2022
by CRC Press
6000 Broken Sound Parkway NW, Suite 300, Boca Raton, FL 33487-2742

and by CRC Press
2 Park Square, Milton Park, Abingdon, Oxon, OX14 4RN

First edition published by CRC Press 1991

CRC Press is an imprint of Taylor & Francis Group, LLC

Library of Congress Cataloging-in-Publication Data

Names: Rhinehart, R. Russell, 1946- author. | Bethea, Robert M., author.
Title: Applied engineering statistics / R. Russell Rhinhart, Robert M. Bethea.
Description: Second edition. | Boca Raton : CRC Press, 2022. |
Revision of: Applied engineering statistics / Robert M. Bethea, R. Russell Rhinehart. 1991. |
Includes bibliographical references and index.
Identifiers: LCCN 2021023519 (print) | LCCN 2021023520 (ebook) | ISBN 9781032119489 (hardback) |
ISBN 9781032119496 (paperback) | ISBN 9781003222330 (ebook)
Subjects: LCSH: Engineering--Statistical methods.
Classification: LCC TA340 .B48 2022 (print) | LCC TA340 (ebook) | DDC 519.502/462--dc23
LC record available at https://lccn.loc.gov/2021023519LC
ebook record available at https://lccn.loc.gov/2021023520

ISBN: 9781032119489 (hbk)
ISBN: 9781032119496 (pbk)
ISBN: 9781003222330 (ebk)

DOI: 10.1201/9781003222330

Typeset in Palatino
by Deanta Global Publishing Services, Chennai, India

Disclaimer

The reader assumes all responsibility for use of the material in this book. Applications are context-dependent, and yours may have aspects not addressed here. The user must choose appropriate tests, methods, etc. Especially, if critical, be sure that your choices are correct.

Contents

Section 3 Applications of Probability and Statistical Fundamentals

Section 4 Case Studies

Appendix Critical Value Tables

Nomenclature

Latin Letters

Capital Latin letters represent populations.

Lowercase Latin letters represent numerical values such as those calculated from a sample taken from the populations.

Lowercase Latin letters $i, j, k, l, m,$ and n represent counting integers.

Lowercase Latin letters $x, y,$ and z typically represent influences to a process or the independent variables in a model. y also is typical of the dependent variable in a model.

Lowercase Latin letters $a, b, c, d, \ldots$ represent model coefficients.

B represents a Belief, or a Condition.

s represents a slope or a sample standard deviation.

Greek Letters

Lowercase Greek letters are used for population parameters (coefficients).

μ represents the population mean.

σ^2 represents the population variance.

σ represents the population standard deviation.

υ represents the degrees of freedom in an analysis.

α represents the probability of a Type-I error, or rejecting a hypothesis that is true, and is termed the level of significance.

β represents the probability of a Type-II error, or accepting a hypothesis that is false, and is termed the power of a test.

$f(x_i) = pdf(x_i)$ is the point probability distribution function of a discretized variable x. It is the expected frequency that a particular value, x_i will occur. It is dimensionless. Note the subscript on x_i, indicating a discrete number, integer, or category.

$F(x_i) = CDF(x_i)$ is the cumulative distribution function of a discretized variable x. It is the probability that a particular value or lower value of x_i will happen. It is dimensionless and bounded between 0 and 1. Note the subscript on x_i, indicating a discrete number, integer, or category.

$f(x) = pdf(x)$ is the probability distribution function of a continuum-valued variable x. It is the expected rate that the probability is increasing a particular value, x. It has dimensions of the reciprocal of the variable x.

$F(x) = CDF(x)$ is the cumulative distribution function of a continuum-valued variable x. It is the probability that a particular value or lower value of x will happen. It is dimensionless and bounded between 0 and 1.

P(Event) or $P(E)$ represents the probability of a particular event.

p is a particular value of P(Event).

N, n, x, or x_i represent the number of items, a count, in a class (group, collection, etc.).

Accents

$\bar{}$, a horizontal overbar, represents the average if on a letter, or "not" if over an event.

$\_$, an underscore, represents a vector.

$\tilde{}$, the squiggle over accent, represents a modeled value.

$\hat{}$, the carat over accent, represents an estimated value.

Subscript

$_i$, the subscript of an integer, representing the ith category or value.

x_i, usually refers to the count in the ith category, but often it also refers to the data value when sorted w.r.t. some other variable or its own value.

Acronyms

w.r.t. means "with respect to".

SS = steady-state.

TS = transient state.

DoE = design of experiments.

DoF = degrees of freedom.

Preface to Second Edition

Although measured data and related calculations possess some level of uncertainty, they are the basis of engineering/business decisions. Engineers must account for that uncertainty for their designs to be both safe and effective. Applied scientists must include uncertainty in experimental design and data analysis for valid processing, and in reporting for communicating uncertainty. Decision-makers must include uncertainty in taking appropriate action. Statistics offers us the tools to do so.

The book is about the practicable methods of statistical applications for engineers, scientists, and business folks. Contrasting a mathematical and abstract orientation of many statistics texts, which express the science/math values of researchers, this book will focus on the application and the interpretation of outcomes (as described with concrete examples). However, the book also presents the fundamental mathematical concepts, and provides some supporting derivations, to show the grounding of the methods.

The science underlying statistics is important, and this book seeks to reveal that. But of greater importance to applications are the choices an investigator makes in defining a hypothesis relevant to a supposition, selecting appropriate confidence levels, selecting a test procedure, and in deciding action from the analysis that is appropriate to the context. These aspects, usually missing in classroom texts, are incorporated here.

The authors each came out of significant engineering practice experience, where we used statistical methods to support legitimacy in decision making. After changing to academic careers, we found ourselves co-teaching the unit operations laboratory course, which required students to use similar statistical techniques as part of their career preparation. We co-authored the first edition of Applied Engineering Statistics, to develop a reference text that would be of utility for the students – in both the course and their engineering careers. We continue to sense the need for the explanation of statistical concepts and application methods for the practitioner, and like the original organization of the book – explain basic concepts, explain commonly applied statistical methods, use examples from industrial practice, and provide case study chapters on advanced applications.

A few things have changed since the original publication: 1) The custom has shifted from critical values and the accept/reject dichotomy to the use of p-values to indicate a degree of confidence. 2) The Big Data and Machine Learning era has introduced a few new techniques on finding associations within data. 3) Computational accessibility has improved the utility of non-linear regression and stochastic approaches for propagating uncertainty. 4) New examples and exercises will take the learner out of the "You are finished after doing a simple word problem" mindset to a more complete analysis of issues and auxiliary aspects of a problem. 5) Statistical tools are widely available and convenient, and this book will reveal those in Excel. 6) Rhinehart has a website that offers Excel/VBA programs to apply some of the procedures – www.r3eda.com.

The book is written for two audiences. One is students in the upper level of an undergraduate program or graduate program in either engineering, science, or business. The other is those in professional life. Our intent is that the text would continue to be useful in professional life after graduation and be appropriate as a self-learning tool for those in their professional practice.

The book is presented in four sections. The first provides the fundamental math/science concepts related to probability, distributions, their characteristics, and hypothesis testing.

The second section provides guidance related to choices that a person makes in statistical analysis. The third section describes many classical statistical applications (such as analysis of variance, regression, design of experiments, model validation, statistical process control, and reliability). Many worked examples, throughout, reveal how to perform the procedures and interpret the outcomes. The fourth section contains case studies which allow the reader to explore a variety of techniques on examples, contrasting the one-aspect application of examples and typical end-of-chapter exercises.

An instructor (college or continuing education) will find that this book serves as the basis for both elementary and intermediate courses.

Although statistical computations can be performed by hand, they can also be programmed on a computer. With the availability of computers and statistical packages at every level (handheld calculator, PC, mini, mainframe), there is something convenient for your specific needs. We detail the computations in most of the examples so that you can easily follow the procedure by hand, but we also demonstrate the use of Excel Spreadsheet and Add-In functions. Many other statistical libraries are available. Some of the applications are supported with open-code software on Rhinehart's website, www.r3eda.com.

Many end-of-chapter exercises are open-ended, asking the reader to make appropriate decisions, not just to apply a recipe to get a numerical value.

We are grateful to Marcel Dekker, Inc. for allowing us to use one example problem and many homework exercises from those included in *Statistical Methods for Engineers and Scientists, Second Edition, Revised and Expanded.* We are also grateful to the American Cyanamid Company, the Institute of Mathematical Statistics, and the American Statistical Association for allowing us to reprint several of the statistical tables as acknowledged in the Appendix. We are also grateful to the American Society for Testing and Materials for data in Tables 21.3 and 21.4.

As well, we are grateful for all those who have provided essential experiences in fundamentals and practice needed for us to be able to compose this book.

Most importantly, we are most appreciative of the support of our families who enabled us to get to this place.

This edition preserves the contributions of both authors to the first edition. However, changes to create this second edition are substantially the work of Rhinehart, who remains very appreciative of the collaboration with Bethea.

R. Russell Rhinehart

Robert M. Bethea

Section 1

Fundamentals of Probability and Statistics

1

Introduction

1.1 Introduction

In this book, the word *statistics* is used in two ways. First, it refers to the techniques involved in collecting, analyzing, and drawing conclusions from data – a procedure, or a recipe. The second, more frequently inferred meaning, is that of an estimated value, a number, calculated from either the data or a proposed theory, that is used for comparative purposes in testing a hypothesis (guess, supposition, etc.) about a parameter of a population – a numerical value. The topics presented in this book have been selected from our experience (and others') to provide you with a set of procedures which are relevant to application work (such as data analysis in engineering, science, and business). Although fundamental concepts are explained and some equations are derived, the focus of this book is on the how-to of statistical applications.

There is a tension between perfection and sufficiency. *Perfection* seeks the truth, which follows the mathematical science viewpoint. Although perfection provides grounding in statistical analysis methods, it is usually a mathematical analysis that is predicated on many idealizations, making it imperfect. By contrast, *sufficiency* seeks utility and functional adequacy, a balance of expediency which is also grounded in mathematical fundamentals. Sufficiency is not sloppiness or inaccuracy. It is appropriate liberty with the idealization, grounded in an understanding of the limitations of the idealization and uncertainty in the "givens". Both perfection and sufficiency are important, and perspectives of both are presented in this text. In this "Applied Engineering" text the balance tends toward sufficiency, rather than unrealistic perfection.

1.2 Deterministic and Stochastic

The term *deterministic* means that there is no uncertainty, there is perfect certainty about a value. Here are some simple examples: What is 3 times 4? Given that the side of a cube is 2.1 cm, what is the surface area? What angle (rounded to three digits) has a tangent value of 0.75? These were very simple calculations, but it is the same with something more complicated, such as: Given a particular heat exchanger and fluid flow rates and associated properties, use the equations in your heat transfer book to calculate the exit temperatures of the fluids. Regardless of the time of day, or location, or the computer type being used, every time the calculation is performed, we get exactly the same answer.

DOI: 10.1201/9781003222330-1

The term *stochastic* means that we get a different answer each time the calculation is performed, or each time the measurement is obtained. Here are some examples: What is the height of the next person you pass on the street? How many grains of sand are in a handful? If the product label indicates that the package contains 40 lbs, what might be the actual weight? If there are the same number of red and green marbles in a box and you draw three, blindfolded, how many green marbles will you have? If you want to compare fertilizer treatments, you will find that the year-to-year variation in weather and insect population, and the location-to-location variation in properties of the earth will cause significant variation in results.

Despite the use of deterministic calculations in teaching concepts and in estimating values, the reality about measurements and samples and predictions is that they have variation. Statistics provides techniques for analyzing and making decisions within the uncertainty.

Sources of variation include the vagaries of weather, the probability of selecting a particular sample, variation in raw material, mechanical vibration, incomplete fluid mixing, prior stress on a device, new laws and regulations, future prices, and many other aspects.

1.3 Treatments, Process, and Outcomes

The term *treatment* refers to the influence on a process. The influence might be how safety training is delivered (video, reading materials, in-person, comically, or seriously). A treatment could be a recipe or procedure to be followed. A treatment could be the type of equipment used (batch or continuous, toaster or microwave). A treatment could be the raw material supplier or the service provider. The treatment might be the operating conditions in manufacturing (flow rates, temperatures, mixing time, etc.).

The *process* is whatever responds to the treatment. It may be a human response to an office lighting treatment. It may be a mechanical spring-and-weight response to treatment by the ambient temperature. It may be a biological process response to a pH (acidity) treatment.

Outcome refers to the response of the process. It may be the time to recover physical health after an infection in response to the medicine dose. It may be the economic response of the nation due to changes in the prime lending rate. It may be the variation in a quality metric due to a particular treatment. It may be the probability of automobile accidents if the speed limit is changed.

Treatments and outcomes are variously termed influences and responses, causes and effects, inputs and outputs, independent and dependent variables, etc.

1.4 Uses of Statistics

You will use statistics in five ways. One is in the *design of experiments* or surveys. In this instance, you need the answers to some questions about an event or a process. An *effective* experiment is one that has been designed so that the answers to your questions will be obtained more often than not. An *efficient* experiment is one that is unbiased (predicts the correct value of the parameter) and that also has the smallest variance (scatter about the true value of the population parameter in question). Efficiency also means that the answers will have been obtained with the minimum expenditure of time (yours, an operator's, a technician's, etc.) and other resources.

The second way you will use statistical techniques is with *descriptive statistics*. This method involves using *sample* data to make an inference about the *population*. The population is the entire or complete set of possible values, attributes, etc. that are common to, describe, or are characteristic of a given experiment or event. A sample is a subset of that data. Descriptive statistics are used for describing and summarizing experimental, production, reliability, and other types of data.

The description can take many forms. The average, median, and mode are all measures of *centrality*. Variance, standard deviation, and probable range are all measures *variation*. The descriptor may be a probability, which refers to the chance an event might happen (such as getting three or more successes in five-coin flips) or the chance that a value might exceed some threshold (the probability of seeing someone taller than 6ft 8in on your next shopping trip).

It is essential that your samples are *random samples* if you are to have any reasonable expectation of obtaining reliable answers to your questions. To obtain a random sample, you must first define, not just describe, the population under consideration. Then you can use the principles of random selection of population values or experimental conditions to obtain the random sample that is essential to statistical inference.

A third statistical use is estimating the *uncertainty* of a value, estimating the possible range of values it might have. The value might be an average from a sample and the question is what range of population means could have generated that sample average. The value might be a predicted outcome from a model when all model coefficient values and influences are not known with certainty.

A fourth use of statistics is in the testing of *hypotheses*. A hypothesis about any event, process, or variable relationship is a statement of anticipated behavior under specified conditions. Hypotheses are tested by determining whether the hypothesized results reasonably agree with the observed data. If they do, the hypothesis is likely to be valid. Otherwise, the hypothesis is likely to be false. Hypotheses could be relatively complex, such as the model matching the data, the design being reliable, or the process being at steady-state.

The fifth use of methods in this book is to obtain quantitative relationships between variables by use of sample data. This aspect of statistics is loosely called "curve fitting" but is more properly termed *regression analysis*. We will use the method of least squares for regression because that technique provides a conventional way to estimate the "best fit" of the data to the hypothetical relationship.

1.5 Stationarity

In statistics, a *stationary* process does not change in mean (average) or variance (variability). It is steady, but any measurement is subject to random variation. The value of the data perturbation changes from sample to sample, but the distribution of the perturbations does not change.

This is in contrast to classic deterministic analysis of transient and steady-state processes. A steady process flatlines in time. The measurement achieves a particular value and remains at that value. When the process is in a transient state the average or mean changes in time.

In statistics the term stationary means that the steady-state process will not deterministically flatline. Instead, the data will be continually fluctuating about a fixed value (mean) with the same variance. In statistics, a stationary process is not in a transient state.

1.6 There are No Absolutes

You should have noticed by now that we have repeatedly used the words "probably" and "likely" in the first part of this chapter. Our use was intentional, as there are NO absolutes in statistics except the following: There are NO absolutes. Every conclusion you reach and everything you say as a result of statistical examination of data is subject to error. Instead of saying, "The relationship between production rate and employee training is ...," you must say, "The relationship between production rate and employee training *probably* is ...." The reason for our statement concerning absolutes is simple. The data from which the estimators were obtained were subject to error. The exact value of any estimator is uncertain, and these uncertainties carry forward every time the estimators are used. This book is devoted to helping you learn how to make qualitative and quantitative statements in the face of the uncertainties.

1.7 A Caution on Statements of Confidence

Level of confidence is a measure of how probable your statistical conclusion is. As an example, after testing raw materials A and B for their influence on product purity, you might be 95% confident that A leads to higher purity. But you *cannot* extend this result to report that you are 95% sure that using raw material A is the better business decision. You have only tested product purity. You have not evaluated product variability, other product characteristics, manufacturing costs, process safety implications, etc. You can only be 95% confident in your evaluation of purity. Be careful that you do not project statistical confidence about one aspect onto your interpretation of the appropriate business action.

1.8 Correlation is Not Causation

Statistics does not prove that some event or value caused some other response. *Causation* refers to a cause-and-effect mechanism. *Correlation* means that there is a strong relationship between two variables, or observations.

As an example, there is a strong correlation to people awakening and the sun rising, but one cannot claim that people awakening causes the sun to rise. The cause-and-effect mechanism for this observed correlation is more akin to the opposite. As another example, there is a strong correlation between gray hair and wrinkles, but that does not mean that gray hair causes wrinkles. The mechanism is that another variable, age, causes both observations.

So, more so than just tempering claims about confidence in taking action from testing a single aspect, be careful not to let indications of correlation dupe you into claiming causation. If you have an opinion as to the cause-and-effect mechanism, and you have correlation that supports it, before you claim it is the truth, perform experiments and seek data that could reject your hypothesized mechanism. State exactly, mechanistically how the treatment leads to the outcome expectations. State what else you expect should be observed, and what should not be observed. State when and where these should be observed. Do the experiments to see if your hypothesized theory is true.

1.9 Uncertainty and Disparate Metrics

Traditionally, statistics deals with the probable outcomes from a distribution. This book is grounded in that mathematical science, and many examples reveal how to describe the likelihood of some extreme value.

But more than this, the basis (the "givens") in any particular application have uncertainty, which is unlike the basis of givens in a schoolbook example. In the real world, to make decisions based on the statistical analysis, the impact of uncertainty needs to be considered. Further, concerns over possible negative choices might not just be about monetary shortfalls. They may be related to disparate issues such as reputation.

This book includes a chapter on propagation of uncertainty, another on stochastic simulation, and frequent discussions on Equal–Concern approaches for combining disparate metrics.

1.10 Takeaway

Statistical statements should be tempered with a declaration related to probability (likelihood, or confidence). Statements about something should only relate to the test data, not an interpretation or extrapolated action. Do not confuse correlation with causation.

1.11 Exercises

1. List several statements that you have recently heard or read and estimate the certainty or qualifications that should be associated with it. Here are example statements: "Men are taller than Women." "I floss each day." "Net wt. 18 oz."
2. List several deterministic and several stochastic processes.

3. List several examples of processes – meaning some influence leads to an outcome. These could be computer procedures, human recipes, physical devices, or human social responses.
4. Sketch data that would come over time from a stationary stochastic process. Sketch another with a change in mean, and another with a change in variance.

2

Probability

2.1 Probability

An *event* is a particular outcome of a trial, test, experiment, or process. It is a particular category for the outcome. You define that category.

The outcome category could be *dichotomous*, meaning either one thing or another. In flipping a coin, the outcome is either a Head (H) or a Tail (T). In flipping an electric light switch to the "on" position, the result is either the light lights or it does not. In passing people on a walk, they either return the smile or do not. These events are mutually exclusive, meaning if one happens the other cannot. You could define the event as a H, or as a T; as the light working, or the light not working.

Alternately, there could be any number of mutually exclusive events. If the outcome is one event, one possible outcome from all possible discrete outcomes, then it cannot be any other. The event of randomly sampling the alphabet could result in 26 possible outcomes. But if the event is defined as finding the letter "T", this success excludes finding any of the other 25 letters.

By contrast, the outcome may be a continuum-valued variable, such as temperature, and the event might be defined as sampling a temperature with a value above 85°F. A temperature of 84.9°F would not count as the event. A temperature of 85.1°F would count as the event. For continuum-valued variables, do not define an event as a particular value. If the event is defined as sampling a temperature value of 85°F, then 84.9999999°F would not count as the event. Nor would 85.00000001°F count as the event. Mathematically, since a point has no width, the likelihood of getting an exact numerical value, is impossible. So, for continuum-valued outcomes, define an event as being greater than, or less than a particular value, or as being between two values.

One definition of *probability* is the ratio of the number of particular occurrences of an event to the number of all possible occurrences of mutually exclusive events. This classical definition of probability requires that the total number of independent trials of the experiment be infinite. This definition is often not as useful as the relative-frequency definition. That interpretation of probability requires only that the experiment be repeated a finite number of times, n. Then, if an event E occurs n_E times out of n trials and if the ratio n_E/n tends to stabilize at some constant as n becomes larger, the probability of E is denoted as:

$$P(E) = \lim_{n\to\infty} n_E / n \tag{2.1}$$

The probability is a number between 0 and 1 and inclusive of the extremes 0 and 1, $0 \le P(E) \le 1$.

DOI: 10.1201/9781003222330-2

Ideally, Equation (2.1) specifies the infinity of data, but certainly reasonable values for an estimate of $\hat{P}(E) = n_E / n_{total}$ can be determined from limited data.

The complement of an event is not-the-event, indicated by $\bar{E}$. Here the overbar does not mean average, it means "not".

$$P(\bar{E}) = \frac{n_{\bar{E}}}{n} = \frac{n - n_E}{n} = 1 - \frac{n_E}{n} = 1 - P(E) \tag{2.2}$$

The value of the event probability can be either derived from fundamental principles, representing an infinite number of experiments, or calculated from a finite number of experiments. These would be called theoretical or experimental values, respectively.

The trials could be of an end-of-process, or batch, nature. For example, during the process of flipping a coin, the coin alternates heads-up and tails-up while in the arc. It may hit the ground and bounce up still spinning. Then it finally settles, and only after it stops do we count H or T.

Alternately, even if the process is in a transient state, one can calculate the probability of an event after a certain point in time. For example, a company might have two versions of a new product and want to know which is more robust. So, they engage 10,000 people to be consumer testers. Half of the testers have one version, and half have the other. They are divided into groups representing equivalent demographics. After each month into a two-year trial period, testers return their product for evaluation, then get it back and continue their use of it. The company may be tracking the point in time that some feature of the product fails. Version A might have 6% failures ($p = 0.06$) after 12 months, and version B might have 11% ($p = 0.11$). In such in-process testing, the probability is not end-of-process, but at a specified time duration.

Once event probabilities are known, we often want to know the probability of composite or compound events. Two types of probabilities are of primary interest to engineers. The first is *a priori* probability, in which we assume that the composite/compound event will happen and that its probability can be predicted or calculated prior to the experiment. The second is *a posteriori*, or conditional, probability, which is calculated after some evidence is obtained.

2.2 Probability Calculations

2.2.1 *A Priori* Probability Calculations

Let us consider that E_1 and E_2 are two user-specified events (results) of outcomes of an experiment. Here are some definitions:

If E_1 and E_2 are the only possible outcomes of the experiment, then the collection of events E_1 and E_2 is said to be *exhaustive*. For instance, if E_1 is that the product meets specifications and E_2 is that the product does not meet specifications, then the collection E_1 and E_2 represents all possible outcomes and is exhaustive.

The events E_1 and E_2 are *mutually exclusive* if the occurrence of one event precludes the occurrence of the other event. For example, again, if E_1 is that the product meets specifications and E_2 is that the product does not meet specifications, then E_1 precludes E_2, they are mutually exclusive, if the outcome is one, then it cannot be the other.

Event E_1 is *independent* of event E_2 if the probability of occurrence of E_1 is not affected by E_2 and vice versa. For example, flip a coin and roll a die. The coin flip event of being a Head is independent of the number that the die roll reveals. As another example, E_1 might be that the product meets specifications, and E_2 might be that fewer than two employees called in sick. These are independent.

The *composite event* "E_1 and E_2" means that both events occur. For example, you flipped a H and rolled a 3. If the events are mutually exclusive, then the probability that both can occur is zero.

The composite event "E_1 or E_2" means that at least one of events E_1 and E_2 occurs. When you flipped and rolled, a H and/or a 3 were the outcomes. This situation allows both E_1 and E_2 to occur but does not require that result, as does the "E_1 and E_2" case.

There could be any number of user-specified events, $E_1, E_2, E_3, \ldots, E_n$.

Two rules govern the calculation of *a priori* probabilities.

2.2.1.1 Rule 1: Multiplication

If $E_1, E_2, E_3, \ldots, E_n$ are independent and not mutually exclusive events having probabilities $P(E_1), P(E_2), P(E_3), \ldots, P(E_n)$, respectively, then the probability of occurrence of E_1 and E_2 and E_3 and … E_n is:

$$P(E_1 \,\&\, E_2 \,\&\, E_3 \,\&\cdots\&\, E_n) = [P(E_1)][P(E_2)][P(E_3)]\cdots[P(E_n)] = \prod_{i=1}^{n} P(E_i) \tag{2.3}$$

The probability of occurrence of *all n* independent events is the product of all their individual probabilities.

2.2.1.2 Rule 2: Addition

If E_1, E_2, and E_3 are independent and not mutually exclusive events with individual probabilities $P(E_1)$, $P(E_2)$, and $P(E_3)$, then the probability of at least one (possibly two or more) of these events is:

$$\begin{aligned} P(E_1 \text{ or } E_2 \text{ or } E_3) &= P(E_1)+P(E_2)+P(E_3) \\ &\quad -P(E_1E_2)-P(E_1E_3)-P(E_2E_3)+P(E_1E_2E_3) \end{aligned} \tag{2.4}$$

This can be more conveniently calculated by considering the not case:

$$P(E_1 \text{ or } E_2 \text{ or } E_3) = 1 - P(\bar{E}_1 \,\&\, \bar{E}_2 \,\&\, \bar{E}_3) \tag{2.5}$$

Here, the overbar on E means "not the event". Assume that $\bar{E}_1$, $\bar{E}_2$, etc. are independent and not mutually exclusive, using the multiplication rule for the "&" conjunction:

$$P(\bar{E}_1 \,\&\, \bar{E}_2 \,\&\, \bar{E}_3) = [P(\bar{E}_1)][P(\bar{E}_1)][P(\bar{E}_1)] \tag{2.6}$$

Then, converting the not-event back to the event:

$$P(E_1 \text{ or } E_2 \text{ or } E_3) = 1 - [1-P(E_1)][1-P(E_2)][1-P(E_3)] \tag{2.7}$$

When Equation (2.7) is algebraically expanded it is the same as Equation (2.4). In general, for n independent and not mutually exclusive events:

$$P(E_1 \text{ or } E_2 \text{ or} \dots \text{or } E_n) = 1 - \prod_{i=1}^{n} \left[1 - P(E_i)\right] \tag{2.8}$$

If the events are mutually exclusive, then the addition rule becomes:

$$P(E_1 \text{ or } E_2 \text{ or} \dots \text{or } E_n) = \sum_{i=1}^{n} P(E_i) \tag{2.9}$$

Also, if the events are not mutually exclusive, but the probabilities are small enough so that the product terms in Equation (2.4) are negligible, then Equation (2.9) is a reasonable approximation. However, if there are many terms in the "OR" conjunction, even if each product is small, the sum of probabilities might add to >1. Equation (2.8) is not prone to such a possibility.

Example 2.1: A person is flipping a fair coin and rolling a fair six-sided die. The desired outcome is a T for the coin and a 4 on the die.

a) What is the probability of E_1 and E_2?
b) What is the probability of E_1 or E_2?

The two outcomes are independent, and not mutually exclusive. Conceptually, the probability of either a H or T is 0.5, and the probability of any particular value on the die is 1/6.

For Part (a) use Equation (2.3):

$$P(E_1 \,\&\, E_2) = \prod_{i=1}^{n=2} P(E_i) = \left[P(E_1)\right]\left[P(E_2)\right] = 0.5\frac{1}{6} = 0.08\overline{333}$$

For Part (b) use Equation (2.8):

$$P(E_1 \text{ or } E_2) = 1 - \prod_{i=1}^{n=2} \left[1 - P(E_i)\right] = 1 - (1 - 0.5)\left(1 - \frac{1}{6}\right) = 0.58\overline{333}$$

There is about an 8% chance that both trials will be successes, and about a 58% chance that at least one of the two will be a success.

Example 2.2: A person is flipping a fair coin and rolling a fair six-sided die. The desired outcome is a H for the coin and a 4 or lower number on the die.

a) What is the probability of E_1 and E_2?
b) What is the probability of E_1 or E_2?

The two outcomes are independent, and not mutually exclusive. Conceptually, the probability of either a H or T is 0.5, and the probability of any particular value on the die is 1/6. The die roll event is a composite event. Either a 1, or a 2, or a 3, or a 4 will represent a success. These outcomes are mutually exclusive. If, for example, the roll shows a 2, it cannot be a 1, 3, or 4. Use Equation (2.9) to calculate the probability of Event 2.

$$P(E_2) = P(1 \text{ or } 2 \text{ or } 3 \text{ or } 4) = \sum_{i=1}^{n} P(E_i) = \frac{1}{6} + \frac{1}{6} + \frac{1}{6} + \frac{1}{6} = 0.666\overline{6}$$

For Part (a) use Equation (2.3).

$$P(E_1 \,\&\, E_2) = \prod_{i=1}^{n=2} P(E_i) = \left[P(E_1)\right]\left[P(E_2)\right] = 0.5\left(0.666\overline{6}\right) = 0.333\overline{3}$$

For Part (b) use Equation (2.8).

$$P(E_1 \text{ or } E_2) = 1 - \prod_{i=1}^{n=2} \left[1 - P(E_i)\right] = 1 - (1 - 0.5)\left(1 - 0.666\overline{6}\right) = 0.833\overline{3}$$

There is about a 33% chance that both trials will be successes, and about an 83% chance that at least one of the two will be a success.

Example 2.3: The results of the rigorous examination of a dozen automobile tires showed the following: One tire was perfect; three had only slight flaws in appearance; two had incompletely formed treads; one had a serious structural defect; and the rest had at least two of these defects. What is the probability that the next set of four tires you buy of this particular brand will be perfect? That they will have at most only undesirable appearances? That they will have less than two defects?

The solution begins with the assumption that the population is adequately represented by the sample of 12 tires.

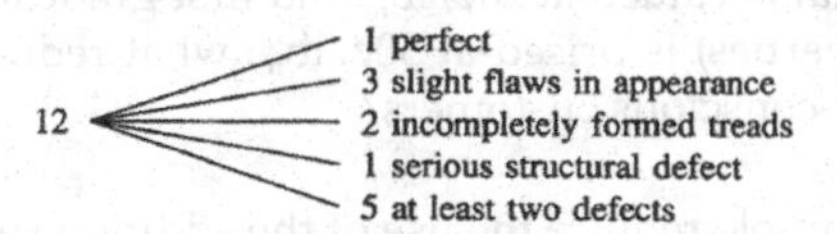

The sample results are displayed fanwise as shown above to help you visualize the probabilities of the events E_i.

The probability of each event, $P(E_i) = n_i/n$ where $n = 12$.

The multiplication rule is appropriate for the solution to Part (a) because we want the probability that all tires in the next set will be perfect, i.e., the first one *and* the second one *and* the third one *and* fourth one.

$$P(4 \text{ perfect tires in next set}) = \left[P(\text{perfect})\right]^4$$

$$P(4 \text{ perfect tires in next set}) = \left(\frac{1}{12}\right)^4$$

$$= 4.82253 \times 10^{-5} \simeq 4.82 \times 10^{-5}$$

For Part (b) the probability of four tires having, at most, undesirable appearance can be obtained from any combination of perfect tires and those with only appearance flaws. Since the sum of the probabilities in any of those cases is 4/12 and the answer to this question again requires the multiplication rule:

$$P(\text{4 tires in next set have, at most, undesirable appearance}) = \left(\frac{4}{12}\right)^4$$

$$= 0.01234568 \simeq 0.0123$$

For Part (c), the probability of any tire having less than two defects is [1 – *P*(any tire having at least two defects)] or *P*(<2 defects) = [1 – *P*(≥2 defects)]. As *P*(<2 defects) = 1 – 5/12 = 7/12, and since each tire must satisfy this criterion:

$$P(\text{4 tires with} < 2 \text{ defects each})$$

$$= \left[P(< 2 \text{ defects})\right]^4$$

$$= \left(\frac{7}{12}\right)^4 = 0.11578897 \simeq 0.1158$$

Example 2.4: In the manufacture of a modified natural resin, the three key product properties are melting point (E_1), color (E_2), and acid number (E_3), each of which is independent of the others. Marketing suggests sales specifications of minimum melting point of 130°C, maximum color of 3 (arbitrary scale), and maximum acid number of 12. Analyses of available data indicate that the probabilities of material being off-grade with respect to each of these properties are $P(E_1) = 0.03$, $P(E_2) = 0.05$, and $P(E_3) = 0.02$. Assuming no process changes,

(a) What percentage of the resin produced will be off-grade?
(b) For customers for whom color is unimportant, what off-grade percentage is to be expected?
(c) If off-grade material is valued at 20¢/lb_m and first grade (on specification with respect to all properties) is priced at 50¢/lb_m, what reduction in price can be given to non-color-conscious customers?

The questions in this example require the use of the addition rule.

Part (a): As all off-grade resin fails to meet the specifications in *at least* one of the three ways, and as these events are not mutually exclusive, we have exactly the situation described by Equation (2.4) or (2.8), where:

$$P(E_1) = P(\text{melting point off-grade})$$

$$= P(\text{MP off}) = 0.03$$

$$P(E_2) = P(\text{color off}) = 0.05$$

$$P(E_3) = P(\text{acid number off}) = 0.02$$

$$P(\text{resin off-grade}) = 0.03 + 0.05 + 0.02$$

$$- 0.03(0.05) - 0.03(0.02)$$

$$- 0.05(0.02) + 0.03(0.05)(0.02)$$

The result is that the probability of resin being off-grade is 0.09693 or about 9.69%

Part (b): If color is unimportant to some customers, the problem reduces to finding the probability that the resin will be off-grade with regard to melting point or acid number, and you can use Equation (2.4) or (2.8) with $P(E_2) = 0$.

$$P(E_1 \text{ or } E_3) = 1 - \left[1 - P(E_1)\right]\left[1 - P(E_3)\right]$$
$$= 1 - (1 - 0.03)(1 - 0.02)$$

Here, $P(E_1 \text{ or } E_3) = 0.0494$ or 4.94%

Part (c): Using a basis: 100 lb_m, and the results of Part (b),

$$4.94\ lb_m\left(20¢/lb_m\right) + 95.06\ lb_m\left(50¢/lb_m\right) = \$48.52$$

As 100 lb_m of on-grade material is worth \$50.00, and 100 lb_m of off-grade material is worth \$48.52, the per lb_m discount to customers who don't care about product color is (\$50.00 – \$48.52)/100 = 1.48¢/lb_m.

2.2.2 Conditional Probability Calculations

In some cases, an event has happened and we wish to determine the probability *a posteriori* (after the fact) that a particular set of circumstances existed based on the results already obtained. Suppose that several factors B_i, $i = 1, n$ can affect the outcome of a specific situation or event, E. The probability that any of the B_i did occur, given that the event or outcome E has already occurred, is a conditional probability. Let's begin with the premise that B_1, B_2, B_3, and B_4 can influence E, an event that has happened. The final event E can take place only if at least one of the preliminary events (the B_i) has already happened. The probability that a particular one of them, e.g., B_3, occurred is $P(B_3 \mid E)$. If one of the B_i, say B_3, had to happen for E to transpire, then B_3 is conditional on E.

These are end-of-process events and beginning-of-process conditions.

Conditional probabilities can be determined by the use of Bayes' theorem. Bayes' theorem is stated in Equation (2.10), where $P(B_i)$ and $P(B_k)$ are the *a priori* probabilities of the occurrences of events B_i and B_k and $P(B_i \mid E)$ and $P(B_k \mid E)$ are the conditional probabilities that B_i or B_k would occur if event E has already occurred.

$$P(B_k|E) = \frac{P(B_k)P(E|B_k)}{\sum_{i=1}^{n} P(B_i)P(E|B_i)} \tag{2.10}$$

Here:

1. E is an event, an outcome.
2. B is a condition (a situation, or an influence).
3. $P(B)$ is the probability of a condition happening.
4. $P(E \mid B)$ is the probability E occurring given that B did.
5. $P(E)\cdot P(E|B)$ is the probability B and it caused E.
6. Σ is the sum of all probabilities of all ways that E could happen.
7. $P(B \mid E)$ is the probability B happening given that E did.

Example 2.5: The game is to roll two dice. The player gets to draw a card if A) either is a 3, or if B) the two numbers are 1 and 6, or if C) none of the numbers are a 1, 3, or 6. Your friend reports a win. If A or B, the chance of drawing a lucky card is 1 in 10. If C, the lucky card chance is 1 in 100. Which condition was most likely?

$$P(A) = P(3 \text{ or } 3) = \frac{1}{6} + \frac{1}{6} - \frac{1}{6}\frac{1}{6} = 0.30555\ldots$$

$$P(B) = P(1 \text{ and } 6) = \frac{1}{6}\frac{1}{6} = 0.27777\ldots$$

$$P(C) = P(\text{not 1 and not 3 and not 6 on either roll}) = \frac{3}{6}\frac{3}{6} = 0.25$$

Situation	*P (Situation)*	*P (Event | Situation)*	Product
A	0.30555	0.1	0.030555
B	0.27777	0.1	0.027777
C	0.25	0.01	0.0025

The sum of all products is 0.060827. Then,

$$P(A \mid E) = \frac{0.030555}{0.060827} = 0.5023$$

$$P(B \mid E) = \frac{0.027777}{0.060827} = 0.4567$$

$$P(C \mid E) = \frac{0.0025}{0.060827} = 0.0411$$

Path C, situation C is very unlikely. Path A is most likely, but B is nearly as probable.

Example 2.6: A certain material is fed to a two-step process. For this process, the probabilities of a malfunction are $P(B_1) = 0.03$ and $P(B_2) = 0.05$, where the factors B_1 and B_2 represent a malfunction in Steps 1 and 2, respectively. A sample of the final product is taken and found to be unacceptable. Our experience over the previous two months indicates that a defective product will be obtained 20% of the time if Section 1 of the process malfunctions and 36% of the time if Section 2 malfunctions. That means $P(E \mid B_1) = 0.2$ and $P(E \mid B_2) = 0.36$. In which part of the process does the fault probably lie?

From Equation (2.10),

$$P(B_1 \mid E) = \frac{P(B_1)P(E \mid B_1)}{P(B_1)P(E \mid B_1) + P(B_2)P(E \mid B_2)}$$

$$= \frac{0.03(0.2)}{0.03(0.2) + 0.05(0.36)} = 0.25$$

And

$$P(B_2 \mid E) = \frac{0.05(0.36)}{0.03(0.2) + 0.05(0.36)} = 0.75$$

From these results, Section 2 of the process is the more likely to need corrective maintenance.

Example 2.7: In the production of polyester fiber to become fabric for consumer use, a molten polyester liquid is forced through hundreds of tiny orifices in a spinneret. The liquid filaments from each orifice are cooled by an air stream, and hence solidified into very thin filaments. These filaments are collated to form tow. The next steps are to stretch the tow (to orientate the polymer to create fiber strength) and to crimp the tow to add bulk. The tow is then cut into short lengths and the filaments separate to become fibers. Samples of the fibers are then routinely inspected before packaging into bales for shipment to the customer.

Among the tests performed on each sample is one to measure fiber tensile properties. If the tow has been insufficiently stretched, the fibers will deform easily, then fabrics made from them will very likely become loose or shapeless in use. If the tow has either been stretched or crimped too much, there will be too many broken or weakened fibers to use in making top-quality garments.

As a polyester production engineer your responsibilities include all aspects of filament and tow production, starting with the arrival of the spinning solution at the spinning heads and ending with the fiber operation. The hourly fiber report has just come in from the lab and indicates that the sample has poor stretch resistance. How can you quickly decide which mechanical operation is probably at fault?

The event E is the occurrence of poor tensile properties. The probability of stretcher–crimper malfunction is known to be $P(B_1) = 0.04$; for clogged spinnerets, $P(B_2) = 0.41$; and for ineffective filament collation, $P(B_3) = 0.12$.

From past experience we know that poor tensile properties occur half the time when the stretcher–crimper malfunctions. So $P(E \mid B_1) = 0.5$. And, from previous observations, we have $P(E \mid B_2) = 0.1$ and $P(E \mid B_3) = 0.8$.

Let us utilize Bayes' formula to calculate the equipment malfunction probabilities $P(B_i \mid E)$ for the different mechanical operations:

$$P(B_1 \mid E) = \frac{0.04(0.5)}{0.04(0.5) + 0.41(0.1) + 0.12(0.8)} = \frac{0.02}{0.157} = 0.1274$$

$$P(B_2 \mid E) = \frac{0.41(0.1)}{0.157} = 0.2611$$

$$P(B_3 \mid E) = \frac{0.12(0.8)}{0.157} = 0.6115$$

Thus, we conclude that we are most probably getting ineffective filament collation.

2.2.3 Bayes' Belief Calculations

Belief is the confidence that you have in making a statement of fact about something, a supposition. Here are some examples of statements that we might make:

"These symptoms are just seasonal allergies."

"The average benefit of Treatment Y is larger than that of Treatment X."

"The process has reached steady-state."

And for each we might claim to be very certain of the supposition (perhaps 99% sure), or somewhat certain (perhaps 80% sure), or even not sure whether it is or is not (perhaps 50% sure).

Bayes' Belief, B, is scaled by 100%, so the value is $0 \le B \le 1$. If you are not so certain about a statement, the belief, B, might be 0.25. If you are very certain, B might be 0.97. If you are not so sure about something, and it could be a 50/50 call, then $B = 0.5$.

Because we act on suppositions, we want to be fairly certain that the statement about what we suppose represents the truth about the reality. When you are not certain, you perform tests, take samples, get other's opinions, etc. to strengthen or to reject your belief in the supposition. But tests are not perfect. There is always some uncertainty about the results. For example, the manufacturer of a particular procedure for detecting the presence of colorectal cancer reports it detects the disease in 92% of the patients with cancer and gives a negative result in 87% of the patients without the disease. (Exact Sciences Laboratories, Cologuard Patient Guide, 2020). The 92% correct positives means 8% false negatives. (The test on 8% of patients with the disease will falsely indicate they do not have it.) Similarly, the 87% correct negatives means 13% false positives. (The test on 13% of patients without the disease will falsely indicate they have it.)

Table 2.1 is a matrix of the probabilities of the medical test giving true and false indications.

Here is another example: A test for steady-state (SS) might look at the past several data points. At SS the time-rate of change, data slope, ideally is zero, $S = 0$. But, because of noise on the data, the slope will not be exactly zero; so, you might accept SS if the test results are $-0.1 \le S \le +0.1$. So, if the test result indicates $S = -0.03$ you say that is just noise, and the test indicates SS. But, at SS, a particular confluence of data perturbations might indicate the local slope is $S = 0.15$, and the test would reject the true condition of SS. Maybe, given a true SS, the test will indicate SS 85% of the time, and reject SS 15% of the time.

On the other hand, if the process is in a transient state (TS), the slope will be much greater than a SS value, the slope will be beyond the $-0.1 \le S \le +0.1$ limits, and the test result will claim TS. However, even in a TS when the process variable is rising, the noise pattern on the past few samples might have a decreasing pattern, and the rate of change might incorrectly indicate SS. Maybe, given a true TS, the test will indicate TS 95% of the time, and SS 5%.

Table 2.2 is a matrix of the probabilities of the test giving true and false indications.

If you somewhat believe something is possibly true, $B = 0.75$, and you do a test which supports your belief (indicates it is probably true), then your belief value rises, perhaps it becomes $B = 0.99$, which may be strong enough belief to take action on it. But, if the test

TABLE 2.1

Probabilities of Correct Diagnosis and False Positive and Negative for a Medical Test

Truth \ Test result	Test is positive, detects cancer	Test is negative, does not detect cancer
Patient has cancer	0.92	0.08
Patient does not have cancer	0.13	0.87

TABLE 2.2

Probabilities of Correct Diagnosis and False Positive and Negative for a Process Test

Truth \ Test result	Test claims SS	Test claims TS
Process is SS	0.85	(1 – 0.85) = 0.15
Process is TS	(1 – 0.95) = 0.05	0.95

indicates it is probably not true, your belief falls. Maybe it falls to 20%, somewhere between zero and your former belief of 70%. If a second test also indicates the hypothesized benefit is probably not true, your belief falls again, perhaps to 1% which may be enough to reject the supposition.

There are two different aspects to knowledge in this analysis. Keep them separate. First, there is the truth about the situation. The patient either has the disease or does not. The process is either at SS or not. Treatment Y is either better than Treatment X, or not. The truth is stated in the first column of the tables. But, the truth is not known. So, we use a test.

The second knowledge aspect is the test result. The test either indicates positive or negative results, or that Y is better than X or it is not. The test outcome is stated in the top row of the tables. But the test is not infallible. The test results may, or may not, represent the truth.

Also, there are two different aspects to probabilities in this analysis. Keep them separate. The first is the entries in the table, the probability of a test giving a correct or wrong outcome.

The second probability aspect is the belief, the personal confidence that something is true. But just because the person wants to believe it is true (or not true) does not make it so. (Humans used to believe that the sun revolved around the Earth, but the belief did not make it so.)

Belief changes with successive test results. The method for updating a belief after test results has the same formulation as Bayes' conditional probability, Equation (2.10). The Bayes' Belief method for updating belief after test results is:

$$B_{\text{after}} = \frac{B_{\text{prior}}\, p(\text{test support of Belief if supposition is true})}{B_{\text{prior}}\, p(\text{support if supposition true}) + (1 - B_{\text{prior}})\, p(\text{support if supposition is false})} \tag{2.11}$$

where B_{prior} is the belief prior to the test results, and B_{after} is the belief after the results. The numerator is the expected outcome if the supposition is true. The denominator is the sum of all such test outcomes that are possible (whether the supposition is true or not true).

The belief could be in either of the two mutually exclusive cases. For example, that Y is better than X. Or, that the patient has cancer or does not have cancer. It does not matter which supposition you choose.

The test results could either support or counter the belief. The term, p(test support of belief if the supposition is true), and its shorter version, p(support if supposition true), is the probability from the table. For example, if you believe you have cancer, and the test result is positive, the probability from Table 2.1 is 0.92. If you believe you are cancer free and the test indicates you have it, the probability from Table 2.1 is 0.13. The complementary probabilities, p(support if the supposition false), is the support of the belief if the supposition is wrong. For example, if you believe that you are cancer-free, but in fact you have the disease, the probability of the test indicating you are healthy, when you are not, is 0.08.

Example 2.8: If the supposition is that the process is at SS, and the belief is 75% that SS is true, and a test affirms that, what is the new belief value?
From Table 2.2

$$p(\text{test supporting belief if supposition is true}) = 0.85, \text{ and}$$

$$p(\text{test supporting belief if supposition is false}) = 0.05$$

So, if the belief is at SS with $B_{\text{prior}} = 0.75$, then

$$B_{\text{after}} = \frac{(0.75)(0.85)}{(0.75)(0.85)+(1-0.75)(0.05)} = 0.980$$

The result is rounded. And, if a second test affirms "at SS" then:

$$B_{\text{after}} = \frac{(0.98)(0.85)}{(0.98)(0.85)+(1-0.98)(0.05)} = 0.999$$

that would be a very strong belief.
By contrast, if the belief is the process is at SS with $B = 0.75$, and a test rejects that supposition, then:

$$B_{\text{after}} = \frac{(0.75)(0.15)}{(0.75)(0.15)+(1-0.75)(0.95)} = 0.32$$

The belief drops. A value of 0.32 means little certainty that the belief is either correct or incorrect. The decision would be to observe more data before taking action.

It works whether the belief is about the situation being either true or false.

Example 2.9: Investigate the complement to Example 2.8. Consider that your belief is that the process is in a TS, with $B_{\text{prior}} = 0.75$, and the test affirms that, then

$$B_{\text{after}} = \frac{(0.75)(0.95)}{(0.75)(0.95)+(1-0.75)(0.15)} = 0.95$$

or, if the test rejects that, indicating SS not TS, then

$$B_{\text{after}} = \frac{(0.75)(0.05)}{(0.75)(0.05)+(1-0.75)(0.85)} = 0.15$$

Example 2.10: There is some evidence to suggest that Treatment B is 10% better than what we are currently using, Treatment A. But we are not sure. B may be just the same as A. Maybe we have a 70% confidence, $B = 0.7$, that B is better. That is not a strong enough belief to switch from A to B but provides an incentive to perform trials. So, we scheduled trials to see if it is true. There is a cost to change from B to A (trials, document updating, etc.), which is equivalent to 30% of the supposed benefit. So, to accept B the trials must indicate that B is greater than 103% of A. For this exercise use the value of A as $1,000/day, then B might be $1,100/day and the needed value of B must be >$1,030/day to justify the change. The trial outcomes have variability, and based on past experience with A, the standard deviation of a single trial observation is about \$50/day. If B is only equivalent to A, any trial has a 0.274 probability of indicating B is better than \$1,030/day. However, if B is \$1,100/day as supposed, then the probability of a trial showing that is better than the required \$1,030/day is about 0.919. Use any standard normal statistics table, or the Excel cell function $p = \text{NORM.DIST}(z,0,1,1)$, where z is the number of standard deviations the value is from the target. If a trial indicates B is better than the threshold of \$1,030, what is the belief?

The probability table is:

Truth \ Test result	Test claims $B \geq \$1,030$	Test claims $B < \$1,030$
$B \geq \$1,030/\text{day}$	0.919	0.081
B is equivalent to A	0.274	0.726

$$B_{\text{after first accept }B\text{ result}} = \frac{(0.7)(0.919)}{(0.7)(0.919)+(1-0.7)(0.274)} = 0.887$$

Because 88.7% is just fairly sure, we'll do another trial to have greater certainty. But with two trials, we'll use the average of the two outcomes, $\bar{B}$, so the standard deviation of the average would be $\$50/\sqrt{n}$ which would change the probabilities in the table. For $n = 2$:

Truth \ Test result	Test claims $\bar{B} > \$1,030$	Test claims $\bar{B} \leq \$1,030$
$\bar{B} \geq \$1,030/\text{day}$	0.976	0.024
$\bar{B}$ is equivalent to A	0.198	0.802

$$B_{\text{after second sequential accept }\bar{B}\text{ result}} = \frac{(0.887)(0.976)}{(0.887)(0.976)+(1-0.813)(0.198)} = 0.975$$

Maybe 97.5% is sure enough to switch to B. But perhaps you want to be 99% sure. So, do another trial.

$$B_{\text{after third sequential accept }\bar{B}\text{ result}} = \frac{(0.975)(0.992)}{(0.975)(0.992)+(1-0.975)(0.149)} = 0.996$$

If, however, the third trial indicated that the average B is worse than the threshold, then:

$$B_{\text{if third result is }\bar{B} < \text{threshold}} = \frac{(0.975)(0.00766)}{(0.975)(0.00766)+(1-0.975)(0.851)} = 0.258$$

which would undermine the growing belief.

Note: After each trial, you have better information about the variability of B and what its average might be. Additionally, if the average of the trial values of the benefits of B are being used, then the standard deviation of the average would be $\$50/\sqrt{n}$ which would also change the probabilities in the table. So, after each trial the probabilities in the table should be updated.

TABLE 2.3

Generic Probabilities of Correct Diagnosis and False Positive and Negative

Truth \ Test result	Test claims it is true	Test claims it is false
Something really is true	p(true outcome, if true)	p(false outcome, if true)
It really is false (not true)	p(true outcome, if false)	p(false outcome, if false)

In general, the table of probabilities is as shown in Table 2.3.

Often the probabilities in the four categories can be determined from extensive and controlled testing on known situations. But often, they can be reasonably estimated from experience. In either case do not think that the probabilities are perfectly known. They have errors. But in our experience the uncertainty on the reasonable values does not undermine the propagation of the belief. Alternate values might lead to needing one more or one less test to provide adequate confidence to make a decision.

What is adequate confidence to take action? If $B = 0.99$ then you are very certain that the supposition is true. If $B = 0.01$ then you are equally confident that it was untrue. But to take action be sure that the consequences of a wrong decision, tempered by the probability of a wrong decision, are acceptable. *Risk* is the probability of an event times the penalty for undesirable consequences. *Benefit* is the probability of an event times the value of the desirable outcome. Set the threshold of the belief to take action by the consequences of taking a right or wrong action (accepting A when it should be B, taking B when it should be A, etc. Reasonable threshold values are in the 0.05 (and 0.95) and 0.001 (0.999) range. If the impact of the consequences of a decision is much greater than the cost of the trials, then use the more extreme thresholds. If the cost of trials is relatively high compared to the impact of a decision, then use the less extreme thresholds. The 0.05 decision is the commonly accepted threshold for statistical testing.

Many accept this as a very good guide to updating belief with sequential results, and then using the belief to make decisions. Alternately, although purists accept the mathematical model, many object to the method because of the uncertainty on the probability values and the initial belief.

Of course, humans might not want to follow this kind of logical rule. Often, when they know something is true, they consider themselves to be absolutely sure, and rather than admit they were wrong, they reject any data that would counter their personal belief. Or when they want something to be true, they reject any opposing data. Here are some examples:

"Win or lose, our sports fans are kind and gracious, but our opponent's fans are disrespectful poor sports."

"My kid is the best looking, smartest, and most athletic in the entire class."

"Reel mowers are better than rotary mowers."

If your boss, or significant other knows the truth, or wants a particular outcome, and the resulting action from an erroneous belief has fewer adverse consequences than the personal cost and effort of proving that person wrong, it might be best to let it go their way. Only martyrs let logic lead to a confrontation against authority. On the other hand, we hope that each of us protects ourselves, our organizations, and society from action based on erroneous beliefs.

2.3 Takeaway

Composite event probability must be within 0 and 1, inclusive. If you are creating your own application, extrapolate to very many or very few events to check that the way you are calculating composite probability does not exceed rational limits.

2.4 Exercises

1. In flipping a fair coin twice, what is the probability of a) getting two Heads, b) getting two Tails, c) getting a Head on the first flip and a Tail on the second, d) not getting any Heads?
2. At a particular summer camp, the probability of getting a case of poison ivy is 0.15 and the probability of getting sunburn is 0.45. What is the probability of a) neither, b) both, c) only sunburn, d) only poison ivy?
3. After rolling three fair six-sided dice, what is the probability of a) getting three ones showing, b) having only one four showing, c) getting a one and a two and a three?
4. If the probability of rain tomorrow is 70% and rain the next day is 50%, then 0% for the next five days, what is the probability of rain a) on both of the next two days, b) on all of the next seven days, c) at least once this week?
5. There are two safety systems on a process. If an over-pressure event happens in the process, the first safety override should quench the source, and if that is not adequate the back-up system should release excess gas to a vent system. Normal control of the process is generally adequate, only permitting an average of about ten over-pressure events per year. The quench system, we are told, has a 95% probability of working adequately when needed, and the back-up vent has a 98% probability of working as needed. What is the probability of an undesired event (the over-pressure happens, and it is not contained by either safety system) in a) the next one-year period, and b) the next ten-year period?
6. There is a belief that Treatment B is better than the current Treatment A in use. The belief is a modest 75%, $B = 0.75$. If B is equivalent to A, not better, then there is a 50/50 chance that the trial outcome will indicate either B is better or worse. However, if B is better, then the chance that it will appear better in the trial is 80%. What is the new belief after the trial if a) the trial indicates B is better, b) if the trial indicates B is not better, and c) how many trials of sequential successes are needed to make the belief that B is the right choice raise to 99%?
7. A restaurant buys thousands of jalapeno peppers per day, of which 5% are not spicy-hot. They use five peppers in each small batch of salsa. If two (or more) of the five are not hot, customers are likely to complain that the salsa is not adequate. What is the probability of making an inadequate batch of salsa? Quantify how larger batch sizes will change the probability.

3

Distributions

3.1 Introduction

Most statistical methods are based on theoretical distributions, described by parameters (such as mean and variance), which can be good approximations to the distribution of experimental data. The mathematical models of the distribution define its shape, but the parameter values define location and variation. Parameters, then, are descriptors of a theoretical distribution. Statistical procedures that state conclusions about such parameters using sample data are *parametric statistics.* Statistical procedures that state conclusions about populations for which the theoretical distribution has not been assumed are termed *nonparametric statistics.* Before you can effectively utilize such methods, you must be able to choose the distribution that best matches the data. This chapter presents key theoretical distributions and the characteristics of the populations involved. It also shows how to describe the distribution of experimental data.

3.2 Definitions

Measurement: A numerical value indicating the extent, intensity, or measure of a characteristic of an object.

Data: Either singular as a single measurement (such as a y-value) or plural as a set of measurements (such as all the y-values). Data could refer to an input-output pair (x, y) or the set $\left(\underline{x}, \underline{y}\right)$.

Observation: A recording of information on some characteristic of an object. Usually a paired set of measurements.

Sample: 1) A subset of possible results of a process that generates data. 2) A single observation.

Sample size: The number of observations, datasets, in the sample.

Population: All of the possible data from an event or process – usually $n = \infty$.

Random disturbance: Small influences on a process that are neither correlated to other variables nor correlated to their own prior values.

DOI: 10.1201/9781003222330-3

Random variable: A variable or function with values that are affected by many independent and random disturbances despite efforts to prevent such occurrences.

Discrete variable: A variable that can assume only isolated values, that is, values in a finite or countably infinite set. It may be the counting numbers, or it may be the digital display values of truncated data.

Continuum variable: A variable that can assume any value between two distinct numbers.

Frequency: The fraction of the number of observations within a specified range of numerical values relative to the total number of observations.

Cumulative frequency: The sum of the frequencies of all values less than or equal to a particular value.

Mean: A measure of location that provides information regarding the central value or point about which all members of the random variable X are distributed. The mean of any distribution is a parameter denoted by the Greek letter μ.

Variance: A parameter that measures the variability of individual population values x_i about the population mean μ. The population variance is indicated by σ^2.

Standard deviation: σ is the positive square root of the variance.

Empirical Distributions: These are obtained from a sampling of the population data. As a result, the models or the parameter values that best fit a model to the data (such as μ and σ) may not exactly match those of the population.

Theoretical Distributions: These are obtained by derivation from concepts about the population. If the concepts are true, then the models and corresponding parameter values represent the population. But nature is not required to comply with human mental constructs.

Category (classification): The name of a grouping of like data, influences, events such as heads, defectives, zero-crossings, integers, negative numbers, green, etc.

3.3 Discrete Distributions

There are two classes of distributions: Discrete and continuous. Discrete distributions are used to describe data that can have only discrete values. Such data have a specific probability associated with each value of the random variable. There are distinct and measurable step changes associated with each value of the variable. Some examples of discrete variables are the size of the last raise you received (it was not in fractions of a cent), the score of the last sporting event you watched, the number of personal protective equipment items available to you on your job, the number of first-quality computer chips on a silicon wafer, the number of defects in a skein of yarn, the energy of electrons in a particular quantum state, the number of raindrops that fall onto a square inch of land, etc.

The variable x_i represents the count of events in the *i*th category. The categories are mutually exclusive, such as alphabet letters, or pass/fail. The value of x_i is an integer number. Looking at this paragraph, if $I = 1$ represents the occurrence of the letter "a" and $I = 2$ that of the letter "b", then the value of $x_1 = 18$ and $x_2 = 4$.

Probability density functions, $pdf(x_i)$ or simply $f(x_i)$, are associated with distributions of discrete variables, x_i represent the probability of possible values of the ith data category. For example, if you flip a coin you expect $k = 2$, two outcomes, Head and Tail, or 0 and 1. If the first classification of x_1 = Head, then $f(x_1) = 0.5$. All such probability functions have the following properties:

1. x_i are the discrete possible values of a variable X, and x_i is the ith of the k finite values of the outcome. Usually, the index i places the x_i values in ascending order.
2. The probability functions are mathematical models of the population, of the infinity of possible samples, not of a finite sample of k number of values.
3. $f(x_i)$ is the frequency, the probability of occurrence that a value x_i will occur. It is positive and real for each x_i. $f(x_i) = \lim_{n\to\infty} \{n_i/n\}$.
4. $\sum_{i=1}^{k} f(x_i) = 1$ where k is the number of categories.
5. $P(E) = \sum f(x_i)$ where the sum includes all x_i in the event E.

These definitions illustrate the notation we use throughout this book. We use capital Latin letters for populations and lowercase Latin letters for particular numerical observation values from the populations. Lowercase Greek letters are used for population parameters. Point and cumulative distributions are identified by f and F (or alternately CDF) respectively. P stands for "probability of …". We are using the conventional notation for discrete distributions: x in the summations of the cumulative distribution functions of discrete distributions sometimes represents the number of items in a class (group, collection, etc.) or at other times, x represents the numerical value that quantifies the class. By using this notation, the formulas in this book are consistent with those you may find in other statistics books. We state this as a warning, because in conventional notation for variables, x means the value of the variable as opposed to the number of occurrences in a category. The cumulative distribution function (CDF) is a function $F(x_r)$ obtained from the probability function and is defined for the values of x_i of the random variable X by

$$CDF(x_r) = F(x_r) = P(X \le x_r) = \begin{cases} 0 & \text{for } X < x_1 \\ \sum_{i=1}^{r} f(x_i) & \text{where } x_1 \le X \le x \\ 1 & \text{for } X \ge x_n \end{cases} \tag{3.1}$$

where x_i is an ordered set $x_1 < x_2 < \ldots < x_{n-1} < x_n$.

The cumulative distribution function is a nondecreasing function with the following properties:

1. $0 \le F(x) \le 1$.
2. $\lim_{x\to-\infty} F(x) = 0$.
3. $\lim_{x\to\infty} F(x) = 1$.

As a result, the probability that X will lie between two values x_i and x_j can be found from the difference of the cumulative probabilities at x_i and x_j or

$$P(x_i < X \le x_j) = P(X \le x_j) - P(X \le x_i) = F(x_j) - F(x_i) \tag{3.2}$$

The mean and variance of a discrete distribution are calculated from

$$\mu = \sum_{i=1}^{\infty} x_i f(x_i) \tag{3.3}$$

$$\sigma^2 = \sum_{i=1}^{\infty} (x_i - \mu)^2 f(x_i) \tag{3.4}$$

Where i is the index representing each of the possible x_i values, and $f(x_i)$ is the weighting factor for the ith x_i value, for all possible i-values.

Note: The terms *CDF* and *F* are used interchangeably. And often, so are *pdf*(*x*) and *f*(*x*).

3.3.1 Discrete Uniform Distribution

When each discrete event has the same likelihood (probability) of occurring, the probability function is given by

$$f(x_i) = \frac{1}{n}, \quad 1 \le i < n \tag{3.5}$$

where n is the number of discrete values for x. For the cumulative discrete distribution function,

$$F(x_i) = P(X \le x_i) = \frac{i}{n} \tag{3.6}$$

where $x_1 < x_2 < x_3 \cdots < x_n$.

A classic example is that of rolling a cubical die. The $n = 6$ categories of possible outcomes are equally probable.

The X in Equation (3.6) may represent either a dimensionless counting number (7 bolts), a category (3 Heads) or a dimensional real number (last raise was \$437.25/month); however, X must be limited to a finite number, n, of discrete values. For the raise example, the discrete values are multiples of 1¢/month. If the maximum possible raise could have been \$600.00/month, then $n = 600.00/.01 + 1 = 60{,}001$ (we cannot exclude the zero-raise event). Consequently, $x_{10{,}000}$ represents the 10,000th value of X, which is \$99.99/month.

Figure 3.1 illustrates the discrete uniform distribution for $n = 5$, and the corresponding cumulative discrete uniform distribution, also for $n = 5$.

Recognizing that each x_i value has the same probability, or frequency of occurring, $f(x_i) = f(x_j)$, the mean and variance of the discrete uniform distribution are

$$\mu = \frac{1}{n}\sum_{i=1}^{n} x_i \tag{3.7}$$

$$\sigma^2 = \frac{1}{n}\sum_{i=1}^{n} (x_i - \mu)^2 \tag{3.8}$$

If the x_i values are also equally incremented between $x_1 = a$ and $x_n = b$, so that $x_{j+1} - x_j = \Delta x = (b-a)/n$ (such as with a die which has sides with values of 1, 2, 3..., 6, where $a = 1$ and $b = 6$) then

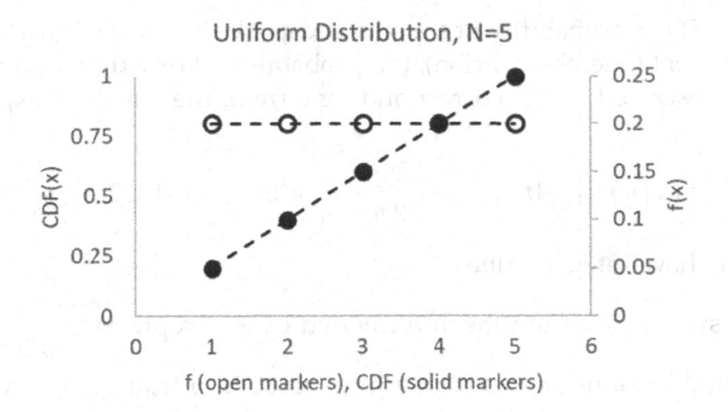

FIGURE 3.1
Discrete uniform distribution: (a) point, (b) cumulative, both for $n = 5$.

$$\mu = \frac{a+b}{2} \tag{3.7a}$$

$$\sigma^2 = \frac{(b-a+1)^2 - 1}{12} \tag{3.8a}$$

Note: In Equations (3.7a) and (3.8a) the μ and σ^2 values are calculated from a theoretical model with a and b known, as opposed to those that would be calculated from observed sample values.

The variable X may be dimensionless, or it could have any units. The parameters μ and σ have the same units as X. The probabilities P and f are dimensionless. The variable X will have k discrete values that need not be uniformly spaced. For instance, if a family has five children, whose heights are 4 ft 2 in., 4 ft 9 in., 5 ft 1 in., 5 ft 5 in., and 5 ft 7 in., we can let X_i be the height of the ith child. Then,

$$\begin{aligned} X_1 &= 4\,\text{ft}\,2\,\text{in.} \\ X_2 &= 4\,\text{ft}\,9\,\text{in.} \\ X_3 &= 5\,\text{ft}\,1\,\text{in.} \\ X_4 &= 5\,\text{ft}\,5\,\text{in.} \\ X_5 &= 5\,\text{ft}\,7\,\text{in.} \end{aligned}$$

If the children are chosen randomly, the probability of selecting a child with a particular height, say 5 ft 5 in. (the fourth child), is $1/n = 0.2$. Also, for instance, the probability of a particular feasible value showing on the roll of a fair six-sided die is $1/n = 1/6 = 0.1\overline{666}$.

3.3.2 Binomial Distribution

A discrete distribution called the binomial occurs when any observation can be placed in only one of two mutually exclusive categories, such as greater-than or less-than-or-equal-to, safe or unsafe, hot or cold, on or off, 0 or 1, pass or fail, Heads or Tails, etc. Although these characteristics are qualitative, the distribution can be made quantitative by assigning the values 0 and 1 to the two categories. The method of assignment is immaterial so long as it is consistent. Customarily, the categories are labeled success (value = 1) and failure (value = 0). If p = probability of success and $q = 1 - p$ = probability of failure in one trial of the experiment (one observation), the probability of exactly x number of successes in n trials can be described by the corresponding term of the binomial expansion, or

$$f(x \mid n) = \binom{n}{x} p^x q^{n-x} \equiv \frac{n!}{x!(n-x)!} p^x q^{n-x}, \ x = 0,1,2,\ldots,n \tag{3.9}$$

where X may only have integer values.

Note: The $\binom{n}{x}$ symbol does not mean n divided by x, it represents $\frac{n!}{x!(n-x)!}$, which is the number of combinations (ways) of having x occur in n trials. If $n = 4$ and $x = 2$ then $\binom{n}{x} = \frac{4!}{2!(4-2)!} = \frac{4\times3\times2\times1}{2\times1(2\times1)} = 6$. The six possible success–fail patterns could be 1100, 1010, 1001, 0110, 0101, and 0011.

Note: The variable x represents the numerical count in a particular category, it is not the value of the category.

Note: When n is large, the factorial terms become large, and direct calculation of either the numerator or denominator can result in digital overflow. Fortunately, the number of integers in the numerator and denominator is equal, there are n digits in each, and a best way to calculate the ratio is to alternate dividing and multiplying. But, many software packages provide convenient functions. In Excel the function is $f(x \mid n) = \text{BINOMIAL.DIST}(x,n,p,0)$.

The binomial cumulative distribution function is

$$F(x_i \mid n) = P(X \le x_i \mid n) = \sum_{k=0}^{x_i} \binom{n}{k} p^k (1-p)^{n-k}, \ i = 0,1,2,\ldots,n \tag{3.10}$$

where X may have only integer values, for selected values of n and p. The notation (something | n) means "something given the value of n". In Excel $F(x_i \mid n) = \text{BINOMIAL.DIST}(x,n,p,1)$.

One can compute other probabilities such as $P(x_i \le X \le x_j)$, indicating the probability that an observation value, X, would be between and including x_i and x_j.

$$P(x_i \le X \le x_j) = P(X \le x_j) - P(X \le x_{i-1}) = F(x_j) - F(x_{i-1})$$

$$= \sum_{k=0}^{x_j} \binom{n}{k} p^k (1-p)^{n-k} - \sum_{k=0}^{x_{i-1}} \binom{n}{k} p^k (1-p)^{n-k} \tag{3.11}$$

The best way to explain the use of Equation (3.11) is by use of a brief example. If you want $P(10 \le X \le 20)$, you need to exclude all values of X which are not in the probability specification. In this case, we want to include only the values X = 10, 11, 12, …, 20. The values of X = 0, 1, 2, …, 9 must be excluded. As $x_1 = 10$, to exclude values below 10, we must use $(x_1 - 1) = (10 - 1) = 9$ as the index.

Specific values, such that the probability will be exactly s successes in n trials, can be found from

$$P(X = s \mid n) = P(X \le s \mid n) - P(X \le (s-1) \mid n) \tag{3.12}$$

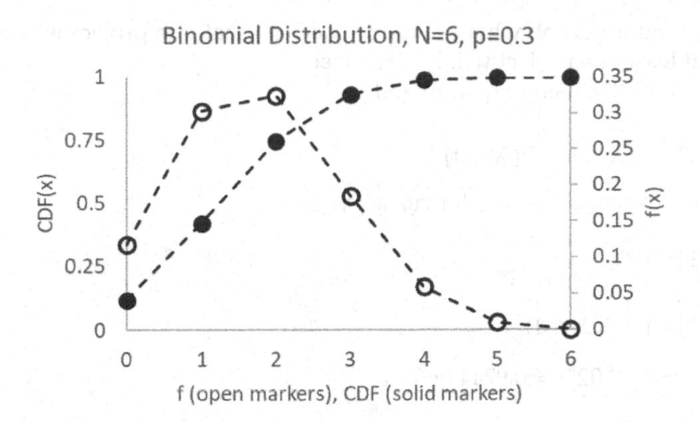

FIGURE 3.2
Typical binomial distribution: (a) point, (b) cumulative for both $n = 6, p = 0.3$.

Depending on the values of n and p, the binomial distribution may have several shapes; however, Figure 3.2, with $n = 6$ and $p = 0.3$, illustrates the characteristic shape.

The mean and variance of the number count of events described by the binomial distribution are

$$\mu_x = np \tag{3.13}$$

and

$$\sigma_x^2 = npq \tag{3.14}$$

where n is the total number of attempts (trials), p is the portion of resulting successes expressed as a numeric or decimal fraction of n, and $q = 1 - p$. Accordingly, μ is the average number of successes, and might not be an integer. If you flip eight coins you expect four to be a head. However, if you flip seven, you expect 3.5 to be a head. Of course, you cannot have a fraction of a count. What that means is if you flipped seven coins millions of times, and counted the number of heads each time, you might have 3, 2, 4, 5, 4, 3, …. Averaging this list is 3.5. Similarly, σ is the standard deviation on the value of the number of successes, not of the probability.

The point probability function f and the cumulative probability function F are dimensionless.

The value of the probability of a particular trial being a success, p, is considered to be the true distribution value, not the chance value one might assign from a few trials. The ideal true value of a coin flipping a head is 0.5. However, if you did not know that value, and flipped a coin seven times to see the head-to-tail ratio you might find three out of seven heads and calculate $p = \frac{3}{7} = 0.42857\ldots$ as the probability. If you do not know the true value for p, you could get a reasonably close approximation by doing many (perhaps thousands) experiments.

Example 3.1: You have submitted four proposals for upgrading the manufacturing facilities in your process area. From past experience you feel that the chance for any one project to be approved by the Finance Committee is 0.6. Accepting that $p = 0.6$ and that selection is a random event, what are the chances (a) that one project will be approved and (b) that at least one project will be approved?

For $n = 4$ and $p = 0.6$, using Equation (3.12):

(a) $P(X=1)=P(X\leq 1)-P(X\leq 0)$

$$=0.1792-0.0256=0.1539 \text{ or } 15\%$$

$$P(X=1)=15\%$$

(b) $P(X\geq 1)=1-P(X\leq 0)$

$$=1-0.0256=0.9744 \text{ or } 97\%$$

$$P(X\geq 1)=97\%$$

Why are these answers different? The first allows the occurrence of only a single event, approval of only one project. The second allows approval of any number of your projects except 0.

3.3.3 Poisson Distribution

The Poisson distribution is concerned with the number of events occurring during a given time or space interval. The interval may be of any duration or in any specified region. The Poisson distribution, then, can be used to describe the number of breaks or other flaws in a particular beam of finished cloth, or the arrival rate of people in a queuing line, or the number of defectives in a paint weathering trial, or the number of defective beakers per line per shift. The Poisson distribution describes processes with the following properties:

1. The number of events, X, in any time interval or region is independent of those occurring elsewhere in time or space.
2. The probability of an event happening in a very short time interval or in a very small region does not depend on the events outside this interval or region.
3. The interval or region is so short or small that the number of events in the interval is much smaller than the total number of events, n.

The point Poisson distribution (point probability) function $f(x)$ can be expressed as

$$f(x)=\frac{\lambda^{x}e^{-\lambda}}{x!} \tag{3.15}$$

where x is the number of events, $f(x)$ is the probability of x events occurring in an interval, λ is the expected average number of events per interval, and $e = 2.7182818\ldots$ is the base of the natural logarithm system.

The cumulative Poisson distribution function $F(x)$ is

$$CDF(x)=F(x)=P(X\leq x)=\sum_{k=0}^{x}\frac{\lambda^{k}e^{-\lambda}}{k!} \tag{3.16}$$

where e is the base of the natural logarithm system.

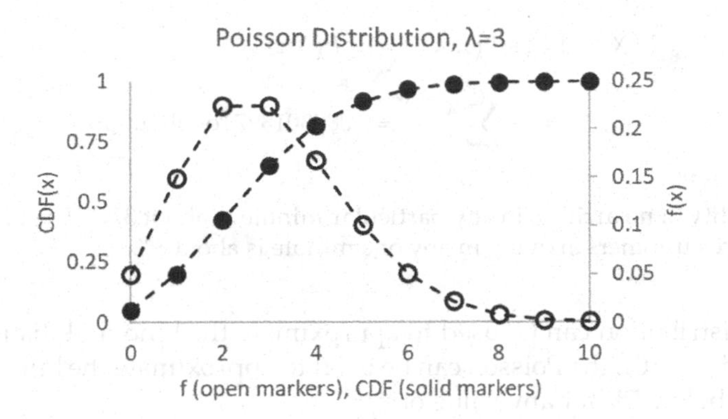

FIGURE 3.3
Poisson distribution: (a) point, (b) cumulative, both for $\lambda = 3$.

Using Excel functions, values of the point Poisson distribution are obtained by $f(x) = \text{POISSON.DIST}(x, \lambda, 0)$, and values for the cumulative Poisson distribution are obtained by $F(x) = \text{POISSON.DIST}(x, \lambda, 1)$.

Figure 3.3 illustrates the shape of the Poisson distribution for $\lambda = 3$ (three events per interval of interest).

Note: The shape of the Poisson distribution is similar to that of the binomial; however, the right-hand tail of the Poisson distribution extends to infinity (although here only graphed up to $n = 10$), whereas there can be at most 6 out of 6 (or n out of n) events in the binomial distribution; even if the expected number of events per unit is 3, the Poisson distribution allows for the rare but possible case in which 9 events occur in the same interval.

The mean and variance of the Poisson distribution are

$$\mu = \lambda = np \tag{3.17}$$

and

$$\sigma^2 = \lambda = np \tag{3.18}$$

where n is the total number of events within the interval, an integer, and p is the probability that each event will occur in an interval. λ may have a fractional value. The variable X is an integer that has the units of number of events or things per unit of time, area, volume, or other specified interval, and λ represents the expected (or average) number of events that will occur within that unit interval. The point and cumulative distributions are dimensionless.

Again, this supposes that you know the true value for λ.

Example 3.2: A branch bank has found that their drive-in customer arrival rate averages one customer per minute during the morning hours. What is the probability that there will be no customer arrivals in any particular minute? What is the probability that three or more customers will arrive in any minute?

From Equation (3.15), $f(x = 0) = \lambda^x e^{-\lambda}/x! = 1^0 e^{-1}/0! = 0.3678944$ or about 37%.

From Equation (3.16),

$$P(X \geq 3) = 1 - P(X < 3) = 1 - P(X \leq 2)$$

$$= 1 - \sum_{k=0}^{2} \frac{\lambda^k e^{-\lambda}}{k!} = 0.080301397 \text{ or about 8\%.}$$

The probability of no arrivals in any particular minute is about 37%. The probability of three or more customers arriving in any one minute is about 8%.

The Poisson distribution can be used to approximate the binomial distribution. In general, if $n \geq 20$ and $p \leq 0.01$, the Poisson can be used to approximate the binomial with errors on $f(x)$ generally below 5% for any value of x.

Example 3.3: In the preparation of sample coupons for a salt-spray corrosion test, 4% of the metal coupons are found to be unusable because of improper cleaning. In a sample of 30 coupons, what is the chance of having two or fewer rejects?

We assume the binomial model as the coupons are initially either clean or not. Furthermore, the cleanliness of any coupon has no effect on whether any of the others in the sample are clean or not. For $p = 0.04$ and $n = 30$, the binomial probability is

$$P_{\text{binomial}}(X \leq 2) = \frac{30!}{2!28!}(0.04)^2(0.96)^{28}$$

$$+ \frac{30!}{1!29!}(0.04)^1(0.96)^{29}$$

$$+ \frac{30!}{0!30!}(0.04)^0(0.96)^{30}$$

$$= 0.883103 \text{ or } 88\%$$

For $\lambda = np = 0.04(30) = 1.2$, the Poisson approximation of the binomial probability is

$$P_{\text{Poisson}}(X \leq 2) = 0.8795 \text{ or } 88\%.$$

Although the Poisson distribution is not the proper descriptor of these events (a single coupon cannot be dirty twice), in the limit of $n = 30$ and low event probabilities the Poisson and binomial distributions are similar.

Example 3.4: If one process has an average of three shutdowns per year, how many would you expect from five similar processes in a month?

For the one process $\lambda = 3[\text{shutdowns per year per process}]$. Five times as many processes would total five times more on average. But, in a month, you expect 1/12 as many on average. So, $\lambda = 3 \cdot 5/12 = 1.25[\text{shutdowns per month per five processes}]$. The individual probabilities of 0, 1, and 2 of shutdowns

$$P(0) = \frac{\lambda^0 e^{-\lambda}}{0!} = 0.2865\ldots$$

$$P(1) = \frac{\lambda^1 e^{-\lambda}}{1!} = 0.3581\ldots$$

$$P(2) = \frac{\lambda^2 e^{-\lambda}}{2!} = 0.2238\ldots$$

To obtain the average

$$\bar{n} = \sum_{i=1}^{\infty} i \cdot P(i) = 1.25[\text{shutdowns per month per five processes}]$$

Expectedly $\bar{n} = \lambda$.

3.3.4 Negative Binomial Distribution

In cases in which the binomial distribution governs the probability of occurrence of one of two mutually exclusive events, we calculated the probability of success exactly s times out of n trials. The negative binomial distribution is used in a complementary way, that is, for calculating the probability that exactly n trials are required to produce s successes. The probabilities of success and failure remain fixed at p and q, respectively. The only way this situation can occur is for exactly $(s - 1)$ of the first $(n - 1)$ trials to be a success, and for the next, or last, trial also to be a success. The probability of $x = n$, the number of trials needed to produce s successful outcomes, then

$$f(x = n \mid s) = \binom{n-1}{s-1} p^s q^{n-s}, \quad s \le n \tag{3.19}$$

is the negative binomial distribution. The cumulative negative binomial distribution is

$$F(x = n \mid s) = P(s \le x \le n) = \sum_{i=s}^{n} \binom{i-1}{s-1} p^s q^{i-s} \tag{3.20}$$

Figure 3.4 illustrates the negative binomial distribution for $p = 0.5$ and $s = 3$.

The mean and variance of the negative binomial distribution are given by

$$\mu = \frac{s}{p} \tag{3.21}$$

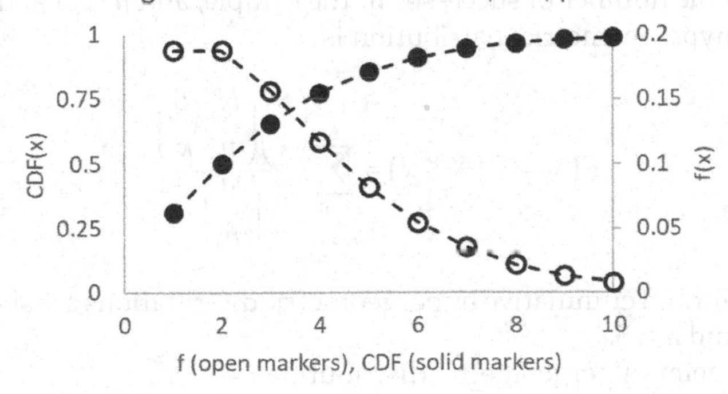

FIGURE 3.4
Negative binomial distribution: (a) point, (b) cumulative both for $p = 0.5$ and $s = 3$.

and

$$\sigma^2 = \frac{sq}{p^2} \tag{3.22}$$

The units on x, s, n, and μ are the numbers of trials. The point and cumulative distribution functions, $f(x_i)$ and $F(x_i)$, are dimensionless. The units on p and q are the probabilities of success or failure.

Example 3.5: Suppose one of your power sources for an analytical instrument in the quality control laboratory has died with a snap and a wisp of smoke. You have finally located the trouble as a faulty integrated circuit (IC). You have been able to find five replacement ICs. You have also found that for this service the chance of failure of an IC is 12%. What is the probability that you will have to use all five ICs before getting one that does not burn out?

Let us define burnout as failure, so $q = 0.12$ and $p = 0.88$. As $x = 5$ and $s = 1$, using Equation (3.19),

$$f(x=5|1) = \binom{4}{0}(0.88)0.12^4 = 1.825\times10^{-4} \text{ or } 0.02\%$$

the probability is less than 0.02% that you will have to try all five of the ICs to repair the power supply.

3.3.5 Hypergeometric Distribution

The hypergeometric distribution is often used to obtain probabilities when sampling is done without replacement. As a result, the probability of success changes with each trial or experiment. The point hypergeometric probability function is

$$P(X=s) = f(s) = \frac{\binom{S}{s}\binom{N-S}{n-s}}{\binom{N}{n}}, \ s \le \min(n, S) \tag{3.23}$$

where N is the population size, n is the sample size, S is the actual number of successes in the population, s is the number of successes in the sample, and $n \le N$ and $(n - s) \le (N - S)$.

The cumulative hypergeometric distribution is

$$F(x) = P(X \le x) = \sum_{k=0}^{x} \frac{\binom{S}{k}\binom{N-S}{n-k}}{\binom{N}{n}} \tag{3.24}$$

Examples of the point and cumulative hypergeometric distributions are shown in Figure 3.5 for $N = 20$, $S = 15$, and $n = 5$.

The mean of the point hypergeometric distribution is

$$\mu = \frac{nS}{N} \tag{3.25}$$

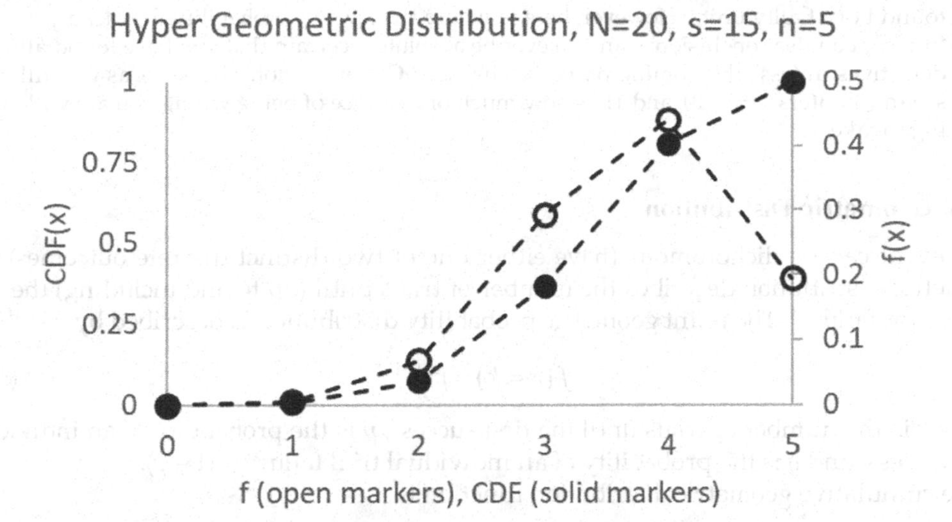

FIGURE 3.5
Hypergeometric distribution: (a) point, (b) cumulative, both for $N = 20$, $s = 15$, $n = 5$.

and the variance is

$$\sigma^2 = \frac{N-n}{N-1} n \frac{S}{N} \frac{N-S}{N} \tag{3.26}$$

The units of μ, s, N, n, and S are the number of items, populations, or successes. The point and cumulative probability functions $f(x)$ and $F(x)$ are dimensionless.

Example 3.6: In the production of avionics equipment for civilian and military use, one manufacturer randomly inspects 10% of all incoming parts for defects. If any of the parts is defective, all the rest are inspected. If 2 of the next box of 50 diodes are actually defective, what is the probability that all of the diodes will be checked before use? This question is really whether the quality control sample of 5 will contain at least one of the defective parts.

For this problem, $N = 50$, $n = 5$, and $s = 2$, as we choose to define success as finding a defective diode. The probability is found from

$$F(0) = P(X \geq 1) = \sum_{k=1}^{2} \frac{\binom{2}{k}\binom{50-2}{5-k}}{\binom{50}{5}}$$

$$= 0.1918367 \text{ or } 19\%$$

With the current sampling procedure, there is approximately a 20% chance of finding a defective part.

If that part would cause failure of the finished item, the 80% chance of not finding that part is inadequate as a quality control measure, considering the application. Either the sample size should be increased (up to 100% inspection of all incoming parts), or a more reliable supplier of diodes should be found! Actually, had you known for certain that 4% of the diodes were defective, you would have tested each one until you had found both faulty units. However, here is where the laws of probability may lead you to accept a false conclusion: Can you ever be absolutely certain that you have found all defectives unless all incoming parts are checked? Of course not. The issue, as we will see in Chapters 5, 6, 7, 10, and 11, is how much of a chance of being wrong you are willing to take.

3.3.6 Geometric Distribution

If an event can be dichotomous (have either one of two distinct discrete outcomes), the geometric distribution describes the number of trials until (up to and including) the first success (or failure). The point geometric probability distribution is described by

$$f(x=k)=pq^{k-1} \tag{3.27}$$

where k is the number of trials until the first success, p is the probability of an individual trial success, and q is the probability of an individual trial failure = $(1-p)$.

The cumulative geometric distribution function is

$$F(x=k)=P(1\le x\le k)=\sum_{i=1}^{k}pq^{i-1} \tag{3.28}$$

Figure 3.6 shows the geometric distribution for $p = 0.6$.

Values for the mean and variance of the geometric distribution are

$$\mu=\frac{1}{p} \tag{3.29}$$

and

$$\sigma^2=\frac{q}{p^2} \tag{3.30}$$

Although the outcome value may have any dimension (ft/sec², kilopascals, ft, etc.) or be a class variable (on or off, wet or dry, smoker or nonsmoker, etc.), the variables p, q, $f(x)$, and

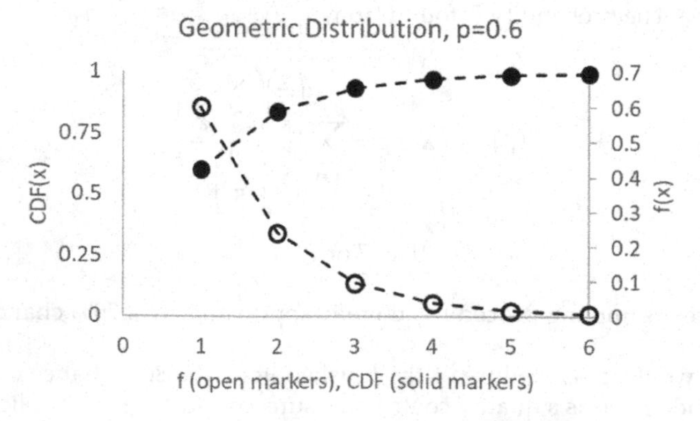

FIGURE 3.6
Geometric distribution: (a) point, (b) cumulative, both for $p = 0.6$.

$F(k)$ are dimensionless. The value of the mean μ is the average number of trials until the first success, and the units on x and k are the number of trials.

The pattern of 5 successive fails and the 6th trial providing a success would be FFFFFS. This is one success in 6 trials. Similar to the geometric distribution, the binomial distribution gives the probability of $x = 1$ with $n = 6$. But the binomial distribution would allow for any sequence that generated that count. For instance, FFSFFF, or FFFFSF. There are 6 possible combinations of one S, and 5 Fs. In the geometric distribution, there is only one of the $\binom{1}{n} = n$ combinations that matters. The geometric distribution is the same as the binomial distribution when $x = 1$ and divided by n to represent that only one of the n combinations is the one sought.

Example 3.7: At a marina, the boat slips either have boats assigned to them or are available. If 100 slips are at the marina and 75 have been rented, what is the probability of having to select 5 slips by random choice of slip number until getting one that is available (4 fails followed by a success is 4 + 1 = 5 trials)?

Since p = probability of success (available empty slip) = 0.25, then q = probability of failure (rented) = 0.75, using Equation (3.27),

$f(x = 5) = 0.25(0.75)^{5-1} = 0.07910156$ or about 8%.

In addition, using Equation (3.29),

$$\mu = \frac{1}{0.25} = 4 \text{ trials before the first success}$$

the probability of having to select five slip numbers at random before finding one empty is about 8%.

On average, you would have to choose four slips at random in order to find one empty.

3.4 Continuous Distributions

In the previous section, the distributions were related to the number count of events. In contrast, many measurements are continuum-valued. Continuous distributions model the probabilities associated with continuous variables, such as those that describe events such as service life, pressure drop, flow rate, temperature, percent conversion, and degradation in yield strength.

That we measure continuous variables in discrete units or at fixed time intervals does not matter; the variables themselves are continuous even if the measuring devices give data that are recorded as if step changes had occurred. A familiar example is body temperature, a continuous variable measured in discrete increments. Think about it: Even if you have a fever, your temperature does not change from 98.6 to 101.2°F in one step or even in a series of connected 0.2°F intervals, just because the thermometer is calibrated that way.

We must acknowledge, however, that the world is not continuous. From an atomic and quantum mechanical view of the universe, no event has a continuum of values. However, on the macroscale of engineering, individual atoms are not distinguishable within measurement discrimination, and so the world appears continuous. For most practical engineering purposes, it is possible to approximate any distribution in which the discrete variable has more than 100 values with a probability density function of a continuous random variable.

A cumulative continuous distribution function $F(x)$ is defined as

$$CDF(x) = F(x) = \int_{-\infty}^{x} pdf(X)\,dX \tag{3.31}$$

where $pdf(X)$ is a continuous probability density function and X is a continuous variable, which could represent time, temperature, weight, composition, etc. x is a particular value of the variable X. The units on x and X are identical and are not a count of the number of events as the x-variable in the discrete distributions. The $F(x)$ is the area under the $pdf(x)$ curve, is dimensionless, and as x goes from $-\infty$ to $+\infty$, $F(x)$ goes from 0 to 1.

$$\int_{-\infty}^{+\infty} pdf(X)\,dX = 1 \tag{3.32}$$

Note, again, the terms *CDF* and *F* are used interchangeably.

Additionally, the terms $pdf(x)$ and $f(x)$ are also used interchangeably. In the discrete distributions, $pdf(x)$ would mean point distribution function, and in continuous functions, it means probability density function.

Although both the continuum $pdf(x)$ and discrete $f(x_i)$ represent the histogram shape of data, they are different. The dimensional units of $pdf(x)$ constitute a major difference between a continuous probability distribution function and the $f(x_i)$ of a discrete point probability distribution. The $pdf(x)$ necessarily has dimensional units that are the reciprocal of the continuous variable. For $F(x)$ of Equation (3.31) to be dimensionless, integrating with dx, the argument of the integral, $pdf(x)$, must have the units of the reciprocal of dx. $pdf(x)$ is often termed a rate, a rate of change of $F(x)$ w.r.t. x. By contrast, in a discrete function $F(x_i)$ is the sum of $f(x_i)$, the fraction of the dataset with a value of x_i, so $f(x_i)$ is dimensionless.

You cannot use a discrete point distribution in Equation (3.31) or a continuous function in Equation (3.1) and expect $F(x)$ to remain a dimensionless cumulative probability. Another difference between discrete and continuous probability density functions is that x is used only to represent values of the variable involved throughout the continuous case. For discrete distributions, x was often the number of events in a particular class (category).

So, whether you are using the term $pdf(x)$ or $f(x)$ take care that you are properly using the dimensionless version for distributions of a discrete variable, and the rate version with reciprocal units of X for distributions of continuum variables.

The mean and variance of the theoretical continuum distributions are:

$$\mu = \int_{-\infty}^{+\infty} x\, pdf(x)\,dx \tag{3.33}$$

$$\sigma^2 = \int_{-\infty}^{+\infty} (x-\mu)^2\, pdf(x)\,dx \tag{3.34}$$

3.4.1 Continuous Uniform Distribution

If a random variable can have any numerical value within the range from a to b and no values outside that range, and if each possible value has an equal probability of occurring, then the probability density function for the uniform continuous distribution is

$$pdf(x) = \begin{cases} \dfrac{1}{b-a}, & a \le x \le b \\ 0 & x > b, x < a \end{cases} \tag{3.35}$$

$$F(x) = P(X \le x) = \begin{cases} 0, & x < a \\ \dfrac{x-a}{b-a}, & a \le x \le b \\ 1, & x > b \end{cases} \tag{3.36}$$

An acronym for data that is uniformly and independently distributed within a range from a and to b is *UID(a,b)*.

Figure 3.7 illustrates the uniform distribution for $a = 2$ and $b = 5$. The mean and variance of the continuous uniform distribution are

$$\mu = \frac{a+b}{2} \tag{3.37}$$

and

$$\sigma^2 = \frac{(b-a)^2}{12} \tag{3.38}$$

Note the parallels and differences between equations for the mean and variance of the continuous uniform and equally incremented discrete uniform distributions.

The random variable X may have any dimensional units which must match that of parameters a and b. The mean, μ will have the same units. Whereas the cumulative distribution function $F(x)$ is dimensionless, the probability density function, $pdf(x)$, has units that are the reciprocal of those of the random variable X.

Again, the population coefficients, a and b, represent the true values. You might not know what they are exactly, but certainly you can do enough experiments to get good estimates for them.

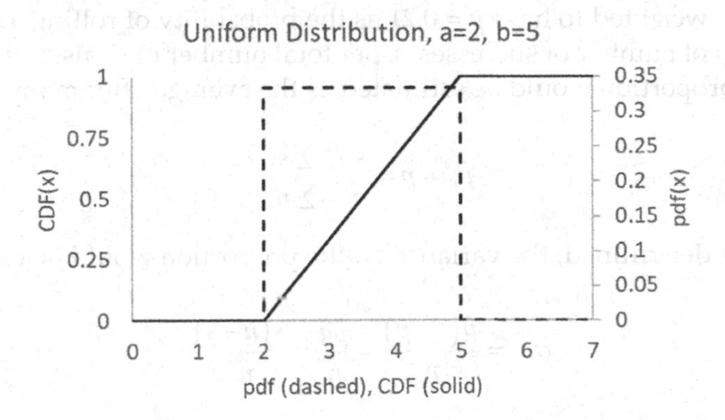

FIGURE 3.7
Continuous uniform distribution: (a) probability density function, (b) cumulative distribution, both for $a = 5$, $b = 10$.

Example 3.8: From observation, it appears that fluid turbulence superimposes a uniformly distributed disturbance of ±2 psi on a differential pressure cell installed across an orifice. If this is true and the average differential pressure is 20 psi, what is the probability of reading a value of more than 21 psi?

From Equation (3.37), we have

$$a = \mu - 2 \text{ psi} = 18 \text{ psi}$$

and

$$b = \mu + 2 \text{ psi} = 22 \text{ psi}$$

From Equation (3.36), we find that

$$P(X \leq 21) = \frac{21-18}{22-18} = 0.75$$

$$P(X > 21) = 1 - P(X \leq 21) = 1.0 - 0.75 = 0.25$$

There is a 25% probability of reading a differential pressure value of greater than 21 psi.

3.4.2 Proportion

A *proportion* is the probability of an event, a fraction of outcomes, and is a continuum-valued variable. In flipping a coin, the probability of a particular outcome is $p = 0.5$. In rolling a die the probability of getting a 5 is $p = 0.1666\overline{6}$. In rolling 10 dice and winning means getting at least one five in the 10 outcomes, the probability is $p = 0.83949441\ldots$. Although the events are discrete, the probability could have a continuum of values between 0 and 1. $0 \leq p \leq 1$.

If the proportion is developed theoretically, then it is known with as much certainty as the basis and idealizations allow. Then the variance on the proportion is 0.

$$\sigma_p{}^2 = 0 \tag{3.39}$$

Alternately, the proportion could be determined from experimental data. For example, a trick die could be weighted to have $p = 0.21$ as the probability of rolling a 5. Here, proportion, p, is the ratio of number of successes, s, per total number of trials, n, $p = s/n$, as $n \to \infty$. Alternately, the proportion would be estimated as the average after many trials.

$$\hat{\mu}_p = \hat{p} = \frac{s}{n} = \frac{\sum s_i}{\sum n_i} \tag{3.40}$$

If experimentally determined, the variance on the proportion would be estimated by

$$\hat{\sigma}_p{}^2 = \frac{\hat{p}(1-\hat{p})}{n} = \frac{\hat{p}\hat{q}}{n} = \frac{s(n-s)}{n^3} \tag{3.41}$$

Note this is similar to the mean and variance of the binomial distribution, but here the statistics are on the continuum-valued proportion. In the binomial distribution, the statistics are on the number count of a particular type of event. The variance of the count of successes of n samples from a population would be given by Equation (3.14).

Example 3.9: What are the mean and sigma when the probability of an event (outcome = 1) is and unknown p, and the probability of a not-an-event (outcome = 0) is $q = (1-p)$? A sequence of n dichotomous events might be

$$0,0,1,1,1,0,1,0,0,1,0,0,1,\ldots$$

Whether we call the event a H or a T, a success or a fail, the {1,0} notation is equivalent. Experimentally, there are $s = 21$ successes out of $n = 143$ trials.

From Equation (3.40) the estimate of p is

$$\hat{p} = \frac{s}{n} = \frac{21}{143} = 0.14685314\ldots$$

From Equation (3.41) the standard deviation on $\hat{p}$ is

$$\hat{\sigma}_p = \sqrt{\hat{\sigma}_p{}^2} = \sqrt{\frac{s(n-s)}{n^3}} = \sqrt{\frac{21(143-21)}{143^3}} = 0.02959957\ldots$$

acknowledging the uncertainty on $\hat{p}$ and $\hat{\sigma}_p$ one might report

$$\hat{p} = 0.148$$

and

$$\hat{\sigma}_p = 0.03$$

3.4.3 Exponential Distribution

The exponential (or negative exponential) distribution describes a mechanism whereby the probability of failures (or events) within a time or distance interval depends directly on the number of un-failed items remaining. It describes events such as radioisotope decay, light intensity attenuation through matter of uniform properties, the failure rate of light bulbs, and the residence time or age distribution of particles in a continuous-flow stirred tank. Requirements for the distribution are that, at any time, the probability of any one particular item failing is the same as that of any other item failing and is the same as it was earlier. Another restriction is that the numbers are so large that the measured values seem to be a continuum. The probability distribution functions are

$$pdf(x) = \alpha e^{-\alpha x}, \quad 0 \le x \le \infty, \quad \alpha > 0 \tag{3.42}$$

and

$$F(x) = 1 - e^{-\alpha x} \tag{3.43}$$

The variable x represents the time or distance interval, not the number (or some other measure of quantity) of un-failed items. The argument of an exponential must be dimensionless,

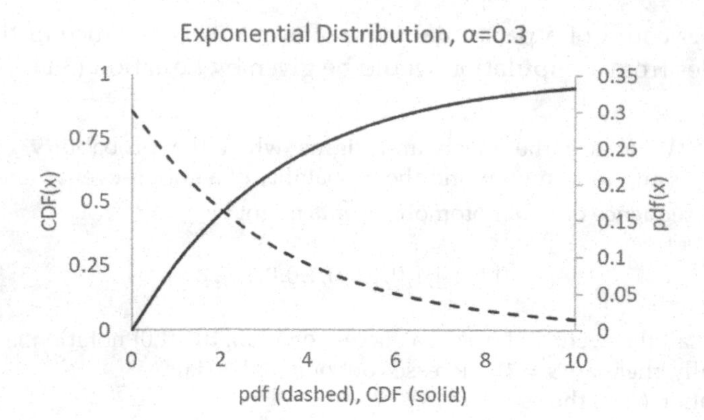

FIGURE 3.8
Exponential distribution: (a) probability density function, (b) cumulative distribution, both for $\alpha = 0.3$.

so the units on α are the reciprocal of the units on x. This requires that the units on $pdf(x)$ are also the reciprocal of the units on x, making $pdf(x)$ be a rate.

Figure 3.8 illustrates the exponential distribution for $\alpha = 0.3$. The mean and variance of the exponential distribution are

$$\mu = \frac{1}{\alpha} \tag{3.44}$$

and

$$\sigma^2 = \frac{1}{\alpha^2} \tag{3.45}$$

The continuous random variable X may have any units. The units on μ will be the same. The units on both α and $f(x)$ are the reciprocal of those of X. $F(x)$ is dimensionless. For a physical interpretation, α represents the fraction of events occurring per unit of space or time. The discrete geometric distribution, in its limit as the number of events is very large and the probability of success is small, approaches the continuous exponential distribution.

Example 3.10: One billion adsorption sites are available on the surface of a solid particle. Gas molecules, randomly and uniformly "looking" for a site, find one upon which to adsorb, which "hides" that site from other molecules. With an infinite gas volume, the rate at which the sites are occupied is therefore proportional to the number of unoccupied sites. If 40% of the sites are covered within the first 24 hours, how long will it take for 99% of the adsorption to be complete? What is the average lifetime of an unoccupied site?

From Equation (3.43),

$$40\% = 0.40 = F(t) = 1 - e^{-\alpha t} = 1 - e^{-\alpha(24)}$$

which gives $\alpha = 0.02128440\ldots$ per hour. From Equation (3.43), 99% = 0.99 = $F(t) = 1 - e^{0.02128\ldots t}$, which gives $t = 216.3636244\ldots$ hours or about 9 days. From Equation (3.44), $\mu = 1/\alpha = 46.9827645$ hours or almost 2 days.

It takes about 9 days for 99% of the sites to become occupied. The average life of an unoccupied site is about 2 days.

Note that although the exact number of initial active sites is immaterial, the fact that it was very large meets the second requirement for the use of the exponential distribution.

3.4.4 Gamma Distribution

The gamma distribution can represent two mechanisms. In a general situation in which a number of partial random events must occur before a complete event is realized, the probability density function of the complete event is given by the gamma distribution. For instance, rust spots on your car (the partial event), may occur randomly at an average rate of one per month. If 16 spots occur before you decide to have your car repainted (the total event), the gamma distribution is the appropriate one to use to describe the repainting time interval. The gamma distribution is

$$pdf(x) = \frac{\lambda}{\Gamma(\alpha)}(\lambda x)^{\alpha-1}\exp(-\lambda x), \quad x \geq 0 \tag{3.46}$$

and where α and $\lambda > 0$ and $\Gamma(\alpha)$ is the gamma function

$$\Gamma(\alpha) = \int_0^\infty Z^{\alpha-1}e^{-z}dZ \tag{3.47}$$

The gamma function has several properties

$$\Gamma(\alpha) = (\alpha-1)\Gamma(\alpha-1) \tag{3.48}$$

and if α is an integer, then

$$\Gamma(\alpha) = (\alpha-1)! \tag{3.49}$$

The variable α represents the number of partial events required to constitute a complete event, and λ is the number of partial events per unit of x (which may be time, distance, space, or item).

If $\alpha = 1$, the gamma distribution reduces to the exponential distribution. For that reason, if an event rate is proportional to some power of x, then the gamma distribution can also be used as an adjusted exponential distribution. Let's look at Example 3.10 again. If adsorption reduces the number of gas molecules available for subsequent adsorption, then the probability of any site being occupied decreases with time. If the frequency with which gas molecules impinge on the particle surface decreases as $(\lambda x)^{\alpha-1}$, then the gamma function describes $f(x)$. However, although close enough for most engineering applications, the power law decrease probably does not describe a real driving force exactly. For such a situation, use of the gamma distribution must be acknowledged as a convenient approximation.

Depending on the values of α and λ, $f(x)$ may have various shapes, some of which are illustrated in Figure 3.9. A general analytical expression for $F(x)$ is intractable. For most α values, to obtain the cumulative distribution function, $f(x)$ must be integrated numerically. Excel provides the function $\text{GAMMA.DIST}(x,\alpha,1/\lambda,0)$ to return the $pdf(x)$ value

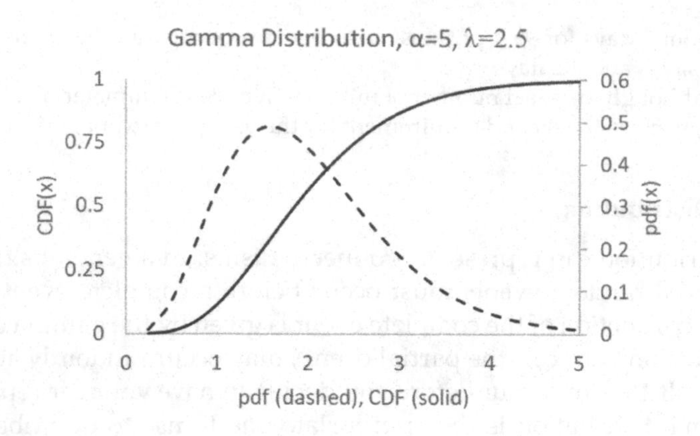

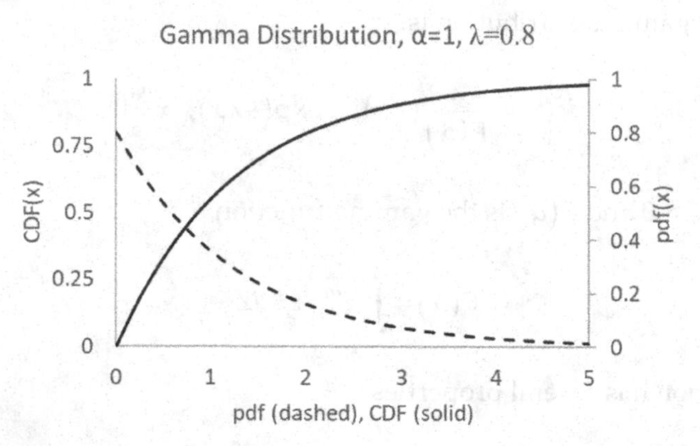

FIGURE 3.9
Gamma distribution: probability density function and cumulative distribution, both for various parameter values – (a) $\alpha = 5, \lambda = 2.5$, (b) $\alpha = 1, \lambda = 0.8$.

and $\text{GAMMA.DIST}(x,\alpha,1/\lambda,1)$ to return the *CDF*(*x*) value. Note that the Excel parameter beta is the reciprocal of λ, here.

The mean and variance of the gamma distribution are

$$\mu = \frac{\alpha}{\lambda} \tag{3.50}$$

$$\sigma^2 = \frac{\alpha}{\lambda^2} \tag{3.51}$$

The units of X are usually count per some interval (time, distance, area, space, or item). Consequently, the units for λ are the fraction of total failures per unit of X. The coefficient, α, is a counting number and is dimensionless, and $f(x)$ has units that are the reciprocal of the units of X.

Example 3.11: There are 8 lightbulbs in a particular chandelier. The lights fail randomly. When three fail, only 5 remain and there is not enough illumination for the room. The illumination is adequate if 0, 1, or 2 lights fail; but when three lights fail, the bulbs must be replaced. On average, we find that the lights need to be replaced every 2 months.

What is the average failure rate of a single light on a monthly basis? What fraction of the lights will last for 4 months without needing replacement?

From Equation (3.50),

$$\lambda = \frac{\alpha}{\mu} = \frac{3(\text{partial per total})}{2(\text{months per total})} = 1.5(\text{partials per month})$$

We expect 1.5 lights to fail each month.

$$P(X > 4) = 1 - P(X \leq 4)$$
$$= 1 - \int_0^4 f(\alpha = 3, \lambda = 1.5, x)dx$$

Using the Excel function GAMMA.DIST$(x, \alpha, 1/\lambda, 1)$ with GAMMA.DIST$(4, 3, 1/1.5, 1)$

$$P(X > 4) = 1 - .938031\ldots = 0.061968\ldots \cong 6.2\%$$

about 6% of the lights will last for four months.

3.4.5 Normal Distribution

The normal distribution, often called the Gaussian distribution or bell-shaped error curve, is the most widely used of all continuous probability density functions. The assumption behind this distribution is that any errors (sources of deviation from true) in the experimental results are due to the addition of many independent small perturbation sources. All experimental situations are subject to many random errors and usually yield data that can be adequately described by the normal distribution.

Even if your data is not normally distributed, the averages of data from a nonnormal distribution tend toward being normal. An average of independent samples will have some values above the mean and some below. The average will be close to the mean, and each sample would represent a small independent deviation. In the limit of large sample size, n, the standard deviation of the average is related to that of the individual data by $\sigma_{\bar{X}} = \sigma_X / \sqrt{n}$. So, when using averages, the normal distribution usually is applicable.

However, this situation is not always true. If you have any doubt that your data are distributed normally, you should use the nonparametric techniques in Chapter 7 to evaluate the distribution. Use of statistics that depend on the normal distribution for a dataset that is distinctly skewed may lead to erroneous results.

An acronym for data that is normally and independently distributed with a mean of μ and standard deviation of σ is $NID(\mu, \sigma)$.

Regardless of the shape of the distribution of the original population, the *central limit theorem* allows us to use the normal distribution for descriptive purposes, subject to a single restriction. The theorem simply states that if the population has a mean μ and a *finite* variance σ^2, then the distribution of the sample mean $\bar{X}$ approaches the normal distribution with mean μ and variance σ^2/n as the sample size n increases. The chief problem with the theorem is how to tell when the sample size is large enough to give reasonable compliance with the theorem. The selection of sample sizes is covered in Chapters 10, 11, and 17.

The probability density function $f(x)$ for the normal distribution is

$$f(x) = \frac{1}{\sigma\sqrt{2\pi}} e^{\left[-\frac{1}{2}\left(\frac{x-\mu}{\sigma}\right)^2\right]}, \quad -\infty < x < \infty \tag{3.52}$$

Note that the argument of the exponentiation, $\left[-\frac{1}{2}\left(\frac{x-\mu}{\sigma}\right)^2\right]$, must be dimensionless. As expected, x, μ, and σ each have identical units. The exponentiation value is also dimensionless. Also, since $f(x)$ is proportional to $1/\sigma$, it has the reciprocal units of x.

As seen in Equation (3.52), the normal distribution has two parameters, μ and σ, which are the mean and standard deviation, respectively. The cumulative distribution function (*CDF*), described by

$$CDF(x) = F(x) = P(X \le x) = \frac{1}{\sigma\sqrt{2\pi}} \int_{-\infty}^{x} e^{-(X-\mu)^2/2\sigma^2} dX \tag{3.53}$$

In Equation (3.53) the variable X is the generic variable, and the lower-case x represents a particular value.

The logistic model, $CDF(x) = F(x) = P(X \le x) = \frac{1}{1+e^{-s(x-c)}}$, is a convenient and reasonably good approximation to the normal *CDF*(x). Convenient: It is computationally simple, analytically invertible, and analytically differentiable. Reasonably good: Values are no more different from the normal *CDF*(x) than that caused by uncertainty on μ and σ. For the scale factor, use $s = \sigma / 1.7$, and for the center, use $c = \mu$ (see Exercise 3.15).

Figure 3.10 shows several possible variations on the distribution. If the variance is the same but the means are different, the shape of the curves are the same, but their locations are different, as shown in Figure 3.10 a and b. However, if the mean is the same and the variance is different, the central part of the curve will be at the same value, but the width of the *pdf* curve will be different, as seen in Figure 3.10 a and c.

Note: In Figure 3.10a, the *pdf* and *CDF* have values at negative x-values. The distribution asymptotically approaches zero at $x = -\infty$, and also at $x = +\infty$. So, even though it is not noticeable in all three figures, the extreme *pdf* and *CDF* values are not exactly zero.

Note: In all cases, the *pdf* distribution is symmetric on either side of the mean, and the peak is at the mean. The *pdf* curves, visually, are effectively zero when x is beyond 3σ of the mean. And the inflection points on the *pdf* curve (where the shape changes between concave to convex) are at the $\pm 1\sigma$ deviation from the mean. When you are sketching normal *pdf* curves, incorporate these features.

Note: The 0.5 *CDF* value of the cumulative distribution corresponds to the mean, as does the inflection point on the *CDF* curve. The *CDF* curves asymptotically approach 0 and 1, but they are visually effectively at 0 and 1 when x is $\pm 3\sigma$ from the mean. When you are sketching normal *CDF* curves, incorporate these features.

The probability of X falling within the range $x_1 < X \le x_2$ is

$$\begin{aligned} P(x_1 < X \le x_2) &= F(x_2) - F(x_1) \\ &= \frac{1}{\sigma\sqrt{2\pi}} \int_{x_1}^{x_2} e^{-(x-\mu)^2/2\sigma^2} dx \end{aligned} \tag{3.54}$$

Because two parameter values must be specified to define the desired normal distribution, it is convenient to consider a single member of the family of distributions, that with μ = 0 and σ = 1. Such a normal distribution is called the standard normal distribution and is characterized by a single parameter z. The *standard normal deviate* z is defined by

$$z = \frac{x-\mu}{\sigma} \tag{3.55}$$

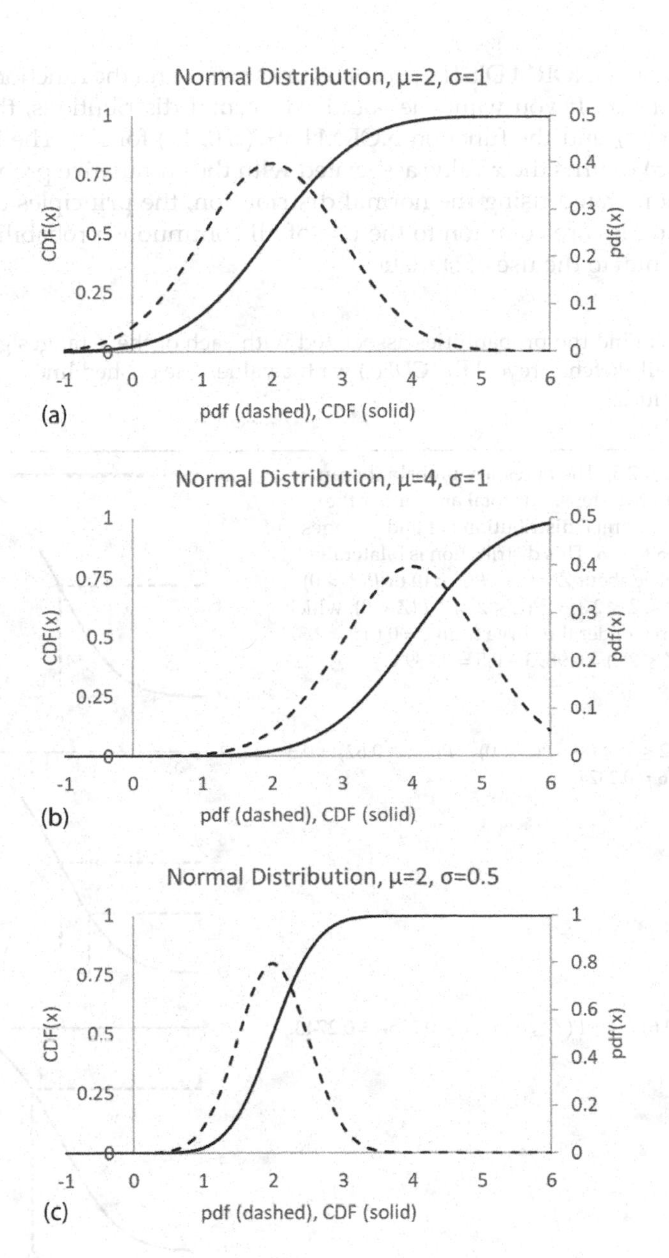

FIGURE 3.10
Normal distribution curves for mean and variance change – (a) $\mu = 2$, $\sigma = 1$, (b) $\mu = 4$, $\sigma = 1$, (c) $\mu = 2$, $\sigma = 0.5$.

The scaled variable z is dimensionless and represents the number of standard deviations that the x-value is away from the mean.

This allows us to scale and translate (standardize) Equation (3.53) to

$$CDF(z) = F(z) = P\left(\frac{x-\mu}{\sigma} \le z\right) = \frac{1}{\sqrt{2\pi}}\int_{-\infty}^{z} e^{-Z^2/2}dZ. \tag{3.56}$$

In this way, we have reduced all normal variables, regardless of parameter values, to a single distribution in z. In Equation (3.56) the variable Z is the generic variable, and the lower-case z represents a particular value.

In Excel, the function NORM.DIST(x, μ, σ, 0) returns $pdf(x)$, and the function NORM.DIST(x, μ, σ, 1) returns $CDF(x)$. If you want the standard normal distributions, then use NORM.DIST(z, 0, 1, 0) for $f(z)$, and the function NORM.DIST(z, 0, 1,1) for $F(z)$. The inverse function NORM.INV(p, μ, σ) returns the x value associated with the cumulative probability, p.

Although demonstrated using the normal distribution, the principles demonstrated in the next two examples are common to the use of all continuous probability distributions and are fundamental to the use of statistics.

Example 3.12: Find the probabilities associated with each of the Z ranges given below. The thumbnail sketches reveal the $CDF(z)$ w.r.t. z-value. The dashed lines represent the numerical values.

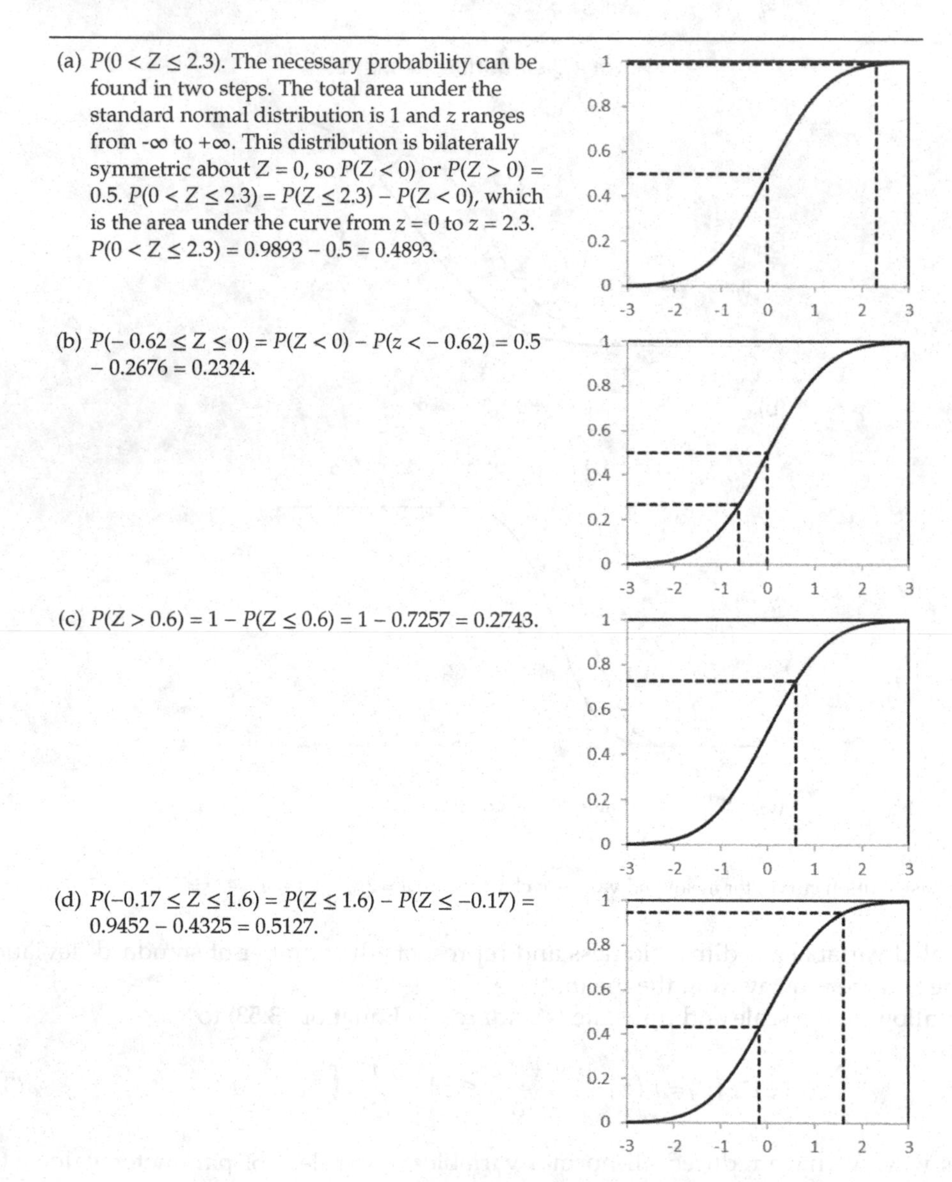

(a) $P(0 < Z \leq 2.3)$. The necessary probability can be found in two steps. The total area under the standard normal distribution is 1 and z ranges from $-\infty$ to $+\infty$. This distribution is bilaterally symmetric about Z = 0, so $P(Z < 0)$ or $P(Z > 0) = 0.5$. $P(0 < Z \leq 2.3) = P(Z \leq 2.3) - P(Z < 0)$, which is the area under the curve from $z = 0$ to $z = 2.3$. $P(0 < Z \leq 2.3) = 0.9893 - 0.5 = 0.4893$.

(b) $P(-0.62 \leq Z \leq 0) = P(Z < 0) - P(z < -0.62) = 0.5 - 0.2676 = 0.2324$.

(c) $P(Z > 0.6) = 1 - P(Z \leq 0.6) = 1 - 0.7257 = 0.2743$.

(d) $P(-0.17 \leq Z \leq 1.6) = P(Z \leq 1.6) - P(Z \leq -0.17) = 0.9452 - 0.4325 = 0.5127$.

(e) $P(0.25 \le Z \le 1.96) = P(Z \le 1.96) - P(Z \le 0.25) =$ $0.9750 - 0.5987 = 0.3763$.

(f) $P(-\infty < Z \le 1.2)$. As the values in the NORM.DIST function are based on the integral in Equation (3.56), the solution can be read directly as the value at $z = 1.2$, and that is 0.8849.

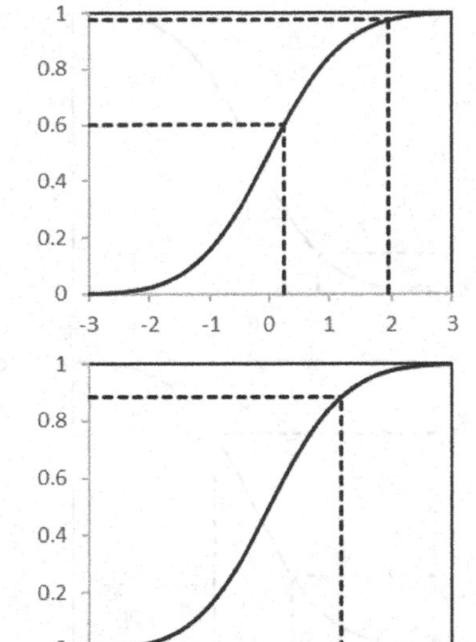

Sometimes, it will be necessary for you to find the values of z_1 and/or z_2 for some given value of $P(z_1 \le Z \le z_2)$. The examples below use the NORM.INV function to determine the limits of Z from the given probability data.

Example 3.13: Determine the z-value associated with each of the probabilities given below.

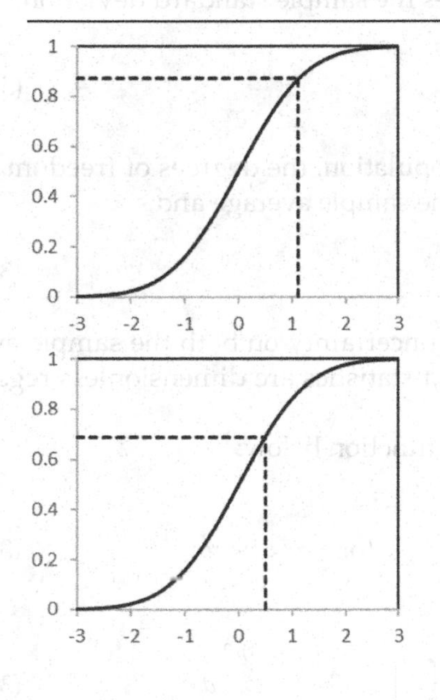

(a) Find the value of z_1 if $P(0 \le Z \le z_1) = 0.37$. As $P(Z < 0)$ is 0.5, we have to find the value of z_1 for which the area under the distribution is $0.5 + 0.37 = 0.87$. We find $z_1 = 1.1264$.

(b) Find the value of z_1 if $P(-\infty < Z \le z_1) = 0.69$. There is only one step needed to answer this problem, as the entire area under the curve from $-\infty$ to z_1 has been specified as 69% of the total. The value of z_1 is 0.49414.

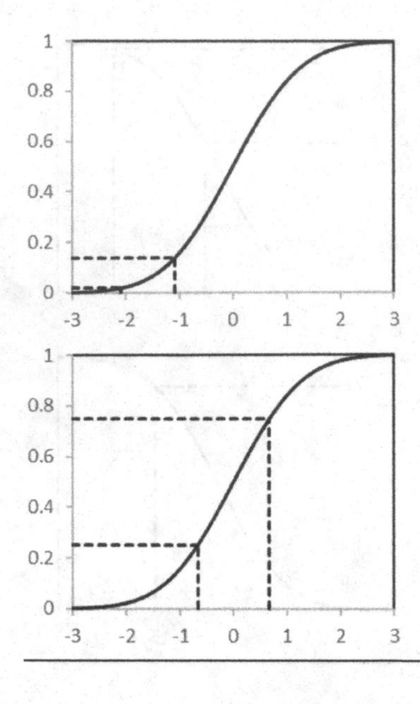

(c) Find the value of z_1 if $P(-2.1 \le Z \le z_1) = 0.12$. The probability interval may be written as $P(Z \le z_1) - P(Z \le -2.1) = 0.12$ or $P(Z \le z_1) - 0.0179 = 0.12$, and the *CDF* of z_1 must be 0.1379. The corresponding value of z is -1.09.

(d) Find the values for z that determine the upper and lower quartiles. For the lower quartile $P(-\infty \le Z \le z_1) = 0.25$. For the upper quartile $P(z_2 \le Z \le +\infty) = 0.25$, which means that $P(-\infty \le Z \le z_2) = 1 - P(z_2 \le Z \le +\infty) = 1 - 0.25 = 0.75$. The z-values are -0.6745 and $+0.6745$.

3.4.6 "Student's" *t*-Distribution

W. S. Gossett, publishing his work under the pseudonym "Student," developed the *t*-distribution. The statistic would become the basis for the *t*-test so widely used for the evaluation of engineering data.

The *t*-statistic is very similar to the standard normal *z*-statistic, but instead of using the true population mean and standard deviation, it uses the sample standard deviation.

$$T = \frac{X - \mu}{s} \tag{3.57}$$

Because it is based on sample data, not the entire population, the degrees of freedom ν is one less than the number of data used to calculate the sample average and s

$$\nu = n - 1 \tag{3.58}$$

Relative to the *z*-statistic, the *t*-statistic includes the uncertainty on both the sample average and sample standard deviation. Both the *z*- and t-statistics are dimensionless regardless of the units on the variable *X*.

The random variable *t* has the probability density function below:

$$f(t) = \frac{1}{\sqrt{\upsilon\pi}} \frac{\Gamma\left((\upsilon+1)/2\right)}{\Gamma(\upsilon/2)} \left(1 + \frac{t^2}{\upsilon}\right)^{-(\upsilon+1)/2} \quad \text{for } -\infty < t < \infty \tag{3.59}$$

$$CDF(t) = F(t) = \frac{1}{\sqrt{\upsilon\pi}} \frac{\Gamma\left((\upsilon+1)/2\right)}{\Gamma\left(\frac{\upsilon}{2}\right)} \int_{-\infty}^{t} \left(1 + \frac{x^2}{\upsilon}\right)^{-(\upsilon+1)/2} dx \tag{3.60}$$

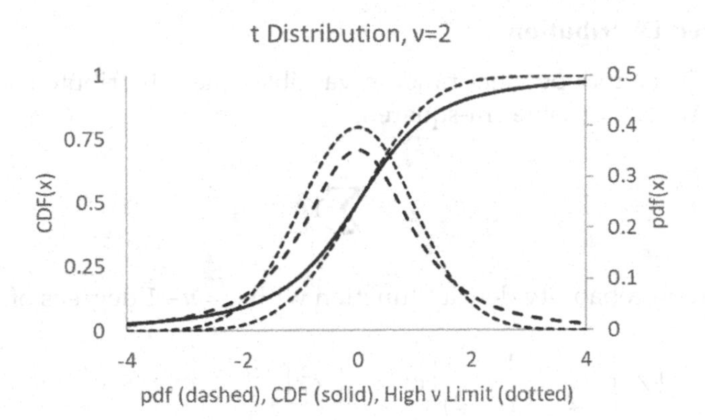

FIGURE 3.11
Characteristic shapes of the "Student's" t-distribution: for $v = 2$, and $v = 200$ (approaching the standard normal).

Note that $\Gamma(v/2)$ is the gamma function. The gamma function is related to the factorial and is *not* the gamma probability density distribution. Like the z-distribution, the distribution of t is bilaterally symmetric about $t = 0$. The t-distribution is illustrated in Figure 3.11 for two values of v, the degrees of freedom. The resulting bell-shaped distribution resembles that of the standard normal. However, more of the area under the t-distribution is in the "tails" of the distribution. In the limit of large n (effectively ν greater than about 150) the t- and standard normal distributions differ in the tenths of a percent.

The use of the t-distribution will be described in subsequent chapters in the sections discussing confidence intervals and tests of hypotheses for the mean of experimental distributions.

The cumulative t-distribution, $F(t)$ from Equation (3.60) can be calculated by the Excel function $T.\text{DIST}(t,v,1)$ where t is calculated from the sample data. Alternately, if you wanted to know the t-value that represents a probability limit then use the Excel function $T.\text{INV}(CDF, v)$ to return a t-value that would represent that CDF value. Alternately, calculate α, the level of significance, the extreme right-hand area, as $\alpha = 1 - F(t) = 1 - CDF$, then use the Excel function $T.\text{INV}(1-\alpha,v)$.

That represented a one-sided evaluation, which considered the area under the t-distribution from $-\infty$ up to a particular t-value. But often, we desire to know either the positive or negative extreme values for t, the "+" or "−" deviations from the central "0" value. You may want to know the range of t-values that includes the central 95% (or some confidence fraction C) of all expected values from sampling the population.

$$P\left(t_{\text{negative limit}} \leq T \leq t_{\text{positive limit}}\right) = C \tag{3.61}$$

Here, the level of significance is again the extreme area. If the 95% interval is desired ($C = 0.95$) then $\alpha = 1 - 0.95 = 1 - C$. Splitting the two tail areas equally, to define the central limits, use $\alpha/2$ to represent both the far right and far left areas in the tails. Then we seek the t-value calculated with $F(t) = 1 - \alpha / 2$. The Excel function $T.\text{INV}(1-\alpha/2,v)$ will return the t-value representing the positive extreme expected value, and $-T.\text{INV}(1-\alpha/2,v)$ will return the negative extreme. This is termed a two-sided (historically a two-tailed) test, because we are seeking the limits of the central area. Alternately, $T.\text{INV}.2T(1-\alpha,v)$ returns the same value.

3.4.7 Chi-Squared Distribution

Let $Y_1, Y_2, Y_3, \ldots, Y_n$ be independent random variables each distributed with mean 0 and variance 1. The random variable chi-squared:

$$\chi^2 = \sum_{i=1}^{n} Y_i^2 \tag{3.62}$$

has the chi-squared probability density function with $v = n-1$ degrees of freedom

$$f(\chi^2) = \frac{1}{2^{v/2}\Gamma(v/2)}\left[e^{-\chi^2/2}\right]\left[\chi^2\right]^{(v/2)-1} \quad for \;\; 0 \le \chi^2 \le \infty \tag{3.63}$$

and cumulative distribution

$$F(\chi^2) = \frac{1}{2^{v/2}\Gamma(v/2)}\int_0^{\chi^2} e^{-Y/2}(Y)^{(v/2)-1}\,dY \tag{3.64}$$

If Y in Equation (3.62) is defined as $(X-\bar{X})/\sigma$ then

$$\chi^2 = \sum_{i=1}^{n} Y_i^2 = \sum_{i=1}^{n}\frac{(X_i-\bar{X})^2}{\sigma^2} = \frac{(n-1)s^2}{\sigma^2} \tag{3.65}$$

Figure 3.12 illustrates the probability density and cumulative chi-squared distributions, respectively. Values of the cumulative chi-squared (χ^2) distribution can be obtained from the Excel function $F(\chi^2) = \text{CHISQ.DIST}(\chi^2, v, 1)$, and the *pdf* by using $f(\chi^2) = \text{CHISQ.DIST}(\chi^2, v, 0)$.

The inverse of the calculation, the value of χ^2 given $F(\chi^2)$ and v can be obtained by the Excel function $\chi^2 = \text{CHISQ.INV}(F, v)$.

Note: Some tables or procedures use χ^2/v. Since Equation (3.62) indicates that χ^2 increases linearly with n, and since degrees of freedom is often $v = n-1$, the scaling makes

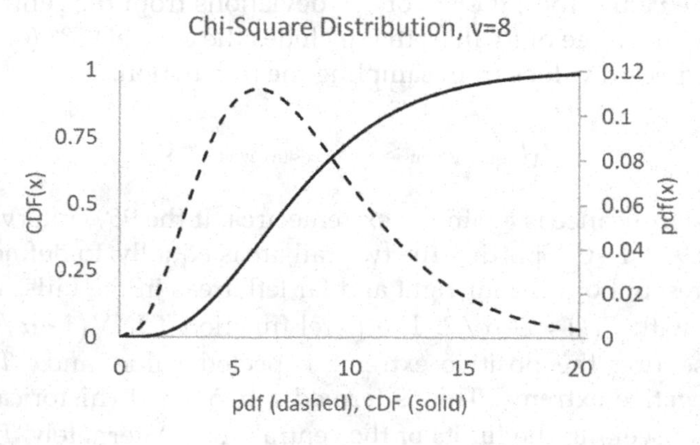

FIGURE 3.12
Chi-squared distribution for $v = 8$.

sense. Mostly, this book will not scale χ^2 by the degrees of freedom. But be aware that the use of either χ^2 / v or χ^2 is common.

The mean and variance of the chi-squared distribution are v and $2v$, respectively.

$$\mu = v \tag{3.66}$$

$$\sigma = 2v \tag{3.67}$$

So, if degrees of freedom is 10, an average-like value of the χ^2 statistic would be about 10. $\chi^2 = 1$ would be an unexpectedly low value, and $\chi^2 = 20$ would be unexpectedly high.

This distribution has several applications, one of which is in calculating and evaluating probability intervals for single variances from normally distributed populations as shown in Chapters 5 and 6. The chi-squared distribution is also used as a nonparametric method of determining whether or not, based on sample data, a population has a particular distribution, as described in Chapter 7. The chi-squared distribution goes from 0 to infinity, or $P\left(0 \le \chi^2 \le \infty\right) = 1$.

The interval

$$P\left(\chi^2_{v,\alpha/2} \le \chi^2 \le \chi^2_{v,1-\alpha/2}\right) = 1 - \alpha \tag{3.68}$$

defines the values for the χ^2-distribution such that equal areas are in each tail. The χ^2-distribution is not symmetric about the mean as are the Z- and t-distributions.

3.4.8 *F*-Distribution

The *F*-distribution (named in honor of Sir Ronald Fisher, who developed it) is the distribution of the random variable F, defined as

$$F = \frac{U / v_1}{V / v_2} = \frac{\chi_1^2 / v_1}{\chi_2^2 / v_2} \tag{3.69}$$

Using Equation (3.65) $\chi^2 = \dfrac{(n-1)s^2}{\sigma^2} = \dfrac{vs^2}{\sigma^2}$

$$F = \frac{s_1^2 / \sigma_1^2}{s_2^2 / \sigma_2^2} \tag{3.70}$$

where U and V are independent variables distributed following the chi-squared distribution with v_1 and v_2 degrees of freedom, respectively. The symbol F in Equation (3.69) does not represent any cumulative distribution but is a statistic, specifically, the ratio of two χ^2 statistics, each scaled by their degrees of freedom. The probability density function of F is

$$f(F) = \frac{\Gamma\left((v_1 + v_2)/2\right)}{\Gamma(v_1/2)\Gamma(v_2/2)} \left(\frac{v_1}{v_2}\right)^{v_1/2} \frac{F^{(v_1-2)/2}}{\left(1 + (v_1/v_2)F\right)^{(v_1+v_2)/2}} \tag{3.71}$$

and the cumulative distribution of F is

$$CDF(F) = \int_0^F f(F)\,dF \tag{3.72}$$

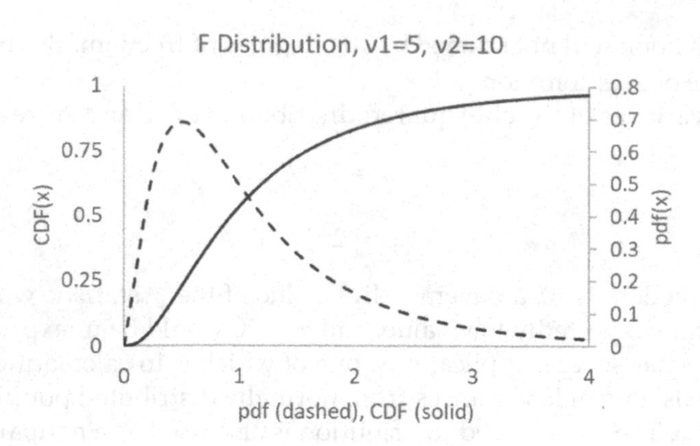

FIGURE 3.13
Characteristic shapes of the *F*-distribution.

The family of *F*-distributions is a two-parameter family in v_1 and v_2. The shape of the *F*-distribution is skewed (more of the area under the curve to the left side of the nominal value, a longer tail to the right), as illustrated in Figure 3.13. The range of all members is from 0 to ∞. This distribution is used to evaluate equality of variances. The *F*-distribution is termed "robust" by statisticians, meaning that the results of such statistical comparisons are likely to be valid even if the underlying populations are not normally distributed. The uses of the *F*-distribution are explained in Chapters 5, 6, and 12.

Values of the *pdf(F)* can be returned by the Excel function $pdf(F) = \text{F.DIST}\left(\chi_1^2/\chi_2^2, v_1, v_2, 0\right)$, and of the cumulative *F*-distribution by $CDF(F) = \text{F.DIST}\left(\chi_1^2/\chi_2^2, v_1, v_2, 1\right)$. The inverse of the distribution returns the chi-squared ratio for a given *CDF* value $\frac{\chi_1^2}{\chi_2^2} = \text{F.INV}(CDF, v_1, v_2)$.

If the chi-squared ratio is 3.58058 and the numerator and denominator degrees of freedom are 6 and 8, then the *CDF* value is 0.95. If, however, you choose to call #1 as #2, then the chi-squared ratio would be 0.279284, and the degrees of freedom would be 8 then 6. With these reversed values the *CDF* value is 0.05 the complement to the first.

3.4.9 Log-Normal Distribution

Many processes (especially particle-creating processes such as prilling, crystal growth, grinding, and attrition) yield a bell-shaped distribution that is skewed to the left (a long tail on the right). Empirically, research has shown that if *f*) is plotted versus Ln *X* or log *X* instead of *X*, the long tail is contracted, and the graph may appear normal in shape. This transformation effect is based on observation, not any theory derived from fundamental phenomena as were the previous distributions. Figure 3.14a illustrates both *pdf* and *CDF* of a skewed distribution from an optimization study. The question the study sought to answer is, "What is the distribution the number of leap-overs (player moves) to convergence, when leap-over distance is randomized?" The figure presents the results from 1,000,000 simulations. Figure 3.14b displays the same data when the abscissa is log transformed. The data is not from the theoretical analysis, but from a finite number of trials (only 10^6), and as a result there are small discontinuities in the curve. (See Ch 35, of Rhinehart, R. R., *Engineering Optimization: Applications, Methods, and Analysis*, 2018, John Wiley & Sons, New York, NY.)

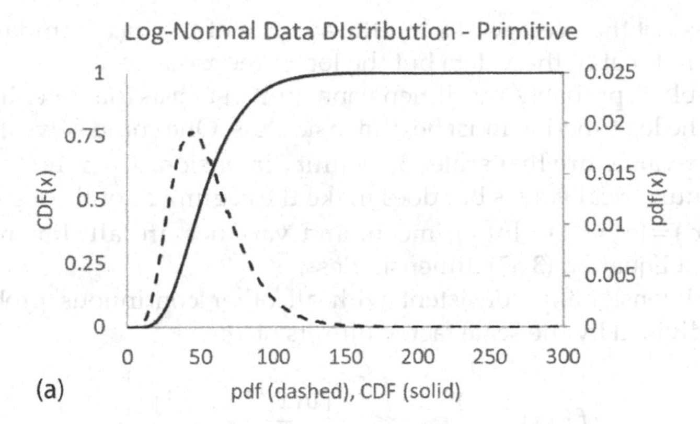

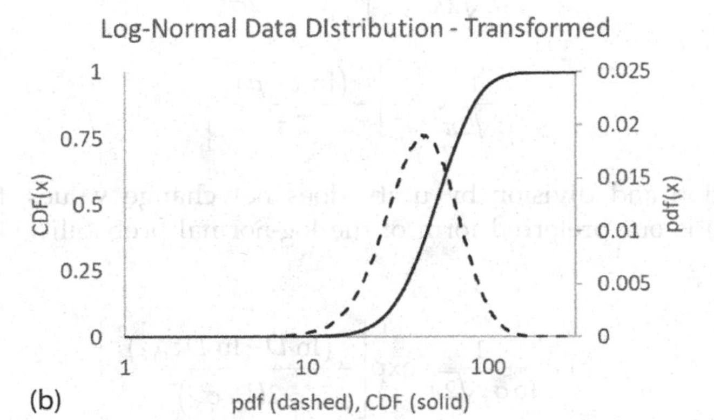

FIGURE 3.14
Log-normal distributions: (a) original data, (b) log-transformed data.

Although the abscissa number of leap-overs is a discrete count, similar shapes and log-transformed normalcy or the distribution is characteristic of many continuum variables, such as particle size (diameter) distribution. One limitation of the log transformation is that the random variable X cannot have any negative values.

There is no universally accepted probability density function for the log-normal distribution. From direct substitution of Ln x into the normal distribution, we obtain

$$f(x) = \frac{1}{\sigma\sqrt{2\pi}} \exp\left[-\frac{(\text{Ln}\, x - \mu)^2}{2\sigma^2}\right], 0 < x < \infty \tag{3.73}$$

where

$$\mu \simeq \overline{\text{Ln}\, x} = \frac{1}{n}\sum_{i=1}^{n} \text{Ln}\, x_i \tag{3.74}$$

and

$$\sigma^2 \simeq S^2 = \frac{1}{n-1}\sum_{i=1}^{n}\left(\text{Ln}\, x_i - \overline{\text{Ln}\, x}\right)^2 \tag{3.75}$$

Note: The mean is not the average of x but the average of the ln(x). Similarly note that the value of sigma is not that of the x-data but the log of the x-data.

Note: The variable X, probably has dimensional units, such as diameter in mm. However, the argument of the log function must be dimensionless. One could solve this issue by considering that the x-values are first scaled by a unit dimension: $x' = x[\text{units}]/1[\text{units}]$. This does not change numerical values but does make the argument of the log dimensionless.

Note: The $\ln(x') = \ln(x/1) = \ln(x)$, mean, and variance are all dimensionless, which would make $f(x)$ in Equation (3.57) dimensionless.

To make $f(x)$ dimensionally consistent with all other continuous probability density functions, $f(x)$ is divided by the scale factor 1 [units of x]:

$$
\begin{aligned}
f(x) &= \frac{1}{1\sigma\sqrt{2\pi}} \exp\left[-\frac{(\ln x/1 - \mu/1)^2}{2\sigma^2}\right] \\
&= \frac{1}{\sigma\sqrt{2\pi}} \exp\left[-\frac{(\ln x - \mu)^2}{2\sigma^2}\right]
\end{aligned}
\tag{3.76}
$$

Since multiplication and division by unity does not change values, the second part of Equation (3.76) is our preferred form of the log-normal probability density function. However,

$$
f(x) = \frac{1}{\ln\sigma_g\sqrt{2\pi}} \exp\left[-\frac{(\ln D - \ln D_{n,md})^2}{2(\ln\sigma_g)^2}\right] \tag{3.77}
$$

where $D_{n,md}$ is the number-median diameter and σ_g is the geometric standard deviation, has also been reported as useful.

See Exercise 3.21 for derivation of a distribution of x assuming the log(x) is normally distributed. The result is

$$
pdf(x) = \frac{1}{x} \cdot \frac{1}{\sqrt{2\pi}\sigma_{\ln(x^*)}} e^{-\frac{1}{2}\left(\frac{\ln(x) - \ln(\bar{x})}{\sigma_{\ln(x^*)}}\right)^2} \tag{3.78}
$$

where

$$
x^* = \frac{x}{\bar{x}} \tag{3.79a}
$$

and

$$
\sigma_{\ln(x^*)} = \frac{1}{1.645} \ln\left(x^*_{0.95}\right) \tag{3.79b}
$$

or

$$
\sigma_{\ln(x^*)} = \frac{-1}{1.645} \ln\left(x^*_{0.05}\right) \tag{3.79c}
$$

Analytic expressions for $F(X)$ are intractable. You will have to calculate $F(X)$ vs ln X by numerical integration. Using the log-transformed data we can estimate the mean of the log of X and the variance of the log of X and use normal (Z) statistics to test hypotheses on log-transformed populations. Since ln X is monotonic with X, if it is found that $\ln x_1 > \ln x_2$ then we may usually accept that $x_1 > x_2$.

We must caution you, however, that the units on the lognormal probability density function may be the reciprocal of the meaningless log of the units on X, the representation is not standardized, and the conclusions concerning ln(X) comparisons may not translate to X comparisons. Graphically, the log-transformed distribution is a convenient visual aid. However, for hypothesis testing, we recommend nonparametric methods.

3.4.10 Weibull Distribution

The Weibull distribution is one of many functions that are heuristically created, as opposed to those derived from a particular probability model. Due to the choice of parameter values, the Weibull distribution is very flexible and is commonly used to describe life distributions in reliability work.

$$f(t) = \frac{\beta}{\eta}\left(\frac{t}{\eta}\right)^{\beta-1} e^{-\left(\frac{t}{\eta}\right)^{\beta}}, \quad t \geq 0 \tag{3.80}$$

$$CDF(t) = 1 - e^{-\left(\frac{t}{\eta}\right)^{\beta}}, \quad t \geq 0 \tag{3.81}$$

where t is the product time to failure, $\eta > 0$ is the characteristic life, and β is the shape factor. If $\beta = 1$, the Weibull distribution reduces to the exponential distribution with $\eta = 1/\lambda$. If $\beta < 1$, the probability of an event w.r.t. time drops faster than an exponential. If $\beta > 1$ the probability of any single event w.r.t. time shifts to a like-normal distribution but slightly skewed with a tail to the left.

Figure 3.15 illustrates some of the Weibull probability density functions for values of β, each with $\eta = 1$.

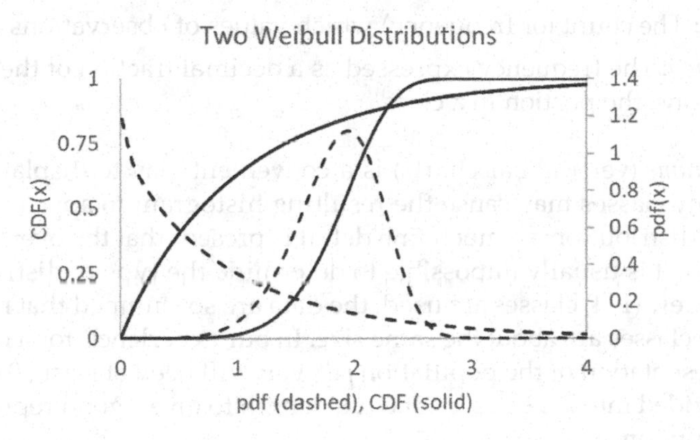

FIGURE 3.15
Weibull distributions; (a) $\alpha = 0.95$, $\beta = 1$ (like the exponential), (b) $\alpha = 6$, $\beta = 2$ (like the normal).

To determine values of η and β that make the Weibull distribution describe your experimental data, simply adjust η and β until the Weibull *CDF* matches the shape of your experimental data.

As a caution, software packages that provide Weibull distribution values, might not use the η and β notation of Equation (3.80). The Excel *pdf* function is WEIBULL.DIST$(x,\alpha,\beta,0)$, in which the Excel alpha is the shape factor, beta, in Equation (3.80) and the Excel beta is the characteristic life, ita, in Equation (3.80). So use WEIBULL.DIST$(x,\beta,\eta,0)$. The Excel Function WEIBULL.DIST$(x,\beta,\eta,1)$ will return the *CDF* value.

3.5 Experimental Distributions for Continuum-Valued Data

All the distributions discussed so far have been the theoretical or semi-theoretical distributions that are commonly used to aid in the interpretation of survey or experimental data. Before we can consider experimental distributions, we need to define a few terms that are commonly used.

Classes: Groups into which data are distributed. These could be linguistic categories (such as "upper-right quadrant", "pass", "dog"), or numerical values. If numerical, these could be discrete values (such as "1, 2, 3, 4, …", "$1\frac{5}{8}, 1\frac{6}{8}, 1\frac{7}{8}, \ldots$) or an interval of continuum valued data. If the linguistic categories or discrete values are known, the histogram is simply the count in each category. What follows is for continuum valued data.

Number of Classes: As a first choice, the number of classes should be about the square root of the number of data.

Class boundary: The numerical value dividing two successive classes. Make these convenient values for a reader to interpret.

Class length: The numerical difference between the boundaries of a class. This should be substantially greater than measurement uncertainty.

Class mark: The mid-value of a class.

Class frequency: The count (or frequency) which values of observations occur.

Relative frequency: The frequency expressed as a decimal fraction of the total number of observations, the portion in a class.

The use of *histograms* (vertical bar charts) is a convenient way to display data. However, choosing too many classes may cause the resulting histogram to appear noisy or even to approach a point distribution; so much fine detail is present that the overall picture is lost. In these situations, it is usually impossible to determine the type of distribution followed by the data. If too few (1–4) classes are used, the data are so clumped that nothing has been gained; often, the classes are about the same size. In our experience, for a histogram to be a meaningful representation of the population *pdf*, you will need at least 50 (and likely more than 100) data divided into 7–12 classes of equal length to give a good representation of the population distribution.

The frequency histogram plots the class boundaries and class marks on the abscissa and the frequency (number in that bin) or relative frequency on the ordinate. Each class is represented by a rectangle centered on the class mark with a height proportional to the corresponding frequency. Empty classes may exist simply because no data fell in those intervals. Having too many open intervals probably indicates that the class length is too small (that too many classes have been used).

Frequency polygons are another convenient way to summarize the behavior of a large collection of experimental data. The upper class boundaries (abscissa) and cumulative frequencies (ordinate) of the corresponding classes locate points that, when connected by a straight line, form the cumulative frequency polygon. This is akin to a *CDF* but discretized by bin intervals. As a result, the percentage of values of the experimental variable expected to be less than or equal to any specified value can be estimated directly from the polygon. As with the *Z* distribution, you can also find values expected to be greater than some predetermined value of the experimental variable or between two values of the variable. A better way to obtain such probabilities is to fit the appropriate theoretical distribution to the relative frequency histogram and to use the corresponding probability function for predictive purposes.

Example 3.14: Air pollution samples are analyzed for both naturally occurring and industrial contaminants. The results of analyses for sulfate (SO_4^{2-}) content in air over a 72-day period are given below expressed as parts per million in air. Prepare a table of grouped frequencies for the data employing suitable class lengths and boundaries. Plot the resulting frequency histogram and the cumulative relative frequency polygon. What distribution seems to best fit the data? What are the upper and lower quartiles and 90% limits? What are the average and standard deviation of the data?

3.14	1.87	1.00	1.12	1.03	5.88	0.91	1.86
0.08	1.95	6.38	2.61	0.10	1.96	0.00	0.00
2.00	5.57	2.96	2.68	2.44	2.62	7.61	0.80
2.11	6.99	9.02	3.24	1.25	0.00	8.06	0.00
3.55	7.80	8.90	1.95	1.85	2.07	0.00	6.85
3.51	10.83	14.71	1.24	1.23	4.11	1.75	0.00
3.69	7.17	1.95	0.63	1.24	1.84	1.78	0.00
0.98	1.93	3.30	0.00	1.66	3.60	0.00	0.00
2.01	2.58	2.01	1.24	2.38	0.00	1.83	1.14

Method: Place the data in a single column in Excel then use the Histogram chart to create a histogram. Right click on the chart to change the bin intervals. Here are three examples:

Note as illustrated in Figures 3.14.1 a, b, and c, the presentation and interpretation of the histogram of only 72 elements is strongly dependent on the user choice of the bin interval. In Figure 3.14.1a, with only three bins, one cannot see the detail and the data appears to be ideally from an exponential distribution. From Figure 3.14.1b, there are too many bins to see the distribution trend. Figure 3.14.1c uses the recommended number of bins about equal to the square root of the number of data $\left(\sqrt{72} = 8.485\ldots\right)$, and adjusted the nominal bin width $\left(\dfrac{(14.71-0)}{8.485\ldots} = 1.733\ldots\right)$ to a convenient value of 2, resulting in 8 bins.

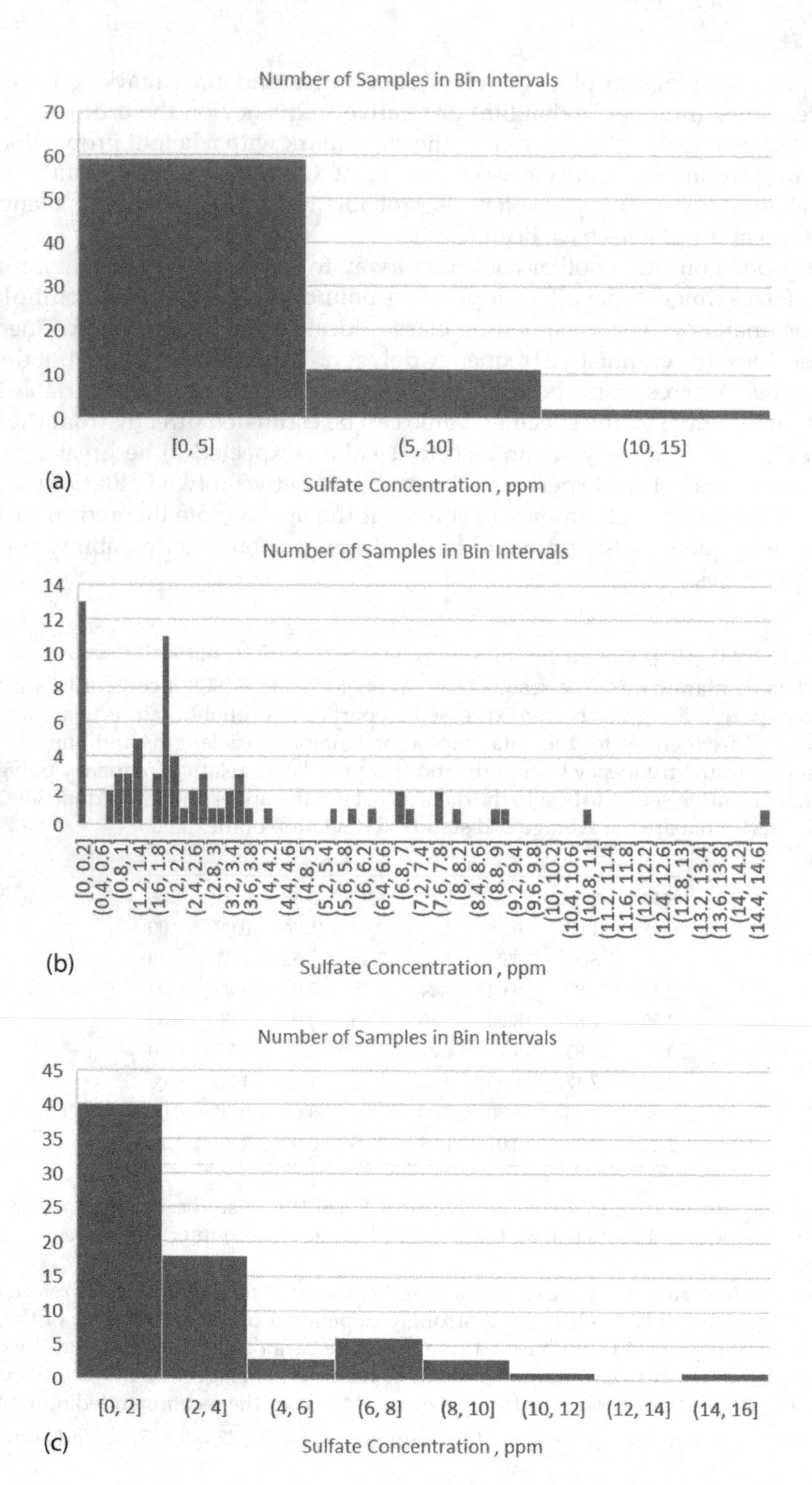

FIGURE EX 3.14.1
a) data histogram with bin interval of 5; b) with a bin interval of 0.2; c) with a bin interval of 2.

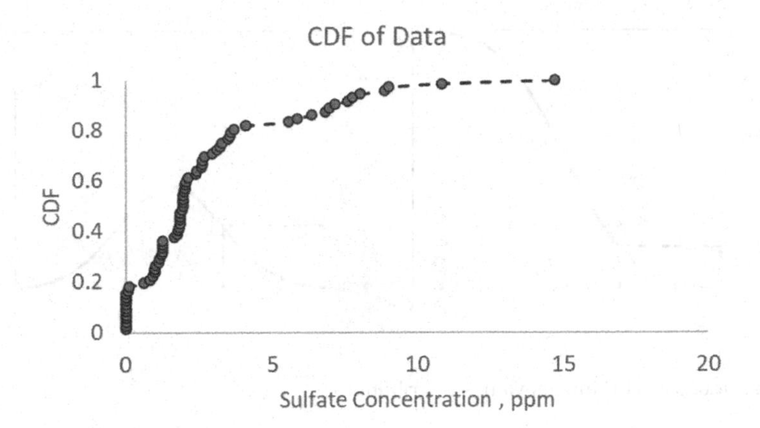

FIGURE EX 3.14.2
Empirical *CDF* (relative frequency polygon) for sulfate concentrations.

Perhaps the data has an exponential distribution, but there is an odd high count of 5 in the bin with a class mark of 7. If $n = 5$ counts is the expected number, the Poisson distribution indicates that the 90% limits are between n = 1 and n = 9. There is the substantial uncertainty the bins with very low counts. So, maybe the distribution is exponential.

To generate an empirical *CDF*, the relative frequency polygon, first sort the data from low to high, then assign a *CDF* value to each data in increments of $1/n$. Then plot the *CDF* value i/n w.r.t. the data value. This is illustrated in Figure 3.14.2. The markers represent the 72 data values, and they are connected by a dashed line to preserve visual order, but not to imply that there is a consistent mechanism or any expectation of between point values.

The quartiles represent the data with *CDF* values of 0.25 and 0.75, which are 1 ppm and 3.3 ppm. Fortunately, some points fall exactly on the quartile values. The upper and lower limits that contain 90% of the data, means that 10% of the data are in the upper and lower extremes. Customarily, this means that 5% of the data are above the upper limit, and 5% are below the lower limit. These would represent data at the 0.05 and 0.95 *CDF* values. However, there are no data points exactly at those *CDF* values; so linear interpolating, the lower 5% limit is 0 ppm, and the upper 95% limit is 8.396. This of course should be rounded to match the implied precision of all the data, reporting an upper limit value of 8.40 ppm.

3.6 Values of Distributions and Inverses

3.6.1 For Continuum-Valued Variables

For continuum-valued variables, *x*, the cumulative distribution function is the probability of getting a particular value or a lower value of a variable. It is the left-sided area on the probability density curve, often expressed as alpha. It is variously represented as $CDF(x) = F(x) = \alpha = p$. Here we'll use the *CDF*(*x*) notation.

For continuous-valued variables, *x*, the probability distribution function, *pdf*(*x*), represents the rate of increase of probability of occurrence of the value *x*. An alternate notation is $pdf(x) = f(x)$.

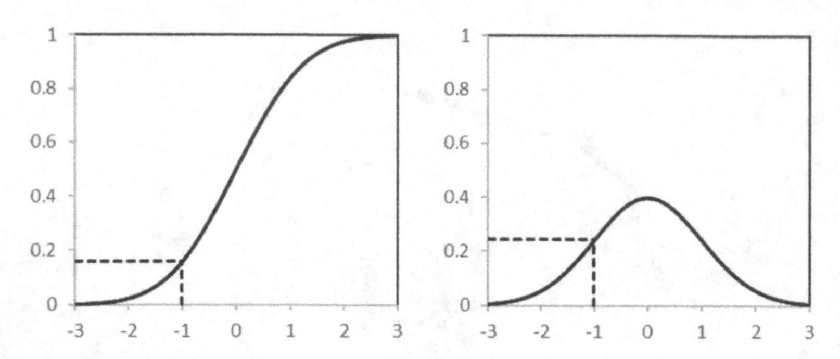

FIGURE 3.16
CDF and *pdf* illustrations for continuous-valued distributions.

The relation between *CDF*(*x*) and *pdf*(*x*) is

$$CDF(x) = \int_{x_{\text{minimum}}}^{x} pdf(x)dx \tag{3.82}$$

Where $x_{minimun}$ represents the lowest possible value for x. In a normal distribution $x_{minimun} = -\infty$. For a chi-squared distribution $x_{minimun} = 0$.

The left-hand sketch in Figure 3.16 illustrates the *CDF* and the right-hand sketch the *pdf* of z for a standard normal distribution (the mean is zero and the standard deviation is unity). At a value of $z = -1$, the *CDF* is about 0.158, and the rate of increase of the *CDF*, the *pdf* is about 0.242. The notations are 0.158 = *CDF* (−1) and 0.242 = *pdf* (−1). In both you enter the graph on the horizontal axis, the *z*-value, and read the value on the vertical axis.

For continuous-valued variables the inverse of the *CDF* is the value of x for which the probability of getting the value of x or a lower value is equal to the *CDF*(*x*).

The inverse would enter on the vertical axis to read the value on the horizontal axis. If the inverse question is, "What *z*-value marks the point for which equal or lower *z*-values have a probability of 0.158 of occurring?" then we represent this inverse question as $z = CDF^{-1}(\alpha)$. In this illustration, $-1 = CDF^{-1}(0.158)$. The inverse of the right-hand *pdf* graph is not unique. If the question is to determine the *z*-value for which the *pdf* = 0.242, there are two values, $z = -1$, and $z = +1$.

3.6.2 For Discrete-Valued Variables

For discrete-valued variables, x, likely a count of the number of events, the cumulative distribution function is the probability of getting a particular value or a lower value of a variable. It is the left-sided area on the probability density curve, often expressed as alpha. It is variously represented as $CDF(x) = F(x) = \alpha = p$. Again, we will use the *CDF*(*x*) notation.

For discrete-valued variables, x, the point distribution function, *pdf*(*x*), represents the probability of an occurrence of the value x. An alternate notation is *pdf*(*x*) = *f*(*x*). Here, *pdf*(*x*) is a probability of a particular value of x, not the rate that the *CDF* is increasing. Unfortunately, the same symbol is used in continuum-valued distribution.

The relation between *CDF*(*x*) and *pdf*(*x*) is

$$CDF(x) = \sum_{x_{\text{minimum}}}^{x} pdf(x) \tag{3.83}$$

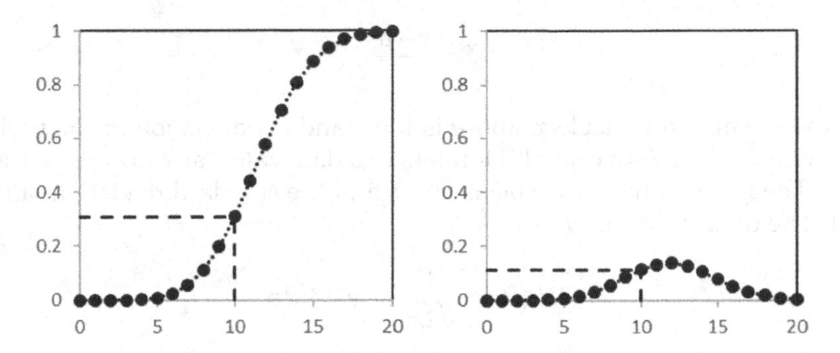

FIGURE 3.17
CDF and *pdf* illustrations for discrete distributions.

where $x_{minimum}$ represents the lowest possible value for x. Normally $x_{minimum} = 0$, the least number of events that could occur.

The left-hand sketch of Figure 3.17 illustrates the *CDF* and the right-hand sketch the *pdf* of s, the count of the number of successes, for a binomial distribution (the number of trials is 40, and the probability of success on any particular trial is 0.3). Note that the markers on the graphs represent feasible values. The light line connecting the dots is a visual convenience. It is not possible to have 10.3 successes. At a value of $s = 10$, the *CDF* is about 0.309, meaning that there is about a 31% chance of getting 10 or fewer successes. The *pdf* is about 0.113, meaning that the probability of getting exactly 10 successes is about 11%. The notations are $0.309 = CDF(10)$ and $0.113 = pdf(10)$. In both you enter the graph on the horizontal axis, the s-value, and read the value on the vertical axis.

For discrete-valued variables the inverse of the *CDF* is the value of s for which the probability of getting the value of s or a lower value is equal to the $CDF(s)$.

The inverse would enter on the vertical axis to read the value on the horizontal axis. If the inverse question is, "What s-value marks the point for which equal or lower counts have a probability of 0.309 of occurring?" then we represent this inverse question as $s = CDF^{-1}(\alpha)$. In this illustration, $10 = CDF^{-1}(0.309)$. The inverse of the right-hand *pdf* graph appears to be not unique. However, it might be. If the question is to determine the s-value for which the $pdf = 0.113$, there is only one value, $s = 10$. It appears that an s-value of about 13.5 could have such a *CDF* value, but the count must be an integer. The *pdf* of $S = 13$ is 0.126, and the *pdf* of $S = 14$ is 0.104.

Although one could ask, "What count value, or lower, has a 30% chance of occurring?" it is impossible to match the 30% $CDF = 0.3\overline{000}$ value. $S \leq 9$ has a *CDF* of about 0.196 which does not include the target $0.3\overline{000}$. $S \leq 10$ has a *CDF* of about 0.309 which does match. $S = 10$ is the lowest value that includes the target *CDF*. One convention is to report the minimum count that includes the target *CDF* value.

3.7 Distribution Properties, Identities, and Excel Cell Functions

3.7.1 Continuum-Valued Variables

3.7.1.1 Standard Normal Distribution

The statistic z is defined as

$$Z = \frac{x - \mu}{\sigma} \tag{3.84}$$

where μ is the mean of a variable x, and σ is the standard deviation of the variable. The z-value is dimensionless. As a capital Z it refers to a data value, as a lowercase z it refers to the variable. The mean of the z-variable is $\mu = 0$, and the standard deviation of the z-variable is $\sigma = 1$. The distribution *CDF* is

$$CDF(Z) = \frac{1}{\sqrt{2\pi}} \int_{-\infty}^{Z} e^{-z^2/2} dz \tag{3.85}$$

The Excel cell functions are

$$Z = \text{NORM.INV}(\alpha, \mu, \sigma)$$

$$CDF(Z) = \text{NORM.DIST}(Z, \mu, \sigma, 1)$$

$$pdf(Z) = \text{NORM.DIST}(Z, \mu, \sigma, 0)$$

where α is the *CDF* value, which is labeled as the probability. The 4th variable in the NORM.DIST function is a trigger to return either the cumulative or the probability distribution value. For the standard normal z-statistic, use $\mu = 0$, and $\sigma = 1$.

The distribution is symmetric. Accordingly,

$$\text{NORM.INV}(\alpha, 0, 1) = -\text{NORM.INV}((1-\alpha), 0, 1)$$

$$\text{NORM.DIST}(Z, 0, 1, 1) = 1 - \text{NORM.DIST}(-Z, 0, 1, 1)$$

3.7.1.2 t-Distribution

The t-statistic is defined as

$$T = \frac{\bar{X} - \mu}{s/\sqrt{n}} \tag{3.86}$$

In which s is the sample standard deviation, not the true population sigma; and $\bar{X}$ is the average, not the true population mean. The degrees of freedom is υ, which is often $n - 1$.

The Excel cell functions are

$$T = T.\text{INV}(\alpha, \upsilon)$$

$$CDF(T) = T.\text{DIST}(T, \upsilon, 1)$$

$$pdf(T) = T.\text{DIST}(T, \upsilon, 0)$$

where α is the *CDF* value, which is labeled as the probability.

The distribution is symmetric. Accordingly,

$$T.\text{INV}(\alpha, \upsilon) = -T.\text{INV}(-\alpha, \upsilon)$$

$$T.\text{DIST}(T, \upsilon, 1) = 1 - T.\text{DIST}(-T, \upsilon, 1)$$

3.7.1.3 Chi-Squared Distribution

The Excel cell functions are

$$\chi^2 = \text{CHISQ.INV}(\alpha, \upsilon)$$

$$CDF(\chi^2) = \text{CHISQ.DIST}(\chi^2, \upsilon, 1)$$

$$pdf(\chi^2) = \text{CHISQ.DIST}(\chi^2, \upsilon, 0)$$

where α is the *CDF* value, which is labeled as the probability, and υ is the degrees of freedom.

The χ^2 distribution is not symmetric. The minimum value is zero.

Since χ^2 nearly increases linearly with degrees of freedom, it is often reported as χ^2/υ.

$$\chi^2/\upsilon = \text{CHISQ.INV}(\alpha, \upsilon)/\upsilon$$

$$CDF(\chi^2/\upsilon) = \text{CHISQ.DIST}(\upsilon(\chi^2/\upsilon), \upsilon, 1)$$

3.7.1.4 F-Distribution

F is a ratio of sample variances scaled by the population variance.

$$F = \frac{s_1^2/\sigma_1^2}{s_2^2/\sigma_2^2} \tag{3.87}$$

The Excel cell functions are

$$F = F.\text{INV}(\alpha, \upsilon_{\text{numerator}}, \upsilon_{\text{denominator}})$$

$$CDF(F) = F.\text{DIST}(F, \upsilon_{\text{numerator}}, \upsilon_{\text{denominator}}, 1)$$

$$pdf(F) = F.\text{DIST}(F, \upsilon_{\text{numerator}}, \upsilon_{\text{denominator}}, 0)$$

where α is the *CDF* value, which is labeled as the probability, and υ is the degrees of freedom for the numerator and denominator values of the sample standard deviations.

The *F*-distribution is not symmetric, the minimum value is zero.

One can choose which sample is labeled 1 (and placed in the numerator) and the other labeled 2 (placed in the denominator). If one ratio is unusually small, then the other will be unusually large, and the extreme changes from the left tail to the right tail. Accordingly,

$$F.\text{INV}(\alpha, \upsilon_1, \upsilon_2) = 1/F.\text{INV}((1-\alpha), \upsilon_2, \upsilon_1)$$

$$F.\text{DIST}(F, \upsilon_1, \upsilon_2, 1) = 1 - F.\text{DIST}(1/F, \upsilon_2, \upsilon_1, 1)$$

3.7.2 Discrete-Valued Variables

3.7.2.1 Binomial Distribution

The cumulative binomial distribution is

$$CDF(S) = \sum_{k=0}^{S} \binom{n}{k} p^k (1-p)^{n-k} \tag{3.88}$$

The Excel cell functions are

$$s = \text{BINOM.INV}(n, p, \alpha)$$

$$CDF(s) = \text{BINOM.DIST}(s, n, p, 1)$$

$$pdf(s) = \text{BINOM.DIST}(s, n, p, 0)$$

Here, s is the number of successes in n number of trials where the probability of a success in any one trial is p. Alpha, α, is the *CDF* value associated with s or fewer successes. All variables are dimensionless. Further s and n must be integers and $0 \le n$, and $0 \le s \le n$.

If you input a non-integer value in the BINOM.DIST function the Excel function truncates either s or n to the integer value. For example, $s = 12.001$ and $s = 12.999$ are both truncated to 12.

If you specify an in-between value for α, the *CDF* value, in the BINOM.INV function Excel will return the next larger s.

3.7.2.2 Poisson Distribution

The cumulative Poisson distribution function is

$$CDF(x) = \sum_{k=0}^{x} \frac{\lambda^k e^{-\lambda}}{k!} \tag{3.89}$$

where x is the number count of events within a time interval (or a distance, area, or space interval, or a per item basis), and λ is the average number for the population. Here $\lambda = \mu$.

The Excel cell functions are:

$$CDF(x) = \text{POISSON.DIST}(x, \mu, 1)$$

$$pdf(x) = \text{POISSON.DIST}(x, \mu, 0)$$

Unfortunately, there does not seem to be an inverse function, but it is fairly easy to use a trial-and-error search (such as interval halving) to determine the count that matches a *CDF* value.

3.8 Propagating Distributions with Variable Transformations

Often, we know the distribution on x-values and have a model that transforms x to y. For instance, $y = Ln(x)$. The question is, "What is the distribution of y?"

Figure 3.18 reveals the case of $y = a + bx^3$ when the distribution on x (on the abscissa) is normal.

Note: For the range of x-values shown, the function is strictly monotonic, positive definite. As x increases, y increases for all values of x. There are no places in the x-range where either 1) the derivative is negative or 2) zero (there are no flat spots in the function).

The inset sketches indicate the *pdf* (dashed line) and *CDF* of x and y, about a nominal value of $x_0 = 2.5$ and the corresponding $y_0 = a + bx_0^3$. Note that the *pdf* of x is symmetric, and that of y is skewed.

The *CDF* of x indicates the probability that x could have a lower value. For any x there is a corresponding y, and since the function is strictly monotonic, the probability of a lower y-value is the same as the probability of a lower y-value. Then

$$CDF\left(y = f(x)\right) = CDF(x) \tag{3.90}$$

Between any two corresponding points x_1 and x_2 separated by $\Delta x = x_2 - x_1$, there are the two corresponding points $y_1 = f(x_1)$ and $y_2 = f(x_2)$ separated by $\Delta y = f(x_2) - f(x_1) \cong \frac{dy}{dx}\Delta x$ for small Δx values (meaning that $\frac{dy}{dx}$ is relatively unchanged over the Δx interval). Since $CDF(y_2) = CDF(x_2)$ and $CDF(y_1) = CDF(x_1)$, the difference is also equal, and by definition:

$$\int_{y_1}^{y_2} pdf(y)\,dy = \int_{x_1}^{x_2} pdf(x)\,dx \tag{3.91}$$

For small Δx intervals, the integral can be approximated by the trapezoid rule of integration, and in the limit of very small Δx,

$$pdf(y) = pdf(x) / \frac{dy}{dx} \tag{3.92}$$

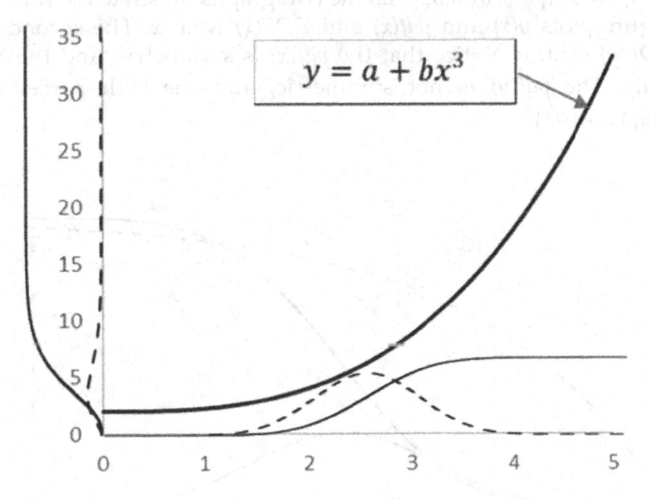

FIGURE 3.18
Illustration of a nonlinear distribution transformation.

To obtain the $CDF(y)$ numerically integrate the $pdf(y)$. Using the trapezoid rule of integration, with y sorted in ascending order.

$$CDF(y_{i+1}) = CDF(y_i) + \frac{1}{2}\left[pdf(y_{i+1}) + pdf(y_i)\right](y_{i+1} - y_i) \tag{3.93}$$

Initialize $CDF(y_{\text{very low}}) = 0$.

This is only true if $y = f(x)$ is a strictly positive definite function (if $\frac{dy}{dx} > 0$ for all values in the range being considered).

Example 3.15: Derive the $pdf(y)$ when $y = Ln(x)$, and $x_0 = 1$, and x is a continuum variable that is uniformly distributed with a range of 0.6.

If the average, or nominal, x-value is 1 and the range is 0.6 then it varies between 0.7 and 1.3. In the uniform distribution $a = 0.7$ and $b = 1.3$, and $pdf(x) = 1.6666\ldots$ for $0.7 \le x \le 1.3$, otherwise $pdf(x) = 0$.

Since $\frac{dy}{dx} = \frac{1}{x}$, then

$$pdf(y) = 1.\overline{66}x, \quad 0.7 \le x \le 1.3$$

$$pdf(y) = 0, \quad x < 0.7 \text{ or } x > 1.3$$

Example 3.16: If $y = a + b\left(1 - e^{-x/s}\right)$ and x is normally distributed, NID(μ_x, σ_x), what is the $pdf(y)$?

The derivative of y w.r.t. x is $\frac{dy}{dx} = \frac{b}{s}e^{-x/s}$. As long as $\frac{b}{s} > 0$ the function is strictly monotonic positive. Then

$$pdf(y) = \frac{1}{\sqrt{2\pi}\sigma_x} e^{-\frac{1}{2}\left(\frac{x-\mu_x}{\sigma_x}\right)^2} \frac{s}{b} e^{x/s}$$

With $a = 1$, $b = 7$, $s = 3$, $\mu_x = 5$, and $\sigma_x = 1.5$ the two graphs illustrate the functions.

The first figure plots $y(x)$ and $pdf(x)$ and $CDF(x)$ w.r.t. x. The second figure plots $pdf(y)$ and $CDF(y)$ w.r.t. y. Notice that the $pdf(x)$ is symmetric, and the 50%ile value of $CDF(x) = \mu_x$. The $pdf(y)$ is not symmetric, and the 50th percentile value of $CDF(y) = a + b\left(1 - e^{-\mu_x/s}\right)$.

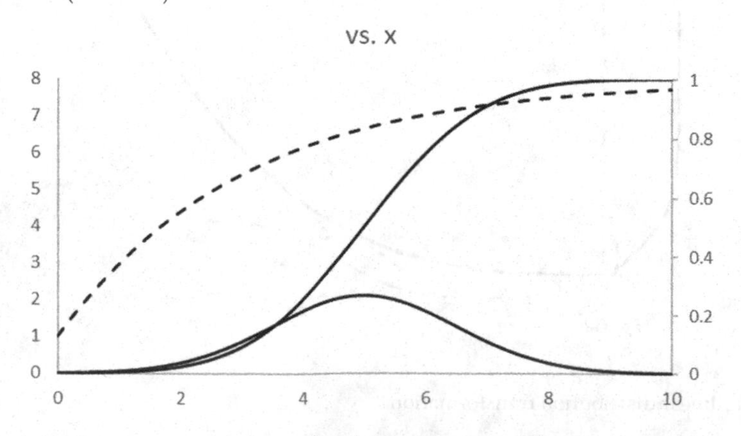

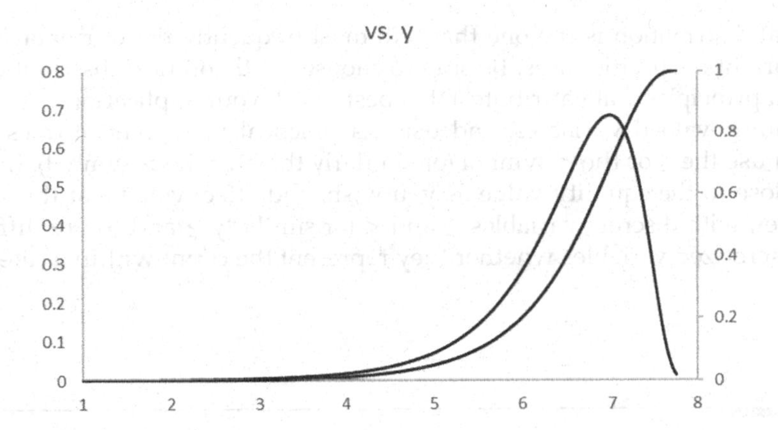

For a strictly negative monotonic function, such as $y = a + \frac{b}{x}$, $x > 0$, or $y = a + bx^2$, $x < 0$, or $y = a - bx$, $b > 0$, the analysis is similar but since larger x-values lead to smaller y-values, the subscripts on y need to be reversed. $y_2 = f(x_1)$ and $y_1 = f(x_2)$. And

$$CDF\left(y = f(x)\right) = 1 - CDF(x) \tag{3.94}$$

$$\int_{y_1}^{y_2} pdf(y)\,dy = -\int_{x_1}^{x_2} pdf(x)\,dx \tag{3.95}$$

For small Δx intervals, the integral can be approximated by the trapezoid rule of integration, and in the limit of very small Δx,

$$pdf(y) = -pdf(x) / \frac{dy}{dx} \tag{3.96}$$

To obtain the *CDF*(y) numerically integrate the *pdf*(y). Using the trapezoid rule of integration, with y sorted in ascending order,

$$CDF(y_{i+1}) = CDF(y_i) + \frac{1}{2}\left[pdf(y_{i+1}) + pdf(y_i)\right](y_{i+1} - y_i) \tag{3.97}$$

initialize $CDF(y_{\text{very low}}) = 0$.

Note: if the data are sorted in ascending order of x, then the y-values are in descending order. Then, Initialize $CDF(y_{\text{very high}}) = 1$.

This is only true if $y = f(x)$ is a strictly negative definite function (if $\frac{dy}{dx} < 0$ for all values in the range being considered).

3.9 Takeaway

Whether the data distribution is normal (Gaussian) or not, it has a mean and a variance. Just because you can get an average and standard deviation from your data does not mean that the data is normally distributed.

The normal distribution is the one that will most frequently fit your data, but the others are important in specific cases. Be sure to choose the theoretical distribution that was derived from principles (data attributes) that best match your application.

For continuum-valued variables, and using a practical viewpoint, it does not matter whether you use the ≤ or the < symbol (or similarly the ≥ or the > symbol). If < or >, you can get as close to the equality value as you wish, and effectively be at the ≤ or ≥ location. However, with discrete variables ≤ and < (or similarly ≥ and >) are different. Take care with discretized variables whether they represent the count within a category or the category.

3.10 Exercises

1. Derive Equations (3.7a) and (3.7b).
2. Derive the mean and variance relations for any one of the distributions of a discrete-valued variable. The uniform, and Poisson distributions are not too difficult.
3. Derive the mean and variance relations for any one of the distributions of a continuum-valued variable. The uniform, and proportion distributions are not too difficult. The exponential might be fun.
4. Over a 20-year driving history, a particular person "earned" two tickets. This represents $\lambda = 0.1[\text{tickets per year per individual}]$. How many tickets per year would be expected in a city of 50,000 individuals with a similar driving style?
5. Derive that the average of z in Equation (3.55) is zero and the standard deviation of z is unity.
6. Show that the sum of all $f(x_i)$ values for a discrete distribution is 1 or show that the integral of a $pdf(x)$ over all x-values is 1.
7. Match your choice of a continuum distribution to the data of Example 3.14 and determine the distribution coefficients that best fit your choice to the data.
8. Compare a normal distribution to a uniform continuum distribution, using the same mean and variance for both.
9. Use the Gaussian *pdf* model to show that the inflection points on the *pdf* curve (where the shape changes between concave to convex) are at the $\pm 1\sigma$ deviation from the mean.
10. Repeat Example 3.15 using $y = a + bx^2$.
11. Repeat Example 3.15 using $y = a + \frac{b}{x}$.
12. If one process has an average of one event per 10 years, and a second process has an average of 0.05 events per year, would you expect the second process, over a two-year period, to have the same number of events as the first?
13. Graph $pdf(x)$ and $pdf(y)$ when $y = ln(x)$. Use the exponential $pdf(x)$.
14. The logistic model is $CDF(x) = 1/\left(1 + e^{-s(x-c)}\right)$, where s is a scale factor and c is the center. Show that if $c = \mu$ and $s = \sigma/1.7$, for your choice of parameter values, that the logistic *CDF* is very similar to the normal *CDF*.

15. Show that the analytical derivation of the inverse of the logistic *CDF* is simply done.
16. Show that the analytical derivation of the *pdf* of the logistic *CDF* is simply done.
17. If one delivery truck in a service has an average of 2 flat tires per year ($\lambda = 2[\text{flats per year per truck}]$), the λ value per year if there are three trucks in the fleet all with the same service is $\lambda = 6[\text{flats per year per fleet}]$. Show that the probability of 0, 1, 2, 3, and 4 flats in a year as calculated for the fleet is the same as that calculated for the three individual trucks. For the fleet $P(x) = 6^x e^{-6} / x!$, and for an individual truck $P(x) = 2^x e^{-2} / x!$. Hint: How can one get 3 flats in the individual trucks? The answer is there are 10 ways: (3,0,0), (0,3,0), (0,0,3), (2,1,0), (2,0,1), (1,2,0), (1,0,2), (0,1,2), (0,2,1), and (1,1,1). The probability of the event (2,0,1) is $P(2 \text{ for Truck } A \text{ AND } 0 \text{ for Truck } B \text{ AND } 1 \text{ for Truck } C) = P(2) * P(0) * P(1)$.
18. Use Figure 3.6 to determine the probability that x will be between 2 and 3 inclusive $P(2 \le x \le 3)$, and also not including the value of 3, $P(2 \le x < 3)$.
19. Use the inverse of Equation (3.92) to derive a model of the distribution of x if the log transformation of x is nearly normal. Here $y = \ln(x)$. Scale x by its average, $x^* = x / \bar{x}$, now x^* is centered on 1, so that $y = \ln(x^*) = \ln(1) = 0$. This means that $\mu_y = 0$, and if the distribution of y is nearly normal then 5% of the area is at $CDF(y = -1.645\sigma_y) = 0.05$. Apply Equation (3.92) twice. First to transform *pdf*(y) to *pdf*(x^*), then *pdf*(x^*) to *pdf*(x).
20. If you drop a marble and lose sight of where it went to hide, then randomly search for it, how many places must you look to be 99% confident of finding it? You decide the appropriate value of the probability of finding it on any particular search. First consider that there are an infinite number of places it could hide. Then consider that there are only a finite number of places.

4

Descriptive Statistics

Descriptive statistics are values of attributes which are calculated from sample data and are used either to describe sample characteristics or to estimate population parameters. The descriptive statistics most often used in engineering are the *mean* and the *standard deviation.* But there are many others.

4.1 Measures of Location (Centrality)

Often termed the arithmetic average, the arithmetic mean is the primary statistic that locates the sample. Usually, it is simply termed the average or mean. If the histogram of sample values is created, the arithmetic mean or arithmetic average of all sample values is the centroid of the distribution. The sample mean is calculated by

$$\bar{X} = \frac{1}{n}\sum_{i=1}^{n} X_i \tag{4.1}$$

where $\bar{X}$ is the average of the n sample members and X_i are the individual sample values. It can be shown that the population mean is the expected value of the sample mean. (If interested in the calculus and proof see the sections on expectations or expected values in any statistical theory or sampling theory text, or Section 4.3 here). But what is important is knowing that the sample mean, $\bar{X}$, is the best estimator of the population mean μ with respect to being consistent, efficient, and unbiased. (An estimator is consistent if the values it predicts become closer and closer to the true value of the parameter as the size of the sample increases. An estimator is unbiased if its expected value is the value of the parameter itself. An efficient estimator is one that not only is unbiased but also has the smallest possible variance. Efficient estimators are often called "best" estimators because of those characteristics.)

If the X_i values can only have discrete values, such as quality points for grades (an A = 4, a B = 3, etc.), then some values of the data are repeated, and you can use a weighted average of a group of numbers, the weighted mean can be found from

$$\bar{X} = \frac{\sum_{i=1}^{k} f_i X_i}{\sum_{i=1}^{k} f_i} \tag{4.2}$$

where the frequencies, f_i, are the weighting factors associated with their respective X_i categories. Here $f_i = n_i / N$, where N is the total number of all data and n_i is the count of items in the kth classification. In this case the denominator term $\sum_{i=1}^{k} f_i = 1$.

DOI: 10.1201/9781003222330-4

Weighted means are useful for far more than just calculating grade point averages. If some of the X_i values are more important than others, you might need a weighted average of a group of numbers. As an example, you might be interested in a characteristic particle size (diameter) that represents the surface area of particles, have screened the particles into size classifications, and have the weight of particles in each screening category. There are more particles in the smaller screening size than in the same weight of a larger size. Here the f_i values would not simply be the weight in each screen category.

The mean might not be a feasible value. Roll a die many times. The average point value per roll will be around a value of 3.5 but that is not a feasible value on any particular roll.

Contrasting the conventional arithmetic mean is the geometric mean, a representation of a characteristic value where the product of attributes is important.

$$\bar{X}_{\text{geo}} = \sqrt[N]{\prod_{i=1}^{N} X_i} \tag{4.3}$$

An alternate yet is the harmonic mean, where reciprocals are used to characterize the feature, as in heat transfer where individual coefficients represent conductance not resistance, but an average value of resistance is desired

$$\bar{X}_{\text{harmonic}} = \left[\sum_{i=1}^{N} X_i^{-1}\right]^{-1} \tag{4.4}$$

The median is value of the middle X_i value if N is odd, or the average of the two middle values if N is even. Sort the data in either ascending or descending order then the middle value is such that half the sample values are larger than the median and the other half, smaller. If N is odd, the median would be a feasible value, because it was one of the sample values; but if N is even, the average of the two middle values might not be feasible.

The mode is simply the most frequently occurring sample value, which could be relevant in describing disparate categories, choices, preferences, etc. As an example of disparate categories, shoppers may buy 4 cans of soup and 2 heads of lettuce. However, the X-values in another example might have consistent units. The mode would be a feasible value.

If the histogram of data is symmetric, then the median and mode are both reasonable representations of the arithmetic average.

Proportion (also termed portion, fraction, probability) is a ratio of the number of events in a particular category to the total number of events, or trials.

$$p = n / N \tag{4.5}$$

Odds is a statistical/probability term for the ratio of the probability of an event, p, to the probability of "not-an-event" $q = (1-p)$.

$$\text{Odds} = \frac{p}{q} = \frac{p}{1-p} \tag{4.6}$$

Almost all of your work will involve the arithmetic mean and proportion, as estimates of centrality or location, and most statistical tests are developed for those measures of centrality, however, some nonparametric procedures are evaluations of the median, not the mean.

4.2 Measures of Variability

You must consider how the data are distributed around that statistic of centrality. The most popular method of reporting variability is an estimate of the population variance, defined in three equivalent forms as follows:

$$S_X^2 = \frac{\sum_{i=1}^{n}(X_i - \bar{X})^2}{n-1} = \frac{\sum_{i=1}^{n} X_i^2 - n\bar{X}^2}{n-1} = \frac{\sum_{i=1}^{n} X_i^2 - \left(\sum_{i=1}^{n} X_i\right)^2 / n}{n-1} \tag{4.7}$$

The *estimated variance* is the sum of the squares of the deviations of the individual data points from the arithmetic mean value of the sample divided by $(n - 1)$. You may wonder why the sample mean uses division by n but the sample variance uses division by $(n - 1)$. The answer is simple, although not obvious: Each statistic is divided by the number of *independent* data points (or *degrees of freedom*) used for its calculation. Because the value of $\bar{X}$ is used in calculating sample variance, that $\bar{X}$ value is presumed to be the truth. Accordingly, you are free to choose any values for $(n - 1)$ of the X_i sample values were used to calculate $\bar{X}$ but the last X_i value is constrained to create the same $\bar{X}$ value. There only $(n - 1)$ independent choices or degrees of freedom involved in the calculation of the sample variance. Note: The summation is over all sample values.

Here, the X_i values represent the n observations from a large population (possibly of infinite number of possible observations). However, if the sample is of all possible data, then divide by n not $(n - 1)$. If n is very large, then it is inconsequential whether n or $(n - 1)$ is used, because the error will be insignificant relative to the inherent uncertainty in S_X^2.

The dimensional units on the number of data, n, is considered to be dimensionless, so the units on the sample variance, S_X^2, are the square of the units on the data.

The standard deviation of a sample is the positive square root of the variance, or

$$S_X = \sqrt{S_X^2} \tag{4.8}$$

Note: The standard deviation has the same dimensional units as the average and as the data.

Another measure of variability often used is the *coefficient of variation* or CV, defined as

$$CV = \frac{S_X}{\bar{X}} \tag{4.9}$$

The coefficient of variation, CV, expresses the standard deviation as a proportion of the mean. CV is dimensionless.

Yet another widely used and misused measure of variability is the standard error. This statistic is obtained by first putting the sample variance on a "per observation" basis and then taking the positive square root as done for the standard deviation. The standard error of the mean is thus defined by

$$S_{\bar{X}} = \sqrt{\frac{S_X^2}{n}} = \frac{S_X}{\sqrt{n}} \tag{4.10}$$

As you can see, the standard error is always smaller than the standard deviation of the individual data, indicating that the sample mean is less variable than the original data from which it was calculated. Consider looking at the values within each sample. One may be large, one small, and another in between. Now consider an average of 5 samples. The only way the average could be large is for each of the samples to be equivalently large, but this is improbable. In 5 samples some will be small and some in the middle. Accordingly, the variability of an average will be less than that of the individual samples. This relation is termed the central limit phenomena and will be derived in Chapter 8.

Again, the dimensional units on sample size, the number of data, n, is considered to be dimensionless; so, the variance on the sample average has the same units as the variance on the individual data. If this is suspicious, consider that the real equation is $S_{\bar{X}}\sqrt{n} = S_X\sqrt{1}$, and since $\sqrt{1} = 1$, it can be removed for convenience.

Do not confuse these two estimates of variability; the sample standard deviation is used to test hypotheses about individual population values, but the sample standard error is used to test hypotheses about the mean of the population from which the sample was drawn. You should distinguish S_X and $S_{\bar{X}}$ for another reason. If the sample data are badly scattered, some people will report the standard error as if it were the standard deviation, thus trying to conceal what they perceive as poor data. When you hear the word "standard" used in describing the variability of data, always ask whether standard deviation or standard error is meant. If in doubt, ask to see the calculations.

If the population is normally distributed (and most of the measuring situations you'll encounter will be), $\bar{X}$ and S_X^2 are the best values for μ and σ^2, the parameters of the normal distribution. However, if the population is not normally distributed, you can estimate the population parameters from the sample characteristics $\bar{X}$ and S_X^2 as described in Chapter 3.

Data Range, R, is also a valid measure of variability, it is the highest value less the lowest value

$$R = \text{MAX}(\underline{X}) - \text{MIN}(\underline{X}) \tag{4.11}$$

where $\underline{X}$ represents the vector (or listing) of all data, and $\text{MAX}(\underline{X})$ indicates the value of the maximum of all values. This two-point calculation of the measure of variation is simpler than the sample standard deviation, but it only uses two data values. The sample standard deviation uses all n values and gives a more representative value.

The X-values in Equation (4.11) could also be paired deviations between data, or residuals between data and a model.

Percentiles are frequently used to represent data variation. These could be values that represent the lowest 25% or upper 75% of values, or lowest 10% or upper 90%, etc. Conceptually, these could be found by sorting the data then finding the particular value. For instance, if there are n = 1,000 data, sorted, then the 250th value would represent the 25th percentile value. Likely, however the value would be interpolated between two neighboring values. If there were 11 data, the 25th percentile value could be interpolated between the second and third in the list. Alternately, if the distribution is presumed to be normal, the percentiles could be estimated from the mean and variance (see Chapter 5).

4.3 Measures of Patterns in the Data

A run is the sequence of ordered data with a like property. The statistic Runs is the number of runs in a dataset of dichotomous, exclusive outcomes. For example, when data are compared to a model, the residual is either + or −. In this sequence of signs $\underline{++}\ \underline{-}\ \underline{+}\ \underline{----}\ \underline{+}$ there are 5 runs (underscored). As another example, here is a coin flip of Heads and Tails sequence, also with 5 runs, H H T H T T T T T H H. Alternately, the categories could be off/on, or 1/0.

The number of runs is often used to determine whether the deviations between model and data are randomly occurring. If there are too few runs, it indicates that there is one or more long sections where the data are on one side of a model, indicating that the model does not match the process that generated the data. The data could be ordered chronologically or with respect to each input or response variable.

In looking at residuals, values will be either + or −, above or below the model. There is a chance, however, that a residual will have a value of zero. In this case it is not a zero-crossing; include the data with the previous run. For example, this set of residuals has 4 runs:

$$\underline{-3, -2, 0, -1}, \ \underline{+4, +1, 0, +2, +1}, \ \underline{-2, -3}, \ \underline{+1, +3, 0}$$

Example 4.1: What is the expected probability of a run of l, if the probability of either outcome is 0.5?

The Geometric distribution generates the probability of a sequence of data with like property which ends with a data of the other property, when the expectation is that the probability of a data having a positive or a negative residual is the same $p = (1-p) = q = 0.5$.

$$f(x=k) = p^{k-1}q = p^k$$

Here k is the number of trials prior to getting one outcome of the other kind. So, using l as the length of a run the distribution of l with $p = 0.5$ is

$$f(l) = p^l = 0.5^l$$

For example:

l	$p(l)$
1	0.5
2	0.25
3	0.125
4	0.0625
5	0.03125

Note: A run cannot have an $l = 0$ value or a value greater than the number of data, N. If there is only one data value, it is a run of 1. If the data are alternating in their property, the number or runs is N. So, there are limits on l.

$$1 \le l \le N$$

Example 4.2: What is the average run length (ARL), if the dichotomous outcomes are equally probable?

$$ARL = \sum_{l=1}^{l=N=\infty} lf(l) = \sum_{l=1}^{l=N=\infty} lp^l = 2[\text{average number of data per run}]$$

It might be fun to confirm that $\sum_{l=1}^{\infty} lp^l = 2$ through simulation.

Example 4.3: Given N data, what is the expected number of runs in that set?

$$n_{\text{expected \# of runs}} = \frac{N[\text{data}]}{ARL[\text{data per run}]}$$

So, if there are $N = 10$ data to compare to a model, and the model was correctly matched the phenomena that generated the data, and the experimental errors on the data were independent, then the expected number of runs in the residuals would be

$$n_{\text{expected \# of runs}} = \frac{10[\text{data}]}{2[\text{data per run}]} = 5[\text{runs}].$$

Although one expects the number of runs to be N/ARL, with a finite sample size, there is a range of possible outcomes that could be obtained. Similarly, if you have a fair coin and flip it 10 times you expect to get 5 Heads (H) and 5 Tails (T). However, of course, because of the event vagaries in a small sample, you don't expect to see exactly the 50/50 outcome expected in the population. Seven H and 3 T would be a reasonable outcome.

Appendix Table A.3 (a and b) provides critical values of the distribution of runs, for finite N values.

Signs is a statistic that is simply the count of the number of "+" or "−" values, without regard to order. If the model goes through the data, then the number of "+" and "−" residuals should be nearly 50/50. Appendix Table A.1 gives critical values of the statistic *signs*.

Correlation between two variables X and Y means that when X is high (or low) Y tends to be high (or low). Or it could be the opposite, when one is high the other tends to be low. Correlation indicates that there a relation between the two variables X and Y. Be aware that it does not mean that there is a cause-and-effect relation. The values of X and Y could both be consequences of a third variable. Correlation also does not mean that the relation is linear. It could be nonlinear.

There is a diverse number of ways to create a correlation statistic. One is

$$r = \frac{\sum_{i=1}^{n}(x_i - \bar{x})(y_i - \bar{y})}{\sqrt{\sum_{i=1}^{n}(x_i - \bar{x})^2}\sqrt{\sum_{i=1}^{n}(y_i - \bar{y})^2}} = \frac{\frac{1}{(n-1)}\sum_{i=1}^{n}(x_i - \bar{x})(y_i - \bar{y})}{s_x s_y} \tag{4.12}$$

The second form of the relation is achieved by dividing the numerator and denominator by $(n - 1)$. The numerator term is called a covariance, a measure of how x and y co-vary. If x_i and y_i are jointly above or below their averages, then the elements in the numerator will all be positive, and the sum will become large. In the opposite case, if y_i is below its average when x_i is above, and vice versa, then the elements in the numerator will all be negative, and the sum will become negatively large. Alternately, if there is no relation between x_i and y_i then half of the elements in the numerator will be positive and half negative, and the sum will tend to be around zero. So, if r is large positive there is a correlation, if r is large negative there is a negative correlation, and if r is small (around zero) there is not a detectable correlation. The quantification of large and small depends on the variation in the variables. So, the numerator is scaled by similar measures of the x_i and y_i variation. If there is perfect, linear positive correlation, $y_i = a + bx_i$, then $\overline{y} = a + b\overline{x}$, and as a result, $r = 1$. If there is perfect linear negative correlation, $r = -1$. And if there is zero correlation, ideally, $r = 0$.

See Chapter 13 for alternate measures.

Autocorrelation means self-correlation. If one deviation is high, the influence that caused that to happen persists, and the next deviation will likely be high also. As an example, on a partly cloudy day the cloud shadows pass by, but they shade one spot on the ground for a minute or so. The shade persists for more than a microsecond. In the shade, the temperature drops. If one temperature measurement is low, then the next (a second later) will probably be low also, until the cloud passes. Alternately, the data could be oscillating high to low, if for instance a controlling mechanism was over-correcting. In those examples the data represents a single variable. Also, it is in chronological order, but if could be ordered by another variable. Autocorrelation would indicate that the data are not independent, but that some influence is persisting. There are many structures for an autocorrelation statistic, and one can look at the adjacent values, or every second value, or third, etc. One autocorrelation statistic is very similar to the correlation statistic above. In this structure it is an autocorrelation of adjacent variables of "lag-1" (adjacent values in an ordered sequence).

$$r_1 = \frac{\sum_{i=2}^{n} r_i r_{i-1}}{\sum_{i=1}^{n} r_i^2} = \frac{\frac{1}{(n-1)} \sum_{i=2}^{n} (x_i - \overline{x})(x_{i-1} - \overline{x})}{s_x^2} \tag{4.13}$$

The variable r_i in the sums is termed a residual, typically it is the difference between model and data, but here it represents the difference between data and average. Note that there are $n - 1$ terms in the numerator sum. Similar to the r in Equation (4.12), $-1 < r_1 < +1$, and if $r_1 \cong 0$ there is no evidence of autocorrelation.

The data might show the same number of values above a model as below, which would be expected if the model was representative of the process that generated the data. But, it could be that many of the + residuals are much greater than the characteristic residual. So, even if the count of signs is nearly 50/50, there is still a skew or bias in the data. A Sign-Rank Sum of Deviations is a common metric to indicate this. Sort the deviations by absolute value, then rank largest to lowest, then sum the ranks with a "+" (or with a "−") deviation. If the sum is too high or too low, it indicates a skew in the residuals (see Chapter 7). Appendix Table A.2 reports critical values of the Wilcoxon Signed-Rank statistic.

Often, we are classifying data (good products vs faulty products, the letter A or B), and have a count of the number of events in all the categories. We might be comparing treatments for a disease, raw material in a process, preferences in age groups, techniques for training a dog, or the success of an Artificial Intelligence algorithm. The treatments may lead to one of two (or more) outcomes. Place the data in a contingency table, Table 4.1. Here, the entry $n_{A,1}$ represents the number of times Treatment A led to Outcome 1.

If the two treatments are identical, if they have the same impact on the outcome, then you expect $n_{A,1} = n_{B,1}$, and $n_{A,2} = n_{B,2}$, but either vagaries in the experiments or unequal numbers of tests will not make them equal. So, calculate an expected value for each of the classifications. For instance, the expected value for $n_{A,1}$ could be based on the total number

TABLE 4.1

An Example of a Contingency Table

	Treatment A	Treatment B
Outcome 1	$n_{A,1}$	$n_{B,1}$
Outcome 2	$n_{A,2}$	$n_{B,2}$

of experiments of Treatment A and the ratio of Outcome 1 to the total number of A and B experiments

$$E_{A,1} = (n_{A,1} + n_{A,2}) \frac{(n_{A,1} + n_{B,1})}{(n_{A,1} + n_{A,2} + n_{B,1} + n_{B,2})}. \tag{4.14}$$

Alternately, the expected values could come from historical data or other experience. The chi-squared (χ^2) statistic is the sum over all categories of the squared deviation of observed from expected, scaled by expected, for each category.

$$\chi^2 = \sum_{i=1}^{k} \frac{(O_i - E_i)^2}{E_i} \tag{4.15}$$

If the value for χ^2 is large, then the two treatments are probably not equal.

Skewness is a measure of nonsymmetry of the histogram or pdf, and kurtosis is a measure of flatness, alternately, of the largeness or the tail area, compared to a normal pdf. These characterizations of distributions are just mentioned here, but of little practical consequence in the authors' experiences.

4.4 Scaled Measures of Deviations

There are several statistics that are used to quantify the magnitude of a deviation, relative to some base situation.

The *t*-statistic is a normalized deviation between the average and true or expected mean (or other averages). It is scaled by the standard error of the average, so it is dimensionless.

$$t = \frac{\bar{X} - \mu}{s / \sqrt{N}} \tag{4.16}$$

The *t*-statistic indicates the number of standard errors that the average is from the mean. You might recognize that this is similar to the CV, except that standard error of the average, not standard deviation of the data is used as the measure of variation.

There are several variations on what to use in the denominator estimate of numerator variation which depend on assumptions about the value of the standard deviation. See Chapter 6 for details on those separate cases.

Chi-squared is a ratio of two variances, with one presumed to be known. $\chi^2 = s^2 / \sigma^2$, usually, and in this book, it is defined as the ratio times the degree of freedom.

$$\chi^2 = \frac{(N-1)s^2}{\sigma^2} \tag{4.17}$$

Take care as to how it is defined. Some sources report $\chi^2/(N-1) = s^2/\sigma^2$.

In a contingency table use, with enough data, the calculated value $\chi^2 = \sum_{i=1}^{k} \frac{(O_i - E_i)^2}{E_i}$ is approximately distributed as the true ratio of variances with one presumed to be known.

The *F*-statistic is a ratio of two variances with neither presumed to be known

$$F = \frac{s_1^2}{s_2^2} \tag{4.18}$$

There is yet another *r*-statistic, r^2. When one set of data, *y*, (expected to be a response or outcome) is plotted with respect to (w.r.t.) another set of data, *x*, (expected to be a cause or influence), an r^2ratio is a conventional measure of how well a linear model, $\tilde{y} = a + bx$, fits the data. If a perfect fit, then $y_i = \tilde{y}_i = a + bx_i$, and each residual, $d_i = (y_i - \tilde{y}_i) = (y_i - a - bx_i)$, the difference between model and data will be zero. Then the sum of the squared deviations (*SSD*) will be zero, $SSD = \sum_{i=1}^{N} d_i^2 = 0$. Alternately, if there is no trend, if $b = 0$, then $a = \bar{y}$, and the sum of the squared deviations will be the $(n - 1)$ times the *y*-data variance, $SSD_1 = \sum_{i=1}^{N} d_i^2 = \sum_{i=1}^{N} (y_i - \bar{y})^2 = (n-1)S_X^2$. Here SSD_1 means the sum of squared residuals from a model with one coefficient. $SSD_2 = \sum_{i=1}^{N} (y_i - a - bx_i)^2$ represents the residual *SSD* from a 2-coefficeint model. The ratio of the reduction in *SSD* of the 2-coefficeint model to the 1-coefficient model is termed r^2

$$r^2 = \frac{SSD_1 - SSD_2}{SSD_1} = \frac{\sum_{i=1}^{N} (y_i - \bar{y})^2 - \sum_{i=1}^{N} (y_i - a - bx_i)^2}{\sum_{i=1}^{N} (y_i - \bar{y})^2} \tag{4.19}$$

The r^2 value ranges between 0 (if there is no relation between *y* and *x*) to 1 (if the linear model perfectly relates *y* to *x*).

4.5 Degrees of Freedom

This is not so much a characterization of the data, as it is an indication of residual flexibility to the data after some attributes have been fixed. DoF $= \upsilon = N - k$, where *N* is the number of data and *k* is the number of model coefficients or characterizations used in the comparison. If an average, $\bar{X}$, is calculated from the data, and if we accept the average is the truth, then all of the data but one could have any value, but the last one must have a value that makes $\bar{X}$ true. Then $N - 1$ of the data values are free to change, and the DoF $= \upsilon = N - 1$. If a model is regressed to the data, and the model has four coefficients, then DoF $= \upsilon = N - 4$.

Example 4.4: Heichelheim obtained values for the compressibility factors for carbon dioxide at 100°C, over the pressure range from 1.3176 to 66.437 atm. (Heichelheim, H. R., The Compressibility of Gaseous 2,2-Dimethyl Propane by the Burnet Method, Ph.D. Dissertation, Library, University of Texas, Austin (1962), with permission.) Calculate the mean, variance, standard deviation, standard error, and the coefficient of variability for the portion of his data listed below.

Compressibility factors		
0.9966	0.9969	0.9971
0.9956	0.9957	0.9960
0.9936	0.9938	0.9940
0.9913	0.9912	0.9915
0.9873	0.9874	0.9980
0.9821	0.9823	0.9829
0.9747	0.9750	0.9758

You should get

$$\bar{X} = \sum_{i=1}^{n} \frac{x_i}{n} = 0.98946667$$

$$S_{\bar{X}}^2 = \sum_{i=1}^{n} \frac{\left(x_i - \bar{X}\right)^2}{n-1} = 0.00005976$$

$$S_X = \sqrt{S_{\bar{X}}^2} = 0.00773022$$

$$S_{\bar{X}} = \frac{S_X}{\sqrt{n}} = 0.00168687$$

$$CV = \frac{S_X}{\bar{X}} = 0.00781251$$

Example 4.5: The following data represent a random subset from a study by one author (Rhinehart) of the academic trend in chemical engineering (ChE) students at Oklahoma State University. The first column represents the science, technology, engineering, math (STEM) grade point average (GPA) of students in their freshman and sophomore years. The second column represents the GPA of the same student in their upper level major ChE courses. The results have been sorted by STEM GPA. The study was intended to be useful in advising students.

Fr. & So. STEM GPA	Jr. & Sr. ChE GPA
1.706	3.020
1.914	2.588
2.143	3.392

(Continued)

Fr. & So. STEM GPA	Jr. & Sr. ChE GPA
2.171	2.891
2.235	3.154
2.382	2.431
2.500	2.500
2.588	2.553
2.600	3.512
2.600	2.569
2.647	3.231
2.676	3.314
2.676	2.471
2.912	3.667
3.086	3.281
3.143	3.261
3.171	3.686
3.257	3.294
3.371	3.922
3.400	3.561
3.412	3.627
3.429	3.314
3.429	3.559
3.471	3.872
3.486	3.391
3.486	3.609
3.500	3.809
3.706	3.882
3.829	3.000
4.000	3.804
4.000	3.769

What are the arithmetic average and standard deviation of the two columns? What is the correlation r-statistic? Segregate the data into four categories: The students below average and above average in STEM GPA, and for each those below and above average in Major GPA. What is the count of number of students in each category? If there was no relation between columns, then there would be (ideally) an equal number of counts in each of the four quadrants. What is the chi-squared statistic value for the observed counts in the four quadrants?

You should find

$$\bar{X}_{\text{STEM}} = 2.9976\ldots$$

$$\bar{X}_{\text{Major}} = 3.2881\ldots$$

$$s_{\text{STEM}} = 0.6167\ldots$$

$$s_{\text{Major}} = 0.4658\ldots$$

$$r = 0.6627\ldots$$

Category	Observed count
High STEM, high GPA	14
High STEM, low GPA	3
Low STEM, high GPA	4
Low STEM, low GPA	10

$$E_{\text{each}} = \frac{14+3+4+10}{4} = 7.75$$

$$\chi^2 = 9.11987\ldots$$

4.6 Expectation

The statistics above have been calculated using sample data. A sampling of the data might have a large number of observations (measurement values), but it is not the entire population; so, values such as sample average and sample variance are not the exact values for the entire population. "Expectation" is the term that means using the entire population to get the descriptive statistics. Of course, one never has the infinite number of measurements that represent the entirety of realizations (possible values) that could be obtained by sampling from the population. But, if one believes that a particular mathematical form of the probability distribution is valid, then one can use it to define population statistics.

The expectation for the average is the f- or pdf-weighted sum. Starting with Equation (4.2) and using the area under the pdf curve to represent the number of X_i values, and recognizing that in the limit of very small dx the sum becomes the integral, and the integral of the pdf is 1

$$E(X) = \bar{X}\Big|_{N=\infty} = \frac{\sum_{i=1}^{k} f_i X_i}{\sum_{i=1}^{k} f_i} = \frac{\sum_{i=1}^{k} (pdf_i\ dx) X_i}{\sum_{i=1}^{k} (pdf_i\ dx)}$$

$$E(X) = \frac{\int_{-\infty}^{\infty} x\, pdf(x)\, dx}{\int_{-\infty}^{\infty} pdf(x)\, dx} = \int_{-\infty}^{\infty} x\, pdf(x)\, dx = \mu \tag{4.20}$$

If you are inclined to enjoy placing the formula for a particular *pdf* from Chapter 3 into Equation (4.20) and integrating, you'll find that $E(X) = \mu$.

In a similar manner, with even more mathematical joy,

$$E\left((X-\mu)^2\right) = \int_{-\infty}^{\infty} (x-\mu)^2\, pdf(x)\, dx = \sigma^2 \tag{4.21}$$

4.7 A Note about Dimensional Consistency

4.7.1 Average and Central Limit Representations

The arithmetic average is represented as

$$\bar{X} = \frac{1}{N}\sum_{i=1}^{N} X_i = \left(\sum_{i=1}^{N} X_i\right) / N \tag{4.22}$$

Here $\bar{X}$ is the arithmetic average of N samples and X_i represents the individual samples in the average. Although the second version is numerically identical to the first, the second is dimensionally inconsistent. N is not dimensionless, a value might be $N = 6$ samples, not just a number, 6. Including units in brackets considering that X represents the weight in lbs in a sack, the equation is

$$\bar{X}[\text{lbs}] = \frac{1[\text{sample}]}{N[\text{sample}]}\sum_{i=1}^{N} X_i[\text{lbs}] \neq \left(\sum_{i=1}^{N} X_i[\text{lbs}]\right) / N[\text{sample}] \tag{4.23}$$

The second version now is obviously dimensionally inconsistent.

The reduction in variance due to the central limit theorem is usually written as

$$\sigma_{\bar{X}} = \sigma_{X_i} / \sqrt{N} \tag{4.24}$$

Here $\sigma_{\bar{X}}$ is the ideal standard deviation of the average of N samples and σ_{X_i} represents the standard deviation of individual samples in the average. This assumes that the variance on each sample is identical to all others.

This is a very powerful and often-used concept. However, it is also dimensionally inconsistent.

To maintain dimensionally consistency, some would prefer to see it written as

$$\sigma_{\bar{X}} = \sigma_{X_i}\sqrt{\frac{1}{N}} \tag{4.25}$$

which now shows the argument of the square root function is dimensionless. However, computationally with or without the 1, the numerical outcome is identical.

4.7.2 Dimensional Consistency in Other Equations (An Aside)

Engineering and science often remove unity values from equations, because when doing the calculation, they have no impact on the result. Other common examples are Newton's First Law $F = ma$, and pressure drop in a fluid system $\Delta P = C_D \frac{1}{2} \rho v^2$. In both, the dimensional unifier g_c is missing. In SI units the value of $g_c = 1\left[\frac{kg - m}{N - s^2}\right]$, and with a unity value it does not matter in a numerical calculation. But it does matter in other systems of dimensional units.

In sizing a flow control valve, the formula is $\dot{Q} = C_v f(x)\sqrt{\frac{\Delta P_v}{G}}$. Here $\dot{Q}$ represents volumetric flow rate and C_v has the same units, representing the flow rate through the fully open valve. Here, G represents specific gravity and $f(x)$ is the fraction of maximum flow rate because of the valve position. Both are dimensionless. However, ΔP_v is the pressure drop across the valve, which might have the units of psig or kPa. The equation is dimensionally incorrect but can be fixed by not removing the unity pressure scale factor, $\xi = 1$, $\dot{Q} = C_v f(x)\sqrt{\frac{\Delta P_v}{G\xi}}$.

The argument of the logarithm must be dimensionless, but it is often simply represented as $y = ln(x)$. With x and y having dimensional units, the equation is incorrect. Since $\ln(x=1)=0$, and subtracting zero to both sides of the equation $y = \ln(x) - \ln(x=1)$ and combining the two log terms, $y = \ln\left(x\left[x_{\text{units}}\right]/1\left[x_{\text{units}}\right]\right)$ the right-hand side (RHS) is dimensionally consistent, but since the log of a value is dimensionless, the equation is still incomplete. However, if the RHS is multiplied by $1\left[y_{\text{units}}\right]$, then $y = 1\left[y_{\text{units}}\right]\ln\left(x\left[x_{\text{units}}\right]/1\left[x_{\text{units}}\right]\right)$. Now the numerical value is identical, and the units are dimensionally consistent.

4.8 Takeaway

The most common measures are arithmetic average (for data centrality) and sample standard deviation (for data variability). These are the best sample statistics to estimate the value of the mean and sigma of the normal (Gaussian) distribution. Most data will be nearly normally distributed, making the average and standard deviation very meaningful. However, a good portion of data that you may be analyzing may not be normally distributed; and just because you can calculate an average and standard deviation for the data does not mean the use of Gaussian model-based analysis is justified. Be aware of the diverse measures of centrality, variability, patterns, and scaled metrics.

Check equations for dimensional consistency.

4.9 Exercises

1. Derive Equation (4.2) from Equation (4.1).
2. If the weighting factor in Equation (4.2) is $f_i = n_i / N$, show that the denominator term $\sum_{i=1}^{k} f_i = 1$.
3. Some liquids have entrained particles, which would plug small orifices in the process line. This is undesirable. Plugging value, X, is a measure of the undesirability. To get the X-value for a batch of liquid, pump a sample through a filter at a uniform flow rate. As particles build up on the filter, the pressure drop across the filter increases. Plugging value is the volume of liquid that causes specified

pressure drop. If volume V of liquid with a plugging value of X_1 is mixed with the same volume of liquid with a plugging value of X_2, show that the mixture plugging value should be calculated as a harmonic mean.

4. Reconsider Exercise 4.3, and define a weighted harmonic mean if the two blended liquid volumes were to be V_1 and V_2.
5. Derive Equations (4.7b) and (4.7c) from (4.7a).
6. Derive the second expression in Equation (4.12) from the first.
7. Derive the second expression in Equation (4.13) from the first.
8. Use the data from Example 4.2 and plot the upper-level major GPA w.r.t. the lower-level STEM GPA. Ask your plotting routine to best fit a linear trend to the data and to display the r-squared correlation value. Compare this value to the square of the correlation r-value in Example 4.5.
9. Use Equations (4.20) and (4.21) to determine the μ and σ values for the continuum uniform distribution where $pdf(x) = \frac{1}{b-a}$ for $a \le x \le b$, or else outside of that range, $pdf(x) = 0$. That is not too difficult an analytical integration exercise.
10. You might want to practice calculating the various descriptive statistics in this chapter on your own data.

5

Data and Parameter Interval Estimation

5.1 Interval Estimation

Given knowledge of the distribution and the parameter values, what value might one sample from the population yield? What is the expected range for possible sample values? This chapter shows how to answer that question.

5.1.1 Continuous Distributions

An example may clarify the question.

Example 5.1: If you knew that the experimental data is normally distributed with $\mu = 8$ and $\sigma = 1.5$, what could be the value of one measurement, x? The sample value could be $x = 500$ or it could be $x = -35$, because the distribution permits values between $\pm\infty$, but such extreme values are improbable. A more useful question is, "What is the expected range for possible sample values?"

The thumbnail sketch shows the upper and lower 10% limits and the corresponding standard normal z range. The upper 10% region will contain 10% of the data with z values greater than about 1.28, similarly, 10% of the data will be in the lower 10% region with z values less than about –1.28. This means that 80% (= 100% – 10% – 10%) of the samples will have z values between about –1.28 and +1.28. In Excel the function NORM.INV(CDF,μ,σ) returns the x-values, or NORM.INV$(CDF,0,1)$ returns the z-values. Here the 1.2815515… value has been truncated to 1.28 to be convenient but not undermine the implied precision of the $\mu = 8$ and $\sigma = 1.5$ values.

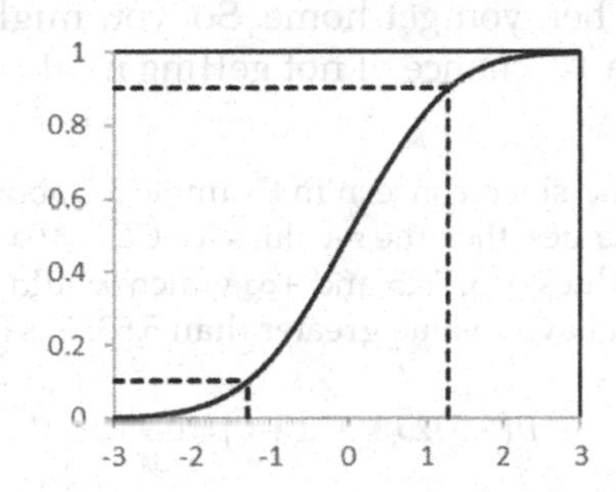

Using the inverse of the definition of z, $x = \mu + z\sigma$, and given the distribution parameter values, the value of x at the upper z-limit is $x = 8 + 1.28(1.5) = 9.92$. And the value of x at the lower z-limit is $x = 8 - 1.28(1.5) = 6.08$. As a result, one would expect that 80% of the sample x-values would be between 6.08 and 9.92.

DOI: 10.1201/9781003222330-5

Note: This example did not assign units to the x, μ, or σ values. The variable might be the time that the mail is delivered, or the high outdoor temperature of your location on 27 April, or the impurity composition of cement. Whatever it is, x, μ, and σ will have the corresponding dimensional units.

Note: Whether you use the inverse function, or a graph, or a table, the procedure is to define the *CDF* limits, then determine the value of the variable at those limits.

This example estimated the x-range that would include 80% of the sample values. But, you might want to have greater surety about what might happen with a sample. So, you might want to know what the 95% limits are. Or, if the outcome is less critical, the 50% limits may be what you are seeking. Where safety and life are involved the 99.99% confidence may be desired. See Chapter 10 for a discussion on choosing appropriate confidence limits.

Example 5.2: Use the normal distribution and $\mu = 8$ and $\sigma = 1.5$ values of Example 5.1 to determine the x-values representing 95% of the possible observations.

These are about at the $z = \pm 1.96$ values, giving an x range of $5.06 \le x \le 10.94$, and the 50% limits are about at the $z = \pm 0.67$ values, giving an x range of $6.995 \le x \le 9.005$. Accordingly, the range of possible values that you might report are strongly dependent on the choice of the *CDF* interval, alternately understood as the confidence interval, probability interval, or percentage chance of happening. So, you cannot just report an interval, you need to include the probability. The probability that the sample will have a value between 5.06 and 10.94 is 0.95.

This could be stated as:

$$P(5.06 \le X \le 10.94) = 0.95$$

There is a 95% chance that the sample value will be included between 5.06 and 10.94.

Further, this example indicated the same level of concern for obtaining extreme high as extreme low values. The confidence interval was the central set of values, with half of the possibility of extreme values in the extreme high region and half in the extreme low region. But you might only have a concern about one side. For instance, if you are risking an investment in a business, any upper extreme return will be acceptable, and you might only be interested in the lower 25% of values. By contrast, if you want to pick up the mail on the way home after work, early mail deliveries are not an issue, but if a very late delivery, it will not be available when you get home. So, you might only be interested in the schedule that results in only a 1% chance of not getting mail.

Example 5.3: Consider a one-sided concern in Example 5.2 above. If one was only interested in the upper 95% of values, then the z-values for $CDF = 0.05$ and $CDF = 1$ are -1.645 and $+\infty$, translating to x-values of 5.5325 and $+\infty$ which would be reported as the probability that the sample will have a value greater than 5.5325 is 0.95.

$$P(5.5325 \le X \le +\infty) = 0.95$$

Both Examples 5.2 and 5.3 are true. But they present different results. So, more than just specifying the confidence interval, one must also specify how it is apportioned.

If not explicitly stated, the custom, is to equally apportion the extreme probabilities, so that a C% level of confidence (c as a fraction) means that $1 - \frac{C}{100} = 1 - c$ is the probability of being in either extreme, so that the *CDF* values that mark the extremes are $(1-c)/2$ for the lower *CDF* value and $1-(1-c)/2$ for the upper *CDF* value.

The Level of Significance, α, is defined as the total extreme area under a *pdf* curve.

$$\alpha = 1 - c \tag{5.1}$$

If the probability of extreme values is apportioned equally

$$\begin{aligned} CDF_{\text{lower}} &= \alpha / 2 \\ CDF_{\text{upper}} &= 1 - \alpha / 2 \end{aligned} \tag{5.2}$$

this means that the probability of getting a value in the lower extreme is CDF_{lower}, and the probability of getting a value in the upper extreme is $1 - CDF_{\text{upper}}$. Certainly, the upper and lower probabilities do not have to be apportioned equally, one can choose any upper or lower probabilities that sum to the desired value of α.

If not equally apportioned, as was the case with Example 5.3, to properly convey the meaning, report each side individually.

Example 5.4: Repeat the Example 5.3, but use the 90% interval, with 1% allocated to the lower and 9% to the upper, then

$$CDF_{\text{lower}} = 0.01$$
$$CDF_{\text{upper}} = 0.91$$

The z-values are

$$z_{\text{lower}} = -2.326$$
$$z_{\text{upper}} = 1.341$$

and the x-values could be presented as

$$P(X \le 4.51) = 0.01$$
$$P(X \ge 10.01) = 0.09$$

Regardless of the continuous distribution model, the procedure to estimate the range that a sample value might provide is:

1. Define the distribution, and parameter values.
2. Determine the confidence interval desired, and the probability allocation to the lower and upper limits. Be sure that the values are appropriate to the context. Use this to assign the CDF_{lower} and CDF_{upper} values.
3. Determine the lower and upper statistic values from the inverse of the distribution.
4. If the statistic is a scaled value, un-scale it to determine the lower and upper X values.
5. Report the lower and upper X-values along with the qualifying givens from Steps 1 and 2.

Example 5.5: The standard deviation of a population is $\sigma = 1.234$ μg/L. What might be the 95% limits on the standard deviation of a sample with $n = 10$ data values?

Assume that the variance will be chi-squared (χ^2) distributed with $\upsilon = n-1 = 9$ degrees of freedom. The 95% limits means that c = 0.95. Using Equation (5.3) $\alpha = 1-0.95 = 0.05$ choosing to split the extreme areas equally, $CDF_{lower} = \frac{\alpha}{2} = 0.025$ and $CDF_{upper} = 1-\frac{\alpha}{2} = 0.975$. In Excel, values for the inverse of the χ^2 distribution are found using CHISQ.INV(CDF, υ). The values are 2.70038… and 19.0227…. Using the definition of $\chi^2 = \upsilon s^2 / \sigma^2$ from Equation (4.16), and solving for $s = \sigma\sqrt{\chi^2/\upsilon}$ the 95% limits on s are 0.67593… μg/L to 1.7940… μg/L.

Assuming that the variance is χ^2 distributed, and rounding to values that are equivalent to the given σ = 1.234,

$$P(s \le 0.676\ \mu g/L) = 0.025$$
$$P(s \ge 1.794\ \mu g/L) = 0.025$$

or with implied equal extreme areas as

$$P(0.676\ \mu g/L \le s \le 1.794\ \mu g/L) = 0.95$$

there is a 95% chance that the standard deviation values could range between 0.676 μg/L and 1.794 μg/L.

Example 5.6: If n = 10 values are sampled from a normal distribution with μ = 8 and σ = 1.5, what could be the 80% range of values of the average of the n measurements?

When data are normal, the standard deviation of an average is calculated as $\sigma_{\bar{X}} = \sigma_X / \sqrt{n}$, the z statistic for the average is then $z_{\bar{X}} = (\bar{X} - \mu)/(\sigma_X/\sqrt{n})$. From z one can calculate the limits on the average $\bar{X} = \mu \pm z_{\bar{X}}\sigma_X / \sqrt{n}$. The 80% range will encompass 80% of the events, leaving 20% (α = 0.2) in the extremes. Centering the extremes are the CDF values of 0.10 and 0.90. The corresponding z-values are ±1.2815515… which translate to about $\bar{X}_{lower} = 7.39$ and $\bar{X}_{upper} = 8.61$.

$$P(7.39 \le \bar{X} \le 8.61) = 0.80$$

There is an 80% chance that the average of 10 randomly sampled values could be included between 7.39 and 8.61.

The procedure is similar for other continuous distributions (uniform, exponential, log-normal, etc.).

5.1.2 Discrete Distributions

The procedure for interval estimation in discrete distributions is the same as that for continuum-valued variables, except that since CDF and X-values can only have particular values, the CDF and/or x-limits need to be truncated (up or down) to the best nearest feasible value.

Example 5.7: Accepting that data are binomially distributed with an individual probability of success as p = 0.7, in n = 15 trials, what is the 70% range of the number of successes?

The thumbnail sketch of the solution is included. Note that abscissa values range from 0 (the minimum possible number of successes) to 15 (the maximum number). The markers, circles, indicate point values, and the dotted line connecting the dots is there as a visual aid to sequence. It does not suggest that in between values are possible. One cannot have 3.456 number of successes. The number of successes must be an integer. The dashed lines represent the *CDF* range and corresponding number of successes.

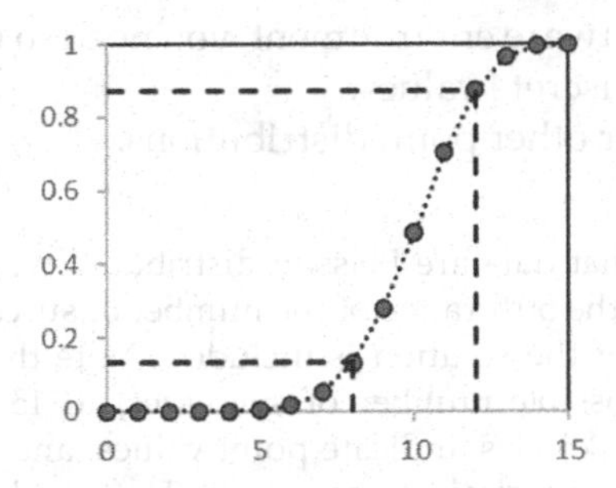

The 70% range requires $\alpha = 1 - \frac{70}{100} = 0.3$. With conventional splitting of the two extreme regions equally, the *CDF* values are $CDF_{lower} = \frac{\alpha}{2} = 0.15$ and $CDF_{upper} = 1 - \frac{\alpha}{2} = 0.85$. As it turns out there are no data points at those *CDF* values. The closest *CDF* values are 0.13114… at a count of 8, and 0.87317… at a count of 12. So the answer is that the range is $8 \le X \le 12$, but the probability of that interval is not the specified 70% of the problem statement, it is the *CDF* difference of 0.87317… − 0.13114… = 0.74203…. As a , the answer is

$$P(8 \le X \le 12) = 0.74$$.

There is a 74% chance that the number of successes will be between 8 and 12, inclusive.

Regardless of the discrete distribution model, the procedure to estimate the range that a sample value might provide is:

1. Define the distribution, and parameter values.
2. Determine the confidence interval desired, and the probability allocation to the lower and upper limits. Be sure that the values are appropriate to the context. Use this to assign the CDF_{lower} and CDF_{upper} values.
3. Adjust the *CDF* values to best match feasible values.
4. Determine the lower and upper statistic values from the inverse of the distribution.
5. If the statistic is a scaled value, un-scale it to determine the lower and upper X-values.
6. Report the lower and upper *X*-values along with the qualifying givens from Steps 1 and 2.

Step 3 is fairly important. In Excel, the function BINOM.DIST$(x,n,p,1)$ returns the *CDF* value. But values for x and n must be integers. Excel truncates a noninteger x and n values. It does not round the value to the next nearest integer. Similarly, the inverse function BINOM.INV(n,p,CDF) returns the x value of the next higher feasible *CDF* value. For example, the *CDF* of $x = 9$, $n = 15$, $p = 0.7$ is 0.2783…, and that for $x = 10$, $n = 15$, $p = 0.7$ is

0.4845.... If you use the inverse to determine an x-value for a *CDF* of 0.2783 (a bit less than 0.2783...) it indicates 9. At 0.2784 (a bit more than 0.2783..., but still far from the 0.4845...) it indicates 10, the next higher value, which should not appear until a *CDF* value of 0.484509.... In Step 3, look at the feasible count and *CDF* values just above and below those indicated from Step 2, and choose the closest to the values of the *CDF* or the count that best match the application context and intent.

If you are using another software environment, you need to understand whether it truncates, rounds, or rounds up discrete values.

The procedure is similar for other point distributions.

Example 5.8: Accepting that data are Poisson-distributed with an average number of successes $\lambda = 7.03$, what is the 50% range of the number of successes?

The thumbnail sketch of the solution is included. Note that abscissa values range from 0 (the minimum possible number of successes) to 15 (but the upper limit is unbounded). The markers, circles, indicate point values, and the dotted line connecting the dots is there as a visual aid to sequence. The number of successes must be an integer. The dashed lines represent the *CDF* range and corresponding number of successes.

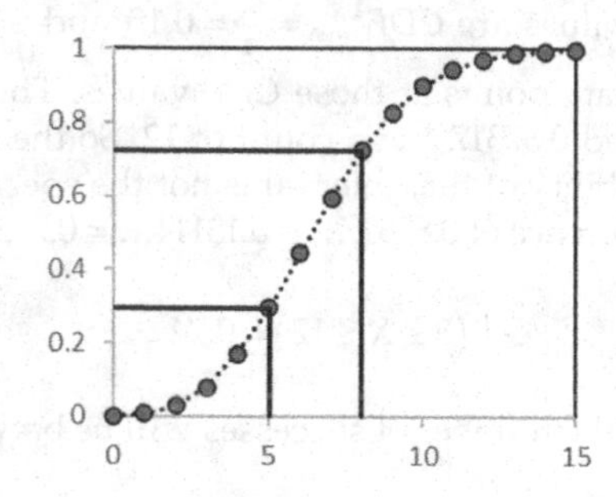

The 50% range requires $\alpha = 1 - \frac{50}{100} = 0.50$. With conventional splitting of the two extreme regions equally, the *CDF* values are $CDF_{lower} = \frac{\alpha}{2} = 0.25$ and $CDF_{upper} = 1 - \frac{\alpha}{2} = 0.75$. These are the quartiles. As it turns out there are no data points at those *CDF* values. The closest *CDF* values are 0.296983... at a count of 5, and 0.725172... at a count of 8. So the answer is that the range is $5 \le X \le 8$, but the probability of that interval is not the specified 50% of the problem statement, it is the *CDF* difference of 0.42827.... As a result, the answer is

$$P(5 \le X \le 8) = 0.43.$$

There is a 43% chance that the number of successes will be between 5 and 8, inclusive.

5.2 Distribution Parameter Estimation

Given knowledge of the distribution and a data value, what range of distribution parameter values might reasonably have generated that data value? Contrasting the question in Section 5.1, here the data has been acquired and the data value is known, the distribution parameter values are unknown.

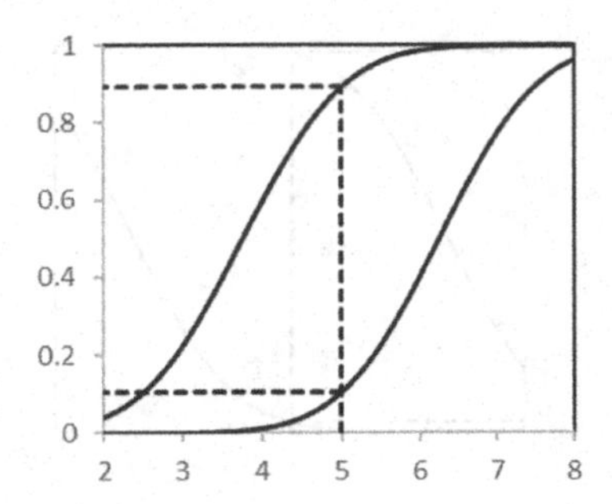

FIGURE 5.1
Illustration of an 80% probable range on the mean.

Figure 5.1 represents data that has occurred with an experimental value of the statistic at 5, on the horizontal axis. The statistic might be a sample average as an estimate of the population mean, a sample standard deviation as an estimate of the population sigma, or a sample proportion as an estimate of the population probability. The dashed vertical line represents the data location. If that experimental value was the true value of the distribution parameter, then you would know the true distribution and would be able to estimate the expected range of a data value that might be generated by the population. The question here is what value of a population parameter could have generated the data. The *CDF* curve to the right has a parameter value (an average value, a 50th percentile value) of about 6.5. It has a high probability of generating data in the 5.5 to 7.5 range and could generate a data value as extreme as about 3 or 9. The lower of the two horizontal dashed lines indicates the probability that it could have generated such an extreme data value as 5 (or lower), is about 10%. The *CDF* curve to the left has a parameter value of about 3.5, and the probability that it could have generated such an extreme data value as 5 (or greater), is about 10%. The objective here is to determine the values of the right and left distributions that could have generated the experimental data value with a desired confidence.

If the center of the right-hand curve were shifted toward the left, there would be a higher probability that the population could have generated the data. Here, with population values between 3.5 and 6.5 it is not unexpected that the data value of 5 could have been generated. Here, between the population values of about 3.5 and 6.5, there is an 80% chance that the population could have generated the data.

In Figure 5.2, the population parameter values are shifted to 3 and 7. There is still a chance that either population could have generated the extreme data value of 5 (or more extreme), but here only 2.5% chance for either. The combined extreme probability is 0.05 = (2.5% + 2.5%)/100%; so, with population parameter range between 3 and 7, there is only a 5% chance that a data value of 5 would be considered extreme. The range of 3 to 7 includes the population parameter values that could have generated an observation value of 5 with less than a 5% chance.

In this section, given the distribution model, and a desired confidence, we'll calculate the population parameter value that could have generated that data.

The procedure is:

1. Choose the population model that best fits the attributes of the data.
2. Choose a confidence interval that best matches the application context, and how to allocate the two extreme probabilities. Here, again, c represents the confidence that either might have generated the data, and $\alpha = 1 - c$ is the combined extreme,

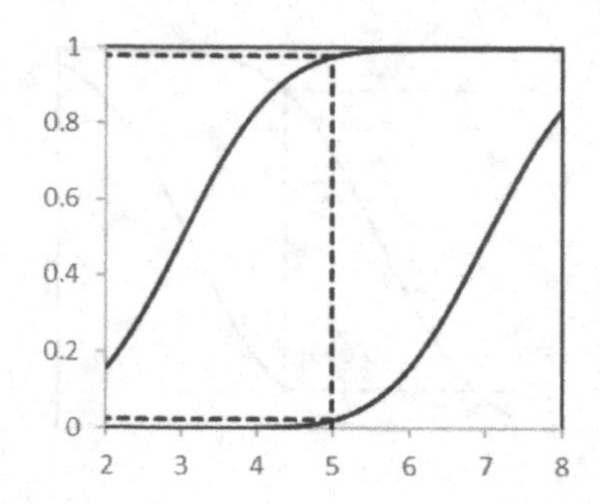

FIGURE 5.2
Illustration of a 95% probable range on the mean.

improbable area. Nominally, $\alpha/2$ would be the area in each tail, defining *CDF* of the right-most curve at the data value x to be $CDF(x) = \alpha/2$, and *CDF* of the left-most curve at the data value x to be $CDF(x) = 1 - \alpha/2$.

3. Determine the value of the population parameter, p, which makes $CDF(x, p_{\text{lower}}) = 1 - \alpha/2$, and the one that makes $CDF(x, p_{\text{upper}}) = \alpha/2$. Or use your alternate choices for the split of the extreme areas. This may need to be a root-finding exercise unless an inverse built-in function is available to do this.
4. If the statistic is a scaled value, un-scale it to determine the lower and upper distribution parameter values.
5. Report the p_{lower} and p_{upper} values along with the assumptions and choices in Steps 1 and 2.

Note: If the distribution is symmetric (such as the normal and *t*-distributions) then assuming that the data value is the true mean and asking what are the extreme values that it might have generated, gives the same answers as the method of this section.

Note: Some nonsymmetric distributions, such as the exponential, $CDF = 1 - e^{-x/\mu}$, are explicitly invertible. Given the value of the sample, x, and a desired *CDF* value, the population parameter can be calculated as $\mu = -x / \left[\ln(1 - CDF)\right]$.

However, if the distribution is not symmetric or not analytically convenient (most are not), then use the method of Section 5.2.1.

5.2.1 Continuous Distributions

We'll explain the procedure with an example.

Example 5.9: The Gulp-a-Cup Coffee Company utilizes spray-drying in their coffee production process. Nozzles employing internal mixing were recently installed for trial runs to determine the entrance pressure. What values of the new gas pressure range correspond to the 99% confidence limit for the population mean? Sample data are:

Entrance gas pressure (psig)									
52.00	51.00	51.80	51.75	51.30	50.85	50.25	49.00	48.65	48.00

The average entrance pressure is 50.46 psig. The sample standard deviation is 1.4331007… psig. The standard error of the mean, $S_{\bar{X}}$, is calculated as $S_{\bar{X}} = \sqrt{S_X^2 / n} = 0.45318625\ldots$

psig. Although, there is suspicion that the data are not normally distributed (note the ending values are all 0 or 5, and the unexpected number of integer values masquerading as decimal-valued numbers), the average of 10 should be about normally distributed. If we knew the population variance, we could use the *z*-statistic to estimate the range, but since the standard deviation is based on the sample, we'll use the *t*-statistic, as the sample has 10 observations, $v = n - 1 = 9$. Desiring the 99% limits, the extreme area is 1%, $\alpha = 0.01$, and splitting the extreme areas equally the two *CDF* values are $CDF_{upper} = \alpha/2 = 0.005$ and $CDF_{lower} = 1 - \alpha/2 = 0.995$.

The objective is to find the values of the *t*-statistics for the upper and lower curves. In Excel we desire $T.\text{DIST}(t_{lower}, 9, 1) = 0.995$ and $T.\text{DIST}(t_{upper}, 9, 1) = 0.005$. One could use any root-finding procedure to determine that the values are $t_{upper} = 3.2498355\ldots$ and $t_{lower} = -3.2498355\ldots$. The Excel Solver Add-In is convenient for root-finding, but it may need reasonable initial guess of the values.

However, in this case, one could also use the Excel inverse functions $T.\text{INV}(.005, 9)$ and $T.\text{INV}(.995, 9)$ to return the values of $\pm 3.2498355\ldots$.

Inverting the *t*-formula to generate the mean, $\mu = \bar{X} - tS_{\bar{X}}$, then approximately

$$P(48.99 \text{ psig} \le \mu \le 51.93 \text{ psig}) = 0.99$$

We can reasonably expect that the average operating pressure will be between these limits if we use any other group (sample) of the same type of nozzles.

Note: Since the *t*-distribution is symmetric, the *t*-values are symmetric about zero $(\pm 3.2498355\ldots)$. Consequently, the mean is symmetric about the average. In this special case one can assume that the data value is the true mean and use the Section 5.1 approach to determine the range of values that it might generate. The range of values will be the same, but the concept of Section 5.2 is different, and equal values for the two procedures only happen of the distribution is symmetric.

If the distribution is not symmetric, not analytically invertible, or an inverse function is not available, then one is required to perform root-finding on the distribution.

Example 5.10: The standard deviation from a sample of $n = 5$ data is 3.05 mm. What might be the 95% limits on the variance of the population that generated the data?

Variance is the square of the standard deviation. From the information presented, the variance of the sample is 9.3025 mm^2, and variance is typically χ^2 distributed. More precisely $V = \upsilon s^2 / \sigma^2$ is χ^2 distributed. We'll divide the extreme areas equally, so that we are seeking *CDF* values of 0.025 and 0.975. The question is what are the $\sigma^2{}_{lower}$ and $\sigma^2{}_{upper}$ values to make the sample s^2 represent the 0.025 and 0.975 limits? The degrees of freedom value is, $\upsilon = n - 1 = 4$.

The operation is to determine $0.025 = \text{CHISQ.DIST}(\chi^2{}_{upper}, \upsilon, 1)$, and $0.975 = \text{CHISQ.DIST}(\chi^2{}_{lower}, \upsilon, 1)$. Using root-finding the values are $\chi^2{}_{lower\,\sigma} = 11.14328\ldots$ and $\chi^2{}_{upper\,\sigma} = 0.484418\ldots$.

Conveniently, Excel also provides the inverse function. Using $\text{CHISQ.INV}(0.975, 4)$ and $\text{CHISQ.INV}(0.025, 4)$ also returns the $\chi^2{}_{lower\,\sigma} = 11.14328\ldots$ and $\chi^2{}_{upper\,\sigma} = 0.484418\ldots$ values.

Translating χ^2 to population variance: $\sigma^2 = \upsilon s^2 / \chi^2$, $\sigma^2{}_{upper} = 4\left(\dfrac{9.3025}{0.484418\ldots}\right) = 76.8137\ldots \text{ mm}^2$, and $\sigma^2{}_{lower} = 4\left(\dfrac{9.3025}{11.14328\ldots}\right) = 3.33921\ldots \text{ mm}^2$.

$$P\left(3.34\ \text{mm}^2 \leq \sigma^2 \leq 76.81\ \text{mm}^2\right) = 0.95$$

Translating variance to population standard deviation:

$$P\left(1.83 \leq \sigma \leq 8.76\right) = 0.95$$

Note: The low and high extreme values are not equidistant from the sample value. The lowest possible value of a variance is zero and the highest is unbounded. The χ^2 distribution is not symmetric.

5.2.2 Discrete Distributions

The procedure is identical to that above when the distribution parameter values are continuum-valued (such as probabilities, or average).

Example 5.11: One mixing experiment over a one-week period indicates that 3 particles bounced out of the tank. The population of rogue particles is hypothesized to be Poisson-distributed. What are the 95% limits on the true mean, λ, on the number of particles bouncing out per week?

The 95% limits will be the center, with the extreme area split on either side. $\alpha = 1 - 95\%/100\% = 0.05$. So, the two *CDF* values are $\frac{\alpha}{2} = 0.025$ and the complement $1 - \frac{\alpha}{2} = 0.975$. The high possible value for the true mean will use the sample data value as its 0.025 *CDF* value, and the low possible value for the true mean will use the sample data value as its 0.975 *CDF* value. The objective is to find values for λ such that $\text{POISSON.DIST}\left(3, \lambda_{\text{upper}}, 1\right) = 0.025$, and $\text{POISSON.DIST}\left(3, \lambda_{\text{lower}}, 1\right) = 0.975$. Using a root-finding algorithm, $\lambda_{\text{lower}} = 1.0898\ldots$ and $\lambda_{\text{upper}} = 8.7672\ldots$

$$P\left(1.09 \text{ particles per week} \leq \lambda \leq 8.77 \text{ particles per week}\right) = 0.95$$

Note: The answer is not symmetric about the data. The high possible average of 8.77 is 5.77 particles/week more than the sample finding of 3 particles/week. If symmetric, on the low side of 3 particles/week would be $3 - 5.77 = -2.77$ particles/week. But the value of the average cannot be less than zero. The low extreme value of 1.09 p/week is about 1.91 away from the sample finding.

Example 5.12: When my rain gauge indicates that we had 1 inch of rain, what might the true average have been?

The nominal rain drop has a size of about 0.08" (inches) diameter. Its volume then is 0.00026808... in^3. My rain gauge has a 1" × 1" opening so when it reads 1" of water, it contains 1 in^3, which means it had accumulated about 3,730 drops of rain.

Modeling the rain drop in a space and time interval as Poisson-distributed, and the experimental sample value as 3,730 drops per square inch of area per one rainfall duration, the question is what is the true population mean (Poisson formula lambda) that could have generated that sample value? Using the 95% interval, the question is, "What low value for λ could have generated the 0.975 upper limit sample value of 3,730, and what high value for λ could have the 0.025 lower limit of 3,730?"

Using the Excel function $\text{POISSON.DIST}\left(3730, \lambda, 1\right)$ and root finding, the values of lambda are

$$\lambda_{\text{lower}} = 3{,}612 \text{ rain drops}$$

$$\lambda_{\text{upper}} = 3{,}851 \text{ rain drops.}$$

Note: The deviations from average are 118 and 122, which is not symmetric because the Poisson distribution is not symmetric.

Returning count of drops to inches of height in the rain gauge,

$$P(0.97 \leq \text{rain height, in} \leq 1.03) = 0.95.$$

Satisfyingly, the 95% interval for the true rainfall amount, which could be about ±3% from my 1" reading, is only about 1/32 of an inch, not visibly detectable.

Note: Often when the number count is high the normal distribution is a reasonable approximation for the Poisson or binomial. If the sample value of 3,730 is the true population mean then the standard deviation of the Poisson distribution is its square root, 61.0737.... Since the standard normal distribution is symmetric, asking "What population mean could have generated the sample average?" is the same as asking, "If the population mean is 3,730, what might the sample value be?" At the 95% confidence, the boundary CDF values are 0.025 and 0975, for which the z-values are ±1.95996.... Then the limits on the mean are $\bar{X} \pm z\sigma = 3370 \pm 1.96 * 61.0737$ for which

$$\lambda_{\text{lower}} = 3{,}610$$

$$\lambda_{\text{upper}} = 3{,}849.$$

These are very close to the values generated by the Poisson distribution.

Example 5.13: They told me it was a fair coin, with a 50/50 chance of winning. I flipped it 20 times and only won 3 times. That is a possible realizable run of bad luck, I imagine; but I also want to know, "Is 3 out of 20, within the 99% range on the probability of winning an individual flip?"

Here, we'll split the extreme area in half. Then $\alpha = 0.01$ and the *CDF* extremes are 0.005 and 0.995. The binomial distribution determines *x*-number of wins in *n* trials where the probability of a particular win is *p*. So, we are solving for the value of *p* to make $CDF_{\text{lower}} = 0.995 = \text{BINOM.DIST}(x, n, p_{\text{lower}}, 1)$, and $CDF_{\text{upper}} = 0.005 = \text{BINOM.DIST}(x, n, p_{\text{upper}}, 1)$. The values are $p_{\text{lower}} = 0.03575\ldots$ and $p_{\text{upper}} = 0.44946\ldots$

$$P(0.035 \leq p \leq 0.449) = 0.99.$$

It appears that the 99% limits on the possible value of the probability of a win do not include 0.5. I'd be inclined to call it a foul coin.

In Excel, the BINOM.INV function returns a count of the outcome associated with the CDF. $x = \text{BINOM.INV}(n, p, CDF)$. If the coin is fair, p = 0.5, and $\text{BINOM.INV}(20, 0.5, .005) = 4$ and $\text{BINOM.INV}(20, 0.5, .995) = 16$. If the coin is fair, one expects the 99% range of outcomes to be within 4 and 16 wins. The 3 -win count is outside this range, corroborating the analysis above, but this inverse function does not provide the population parameter value.

Example 5.14: The developers of a new, less expensive, method of manufacturing compression rings for 1/4 inch copper refrigeration tubing claim that their procedure has a failure rate of only 0.1%, that is, 1 failure per 1,000 rings. You didn't believe their claim, so you had your purchasing agent order a sample of 1,000. Of that sample, three rings were defective. Construct the 99% confidence limit for the expected proportion of failures.

We'll choose to split the extreme area equally, $\alpha = 1 - 0.99 = 0.01$, so that we are solving for the value of p to make $CDF_{\text{lower}} = 0.995 = \text{BINOM.DIST}(x, n, p_{\text{lower}}, 1)$, and $CDF_{\text{upper}} = 0.005 = \text{BINOM.DIST}(x, n, p_{\text{upper}}, 1)$. The p-values are 0.0006729… and 0.01093377…

$$P(0.00067 \le p \le 0.0109) = 0.99.$$

As the interval for p contains 0.001, corresponding to the fraction of defective rings claimed in the new manufacturing method, you'll have to conclude that the developers' claim may be legitimate. At this point, you have no way to tell whether the new method is better than the standard method of manufacturing the compression rings at a 99% confidence level.

5.3 Approximation with the Normal Distribution

We can often take advantage of the *central limit theorem*, which states that as the number of samples increases, the resulting distribution of their means, $\bar{X}$, approaches the normal distribution with mean μ and variance σ^2. Thus, the random variable for proportion, x/n, the number of successes per number of trials, has an approximate normal distribution with mean p and variance pq/n. For large n and not near extreme values of either p *or* q or we can use the Z-statistic to construct an approximate confidence interval for p if Z is expressed as

$$Z = \frac{\text{P} - p}{\sigma_p} \tag{5.3}$$

Substituting, we have

$$Z = \frac{\text{P} - p}{S_p} = \frac{\text{P} - p}{\sqrt{\text{pq}/n}} \tag{5.4}$$

Proceeding in the usual manner, of equal allocation of the extreme probabilities, we define the confidence interval for the proportion p as

$$P\left(z_{\alpha/2} < Z < z_{1-\alpha/2}\right) = 1 - \alpha \tag{5.5}$$

Proceeding, to transform Equation (5.5), we have the desired *approximate confidence interval on the proportion* P:

$$P\left(\text{P} - z_{1-\alpha/2}\sqrt{\text{P}\frac{(1-\text{P})}{n}} < p < \text{P} + z_{1-\alpha/2} \times \sqrt{\text{P}\frac{(1-\text{P})}{n}}\right) = 1 - \alpha \tag{5.6}$$

where P = x/n and Q must be neither 0 nor 1.

Example 5.15: Repeat Example 5.14 but use the normal approximation for the distribution.

The sample values are $Q = 0.003$, $P = 0.997$. The sample size of $n = 1000$ is large and neither P nor Q is near 0 or 1. Using the Excel function NORM.INV(CDF, μ, σ) with NORM.INV$(0.995, 0, 1)$ to return $z_{0.995}$, we find $z_{0.995} = 2.575829\ldots$ and $z_{0.005} = -2.575829\ldots$. Now we can construct the interval for the proportion of failures:

$$\mathrm{P}\left(0.003 - 2.575\sqrt{\frac{0.003(0.997)}{1000}} < q < 0.003 + 2.575 \times \sqrt{\frac{0.003(0.997)}{1000}}\right) = 0.99$$

$$P(-0.0014533 < q < 0.0074533) = 0.99$$

Again, as the interval for q contains 0.001, corresponding to the fraction of defective rings in the new manufacturing method, you'll have to conclude that the developers' claim may be legitimate.

Note: The normal distribution is unconstrained. Values can range from $-\infty$ to $+\infty$, but the proportion is constrained. The p-value and the q-value can only be between 0 or 1 (inclusive). In using the normal distribution as an approximation to the binomial, near to the p or q limit the normal distribution will permit out-of-range values, such as $q = -0.0014533$. Although the normal approximation to a particular distribution can often be justified, our preference is to use the distribution that best matches that expected for the data population.

5.4 Empirical Data

It could very well happen that you have a bunch of data (perhaps more than 1,000) representing a process, do not have a model for the distribution, and are asked to determine the upper and lower limits for either the data of the range of the distribution parameter values.

5.4.1 Data Range

To estimate the confidence limits for the data, first sort the data from low to high value. There will be a high and a low value, representing the particular sample, not the population. You could report the data range, $R = x_{\text{high}} - x_{\text{low}}$, but this depends on the vagaries in the particular sample and is not the 100% possible range. Further, it uses only two of all of the values you collected and ignores the information in the other values.

A better measure is to use the empirical *CDF* of the data to estimate the 90% range or quartiles or suchlike. In the sorted data, there will be n data values. Assign to them index numbers $1 \le i \le n$. Assign a *CDF* value to each data using $CDF = i/n$. As usual, choose the desired confidence interval, and its allocation to the two extreme limits, then search for the sorted data for the i/n values representing the CDF_{lower} and CDF_{upper} values. Probably the i/n values of your data will not exactly match the desired lower or upper *CDF* values, so interpolate.

Both of those approaches are simple to implement. But still, those methods use just a few data values in the sample and ignore all of the information in the rest of the data. So, a preferred technique for more comprehensive analysis is to fit a best matching distribution to the data, then use the techniques of Section 5.1 to determine the limits on the presumed distribution model.

If you can expect the population to be normally distributed, then calculate the average and standard deviation of the sample, presume this to be the population mean and sigma, then use the techniques of Section 5.1 to determine the limits on the presumed distribution model.

Example 5.16: Use the data from Example 3.14, presume that the population is normally distributed, and estimate the 95% range on the data that might come from the population.

The average and standard deviation of the sample are approximately 2.785417 ppm and 2.910418 ppm, respectively. The 95% interval means that the extreme area is $\alpha = 1 - \frac{95}{100} = 0.05$, and assigning equal probabilities to the extreme low and extreme high values means that we are seeking the population values representing the CDF_{lower} = 0.025 and CDF_{upper} = 0.975 values. Since the sample parameters have been estimated from the data, we'll use the *t*-distribution, with $\upsilon = n - 1 = 72 - 1 = 71$ degrees of freedom.

Using the Excel function $T.INV(CDF, \upsilon)$ the *t*-values are about ±1.9939434 and using the definition of the t-statistic to calculate the *x*-values, $x = \mu + t\sigma$, the 95% limits are estimated to be

$$P(-3.02 \text{ ppm} \le X \le 8.59 \text{ ppm}) = 0.95.$$

Note: The limits are rounded to match the implied precision of the data.

Note: The procedure is correct, but the negative concentration value is not feasible. A result that is not reasonable should be an indication that the assumptions need to be revisited. In Example 3.14, the experimental distribution of the data suggested that the data might be exponentially distributed, not normal.

Note: Just because you can calculate an average and standard deviation of a sample, does not mean that the population is normal. Use a distribution that seems to best match the attributes of the process that generated the data, or best match the shape of the empirical distribution.

Example 5.17: Repeat Example 5.16, but presume that the population distribution is exponential, as is suggested by the empirical *CDF* of Example 3.14.

In an exponential distribution, the population mean and standard deviation are both the reciprocal of the exponential factor, α. This alpha is not the level of significance for the confidence interval, but the multiplier for the *x*-value in the exponential distribution $CDF(x) = 1 - e^{-\alpha x}$. The reciprocals of the sample average and sample standard deviation are about 0.3590313 ppm and 0.343593 ppm. The closeness of these two values to the expected one α value is a reassuring check on the presumption that the data come from an exponential population. We'll use an average of the two as a best estimate of the population parameter value. $\alpha = 0.351303/\text{ppm}$. With the same CDF_{lower} = 0.025 and CDF_{upper} = 0.975 values, the inverse of the distribution, $x = -\ln(1 - CDF)/\alpha$, can be used to solve for the extreme *x*-values.

$$P(0.07 \text{ ppm} \le X \le 10.50 \text{ ppm}) = 0.95$$

Note: Using the exponential distribution, which appears to better represent the empirical *CDF* returns, more reasonable values for the 5% extreme boundaries of the data that could be generated by the process that generated the experimental data.

5.4.2 Empirical Distribution Parameter Range

The question in this section is not what might be the data range that the population could generate, but what range of population parameter values might have generated the dataset. One has the dataset.

A preferred method is called bootstrapping.

1. Consider that the sample represents all aspects of the population. The sample size, n, is large enough to reveal representative extreme values, and all the vagaries that might be expressed by the data-generating experiment. We will call the sample of n data the surrogate population.
2. Assume a population distribution.
3. Specify the confidence limits desired for the population parameter values.
4. Randomly sample n number of data from the surrogate population. Sample with replacement, meaning that if a particular data is taken from the surrogate population, it remains in the surrogate population and might be sampled again. For example, if the four data values in the surrogate population are 1, 2, 3, and 4, then a random sampling with replacement might generate the sampling 3, 2 3, and 1. Note that the data value of 4 is missing from the sample, that the data value 3 is repeated, and the data of the sample are not in the same order of the surrogate population. This sample is termed a realization, and it represents a dataset of n values that could have been generated by the surrogate population.
5. Use the data in the realization set of Step 4 to generate the population parameter values (such as average, standard deviation, proportion, etc.) of the assumed population from Step 2. These might be directly calculated from the data, or by best fitting the presumed distribution to the data. This one realization of the population parameter values.
6. Record the parameter values for that realization.
7. Repeat Steps 4–6 many times. Probably this means 100, or more, realizations. Now you have 100 or so realization values for the population parameters.
8. Create an empirical *CDF* of the parameter values and characterize the distribution of those values. Most likely they will appear normally distributed, but it may be exponential, log-normal, or other. The distribution of parameter values will not necessarily have the same distribution as that of the surrogate data in Step 2.
9. Use the technique of Example 5.15 or 5.16 to generate the confidence limits on the distribution parameter values from Step 7.

How many realizations are needed? A nominal number is 100, as stated in Step 7. Fewer may be fully functional, or more may be needed. It all depends on the vagaries in the surrogate population data, the number of data, and the precision needed on the parameter values. Once the bootstrapping procedure is automated, it is a trivial exercise to run it again. If a new randomized bootstrapping procedure gives the same results (within desired precision) as the prior one, then 100 realizations was enough. If the results are not equivalent, then increase the number of realizations, perhaps by a factor of 5, and repeat.

5.5 Takeaway

Interval estimation has two aspects, and for each there are two situations. When you are interested in a single data value: In one aspect you are given the distribution and associated parameter values and are asked to determine the range of values that a sampling might generate. In the second aspect, you are given the distribution, but not the parameter values, and you are given a data value, and you wish to determine the range of parameter values that could have generated the data. Don't get them mixed up.

When you have a set of data, again you could be either seeking the possible limits that the population might generate or the range on population parameter values that might have generated the data. Don't get them mixed up.

In either case, you need to specify a level of confidence on the interval, as well as the allocation of the upper and lower extremes. This is specific to a particular context. Traditionally, the default is the 95% confidence interval with equal allocation of the extremes. But that does not make it right for your particular application. Chapter 10 discusses choices for confidence interval.

If the number of samples is high and the values are far from the constraints, then the normal distribution is a reasonable approximation to your data. If you do not know what the underlying distribution is, using the normal may be the best you can do. But seek to use the distribution that best matches that revealed by the empirical data. If your analysis leads to infeasible values, you probably need to change the distribution model.

5.6 Exercises

1. Data taken from the filter located in one section of the pilot plant are used to determine the specific cake resistance of a slurry. Several values of the variable, expressed in ft/lb$_m$, have been calculated from data taken during the past month. Based on these values, what is the 95% confidence interval for the variability of the specific cake resistance?

Specific cake resistance (ft/lb$_m$)	
2.49×10^{11}	2.67×10^{11}
2.40×10^{11}	2.60×10^{11}
2.43×10^{11}	2.50×10^{11}
2.30×10^{11}	2.54×10^{11}
2.53×10^{11}	2.55×10^{11}

2. Show the details of the calculations leading to the results in Examples 5.3 and 5.4.
3. Repeat Example 5.12 but consider that the rain gauge has a 1/3 inch diameter opening. What is the uncertainty on the rainfall if the amount in the gauge reads 1 inch?
4. Use the bootstrapping technique of Section 5.4.2 on the data of Example 3.14 to calculate the 95% range on the average. Compare your results to Example 5.17.

6

Hypothesis Formulation and Testing – Parametric Tests

6.1 Introduction

A hypothesis is based on a supposition or claim about something. The supposition might be, "This is a fair coin." Or, "Treatment A gives a yield that is 10% better than Treatment B." Or, "Including "ratio" to the controller strategy reduces process variability." Or, "The moon is the same size as the sun." Just because something is supposed does not mean it is true.

But if the supposition is true, then it might be manifest in some attribute of the data. The hypothesis is a statement about what you expect to see in the data.

The hypothesis is about the attributes of the population, not about the attribute of the sample. For example, Sample 1 data are 3, 4, and 5; and Sample 2 data are 4, 5, and 6. The averages are $\bar{X}_1 = 4$ and $\bar{X}_2 = 5$. There is no question that the average of Sample 2 is greater than the average of Sample 1. There is no question that the averages are not equal. But these are samples from a population, not the truth about the population. The question is not about the sample. The question would be about the population. The related question may be, "Are the population means equal?" The two sample sets could come from the same population. Just because the sample averages are different does not mean that the population means are different.

The hypothesis must be tested. You consider what attributes the supposition might express, then do experiments and collect data that could be used to test the hypothesized aspect of the data. For example, if it is a fair coin, then after a bunch of flips, you expect the number of Heads (H) to equal the number of Tails (T). But the experimental data is a sample of the population, and it does not represent the definitive truth about the population. There will be a range of the experimental results that will support the hypothesis, and a contrasting range that will indicate that the hypothesis is not likely true.

Consider the fair coin supposition as an archetypical illustration to explain the method of this chapter: How could one test the fair coin supposition? An answer is to flip a coin many times and record the number of H and T, if the results are 50/50, the coin is likely fair. The hypothesis is that the number of H is equal to the number of T. In an experiment of 20 flips, we might get 8 H and 12 T. It is not a 50/50 distribution, but this does not mean the coin is not fair. This is a likely outcome of a fair coin with only 20 trials.

Figure 6.1 illustrates the number of successes from the binomial distribution with n = 20 trials and $p = 0.5$ (a fair coin). The horizontal axis is the possible count of the number of successes (either H or T, depending on your choice of what constitutes a success), and the vertical axis is the binomial *CDF*. The dashed lines indicate that experiencing fewer than

DOI: 10.1201/9781003222330-6

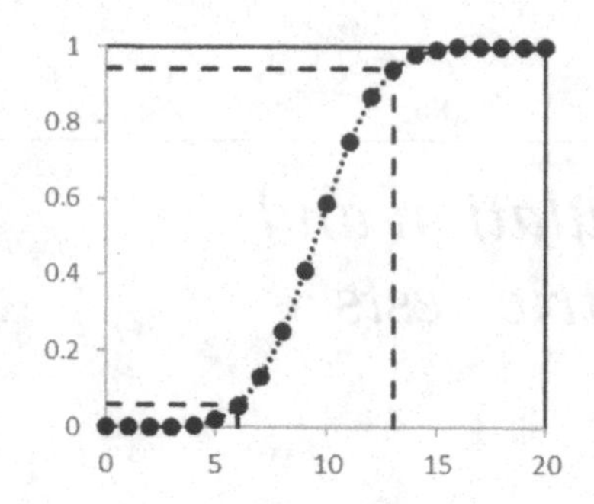

FIGURE 6.1
CDF of the number of successes from a binomial distribution with $n = 20$ and $p(H) = 0.5$, and roughly 90% limits.

6 successes only has about a 5% chance of happening, and 14 or more successes about an equal chance. Between, the values of 6 through 13 successes are within the about 90% possibility of occurrence with a fair coin. Only 8 wins in 20 flips is not an improbable event. It does not provide strong evidence that the coin is foul.

Because of random events and the vagaries associated with experimental testing, the sample data will not reveal the true mean or true variance of the population. So, we need to see if the experimental statistic is so extreme that it probably would not be from the hypothesized situation. This will use techniques in Chapter 5, which means that a probability distribution that is representative of the population of outcomes needs to be chosen. Also, so does a level of confidence.

In the coin flipping illustration, the distribution of successes should be the binomial. If one chooses the 50% level of confidence, the level of significance, the extreme areas is $\alpha = (1-0.5) = 0.5$. And equally splitting the extremes, the range of extreme outcomes represents the quartiles (*CDF* = 0.25 and *CDF* = 0.75). The hypothesis is *P*(*H*) = 0.5 and using n = 20 the binomial distribution model reveals that the probability of the number of heads being between 9 and 11 is 50%, $P(9 \le x \le 11) = 0.5$; and since the experimental outcome of 8 H is outside of that range one might claim "Foul!". But should we reject the hypothesis, accuse the flipper with misconduct, and presume to extract justice for being defrauded, when there is only 50% confidence? Some people might claim "Foul!" if they lost on a single flip, but that is not adequate evidence to be confident in the accusation. What level of confidence is appropriate? In the US legal system, for criminal cases, it is the intuitive "beyond a reasonable doubt". In conventional economic business decisions, it is 95%. If 95% on the coin flipping with n = 20, a fair coin could provide between 5 and 14 successes, $P(5 \le x \le 14) \cong 0.95$. And since the experimental outcome of 8 H is not outside of that range, one cannot claim "Foul!" with a 95% confidence. One cannot reject the hypothesis at the 95% confidence level.

Note: Not rejecting the hypothesis, does not mean the hypothesis is true. It may be a trick coin that has a 0.45 probability of flipping a T. It may mean that the number of samples is not enough to see the level of detail to be able to confidently reject the hypothesis. Also, it may mean the statistic and test chosen is not able to provide a legitimate test.

We cannot use experimental evidence to claim that a hypothesis is true. Here are some examples from human history: They once hypothesized that the aether conveyed light and other forms of electromagnetic radiation through space, and based on that magical substance, the Maxwell Equations modeled electromagnetic transmission. The equations worked, and remain useful, but that evidence does not prove the presence of the aether.

They once hypothesized that heat was a fluid-like substance called caloric, and the differential equations to describe how this mystical fluid flowed within material remain useful today. Indirect supporting evidence cannot prove that the aether or caloric exist.

If for instance, the coin was not fair, with a probability of flipping a H as $p(H) = 0.45$, then getting between 6 and 13 successes is not unlikely (it is within the 95% of possible outcomes). Figure 6.2 compares the distribution with $p = 0.50$ to that of $p = 0.45$. The $p = 0.45$ value shifts the distribution slightly to the left.

One could get a 50/50 split (10 each) with 20 flips of a trick coin with a $p = 0.45$. Getting the expected outcome does not prove that the fair-coin hypothesis is true. Getting an 8/12 split does not prove that the fair-coin hypothesis is false. There are no absolutes.

Accordingly, the hypothesis testing will either "reject the hypothesis at the specified confidence" or "not reject the hypothesis at the specified confidence". The "not reject" claim does not mean we have proved that the hypothesis is true. It just means either of two things: 1) The hypothesis might be true. It also could mean: 2) The hypothesis might not be true, and we have insufficient evidence to confidently claim it is not true.

The parallel in the legal system in the US is to begin with the hypothesis that the accused is innocent. The jury sees the evidence, then either declares "guilty" (meaning there was enough evidence to confidently claim that the accused did commit the crime) or "not guilty" (meaning that there was not enough confidence to claim guilty). Note that "not guilty" does not mean "innocent" (although the accused might want to claim that a "not guilty" verdict proves innocence).

Unfortunately, the tradition in statistical hypothesis testing, uses "accept" instead of "not reject". If the hypothesis cannot be confidently rejected, if the statistical decision is "not reject", the statistical term is to "accept" the hypothesis. However, accepting the hypothesis does not prove it is true.

The hypothesis being tested is usually the null hypothesis, meaning that there is no difference between a population and its expected value, or between the population parameters of two or more treatments. The symbol is H_0. For every hypothesis, H, there is an alternate hypothesis, H_A, which is the complement, or opposite, of the hypothesis. Rejection of the hypothesis automatically requires acceptance of the alternate hypothesis (but again, statistical "accept" does not mean proof that H_A is true). No matter how careful you are, the probability of an erroneous claim exists. Uncertainty in hypothesis acceptance/rejection will be reflected in either a confidence interval about a parameter estimate, or confidence level in the acceptance/rejection of the hypothesis.

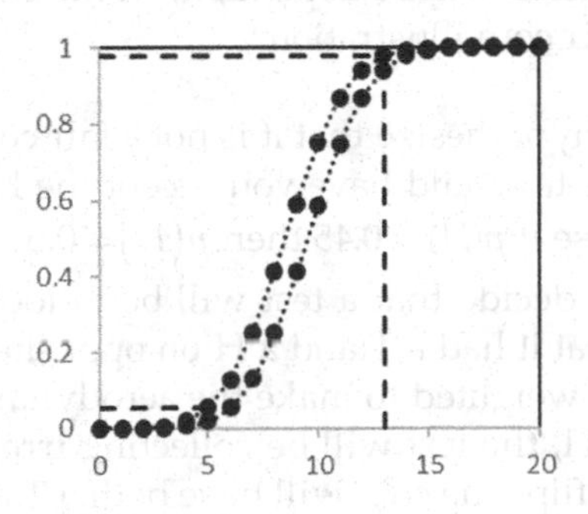

FIGURE 6.2
Comparing the binomial *CDF* with $n = 20$, and $p(H) = 0.5$ (right-most *CDF*) and $p(H) = 0.45$ (left-most *CDF*). The dashed lines at 5 and 13 successes indicate the 95% confidence interval for the trick coin.

6.1.1 Critical Value Method

Here is an outline of the procedure for statistical hypothesis testing based on critical values:

1. Based on your supposition, hypothesize some expected feature of the data, and the alternate to that. The hypothesis will be about the population, not a sample.
2. Expect what an experiment might reveal if the hypothesis is true and define an experimental procedure with a measurable outcome that could either reveal or counter the expected outcome.
3. Define the statistic for the test data. It might be any one of the several descriptive statistics described in Chapter 4 – a count, or a ratio of deviation from expected scaled by the standard deviation.
4. Choose the probability distribution that matches the characteristics (properties) of the test data associated with the statistic. You could use theoretical analysis, intuition, previous experience, or a match to the empirical data histogram or *CDF* as your basis for selecting the distribution.
5. Choose a level of confidence that is appropriate for the context. It might be 50% if the consequences of rejecting hypothesis that is true is not critical, or 99.9% if you want to be very sure of only rejecting a hypothesis that is false. It might equally split the extreme regions, or only consider one side as extreme. In the absence of special context, the 95% limit is the traditional level of confidence for economic/business decisions, and the extreme areas are split equally.
6. Do the experiment and collect data.
7. Use the data to formulate the statistic (average, count, z, t, χ^2, F, etc.).
8. Use the chosen distribution and confidence to determine the expected range on the sample statistic corresponding to the confidence interval if the hypothesis would be true.
9. If the sample statistic value is beyond the probable range of the distribution statistic, reject the hypothesis. If it is within the probable range, accept the hypothesis.
10. Report the decision in a manner that clearly reveals the choices you made in Steps 1, 2, 3, 4, and 5, and in a manner that does not cause the audience to misinterpret the accept/reject conclusion.

Note: As a caution, the conclusion will be dependent on the choices you make in Steps 1, 2, 3, 4, and 5. Continuing the fair coin illustration:

a. Suppose in Step 1, you hypothesize that it is not a fair coin, that the chance of a H is only 0.45. Then the data would have you accept the hypothesis and claim that the coin was foul because if $p(H) = 0.45$ then $p(H) \neq 0.5$.

b. Suppose in Step 2, you decide that a test will be to look at the bottom side of a flipped coin to check that it had a T and a H on opposite sides. Then if the H and T coin was beveled and weighted to make the aerodynamics of a flip have a mere 10% chance of being a H, the test will be collecting irrelevant data. The test will find that in 100% of the flips the coin will have both a T and an H. But having two sides does not mean that they are equally probable. Be sure that the hypothesis is relevant to the supposition.

c. Suppose in Step 3, that you choose a normal distribution, not the binomial. Then the basis for any conclusion will not be grounded in the ideal truth about the events.
d. Suppose in Step 4, you choose a 5% confidence, then the lower 47.5% and upper 52.5% of expected outcomes are both 10, and any reasonably expected fair outcome other than $x = 10$ would lead you to reject the fair coin hypothesis.
e. Finally, in Step 5, if you wanted to show that the coin was foul, you could contrive an experimental protocol so that the flipping is not exactly unbiased. It may be a mechanical flipper designed to only make one flip during the path, with each flip starting with a H-up coin orientation. Be sure that the experimental process is fair.

Critically review your choices in Steps 1–5 to ensure that your choices are legitimate.

6.1.2 *p*-Value Assessment

The critical value procedure is one approach to hypothesis testing. There are alternate approaches to rejecting the hypothesis, to assessing the degree of violation or improbability of a data outcome if the hypothesis is true. The issue with the critical value approach is: If the statistic value is very near to, but does not exceed, the critical value, the statistical statement is "accept the hypothesis". For instance, if the expected number of successes in flipping a fair coin is 1,000 (out of 2,000 flips), and the 99% confidence is chosen, the two critical values are 942 (or fewer) and 1,058 (or more) successes. If there are only 943 successes, the conclusion would be to accept the fair coin hypothesis because there is not enough evidence to reject the hypothesis at a 99% confidence, because 943 is not in the critical region. Instead of being directed to think the coin is fair, at least, the audience should be informed of how close the statistic is to the critical values. Although the statement is "accept the fair coin hypothesis", the trial outcome was one event out of 2,000 away from being rejected. The trial outcome of 943 successes is very close to the critical value of 942 or fewer which would lead to the claim "reject with a 99% confidence". This closeness to the reject decision needs to be revealed. The *p*-value approach is one.

The *p*-value is the probability that the data could have an outcome as (or more) extreme if the hypothesis is true.

For instance, in the coin flipping situation of 943 successes out of 2,000 trials, the *CDF* associated with that is 0.005747…. In Excel use the $\text{BINOM.DIST}(s,n,p,1)$ function, $\text{BINOM.DIST}(943,2000,0.5,1) = 0.005747\ldots$. There is a 0.005747 probability of getting 943 successes or fewer. Since the fair coin hypothesis is a two-sided test that is rejected if either too few or too many successes, the *p*-value is twice that value or ~0.0115. There is only a 1.15% chance that a fair coin could have such an extreme outcome. So, the qualification to the "accept the fair coin hypothesis" statement could be extended to include, "But, the *p*-value, the probability of a fair coin producing such a lopsided outcome is 0.0115 – a 1.15% chance. It is fairly improbable that the coin is fair."

For a one-sided t-test the *p*-value is $T.\text{DIST}(T,\upsilon,1)$ if $T < 0$, or $1 - T.\text{DIST}(T,\upsilon,1)$ if $T > 0$. Since the t-distribution is symmetric, for either case

$$\text{pvalue}_{1\text{-sided},\, t\text{ distribution}} = 1 - T.\text{DIST}(|T|,\upsilon,1) \tag{6.1}$$

For a two-sided t-test the p-value is

$$pvalue_{2\text{-sided},\, t\text{ distribution}} = 2\left[1 - T.\text{DIST}\left(\left|T\right|, \upsilon, 1\right)\right] \tag{6.2}$$

Note the absolute value of T is used for both.

Here is a modified outline of the procedure for statistical hypothesis testing which uses p-values to determine the improbability of the hypothesis:

1. Same.
2. Same.
3. Same.
4. Same.
5. Same.
6. Same.
7. Same.
8. Use the chosen distribution and determine the probability (p-value) that the population could have generated such an extreme value of the sample statistic if the hypothesis is true.
9. If the p-value is improbable relative to the choice in Step 5, reject the hypothesis. If it is within the permissible range, accept the hypothesis.
10. Report the decision and p-value in a manner that clearly reveals the choices you made in Steps 1, 2, 3, 4, and 5, and in a manner that does not cause the audience to misinterpret the accept/reject conclusion.

Note: Again, as a caution, the conclusion will be dependent on the choices you make in Steps 1, 2, 3, 4, and 5.

The advantage of the p-value approach is that it reveals the magnitude of the deviation of the hypothesis. It conveys more than just the accept/reject decision. For instance, the critical value for a test statistic might be 12.34, and any larger value would reject the hypothesis. If the test statistic value is 12.33, nearly rejecting the hypothesis, but not, it leads to the "accept" decision. This black/white accept/reject dichotomy does not relay to the audience how close the decision is. The p-value would report that such an extreme value of the test statistic has a probability of 0.051, indicating it is close to the standard 95% confidence interval limit. Further, a test statistic of 12.35 gives the same "reject" decision as one that might have a value of 278.3. Reporting a "reject" decision does not reveal that the second value is not even close. However, its p-value might be 0.000001.

Reporting the p-value indicates the strength of justification for rejecting (or accepting) the hypothesis.

6.1.3 Probability Ratio Method

For an equality hypothesis, such as $H: \mu_B = \mu_A + k$, which can be expressed as a null hypothesis $H: \mu_B - \mu_A - k = 0$, the test has a two-sided rejection region. Representing the means would be the averages, and the statistic would have $\bar{X}_B - \bar{X}_A - k$, perhaps as a t-statistic scaled by the standard deviation on the difference of the averages, $T = \dfrac{\bar{X}_B - \bar{X}_A - k}{s_{\bar{X}_B - \bar{X}_A}}$.

If the hypothesis is true, the ideal *t*-value is zero. Reject if the *t*-value is too extreme, either "+" or "−". If the distribution is symmetric, and the hypothesis is true, then the chance of getting a $T > 0$ is the same as a $T < 0$, 50%. $P(T > 0) = 0.5$, and $P(T < 0) = 0.5$.

If, however, the hypothesis is not true, then there will be a higher probability of getting either "+" or "−" values. For example, if the data indicates $\bar{X}_B - \bar{X}_A - k = +5$, and that value better represents $\mu_B - \mu_A - k$ than a value of 0, then the chance of getting a negative value from the experiment is lower than a 50% chance.

Figure 6.3 Illustrates the two cases with the *CDF* of *t*-statistic distributions. The dashed curve represents the hypothesis $\mu_B - \mu_A - k = 0$. It is symmetric about the *T*-value of zero, and the *CDF* value of $T = 0$ is 0.5, half of the expected t-values from experiments should be >0 and half <0. The solid curve is centered on the experimental *T*-value, here illustrated at a value of about 1.2. If the null hypothesis is true, then the chance of having a *T*-value equal or greater than 1.2 is about 1 − 0.87 = 0.13. Because the distributions are symmetric, if the data is true, if $\mu_B - \mu_A - k = 1.2$, then the chance of getting a *t*-value less than zero is also 0.13. (Read this from the *CDF* value of the solid curve at a *t*-value of zero.)

The probability of getting a statistic value on the other side of 0 can be obtained using the Excel function $T.\text{DIST}(T, \upsilon, 1)$. If the experimental $T > 0$, and $\mu_B - \mu_A = \bar{X}_B - \bar{X}_A$, then $P(T < 0) = 1 - T.\text{DIST}(T, \upsilon, 1)$. If the experimental $T < 0$, and $\mu_B - \mu_A = \bar{X}_B - \bar{X}_A$, then $P(T > 0) = 1 - T.\text{DIST}(-T, \upsilon, 1)$. These can be generalized using the absolute value of the experimental T. If $\mu_B - \mu_A = \bar{X}_B - \bar{X}_A$, then $P(T \text{ having the other sign}) = 1 - T.\text{DIST}(|T|, \upsilon, 1)$.

The ratio of $P(T \text{having the other sign} \mid \mu_B - \mu_A - k = 0) = 0.5$ to $P(T \text{having the other sign} \mid \mu_B - \mu_A = \bar{X}_B - \bar{X}_A) = 1 - T.\text{DIST}(|T|, \upsilon, 1)$ for testing the null hypothesis of a symmetric distribution is the probability ratio.

$$P_r = \frac{0.5}{1 - T.\text{DIST}(|T|, \upsilon, 1)} \tag{6.3}$$

If the experimental *T*-value is close to the hypothesized zero, then the *CDF* value will be close to 0.5, and the probability ratio will be around unity. However, if the experimental *T*-value is not close to the hypothesized zero, then the denominator will have a low value and the probability ratio will have a large value.

As illustrated in Figure 6.3, $P_r = \dfrac{0.5}{1 - .87} \cong 3.8$. the chance of getting a t-value of 1.2 if the hypothesis is true is about $2 \times 0.13 = 0.26$, which is far greater than the normal 0.05 criteria

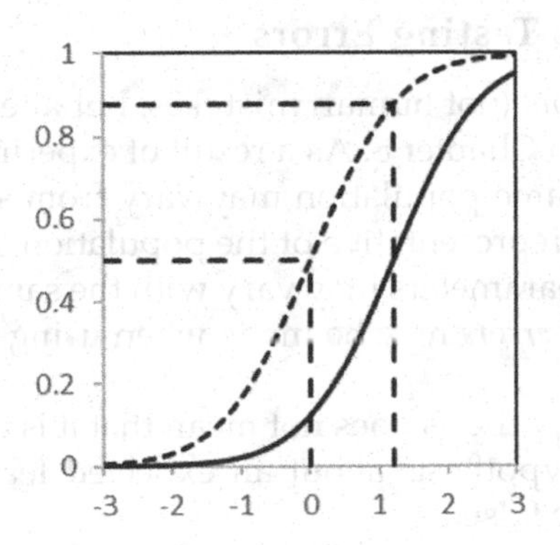

FIGURE 6.3
Illustrating the probability of getting a t-value on the other side of zero.

for rejecting the hypothesis. So, a probability ratio of 3.8 is not adequate cause to reject. However, if the probability ratio is 20:1, it means that there is 20 times greater chance of getting the sign of T than if the null hypothesis were true. Perhaps 20:1 odds is strong enough evidence to reject.

Not unexpectedly, the three criteria are related. For the two-sided t-test, the probability ratio is the reciprocal of the p-value.

$$\text{pvalue}_{\text{2-sided}, t \text{ distribution}} = 1/P_r = \frac{\left[1 - T.\text{DIST}\left(|T|, \upsilon, 1\right)\right]}{0.5} = 2\left[1 - T.\text{DIST}\left(|T|, \upsilon, 1\right)\right] \quad (6.4)$$

The probability ratio is valid when considering the two-sided null hypothesis for a symmetric distribution. The P_r is not the traditional statistical metric, but the odds that it represents seems to be a more familiar metric to most people than either the critical value or the alpha level of significance.

6.1.4 What Distribution to Use?

The normal distribution describes most data, and more so the average of data, and is therefore the most widely used model for comparing the mean of continuous distributions. However, the normal distribution is not the appropriate choice for all datasets. Chapter 7 provides test procedures to determine whether or not the normal distribution is a valid model.

When comparing averages to an expected mean or when comparing averages to averages, when sigma is known, the standard normal distribution is probably the correct choice. When sigma is estimated from the data, the t-distribution is probably the correct choice.

When comparing sample standard deviation to an expected sigma the chi-squared distribution is probably the correct choice. When comparing two sample standard deviations, the F-distribution is probably the correct choice.

6.2 Types of Hypothesis Testing Errors

We discuss the *sources of errors* (not human mistakes, but alternately, naturally occurring deviations, vagaries, noise) in Chapter 8. As a result of experimental errors, the characteristics of a sample from the same population may vary from sample to sample. Although the sample is expected to be representative of the population, it may not always be. As the estimates of the population parameters may vary with the sample, it is possible that one of two *types of hypothesis testing errors* may be made when using sample data to evaluate the hypothesis.

Just because you have a hypothesis does not mean that it is true. We suppose that something is true and create a hypothesis about an expected feature of the test outcome. It might be true, but it might be false.

A *Type 1 error* occurs when a true hypothesis, H, is rejected. We use the Greek letter α to indicate the probability of committing a Type 1 (T-I) error, i.e., rejecting a hypothesis when it is in fact true. The probability of *not* making a T-I error is $(1 - \alpha)$ and is the basis of the confidence intervals in Chapter 5. α is also called *level of significance*, indicating the chance we are willing to take of committing a T-I error.

As an example, Figure 6.1 shows the *CDF* of number of successes (flipping a H) for a fair coin $p(H) = 0.5$ and $n = 20$ trials. There is a possibility of only getting 2 heads, or only 4, or even 15, but these are extremely rare outcomes. The probability of getting 6 or fewer wins (successes) or 13 or more ($s \le 6,\ s \ge 13$) is roughly 10%. In that example $\alpha \cong 0.1$. If the fair coin acceptance region is from 7 to 12 heads inclusive ($7 \le s \le 12$), then there is about a 10% chance, $\alpha \cong 0.1$, of rejecting the fair-coin hypothesis when it is true. That is about a 10% chance of making a T-I error.

A *Type 2 error* occurs when we accept a hypothesis, *H*, when it is actually false. The Greek letter β represents the probability of committing a Type 2 (T-II) error. The power of a test is defined as $(1 - \beta)$, which is the probability of not rejecting a false hypothesis. The probability of committing a T-II error depends, partly, on the degree of wrongness in the false hypothesis. For example, if the hypothesis is that the coin is fair, but in reality $p(H) = 0.499$, the fair-coin hypothesis is false. But it would take hundreds of thousands of flips to differentiate the outcome from the hypothesized $p(H) = 0.5$. So, in all practicality, after a reasonable number of flips, the data would reveal $7 \le s \le 12$, and the false hypothesis would be accepted. That is a T-II error. By contrast, if the coin $p(H) = 0.9$, then in just a dozen flips the fair coin hypothesis would be rejected.

As an illustration, see Figure 6.4. The *CDF* of the fair coin, $p(H) = 0.5$, is to the right and the *CDF* of a trick coin, $p(H) = 0.4$, is to the left. The 90% fair-coin hypothesis acceptance region is a number of successes $7 \le s \le 12$, roughly $\alpha = 0.1$. However, the trick coin will also generate successes between 7 and 12, inclusive, and the rejection area for 6 or fewer is about 13% for the trick coin and for 13 or more is about 1%, so the probability of rejecting a trick coin as being fair is about 14% or accepting the trick coin as a fair coin is about 86%. Here the chance of making a T-II error is 0.86, $\beta = 0.86$.

The probability of committing a T-II error depends on the choices of both α and n as well as the magnitude of the deviation from the hypothesis. In Figure 6.5 for the trick coin $p(H) = 0.2$ (left-most *CDF*), and $\alpha = 0.1$ and $n = 20$ are as before. With the larger deviation from the hypothesis there is only a 9% chance that the trick coin will give a count in the ≤ 6 or ≥ 14 rejection region. Here $\beta = 0.09$.

In Figure 6.6 for the trick coin $p(H) = 0.45$ (left-most *CDF*) and $\alpha = 0.1$, as in Figure 6.4, but here $n = 200$. With the larger number of samples there only a 42% chance that the trick coin

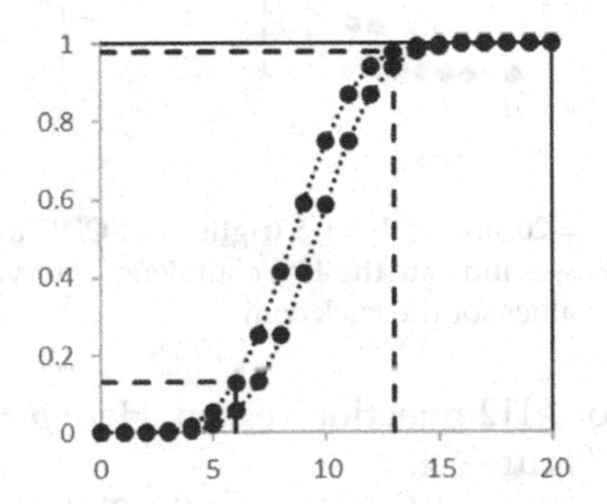

FIGURE 6.4
Comparing the binomial *CDF* with $n = 20$, and $p(H) = 0.50$ (right-most *CDF*) and $p(H) = 0.45$ (left-most *CDF*). The vertical dashed lines at 6 and 13 successes indicate the 90% confidence interval for the fair coin, but the horizontal lines are the corresponding CDF values for the trick coin.

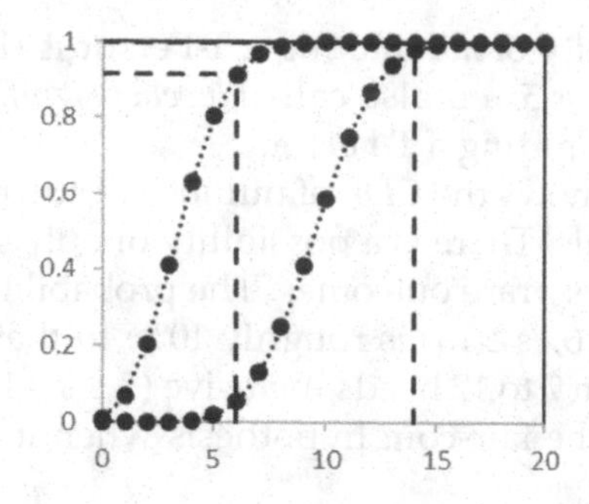

FIGURE 6.5
Comparing the binomial *CDF* with $n = 20$, and $p(H) = 0.5$ (right-most *CDF*) and $p(H) = 0.2$ (left-most *CDF*). The vertical dashed lines at 6 and 14 successes indicate the 90% confidence interval for the fair coin, but the horizontal lines are the corresponding *CDF* values for the trick coin.

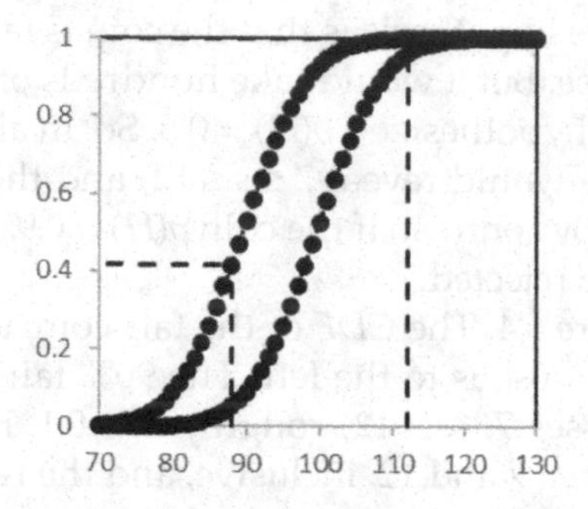

FIGURE 6.6
Comparing the binomial *CDF* with $n = 200$, and $p(H) = 0.5$ (right-most *CDF*) and $p(H) = 0.2$ (left-most *CDF*). The vertical dashed lines at 88 and 112 successes indicate the 90% confidence interval for the fair coin, but the horizontal lines are the corresponding *CDF* values for the trick coin.

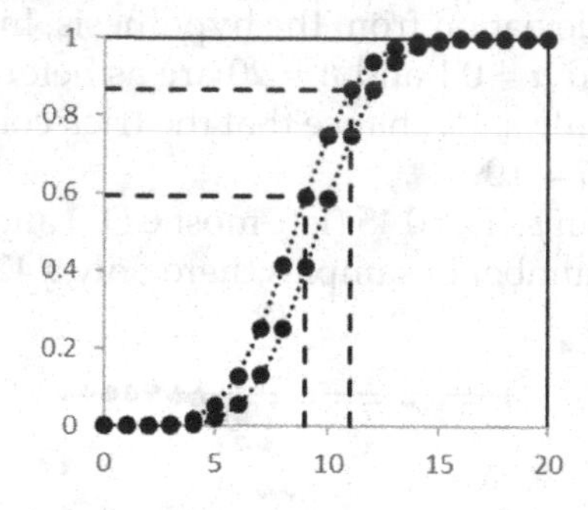

FIGURE 6.7
Comparing the binomial *CDF* with $n = 20$, and $p(H) = 0.5$ (right-most *CDF*) and $p(H) = 0.45$ (left-most *CDF*). The vertical dashed lines at 9 and 11 successes indicate the 40% confidence interval for the fair coin, but the horizontal lines are the corresponding *CDF* values for the trick coin.

will give a count in the ≤88 or ≥112 rejection region. Here $\beta = 0.58$, an improvement over $\beta = 0.86$ for the $n = 20$ case in Figure 6.4.

As a final exploration of the impact of choices on the T-II error, in Figure 6.7 for the trick coin $p(H) = 0.45$ (left-most *CDF*) and here $n = 20$, as in Figure 6.4, but here $\alpha = 0.6$. With the larger α, a greater chance of making a T-I error, a lower confidence of 40%, there only a 28% chance that the trick coin will give a count in the ≤9 or ≥11 rejection region. Here $\beta = 0.28$.

Further, the probability of committing a Type-II error is not the complement of the probability of committing a T-I error, i.e., $\beta \neq (1-\alpha)$. Consider two populations, X_1 and X_2, for which μ_1 is really less than μ_2. However, when the samples are collected and the data averaged, you find that $\bar{X}_1 > \bar{X}_2$. If the hypothesis is H: $\mu_1 \leq \mu_2$, the data may cause us to reject H, committing a Type-I error. Conversely, the data might have been such that $\bar{X}_1 \approx \bar{X}_2$. Depending on the variances involved, we might accept H: $\mu_1 = \mu_2$, thus making a T-II error. Unfortunately, no test of a null hypothesis has yet been devised that simultaneously minimizes both types of error. However, increasing the number of samples can, but it increases experimental cost.

Hypotheses are accepted or rejected based on the choice of α, the probability of a T-I error. Various considerations to guide that choice are given in Chapters 10 and 17.

However, β, the probability of a T-II error is also important. The value of β depends on the degree that the hypothesis might be wrong, the number of data, and the alpha-value that sets the acceptance limits. Economic considerations for these aspects are also given in Chapter 10.

6.3 Two-Sided and One-Sided Tests

There are several basic choices for the hypothesis and its alternate. Using the population mean μ as the example of a continuum-valued statistic, the four choices could be represented as:

$$H_0 : \mu = \mu_0 \quad \text{vs.} \quad H_A : \mu \neq \mu_0 \tag{6.5}$$

$$H : \mu \leq \mu_0 \quad \text{vs.} \quad H_A : \mu > \mu_0 \tag{6.6}$$

$$H : \mu \geq \mu_0 \quad \text{vs.} \quad H_A : \mu < \mu_0 \tag{6.7}$$

$$H : \mu \neq \mu_0 \quad \text{vs.} \quad H_A : \mu = \mu_0 \tag{6.8}$$

The first hypothesis says that you expect the population mean to have a particular value, μ_0. The alternate hypothesis in this case has two possibilities: H_{A1}: $\mu < \mu_0$ and H_{A2}: $\mu > \mu_0$. Because there are two options, evaluation of the null hypothesis given in Equation (6.5) requires a two-sided, or two-tailed, test. So does the hypothesis of Equation (6.8). On the other hand, the hypotheses of Equations (6.6) and (6.7) are one-sided, any statistic to one side would lead to acceptance, and only one extreme side would lead to rejection.

Note: In this discussion, the statistic does not have to be the mean, as illustrated in Equations (6.5) to (6.8). It could be a variance, or a scaled statistic such as t, χ^2, or F.

There are several other hypothesis choices that are essentially equivalent to those four. These two are essentially equivalent to Equations (6.6) and (6.7).

$$H_0 : \mu < \mu_0 \quad \text{vs.} \quad H_A : \mu \geq \mu_0 \tag{6.9}$$

$$H_0 : \mu > \mu_0 \quad \text{vs.} \quad H_A : \mu \leq \mu_0 \tag{6.10}$$

If the possible values for μ were discretized values, perhaps counting integers, and μ_0 = 5 then the hypothesis $\mu < \mu_0$ could only permit numbers 0 through 4. While the nearly equivalent hypothesis $\mu \le \mu_0$ could permit numbers 0 through 4 and 5. These test procedures are equivalent, because the test for $\mu < 5$ has the same rejection/acceptance areas as the test for $\mu \le 4$. Although, mathematically for continuum-valued variables the $\mu \le \mu_0$ condition includes the μ_0 point value, and the $\mu < \mu_0$ condition excludes the point value, the $\mu < \mu_0$ condition permits being infinitesimally close to that point. However, in any practical application with continuum-valued variables, numerical truncation on the digital values will exceed that infinitesimal amount; and further, error on the data values and the mismatch between the ideal distribution and the true sample-generating phenomena will be even larger. In theory, we can differentiate, less-than from less-than-or-equal-to; but in reality, we cannot.

There are four more possible hypotheses! Hypotheses could also use not-greater-than and not-less-than, but these are identical to less-than-or-equal-to and greater-than-or-equal-to. And, not-less-than-or-equal-to and not-greater-than-or-equal-to are equivalent to greater-than and less-than.

Equations (6.6) and (6.7) are complementary to each other. The method of rejecting/accepting the less-than hypothesis is the same as rejecting/accepting the greater-than hypothesis, except for using the rejection region to the right, instead of to the left. These use the same procedures.

Equation (6.8), the not-equal hypothesis, is not practical. Except in mathematical concept, nothing is equal. We cannot make things equal. The can of juice indicates it is 8 oz. That is a nominal or target value. To have the contents be exactly 8 oz means that it cannot be over or under by one molecule. Nothing is truly equal. In reality, for continuum-valued variables, the hypothesis $H : \mu \ne \mu_0$ is effectively always true. Chance might make it appear true for discrete variables.

So, essentially there are only two procedures that need to be considered. A two-sided test for Equation (6.5), and the identical (but right vs left) one-sided test for Equations (6.6) and (6.7).

After selecting the significance level of a two-sided test, it is customary to divide the chance of committing a T-I error into halves and to assign one half to each of the alternate hypotheses. These halves define the *critical regions* for the test of the null hypothesis. The values of the test statistic defining the inner limit between each too-extreme region, the acceptance region, are called the *critical values.* The too-extreme regions are termed critical regions. If the value of the test statistic calculated from sample data falls in either critical region, we reject the hypothesis. It is possibly false, and we accept the alternate hypothesis as possibly true. For this reason, the critical regions are also termed *rejection regions.* In the case of a two-sided test, the area under the *pdf* curve in each rejection region is conventionally set to $100(\alpha/2)$ percent of the total area under the curve. The *CDF* values of the critical region are $\alpha/2$ and $(1-\alpha/2)$. All the remainder of the area is the *acceptance region.* If values of the test statistic fall in the acceptance region, we accept the null hypothesis as possibly true. Figure 6.1 shows the rejection (outside the dashed lines) and acceptance regions (between the dashed lines) for a two-sided test of a hypothesis of the type in Equation (6.5) for the hypothesis $\mu = \mu_0$.

Alternately, if the hypothesis is that of Equation (6.8), $\mu \ne \mu_0$, then the rejection region would be between the dashed lines and the acceptance region the more extreme values beyond either line.

Figure 6.8 illustrates the one-sided rejection region associated with Equation (6.6) for roughly a 90% confidence. If the hypothesis is $\mu < \mu_0$ then the acceptance region is below the $CDF = 0.9$ (or for a statistic value to the left of the dashed line on the horizontal axis), and the rejection region is above and including (or to the right). If the hypothesis is or $\mu \le \mu_0$, then the acceptance region is below and inclusive of the $CDF = 0.9$ value. Again, the

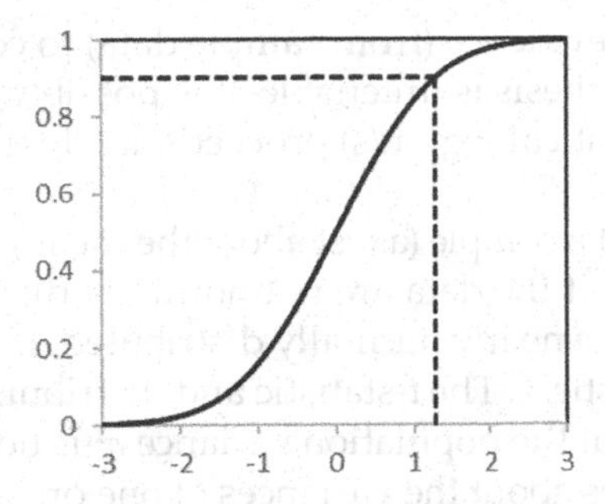

FIGURE 6.8
Illustrating a one-sided test with acceptance region with lower *CDF* values and a statistic to the left. Representing a continuum-valued statistic which is unbounded on either side, such as the t-statistic.

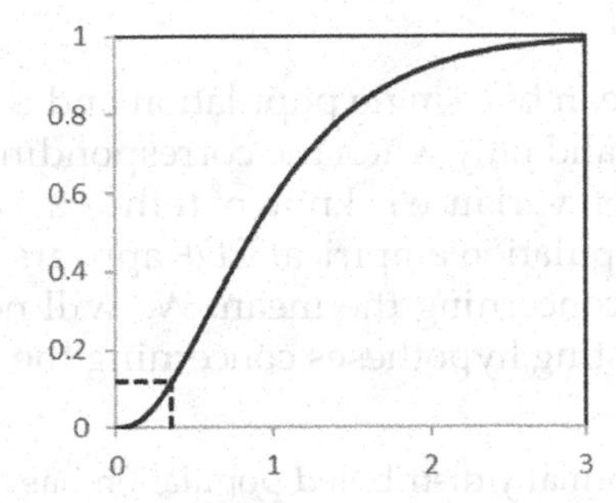

FIGURE 6.9
Illustrating a one-sided test with acceptance region with higher *CDF* values and a statistic to the right. Representing a continuum-valued statistic which is bounded on one side such as the χ^2 or *F*-statistic.

statistic does not have to be the value of the mean, it could be a count, average, variance, t, χ^2, etc.

The complement situation is used for the hypothesis $\mu > \mu_0$. Again, the statistic does not have to be the value of the mean, it could be a count, average, variance, t, χ^2, etc. In Figure 6.9 the illustration is for the χ^2 statistic, $\chi^2 = \frac{\upsilon s^2}{{\sigma_0}^2}$, and the hypothesis $\sigma > \sigma_0$. The acceptance region is above the $CDF = 0.1$ (or for a statistic value to the right of the dashed line on the horizontal axis), and the rejection region is below (or to the left) and inclusive of the critical value.

Thinking about permissible values of μ or σ for the hypothesis will reveal whether the *CDF* for 90% confidence should be 0.9 or 0.1. In Figure 6.9 the hypothesis is $\sigma > \sigma_0$. If the ratio s / σ_0 is large, or very large, or infinity, then you accept that $\sigma > \sigma_0$ is possibly true. In contrast, if the ratio s / σ_0 is small, or the extreme of zero, then you have confidence in rejecting the $\sigma > \sigma_0$ hypothesis. The rejection region for the hypothesis $\sigma > \sigma_0$ is unusually small values, the low *CDF* region. Some use the mnemonic that the less-than or greater-than symbol in the alternate hypothesis points to the rejection region.

We use the terms "accept" and "reject" with regard to hypothesis testing throughout this and subsequent chapters to be consistent with the terminology you will find in practice. The actual meaning of "accept" is "there is insufficient evidence (from the sample) to confidently reject the hypothesis." We stress that acceptance of either the hypothesis or its alternate does not imply that that one is absolutely true and the other is absolutely false. Acceptance indicates only that the hypothesis could be true. The actual meaning of "reject" is "there is sufficient evidence (from sample data) to confidently reject the hypothesis, we believe that the hypothesis is untenable, it is possibly false."

The determination of the critical region(s) proceeds as illustrated above whether the test statistic is t, χ^2, F (or others).

In a test using the average of a sample (a test about the mean) it is usually acceptable to use the normal distribution. Even if the data are not normally distributed, the average of more than several data values will be nearly normally distributed. If the variance value is known, use the standard normal statistic, z. The t-statistic and distribution is used when hypotheses on the mean must be tested but the population variance σ^2 is not known. The χ^2 and F statistics are used to test hypotheses about the variances of one or two populations, respectively.

6.4 Tests about the Mean

The Z tests concerning the mean of a single population and those involving the means of two populations are strictly valid only when the corresponding populations are normally distributed, and the population variance is known. If the variances of the populations are estimated by S^2 (and if the population empirical *CDF* appears approximately normal), the t-test must be used for tests concerning the mean. We will now discuss some examples illustrating the concepts of testing hypotheses concerning the mean.

Example 6.1: Suppose a normally distributed population has a variance of 6. We don't know the population mean μ, but we expect its value to be $\mu = 2$. By experiment, we collect a sample of 20 items for which $\bar{X} = 3.1$. We want to test the null hypothesis H_0: $\mu = \mu_0 = 2$ against the alternate hypothesis H_A: $\mu \neq 2$.

This test is two-sided because two rejection regions are defined by H_A. Since the variance is known, we will use the standard normal distribution to test the hypothesis. We select $\alpha = 0.05$ as the significance level for the test, and dividing the two rejection regions in equal sizes, the critical *CDF* values are $\alpha/2 = 0.025$ and $1-\alpha/2 = 0.975$. The associated critical values of Z are $z_{\alpha/2} = -1.95996...$ and $z_{1-\alpha/2} = 1.95996....$ In Excel these values can be obtained from the function $\text{NORM.INV}(CDF, 0, 1)$. This can be stated as:

$$P(z_{\alpha/2} < Z < z_{1-\alpha/2}) = 1-\alpha$$

$$P(-1.96 < Z < 1.96) = 0.95$$

Where the rounded value 1.96 is used instead of 1.95996....

The symbolic representation means the 95% limits of the z-statistic value are between −1.96 and +1.96.

The thumbnail sketch of the standard normal *CDF* shows these limits.

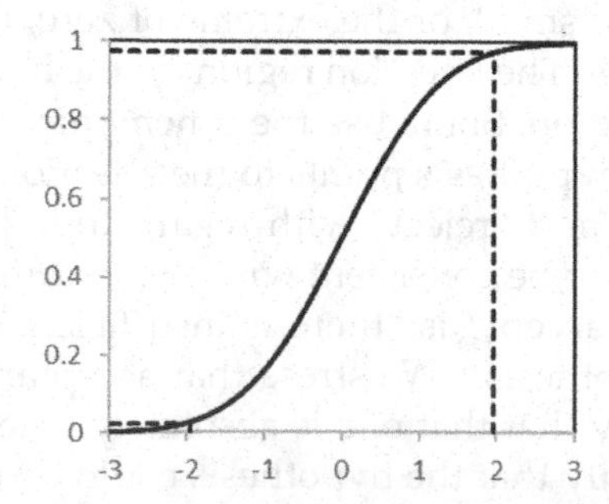

We next calculate the value of the data statistic, *Z*, based on sample data:

$$Z = \frac{\overline{X} - \mu_0}{\sqrt{\sigma^2 / n}} = \frac{3.1 - 2}{\sqrt{6 / 20}} = 2.0083\ldots$$

Because 2.008 > 1.96, the result is outside the 95% confidence limit on Z. The experimental data is too extreme to be believed probably from data generated by a $\mu = 2$ population. Accordingly:

We therefore reject H_0, accept H_A, and conclude with 95% confidence from the *Z* test that μ is probably not 2.

To support this with a p-value, first determine the *CDF* value that $z = 2.0083$ represents. The Excel function NORM.DIST$(z,0,1,1)$ returns 0.977694, meaning that the chance of getting $z = 2.0083$ or higher values is 1 − 0.977694 = 0.022306. But since this is a two-sided test, the chance of such an extreme value is twice that.

The *p*-value for the test statistic is 0.04461. There is only a 4.5% chance of the experimental outcome being so extreme if the hypothesis is true.

The probability ratio is 0.5/0.022306 = 22.4. If the data represents the truth, the chance of having a such an extreme "+" deviation on the statistic is 22.4 times more probable than if the null hypothesis were true.

This *p*-value of 0.044 is marginally less probable than the 95% specification of 0.05. If one chose a 96% level of confidence, the test would accept the hypothesis.

We could have tested the hypothesis by calculating the 95% confidence interval on the population mean about the sample average using the inverse of the t-formula. For this situation, $\sigma_{\overline{X}} = \sigma / \sqrt{n} = 0.5772256$, $z = \pm 1.96$ as above, and

$$P\left(\overline{X} + z_{\alpha/2}\sigma_{\overline{X}} < \mu < \overline{X} + z_{1-\alpha/2}\sigma_{\overline{X}}\right) = 1 - \alpha$$

$$P\big(3.1 - 1.96(0.5477256) < \mu < 3.1$$

$$+1.96(0.5477256)\big) = 0.95$$

$$P(2.0264 < \mu < 4.1735) = 0.95$$

Again, the population mean μ is not inside the 95% confidence interval (CI) about $\overline{X}$, so we reject H_0 and conclude that the population mean is probably not 2. The result is the same, as you should have expected.

Note: Rounding the several values in this example did not affect the conclusion. As a rule of thumb when you round values, keep two more decimal digits than those that might have an impact on the result.

Example 6.2: A particular normal population has a variance of 5. Is it reasonable to expect that $\mu \leq 10$? To answer this question, a 16-member sample was taken from the population and was found to have a mean of 12.46.

In this instance, $\mu_0 = 10$. For this example, let us select $\alpha = 0.02$ representing the 98% confidence. The hypothesis is H: $\mu \leq \mu_0 = 10$. The corresponding alternate hypothesis is H_A: $\mu > \mu_0 = 10$. This is a one-sided test because only one rejection region is defined by H_A. For $\alpha = 0.02$ as the significance level, we want $Z_{1-\alpha} = Z_{0.98}$ to define the critical region. The value can be found from the Excel function NORM.INV$(0.98,0,1) = 2.053748\ldots$. The rejection region (critical region) is that with a *CDF* > 0.98, which is to the right of $z =$ 2.053748….

The confidence interval for testing this hypothesis is one-sided. We will reject H if $Z > z_\alpha$

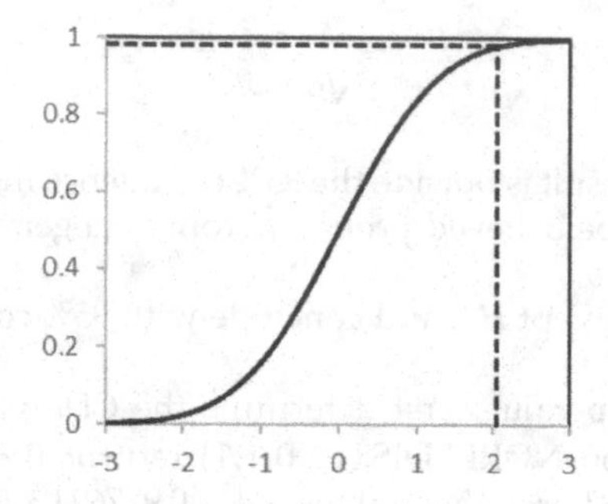

Equivalently, the confidence interval for μ is obtained by substituting $Z = \left(\bar{X} - \mu\right)/\sigma_{\bar{X}}$ into the confidence interval for Z. We will reject H if $\bar{X} - z_\alpha \sigma_{\bar{X}} > 10$.

From the data, we calculate $\sigma_{\bar{X}} = \sqrt{5/16} = 0.559017$ and $Z = \left(\bar{X} - \mu\right)/\sigma_{\bar{X}} = 4.40058$. We have $Z_{data} = 4.40058 > Z_{critical} = 2.0538$, or equivalently, $12.46 - 1.1481 > 10$ for μ.

As either Z and μ fall within their respective rejection ranges, the hypothesis H: $\mu \leq 10$ is rejected by the Z test as possibly false at the 2% significance level.

But this does not relay whether the rejection was overwhelming or just close. The p-value is found by determining the CDF associated with the test statistic and converting that to the probability of that or a more extreme value. $CDF = \text{NORM.DIST}\left(4.40058, 0, 1, 1\right) = 0.9999946$. Since the rejection area is one-sided, the p-value is $1 - 0.9999946 = 5.4 \times 10^{-6}$. The justification for rejecting the hypothesis is very strong!

Note: Neither Example 6.1 nor 6.2 used data values that had dimensional units. Most likely your data will have units, and the units on $X, \bar{X}, \mu_0$, and σ will all have the same dimensional units. The Z-statistic will be dimensionless.

Note: Examples 6.1 and 6.2 might represent the common business situation of qualifying a new supplier of a raw material, device, or test procedure, or new operator. The mean and variance of the hypothesized population would represent the traditional treatment, and the question in Example 6.1 is, "Is the new treatment equivalent?" The data could represent any quality metric (product yield, completion time, manufacturing cost, etc.). Whether one test, based on one hypothesis on one of several important metrics, accepts or rejects, you should also test the other key performance metrics.

To test hypotheses on the population mean μ when the value of σ^2 is unknown and estimated by sample standard deviation, we must use the test statistic $T = \dfrac{\bar{X} - \mu_0}{S_{\bar{X}}} = \dfrac{\bar{X} - \mu_0}{S/\sqrt{n}}$ with $v = \left(n - 1\right)$ degrees of freedom.

Example 6.3: In a quality test, painted surfaces are subjected to intense light and heat to accelerate the rate of appearance of paint defects. The specification requires the mean time to first appearance of a defect to be greater than 10.5 hrs, $\mu > 10.5$ hrs. After the supplier of one component in the paint formulation changed, the company sampled 10 items and the quality test showed an average of 10.21 hrs. The standard deviation of the sample data was 0.74 hrs. The sample average is less than 10.5 hrs, but could the population mean still meet specification? Certainly, some samples will have lower values and some higher. Maybe this set of 10 samples just happened to come out with more of the low values. Can we claim that we are still on-specification?

The supposition is that we are still on-spec. If so, the hypothesis is that the new population mean is greater than 10.5 hrs: $H: \mu > 10.5\,\text{hrs}$. We'll use the conventional 95% confidence, meaning that $\alpha = 0.05$. Since the sample provides the standard deviation, we'll use the t-statistic with $\upsilon = n - 1 = 10 - 1 = 9$ degrees of freedom. This is a one-sided test; we only reject the hypothesis if it is improbable that the population mean is above 10.5. The experimental statistic value is

$$T = \frac{\overline{X} - \mu_0}{S_{\overline{X}}} = \frac{\overline{X} - \mu_0}{S/\sqrt{n}} = \frac{10.21 - 10.5}{0.74/\sqrt{10}} = -1.2392709\ldots$$

The one-sided t-critical value can be found using the Excel T.INV function. If the sample average is too small, then its population mean could not be above the spec. So, we reject the hypothesis at the too-small, left extreme where the $CDF = \alpha$

$$t_{\alpha,v} = t_{0.05,9} = T.\text{INV}(0.05, 9) = -1.833112\ldots$$

We see that $T_{\text{data}} = -1.2392709\ldots \geq t_{\text{critical}} = -1.833112\ldots$.

Since the experimental *t*-value is not beyond the 95% confidence limit, we accept the hypothesis that the population mean with the new supplier could still be above 10.5 hrs.

Although there is not enough evidence to reject the hypothesis at a 95% level of confidence, the new average of 10.21 hrs is cause for suspicion. At the 85% confidence level, the critical *t*-value is –1.0997... and the 10-sample results would cause the hypothesis to be rejected. Alternately, if increased testing of 20 samples gave the same average and standard deviation, the experimental *T*-value would be –1.75259... and the *t*-critical would be –1.72913..., and the hypothesis would be rejected. So perhaps an action should not be to unconditionally approve the new supplier, but to perform more tests.

Statistics should support rational decision-making, not be the decision.

The *p*-value associated with the experimental $t = -1.2392709\ldots$ and $\upsilon = 9$ is $T.\text{DIST}(t, \upsilon, 1) = 0.12329\ldots$ indicating only a 12% chance that the sample could have come from a population with a mean above 10.5 hrs. It is possible, but there is only a 12% chance. Perhaps not enough of a possibility to make a blanket acceptance of the new supplier.

Example 6.4: A new vertical elutriator was calibrated upon receipt before use in cotton dust sampling. The flow rates, in liters per minute, are shown below. Our concern is whether this elutriator complies with the standard flow rate range of 7.2 to 7.6 L/min as specified by 29 CFR 1910.1043.

7.66	7.43	7.32	7.34
7.79	7.55	7.40	7.67

The supposition is that the new device is compliant.

This example will explore three hypotheses about the data, statistical tests. First, we'll determine whether the population mean of these calibration data (sample) might be 7.4 L/min, the midpoint of the allowed range.

1. We assume that the population is approximately normal with unknown variance.
2. The hypothesis is H_0: $\mu = \mu_0 = 7.4$ L/min and H_A: $\mu \neq 7.4$ L/min.
3. As σ^2 is unknown, T will be the test statistic.
4. Choose $\alpha = 0.05$ (the 95% confidence). There are two rejection regions for the extreme large and extreme small values of the possible mean. So, the *CDF* values of $\alpha/2 = 0.025$ and $1 - \alpha/2 = 0.975$ will define the limits.

5. T is distributed with $n-1=\nu=7$ degrees of freedom and the values of t from the Excel function $T.INV(CDF, v)$, are ±2.3646... for the two-tailed test involved.
6. For the sample data, $\bar{X}=7.52$, $S_x = 0.173534...$, and $s_{\bar{X}} = 0.0613537...$ (all in L/min); and T is calculated as

$$T=\frac{\bar{X}-\mu}{S_{\bar{X}}}=1.9558...$$

As $-2.3646 < T = 1.9558 < 2.3646$, we accept H_0 as a result of the t-test and conclude with 95% confidence that μ could be 7.4 L/min.

However, the specification was not about the mean of the population being at the mid-point of the range, it was about whether the mean might exceed the range. What we really wanted to know was whether the allowed flow rate limits are likely to be exceeded. We should be confident that μ is not less than 7.2 L/min and that μ is not greater than 7.6 L/min. That range is ±0.2 L/min. For the second test, also statistically legitimate but incompatible with the situation, we'll see if the range for the population mean is less than ±0.2 L/min.

The possible range on μ is found from $\bar{X} \pm t_{v,1-\alpha/2} S_{\bar{X}}$

$$= \pm 2.3646...(0.0613537..) \cong \pm 0.145 \text{ L/min}$$

The estimated 95% tolerance on the population mean is smaller than the number allowed (±0.2 L/min).

However, that second claim also does not address the issue. With a possible 95% interval on the population mean being 7.52 ± 0.145 L/min, the upper 95% value is 7.665 L/min, which exceeds the 7.6 L/min limit. This third analysis leads to the conclusion:

The elutriator is unacceptable because there is a significant probability that the flow rate is either <7.375 or >7.665 L/min, and it is that the upper range exceeds the CFR limits.

In this example, we show the potential for the inappropriate use of statistics. We have demonstrated that it is possible to choose a statistical hypothesis that does not correspond to the engineering/business question that must be addressed. You must always be careful to select the proper hypothesis and to use the correct test for the evaluation of that hypothesis.

Our primary concern in this situation was whether the elutriator complied with the 7.2 to 7.6 L/min range allowed by OSHA. Since the mean might be greater than 7.665 L/min, the answer must be "no". The performance specification might not be met. Although, the t-test of the mean told us that the elutriator flow rate was "acceptably close" to the tolerance midpoint, in this case that does not matter. The second analysis indicated that the 95% confidence interval on the mean is less than the specified range. Although that is true, it also does not address the question, "Does the probable flow rate of the elutriator match the allowed range?"

The *CDF*-value associated with the mean violating the lower limit is 0.0004..., which indicates there is hardly any chance of that. However, the *CDF*-value associated with the possibility of the mean violating the upper limit is 0.18..., which is too large a possibility to accept the hypothesis.

Appropriate action must be determined. Here are some options: 1) Do more tests. More tests might shift the average so that the limits are not exceeded, so take more samples. 2) Redesign the device so that the average flow rate is a bit smaller.

However here are some other actions: 3) Choose to use the 80% confidence. This will permit acceptance. 4) Claim that the 7.79 sample value is an outlier that can be rejected. This leaves 7 samples and shifts the average lower. Now the 7-sample data meets spec. (But those last two actions are just being gamey and tend to violate engineering ethics. Don't shape the test so that the outcome seems to support a desired supposition.)

6.5 Tests on the Difference of Two Means

Three cases must be considered when comparing two means. They are the cases for which variances 1) known, 2) unknown but presumed equal, and 3) unknown with no reason to presume them equal. The most common confidence intervals and corresponding tests are presented below.

6.5.1 Case 1 (σ_1^2 and σ_2^2 Known)

For random samples X_{1i} and X_{2i} from different populations with known variances σ_1^2 and σ_2^2, we formulate H_0: $\mu_1 = \mu_2 + k$ (or $\mu_1 - \mu_2 = k$) and H_A: $\mu_1 - \mu_2 \neq k$. The appropriate test statistic is

$$Z = \frac{\bar{X}_1 - \left(\bar{X}_2 + k\right)}{\sqrt{\sigma_1^2 / n_1 + \sigma_2^2 / n_1}} \tag{6.11}$$

where n_1 and n_2 are the sample sizes. If you are hypothesizing that the two populations that generated the sets of samples that resulted in the two averages, $\bar{X}_1$ and $\bar{X}_2$, have equal means then $k = 0$. If you are hypothesizing that the means are equal, $H: \mu_1 = \mu_2$, or that they differ by the amount k, $H: \mu_1 = \mu_2 + k$, then use a two-sided test. Reject if either the data-based z-value is less than $z_{\alpha/2}$ or greater than $z_{1-\alpha/2}$. If you are hypothesizing that one population has a larger or smaller mean, then use a one-sided test. If $H: \mu_1 > \mu_2 + k$ then reject if the data-based z-value is less than the z_α. If $H: \mu_1 < \mu_2 + k$ then reject if the data-based z-value is greater than the $z_{1-\alpha}$.

Since the z-distribution is symmetric, $z_\alpha = -z_{1-\alpha}$ and $z_{\alpha/2} = -z_{1-\alpha/2}$.

Example 6.5: In a beginner-level competition, female gymnasts perform two routines on each of four apparatus. At the end of a meet the all-around winners are based on the sum of their 8 scores. At the beginner level, an informal competition level, routines are often judged by two judges on a maximum basis of a score of 10 in increments of 0.1. The gymnast's score for a routine is the average of the two judges scores. Judging follows rules, but the judges' opinions of elegance or amplitude, might be a bit different; and more importantly, at the lower levels, there are so many deductions happening so rapidly that judges miss things. If the judges' scores of a routine differ by more than 1 point, they converse to normalize, and re-score. If the top two all-around scores, the sum of the 8 routine scores, are 56.85 and 55.25, the girl with 56.85 points gets the first-place blue ribbon and the girl with 55.25 points gets the second-place red ribbon. Can we claim that the one gymnast was better than the other at that meet, or is that difference within judging uncertainty (and an appropriate claim is the two girls had equivalent performance)?

If the 1-point allowable difference in judge's scores represents the range which has uniformly distributed values, then from Chapter 3, the standard deviation on an individual score is $\frac{1}{\sqrt{12}} = 0.28868$ points, and then the standard deviation on the routine score, the average of the two judges scores, is $\frac{0.28868}{\sqrt{2}} = 0.204124$ points. Since the girls have similar scores and scores from the same judges, we'll assume that this is a case of equal variances, known. Propagating uncertainty of a sum, the 8-score total has a standard deviation of $0.204124\sqrt{8} = 0.57735$ points.

The Z-statistic value from the data is $\dfrac{56.85 - 55.25}{0.57735 / \sqrt{\dfrac{1}{1} + \dfrac{1}{1}}} = 1.959592.$

For the hypothesis that the two scores are equal, at the 95% confidence level the two-sided critical value of z is ±1.95996, as illustrated in the sketch. Since the experimental z-value is within that range, we accept, at the 95% confidence that the two scores come from the same population, that the girls' performances were equivalent. If this is the case, then assignment of the blue and red ribbons might not be based on the girls' performance but because of the vagaries of sampling.

Note: the data z-value is very near to the reject value. Although the statement is "accept the null hypothesis" the data is nearly at the reject value.

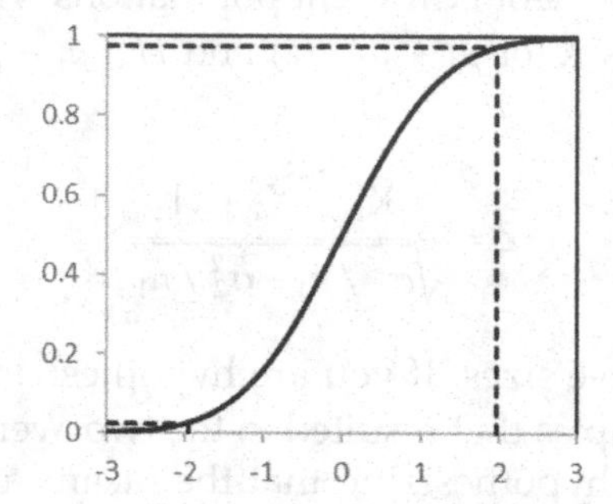

Alternately, one could ask if the 56.85 score represented a population of scores (the truth about one girl's performance) that was better than the other. For the hypothesis that the 56.85 represents a population with a higher mean than the 55.25, at a 95% confidence level, the one-sided z-critical is −1.64485, as illustrated in the sketch. Since the actual Z-score of 1.959592 is not beyond that value, we can accept the hypothesis that the one girl outscored the other.

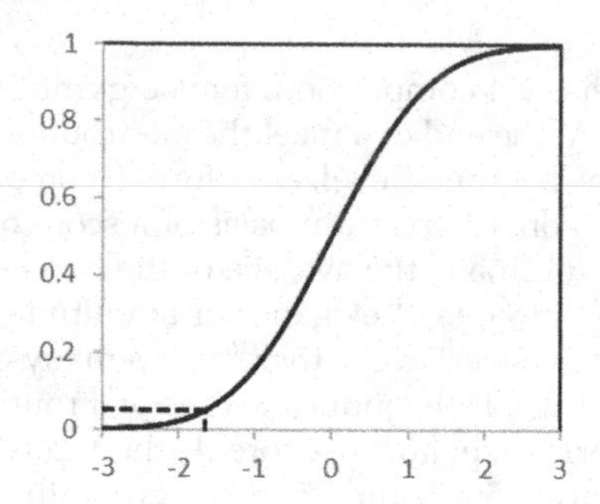

This may seem to be a contradiction. We have accepted both the $\mu_1 = \mu_2$ and $\mu_1 > \mu_2$! But realize that accepting the $\mu_1 = \mu_2$ hypothesis does not mean it is true, it might be, but there is not enough evidence to reject it. Similarly, accepting the $\mu_1 > \mu_2$ hypothesis does not mean it is true, it might be, but there is not enough evidence to reject it.

The degree to which the accept result is made can be represented by the p-values, the probability of the data T being that or more extreme. For the $H: \mu_1 = \mu_2$ the p-value is 0.0500435, is nearly at the reject level, and the probability ratio is 19.98, also a very suspicious value should the null hypothesis be true.

For the $H: \mu_1 > \mu_2$ the p-value is 0.025021, also nearly at the reject level. So, although the statistical decision was to accept the greater than hypotheses, it is nearly rejectable.

Alternately, one could hypothesize that the lower scoring girl was actually the better and appears worse because of the vagaries of judging. $H: \mu_1 < \mu_2$. At the 95% level the *z*-critical is +1.644854. Since the experimental *T*-value of +1.959592 is beyond that value, we can reject the hypothesis. But, again, the action does not mean that the hypothesis is not true. It simply means there is less than a 5% chance of it being true.

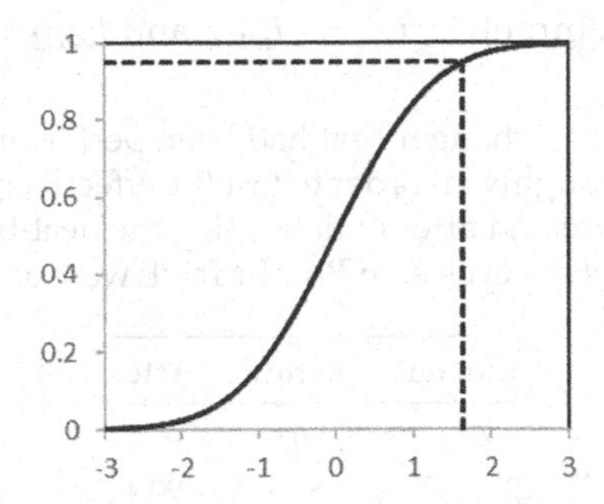

So, we accept both $H: \mu_1 = \mu_2$ and $H: \mu_1 > \mu_2$. What action to take? If we choose action based on $H: \mu_1 = \mu_2$ we might give two first-place awards and elevate the third highest gymnast to second place. I think this best represents the truth of the situation, but it is not practical. Where would we find gymnastic meet organizers who are also competent in statistical analysis? And think of how the youth will perceive adults who are admitting that judges and referees do not own the truth. It would create a chaos of parenting challenges! It would diminish the incentive to practice harder. So, the practical action is to pretend that although a score may not match what the gymnast actually did, it is what happened at the meet when judges are included in the outcome. And then, a 55.31 even beats a 55.30!

Again, statistical analysis is not the decision of what to implement. Use it to guide and support action but ground the action in all aspects of the situation.

6.5.2 Case 2 (σ_1^2 and σ_2^2 Both Unknown but Presumed Equal)

In this situation, the sample variances are pooled. The result is a pooled estimate of the sample variance S_p^2, where

$$S_p^2 = \frac{\sum_{i=1}^{n_1} \left(X_{1i} - \bar{X}_1\right)^2 + \sum_{i=1}^{n_2} \left(X_{2i} - \bar{X}_2\right)^2}{n_1 + n_2 - 2} = \frac{(n_1 - 1)S_1^2 + (n_2 - 1)S_2^2}{n_1 + n_2 - 2} \tag{6.12}$$

The corresponding test statistic is

$$T = \frac{\bar{X}_1 - \left(\bar{X}_2 + k\right)}{S_P\sqrt{1/n_1 + 1/n_2}} \tag{6.13}$$

The statistic is described by the *t* distribution with $v = n_1 + n_2 - 2$ degrees of freedom.

Again, with the equality condition $\mu_1 - \mu_2 = k$, if you are hypothesizing that the two populations that generated the sets of samples that resulted in the two averages, $\bar{X}_1$ and $\bar{X}_2$, have equal means then $k = 0$. If you are hypothesizing that the means are equal, $H: \mu_1 = \mu_2$, or that they differ by the amount *k*, $H: \mu_1 = \mu_2 + k$, then use a two-sided test. Reject if either the data-based *t*-value is less than $t_{v,\alpha/2}$ or greater than $t_{v,1-\alpha/2}$. If you are hypothesizing that one population has a larger or smaller mean, then use a one-sided test. If $H: \mu_1 > \mu_2 + k$ then reject if the data-based t-value is less than the $t_{\alpha,v}$. If $H: \mu_1 < \mu_2 + k$ then reject if the data-based t-value is greater than the $t_{v,1-\alpha}$.

Since the t-distribution is symmetric, $t_{v,\alpha} = -t_{v,1-\alpha}$ and $t_{v,\alpha/2} = -t_{v,1-\alpha/2}$.

Example 6.6: A certain heat exchanger that had been performing poorly was taken out of service and cleaned thoroughly. In order to test the effectiveness of the cleaning, measurements were made before and after to determine the heat-transfer coefficient. These results from ten use-then-clean cycles, in Btu/hr ft² °F were as follows:

Run no.	Before	After
1	90.5	93.4
2	87.6	90.4
3	91.3	99.6
4	93.2	93.7
5	85.7	89.6
6	89.3	88.1
7	92.4	96.7
8	95.3	94.2
9	90.1	98.6
10	83.2	91.1

Did the cleaning of the heat exchanger significantly improve the heat-transfer coefficient?

Note: It was not necessary to have equal numbers of observations for the "before" and "after" data.

1. Assume normal populations (before cleaning = 1, after cleaning = 2).
2. The hypothesis is that cleaning increased the heat transfer performance: H: $\mu_1 - \mu_2 < 0$ vs $H_A = \mu_1 - \mu_2 \geq 0$. For this test, $k = 0$.
3. The test statistic is T as given by Equation (6.13) with $v = 10 + 10 - 2 = 18$.
4. Choose $\alpha = 0.05$, the conventional value.
5. The critical value of T is $t_{18,0.95} = +1.73406$. We will reject the hypothesis if the data T is greater than this value.
6. The sample averages are 89.86 and 93.54 and the standard deviations are 3.6028 and 3.8546 (all in Btu/hr ft² °F). The similarity of the standard deviations supports the Case 2 condition (variances are presumed equal).
7. Then $S_P = 3.73086$ and $T = -2.20558$.
8. The thumbnail sketch illustrates the *CDF* of t and the critical value. As $T < t_{18,0.95}$, T is not in the rejection region. The hypothesis ($\mu_1 - \mu_2 \leq 0$) is thus accepted as possibly true with 95% confidence.

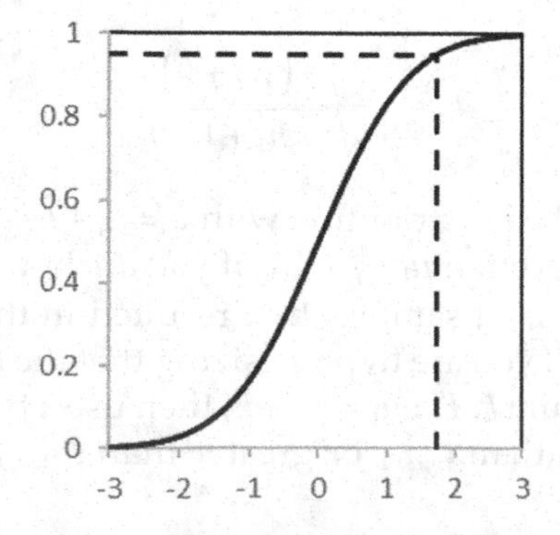

The heat-transfer coefficient may have been improved by cleaning.

Alternately the hypothesis could have been that the cleaning did not affect the heat transfer coefficient, $H: \mu_1 = \mu_2$. In this case, at the 95% confidence limits, the two critical t-values, $t_{v=18,\frac{\alpha}{2}=0.025}$ and $t_{v=18,1-\frac{\alpha}{2}=0.975}$ are ±2.10092. Since the T-value from the data, −2.20558 exceeds one limit, we reject the hypothesis that the before and after performances are equal.

However, the data T-value is almost at the critical value, so your really should acknowledge that the no-cleaning-effect hypothesis was barely rejected, nearly accepted. Looking at a p-value of getting such an extreme value as ±2.20558 is 0.040653, very nearly the level of significance associated with the 95% confidence. If a 96.1% confidence had been chosen, just a bit more desired surety, the null hypothesis would have been accepted.

Although the after-cleaning average heat transfer coefficient is better than before cleaning, the improvement is not statistically overwhelming. Moreover, what might be more important is the cost impact that cleaning has on production, not whether a test outcome is statistically significant or not.

Example 6.7: Supplier B is being compared to the currently used Supplier A. Since all product quality aspects are identical, the comparison metric will be production utility cost in \$k/month. There are costs associated with the proposed change from A to B which includes product testing, operator training, records revision, and risk associated with the unexpected from something new. The company wants a two-year payback on the costs of any investment in manufacturing, which, in this A-to-B change translates to an \$8k/month necessary reduction in production cost. Employee X has been bragging about supplier B, and claims the switch is worth the costs. Employee Y says there might be some benefit of B, but the cost to change over is not worth the utility cost savings.

So that you know, the true mean for A is \$65k/month and for B is \$55k/month, and the sigma for each is \$3k/month. So, the savings from a switch from A to B (\$10k/month) exceeds the \$8k/month threshold. But the true mean is unknowable, so the company performs experiments to compare B to A.

Here is the data:

Supplier A \$k/month	**Supplier B \$k/month**
64.39199	55.32395
64.04099	59.40266
65.04488	54.36507
64.30435	58.65597
68.34815	57.34935
66.03006	
69.88677	
66.42114	
67.22042	
66.27281	

The sample average, standard deviation, and count for the two treatments are

	Supplier A	**Supplier B**
Average, \$k/month	66.19616	57.0194
Standard deviation, \$k/month	1.891179	2.144015
n	10	5

The hypothesis is to switch if the benefit is greater than k = $8k/month. Employee X, who favors the switch, tests the hypothesis $H: \mu_B < \mu_A - k$ (the utility cost of using B is lower), but Employee Y, who opposes the switch tests the hypothesis $H: \mu_B > \mu_A - k$. For either set of tests, standard deviations appear to be equivalent, indicating a Case 2 t-test. The t-statistic for either hypothesis uses the same numerator $\mu_B - (\mu_A - k)$, the same degrees of freedom $\nu = 10 + 5 - 2 = 13$, and the same pooled standard deviation of 1.97243 $k/month. The t-statistic is 1.089242.

Employee X uses the 99% confidence (to be very sure of the conclusion), which would reject $H: \mu_B < \mu_A - k$ if the t-value were greater than the critical value of 2.65031. Since the experimental t of 1.089242 is in the acceptance region, Employee X, triumphantly accepts the hypothesis and claims with 99% confidence "We should switch to Supplier B. Told ya!"

Employee Y, not to be outdone, uses the 99.9% confidence, which would reject $H: \mu_B > \mu_A - k$ if the t-value were less than the critical value of −3.85198. Since the experimental t of 1.089242 is in the acceptance region, Employee Y, triumphantly accepts the hypothesis and claims with 99.9% confidence, and emphases his triumph with a drop the mic gesture, "Keep using Supplier A."

Of course, the statistical term for "not reject" is "accept", but "accept" does not mean the hypothesis is true. "Accept" just means that there is not enough information to confidently reject it. And, the greater the confidence in the statistical test does not mean greater surety of the truth of the hypothesis it means greater surety of rejecting it. Employees X and Y both are misusing statistics.

Since this is a standard economic decision, Employees X and Y should be using the 95% interval which has the one-sided t-critical of ± 1.770933. Since the experimental t of 1.089242 is within the acceptance region there is not enough evidence to confidently reject either hypothesis. Either hypothesis may be true. However, the p-value for $H: \mu_B < \mu_A - k$ is 0.852, and the p-value for $H: \mu_B > \mu_A - k$ is 0.148, suggesting that rejecting $H: \mu_B > \mu_A - k$ may be more probable than rejecting $H: \mu_B < \mu_A - k$. Although not statistically definitive, Supplier B seems be the better choice. If one wants to be more certain, do more trials.

After more trials, a total of 15 trials each, sample average, standard deviation, and count for the two treatments are:

	Supplier A	Supplier B
Average, $k/month	65.35417	55.44148
Standard deviation, $k/month	2.609777	2.364961
n	15	15

The Case 2 t-value is 2.103342, which exceeds the 95% critical value $t_{0.95,28} = +1.701131$ of X's hypothesis, $H: \mu_B < \mu_A - k$, and does not exceed the 95% critical value $t_{0.05,28} = -1.701131$ of Y's hypothesis, $H: \mu_B > \mu_A - k$. So, the hypothesis that Supplier B does not have enough economic advantage is confidently rejected. The p-value for rejection was 0.0223, there was only a 2.23% chance that it was an erroneous decision.

Accept the switch to B.

As you might suppose, the Case 2 t-test is the most general case. It is commonly used in the evaluation of the effects of maintenance, production improvements, operating changes, and raw material/parts source changes.

6.5.3 Case 3 (σ_1^2 and σ_2^2 Both Unknown and Presumed Unequal)

This case is often used when major changes in manufacturing methods have been made. A new type of raw material, replacement of a major piece of processing equipment with a unit from a different source, and other factors all lead to the situation in which you have no reason to expect the unknown variances to be equal. This situation requires the use of a modified statistic, such as Satterthwaite's statistic T_f, which is approximately distributed as Student's t. The test statistic is calculated like Z in Case 1 as if $S_1^2 = \sigma_1^2$ and $S_2^2 = \sigma_2^2$. Use

$$T_f = \frac{\overline{X}_1 - \overline{X}_2 - (\mu_1 - \mu_2)}{\sqrt{S_1^2 / n_1 + S_2^2 / n_2}} \tag{6.14}$$

to test hypotheses involving $\mu_1 - \mu_2$. As usual, n_1 and n_2 and S_1^2 and S_2^2 are the sample sizes and variances of the two samples involved. In Satterthwaite's method, the degrees of freedom cannot be calculated exactly but are approximated by

$$f = \frac{\left(S_1^2 / n_1 + S_2^2 / n_2\right)^2}{\left(S_1^2 / n_1\right)^2 / (n_1 - 1) + \left(S_2^2 / n_2\right)^2 / (n_2 - 1)} \tag{6.15}$$

It seems that this is also termed Welch's method or the Welch–Satterthwaite method.

Example 6.8: In the manufacture of a synthetic fiber, the polymer material, still in the form of continuous monofilaments, is subjected to high temperatures under tension to improve its shrinkage properties. The shrinkage test results for fibers from the same source, treated at two different temperatures, are given below. Is the shrinkage after treatment at 140°C less than that after 120°C?

Percent Shrinkage	
140°C	**120°C**
3.45	3.72
3.64	4.03
3.57	3.60
3.62	4.01
3.56	3.40
3.44	3.76
3.60	3.54
3.56	3.96
3.49	3.91
3.53	3.67
3.43	

1. Assume normal populations (1 = 140°C, 2 = 120°C) with unequal variances.
2. $H: \mu_1 - \mu_2 < 0$ *vs* $H_A: \mu_1 - \mu_2 \geq 0$.
3. Use $T_f = -3.1553$ and $f = 10.93 \approx 11$ from Equations (6.14) and (6.15), respectively. Note: It is probably more conservative when calculating f to round up to the next higher integer, thus decreasing the acceptance region.
4. Choose $\alpha = 0.01$.
5. The critical value of T_f is approximately $t_{11,0.99} = 2.7181$.
6. From the data, the means are $\overline{X}_1 = 3.5355$ and $\overline{X}_2 = 3.760$. The variances are $S_1^2 = 0.005427$ and $S_2^2 = 0.045689$. From these values, the calculated value of T_f is -3.1553.

7. As $T_f = -3.1553 < t_{11,0.99} = 2.7181$, we will accept (with 99% confidence) the hypothesis as possibly true based on the t-test.

We now believe that the shrinkage after treatment at 140°C is probably less than that subject to 120°C.

Example 6.9: The percent conversion data below were obtained with a spacecraft (zero-gravity) reactor for contaminant control. Two different catalysts (MnO_2 and CuO) were used for the oxidation of organic materials.

MnO_2:	55,	62,	64,	63,	58,	61,	60,	62,	64
CuO:	50,	57,	52,	55,	57,	54,	56,	51,	55

Because MnO_2 is more expensive than CuO, it will be selected only if its efficiency is clearly superior to that of CuO. It has been decided that superiority can be adequately demonstrated if the conversion when using MnO_2 is at least 4% higher than that attainable with CuO. A significance level of 0.01 is required.

Should $Mn0_2$ be specified for the catalytic oxidizers?

1. Assume that the populations (1 = MnO_2, 2 = CuO) are approximately normally distributed.
2. $H_0: \mu_1 - \mu_2 \geq 4$ vs. $H_A: \mu_1 - \mu_2 < 4$.
3. The test statistic will be t with $9 + 9 - 2 = 16$ degrees of freedom.
4. $\alpha = 0.01$.
5. The critical region for t must be determined after the decision regarding Case 2 (equal variances) or Case 3 (unequal variances) is reached.
6. $F_{calc} = S_1^2 / S_2^2 = 8.75 / 6.6111 = 1.3235$ is within the 99% CI described by $F_{8,8m}$; so, Case 2 (equal variances) will be used for the t-test.
7. The critical region is $t_{16,0.99} = -2.5835$.
8. From Equation (6.13), $T = \dfrac{(61 - 54.111) - 4}{1.306441779} = 2.2113$.

As $T \nless -t_{16,0.99}$, the calculated T value is in the acceptance region for the one-tailed t-test. MnO_2 should be recommended as the catalyst with 99% confidence.

The conversion with MnO_2 is probably at least 4% higher than that when CuO is used.

6.5.4 An Interpretation of the Comparison of Means – A One-Sided Test

Consider two normally distributed variables as illustrated in Figure 6.10. The distribution A on the left represents a population with a mean of 2 and a sigma of 1. The other, distribution B, has a mean of 3 and a sigma of 2. It is possible for both to generate values of about 1 (within the region included between the vertical lines). Distribution B, with the higher mean, generally generates values higher than A, the one with the lower mean; but it also has a higher probability of generating values lower than the population with the lower mean. If you sample one time from each, you could get a $x = 3$ from A and a $x = 1$ from B, and mistake which is greater on average. The question is: "If you sample a value from each, what is the probability that one is greater than the other?"

The solid vertical lines indicate an interval of Δx about the center of $x = 1$. In general, call the bin center c, $x = 1 = c$. The dashed vertical line indicates the midpoint of the interval. The probability of a sample from population A falling in the $x = c \pm \Delta x / 2$ interval,

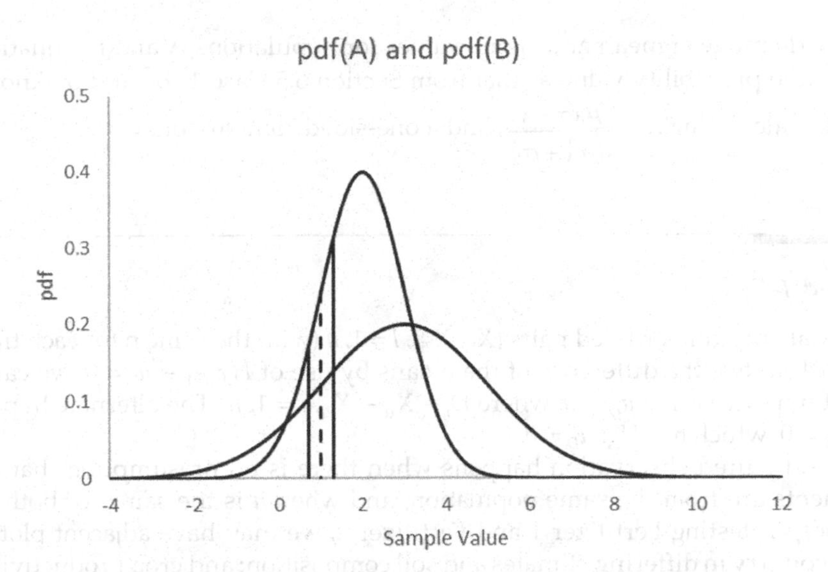

FIGURE 6.10
Interpretation of the one-sided z comparison.

$p_A\left(c-\frac{\Delta x}{2}<x\leq c-\frac{\Delta x}{2}\right)$ is the area under the *pdf* curve. This can be estimated in any number of ways. One simple approach is to use the rectangle rule with the midpoint *pdf* value $A=\Delta x\ \text{pdf}_A(c)$, a more accurate approach would be to use the difference in *CDF* values $A=CDF_A\left(c+\frac{\Delta x}{2}\right)-CDF_A\left(c-\frac{\Delta x}{2}\right)$. The area under the $pdf_B(x)$ curve to the right of that x-value is the probability that population B could have generated a greater value, $p_B(c>1)$. This can easily be evaluated as $1-CDF_B(c)$.

The probability of 1) sampling from distribution A in the Δx interval AND 2) getting a greater sample value than the interval midpoint from B is the product of the two probabilities: $p_A\left(c-\frac{\Delta x}{2}<x\leq c-\frac{\Delta x}{2}\right)p_B(x>c)=\Delta x\ pdf_A(c)\left[1-CDF_B(c)\right]$.

The probability of population B giving a higher value than A for any possible c-value is the probability for $x=c=1$ OR $x=2$ OR $c=3$ OR…. It is the sum of all probabilities for all possible x-values.

$$P(x_B>x_A)=\sum_{x=-4}^{12,\text{ in }\Delta x\text{ increments}}\Delta x\ pdf_A(x)\left[1-CDF_B(x)\right] \tag{6.16a}$$

$$P(x_B>x_A)=\int_{-\infty}^{+\infty}pdf_A(x)\left[1-CDF_B(x)\right]dx \tag{6.16b}$$

In truth, the sum in Equation (6.16a) should go from $x-\infty$ to $x+\infty$, but as Figure 6.10 indicates effectively values are zero beyond the −4 to +12 range. In the limit, as $\Delta x\to 0$, the rectangle rule of integration becomes the integral of Equation (6.16b).

Over a wide range of mean and sigma values for populations A and B, Equation (6.16a) gives the same probability value as that from Section 6.5 Case 1 (σ_1^2 and σ_2^2 known) with one sample, calculating $Z = \frac{\mu_B - \mu_A}{\sqrt{\sigma_A^2 + \sigma_B^2}}$, and a one-sided comparison.

6.6 Paired *t*-Test

When the data occur in ordered pairs (X_{1i}, X_{2i}), $I = 1, n$ (with the same n for each treatment), and we wish to test the difference of the means by use of H_0: $\mu_1 - \mu_2 = 0$, we can use the equivalent hypothesis H_0: $\mu_D = 0$, where $D_i \equiv X_{1i} - X_{2i}$, $i = 1, n$. The alternate hypothesis is H_A: $\mu_1 - \mu_2 \neq 0$, which has H_A: $\mu_D \neq 0$.

This sort-of-paired observation happens when there is no presumption that all of the measurements are from the same population, and when n is the same for both samples. For instance, in testing Fertilizer 1 and Fertilizer 2, we may have adjacent plots of land across the country in differing climates and soil composition; and crop productivity will be affected by the soil and climate as well as the fertilizer. In another example, male Gymnasts A and B perform routines on parallel bars, rings, pommel horse, high bar, etc. Each apparatus has its own attributes requiring different abilities (balance, speed, strength, flexibility, etc.). The apparatus is the same for each gymnast, but the compatibility of the boys' skills is not identical on every apparatus. Again, individual scores do not all come from the same population. As a final example you may have two designs for a product, and ask possible customers to rate each design. Some people will favor sleek over strong, blue over green, or utility over aesthetics; so, each persons' rating of any design will not come from the same population. In these cases, the average of all the scores might remain a best indication of the difference, but the standard deviation of all scores will not represent a single population. Accordingly, we'll use the paired differences as the data. Now the average of the differences will be the same as the difference of the averages, but the standard deviation of the differences will represent the population variation.

Here is a diversion example. When my grandkids come to visit we like to measure their heights on the measuring stick. I hold the stick against the wall, the kid backs up to it, and I use a pencil, level on the top of their head, to mark their height on the stick. Perhaps my measurement technique has an error range of about 1/8 inch ("). Once, their heights were 52", 49", 48", 40", 35", 33", and 27". The standard deviation of the heights is 9.4", and if the heights all came from the same population the 95% range would be about 30". But none of my measurements has an error so large. The grandkids' heights are not from the same population. In calculating a standard deviation, the data must come from the same population.

In paired data, it is not presumed that all values are from the same population, but it is assumed that the differences reflect the treatment effects. We assume that the differences represent a consistent population.

Testing hypothesis about the differences follows the same rules as for testing data against an individual mean. Use a *t*-statistic on the average difference and the sample standard deviation of the differences,

$$T = \frac{\bar{D} - \mu_D}{S_D/\sqrt{n}} \quad \text{with} \quad S_D^2 = \sum_{i=1}^{n} \frac{(D_i - \bar{D})^2}{n-1} \quad \text{and } v = n - 1 \tag{6.17}$$

Example 6.10: Reconsider Example 6.5. In a competition, female gymnasts perform two routines (a compulsory, C, and an optional, O) on each of four apparatus (uneven bars, balance beam, vault, and floor exercise). At the end of a meet the all-around winners are based on the sum of their 8 scores. If the top two all-around scores are 56.95 and 55.25, the girl with 56.95 points gets the first-place blue ribbon and the girl with 55.25 points gets the second-place red ribbon. Can we claim that the one gymnast performed better than the other, or is that difference within judging uncertainty?

Here is the data. Gymnast A has the slightly higher average score. The hypothesis is that $\mu_A > \mu_B$

	Gymnast A	Gymnast B
Bars C	7.45	7.05
Bars O	5.7	5.85
Beam C	7.85	7.15
Beam O	6.95	6.65
Vault C	6.65	6.8
Vault O	7.5	6.9
Floor C	7.95	7.7
Floor O	6.9	7.15

These are paired differences, the beam especially requires balance, bars require strength, and the vault goes very fast. The events are very different. The scores for any one gymnast do not come from the same population. So, use paired differences.

The difference is defined as $d = X_A - X_B$, and the average difference is 0.2125, the standard deviation of the differences is 0.36037, $n = 8$, and $\upsilon = 7$. The data T-value is 1.668133. At a 95% confidence, the t-critical is −1.89458. We reject the hypothesis that A beats B if the data T is outside of t-critical. Since it is not, we accept that A is better than B.

Note: If the question was, "Were the two gymnasts equivalent?", the two-sided t-critical is ±2.36462, and the experimental T of 1.668133 is not outside of either limit. So, we accept the Hypothesis that the population of differences could be zero, the girls could be equivalent, and the apparent difference may be because of judging vagaries.

In this case we would accept either hypothesis, as well as the hypothesis that B could have been the better gymnast. Accordingly it is useful to report p-values. If the two gymnasts are presumed equivalent, the p-value for a T-value of 1.668133 is 0.13922, and the p-value for A being the better gymnast B is 0.06961. The first is marginally acceptable. The second is near to the classic rejection criterion.

Example 6.11: A synthetic lipid was applied at a uniform rate to soil samples to determine the effectiveness in reducing evaporative water losses. All the soil samples were made up from the same batch of well-mixed materials. Twelve soil samples were available. Half were sprayed with a wetting agent prior to application of the lipid; the other 6 samples were not. The results obtained in grams of water lost per square decimeter per minute for a particular set of temperature and humidity conditions were as follows:

Sample	1	2	3	4	5	6	
Lipid	11.5	13.4	14.0	13.6	11.6	14.6	X_{1i}
Wetting agent + lipid	10.8	10.8	12.5	12.1	12.1	13.5	X_{2i}
Difference	0.7	2.6	1.5	1.5	−0.5	1.1	D_i

Did the inclusion of the wetting agent significantly affect the water loss rate?

1. Assume that population of differences is approximately normally distributed.
2. H_0: $\mu_D = 0$ *vs* H_A: $\mu_D \neq 0$.
3. The test statistic is T with $6 - 1 = 5$ degrees of freedom.
4. We select $\alpha = 0.02$ as we are dealing with samples not totally within our control and want a strong confidence should we reject the hypothesis. (Since a significant stakeholder wants them to be equivalent, you need strong evidence to reject the hypothesis.)
5. The critical value of T occurs at $\pm t_{5,0.99} = \pm 3.3649$.
6. The values of $\bar{D}$ and $S_{\bar{D}}$ are 1.15 g/dm² min (which seems to indicate a difference of about 10% of the nominal values) and 0.419325 g/dm² min, respectively. The resulting value of the experimental T is 2.7425.
7. As T is within the 98% Cl for the t-test, we accept H_0: $\mu_D = 0$ and conclude:

The addition of the wetting agent was probably ineffective.

Let's say that you work for the supplier of the wetting agent, and you want to show that it probably is effective. You may think that all you would have to do is change the significance level to $\alpha = 0.05$ so that the critical value of t for the two-tailed null hypothesis is 2.571. Then the calculated value of T would be in the rejection region, and you could conclude that the wetting agent probably does make a difference for the better. To guard against possibly unethical behavior, you should set the significance level first, independent of the test outcome you might desire, not after you see how the choice could influence the outcome.

Alternately, to avoid an unqualified accept/reject conclusion, which could be a distortion of the reality for an audience, include a p-value in a statement. In this case, the p-value associated with the null hypothesis is $2\left(1 - T.\text{INV}(2.7425, 5, 1)\right) = 0.04067\ldots$. There is only a 4% chance that the experimental difference could be so great between the two treatments if the means were the same. The probability ratio is 24.6, again providing high odds that the two treatments are not the same.

Although not absolutely definitive, the data provides strong evidence that the treatments are different.

Example 6.12: Two gaskets were cut from each of eight different production runs of a common gasket sheet stock material. One gasket of each pair was randomly selected for use in dilute HCl service. The other gasket of each pair was for concentrated HCl service. All gaskets were subjected to accelerated life tests for their respective intended uses. From the data so obtained, the estimated values of the average service life in weeks are given below:

Run no.	Dilute HCl	Concentrated HCl
1	35	30
2	40	32
3	27	28
4	25	27
5	36	33
6	48	38
7	53	41
8	48	39

Use the data to test the hypothesis that the service life of the gaskets is independent of the HCl acid strength.

Use a paired *t*-test. From the data $\bar{D} = 5.625$, $S_{\bar{D}} = 1.82247869$ (both in weeks), and $T = 3.02$. As $t_{7,0.995} = 3.4995$, for a two-sided test with $\alpha = 0.01$, we do not reject H_0: $\mu_D = 0$ at the 99% confidence interval and conclude:

Acid concentration may not affect the service life.

However, a different conclusion would be reached using a 95% confidence, at $\alpha = 0.05$, the *t*-critical values are ± 2.3646, which is exceeded by the data *T*-value of +3.02. So, at the 95% confidence level (pretty sure, but not really-really sure) we would claim:

Acid concentration may actually affect the expected life.

The *p*-value for rejecting the null hypothesis is 0.019386, which should be reported to help a reader understand the test conclusion. There is less than a 2% chance that the average difference could be so large if the means were identical. The probability ratio is nearly 52. Very strong odds.

It seems that there is fairly strong evidence to claim that acid concentration might affect gasket life.

Note: If all of the samples do come from the same population, and treatment A and B are being compared, and $n_A = n_B$ then either the paired *t*-test on the data differences, or the normal *t*-test could be used. However, with the paired *t*-test $v = n_A - 1 = n_B - 1$, and for the normal *t*-test $v = n_A + n_B - 2$. The numerator value in the two *t*-statistics is the same, since $\bar{D} = \bar{X}_B - \bar{X}_A$, although, the variance on the difference in the paired test is larger than the variance for either individual average, since $n_A = n_B$ the denominator values of the *t*-statistic will be nearly the same. So, the *t*-value for the paired test will be nearly same as the *t*-value for the unpaired test. However, since the degree of freedom for the unpaired test is larger, it will be more sensitive than the paired test. Only use the paired test if the paired samples come from populations that cannot be presumed to have the same mean.

6.7 Tests on a Single Variance

Any one of the several hypotheses may be used to test a single population variance. These tests may be formulated in terms of the confidence interval on chi-squared, Equation (6.18), or the confidence interval on the unknown variance σ^2 of the population, Equation (6.19). These equations are for the null hypothesis $H_0 : \sigma^2 = \sigma_0^2$ vs. $H_A : \sigma^2 \neq \sigma_0^2$:

$$P\left(\chi^2_{n-1,\alpha/2} \leq \chi^2 \leq \chi^2_{n-1,1-\alpha/2}\right) = 1 - \alpha \tag{6.18}$$

and

$$P\left(\frac{(n-1)S^2}{\chi^2_{n-1,1-\alpha/2}} \leq \sigma^2 \leq \frac{(n-1)S^2}{\chi^2_{n-1,-\alpha/2}}\right) = 1 - \alpha \tag{6.19}$$

The only restriction is that the sample was drawn from a normal population. In that case, χ^2 calculated by $(n-1)S^2/\sigma_0^2$ has a chi-squared distribution with $\upsilon = (n-1)$ degrees of freedom.

$$\chi^2 = \frac{(n-1)S^2}{\sigma_0^2} \tag{6.20}$$

Note: Although the population variance will likely have dimensional units, χ^2 is dimensionless.

Note: χ^2 is a ratio of variances, with one known. Contrasting the F-statistic a ratio of variances with neither known.

The appropriate acceptance and rejection regions for testing hypotheses about a single variance are shown in Table 6.1.

TABLE 6.1

Testing Hypotheses Involving a Single Variance

$$\textbf{Statistic: } \chi^2 = \frac{(n-1)S^2}{\sigma_0^2}, \upsilon = (n-1)$$

Null hypothesis	Alternative hypothesis	Rejection region
$H_0: \sigma^2 = \sigma_0^2$	$H_A: \sigma^2 \neq \sigma_0^2$	$\chi^2 > \chi^2_{n-1,1-\alpha/2}$ or $\chi^2 < \chi^2_{n-1,\alpha/2}$
$H_0: \sigma^2 \leq \sigma_0^2$	$H_A: \sigma^2 > \sigma_0^2$	$\chi^2 > \chi^2_{n-1,1-\alpha}$
$H_0: \sigma^2 \geq \sigma_0^2$	$H_A: \sigma^2 < \sigma_0^2$	$\chi^2 < \chi^2_{n-1,\alpha}$
$H_0: \sigma^2 \neq \sigma_0^2$	$H_A: \sigma^2 = \sigma^2$	$\chi^2_{n-1,\alpha/2} \leq \chi^2 \leq \chi^2_{n-1,1-\alpha/2}$
$H_0: \sigma^2 < \sigma_0^2$	$H_A: \sigma^2 \geq \sigma_0^2$	$\chi^2 \geq \chi^2_{n-1,1-\alpha}$
$H_0: \sigma^2 > \sigma_0^2$	$H_A: \sigma^2 \leq \sigma_0^2$	$\chi^2 \leq \chi^2_{n-1,\alpha}$

Example 6.13: The long-term average variance of the heat transfer coefficient for the drying of plywood is 0.2 (Btu/hr ft^2 °F)^2. Recent quality control data indicate that the plywood is not bonding properly in the kiln. Has the variance of the heat transfer coefficient changed significantly? A larger variance would indicate more drying variability and poorer bonding in the undried sheets. Test values are below in Btu/hr ft^2 °F.

4.68	4.52	4.70
4.73	4.79	4.67
4.65	4.70	4.63
4.69	4.57	4.58

1. Assume population is normally distributed.
2. H_0: $\sigma^2 = 0.2$ vs $\sigma^2 \neq 0.2$.
3. The test statistic is χ^2 with 12 – 1 = 11 degrees of freedom.
4. Choose $\alpha = 0.05$.
5. The critical regions are $\chi^2 > \chi^2_{11,0.975} = 21.9200$ and $\chi^2 < \chi^2_{11,0.025} = 3.8157$. In Excel these are calculated by the cell function CHISQ.INV(CDF,υ).
6. From the data, $S^2 = 0.074767$ and $\chi^2 = 4.112$.
7. The thumbnail sketch is the distribution for χ^2/υ, for which the critical values are 21.9200/11 = 1.9927… and 3.8157/11 = 0.34688….

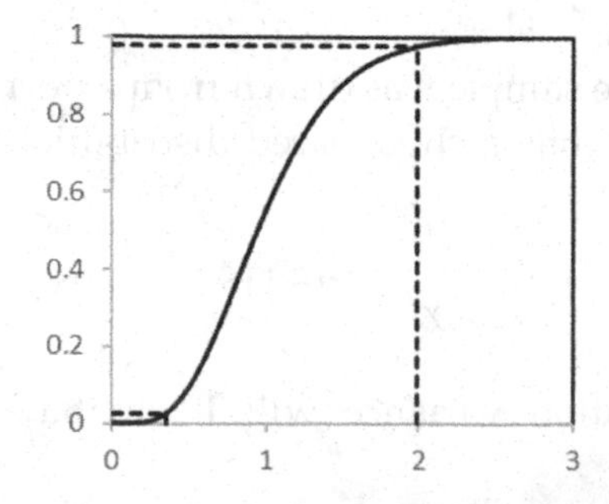

8. As the calculated value of $\frac{\chi^2}{\upsilon} = \frac{4.112}{11} = 0.3738\ldots$ is within the 95% confidence interval, we accept H_0: $\sigma^2 = 0.2$ and conclude as a result of the χ^2 test.

We accept the null hypothesis that heat transfer variance has not changed at the 95% confidence. The heat transfer variance may not be the source of poor plywood bonding.

However, notice that the calculated $\chi^2 = 4.112$ is very close to the lower critical value of $\chi^2_{11,0.025} = 3.8157$, suggesting that heat transfer variance may have actually decreased. The *CDF* value of $\chi^2 = 4.112$ is 0.033475. In Excel this is calculated by the cell function CHISQ.DIST$\left(\chi^2, \upsilon, 1\right)$. The *p*-value associated with the null hypothesis, H_0: $\sigma^2 = 0.2$ is twice 0.033475 = 0.06695 (because it includes the equal probability of both extremes). Since 0.06695 is near to the level of significance for the test, 0.05, one should report more information than just "accept the null hypothesis": The heat transfer variance may have changed, and actually may have reduced, not increased.

By considering a new hypothesis *H*: $\sigma^2 > 0.2$, the 95% one-sided critical value is $\chi^2_{11,0.05} = 4.5748$. Since the data value of 4.1224 is more extreme, we reject the hypothesis that variance has increased with a 95% confidence. The *p*-value for the $\sigma^2 > 0.2$ hypothesis is 0.033475, again a value less than the conventional 0.05 level of significance, and support to reject the hypothesis.

6.8 Tests Concerning Two Variances

The most common two-variance test is that of equality under the null hypothesis $H_0: \sigma_1^2 = \sigma_2^2$ vs. $H_A: \sigma_1^2 \neq \sigma_2^2$. The null hypothesis is usually stated $H_0: \sigma_1^2/\sigma_2^2 = 1$ vs. $H_A: \sigma_1^2/\sigma_2^2 \neq 1$. This particular test is of considerable importance when examining the variability of production lines, the effect of changing production parameters such as a raw material source, the efficiency of different machines, etc.

For two independent normal populations $X_{1,i}$, $i = 1, n_1$, and $X_{2,j}$, $j = 1, n_2$,

$$F = \frac{S_1^2/\sigma_1^2}{S_2^2/\sigma_2^2} = \frac{\chi_1^2/(n_1-1)}{\chi_2^2/(n_2-1)} \tag{6.21}$$

is the appropriate statistic for testing any hypothesis about two population variances. If the equality hypothesis is true, $\sigma_1^2 = \sigma_2^2$ and then $F = S_1^2/S_2^2$ is distributed with $\upsilon_1 = n_1 - 1$ and $\upsilon_2 = n_2 - 1$ degrees of freedom, commonly referred to as the numerator and denominator degrees of freedom, respectively. The expected value for the data-calculated *F*-value is ~1. The corresponding confidence interval for *F* is

$$P\left(F_{n_1-1,n_2-1,\alpha/2} \leq F \leq F_{n_1-1,n_2-1,1-\alpha/2}\right) = 1-\alpha \tag{6.22}$$

Table 6.2 shows the appropriate hypotheses and their corresponding acceptance and critical regions for examining relations between two variances. The rejection regions for the calculated values of *F* follow the same pattern as χ^2 in Table 6.1.

TABLE 6.2

Testing the Equivalent Hypotheses Involving Two Variances

Satatistic: $F = \frac{S_1^2}{S_2^2}$, $v_1 = n_1 - 1$, $v_2 = n_2 - 1$		
Null hypothesis	**Alternative hypothesis**	**Rejection region**
$H_0: \sigma_1^2 = \sigma_2^2$	$H_A: \sigma_1^2 \neq \sigma_2^2$	$F > F_{v_1,v_2,1-\alpha/2}$ or $F < F_{v_1,v_2,\alpha/2}$
$H_0: \sigma_1^2 \leq \sigma_2^2$	$H_A: \sigma_1^2 > \sigma_2^2$	$F > F_{v_1,v_2,1-\alpha}$
$H_0: \sigma_1^2 \geq \sigma_2^2$	$H_A: \sigma_1^2 < \sigma_2^2$	$F < F_{v_1,v_2,\alpha}$

Example 6.14: Pilot plant runs were made on two variations of a process to produce crude naphthalene. Product purities for the several runs are given below. In each series, all conditions were controlled in the normal manner, and there is no evidence from the log sheets of bad runs.

	Product purity (% naphthalene)						
Conditions A	76.0,	77.5,	77.0,	75.5,	75.0		
Conditions B	80.0,	76.0,	80.5,	75.5,	78.5,	79.0,	78.5

The development engineer reports that, on the basis of these data, Conditions B give better purities, but uniformity is poorer at these conditions. Do you agree?

To answer the question about product purity, you should use a one-tailed *t*-test of H_0: $\mu_B > \mu_A$ vs H_A: $\mu_B \leq \mu_A$. We address the claim that uniformity is worse for B by examining the ratio of variances.

1. Assume that both populations are normally distributed.
2. $H: \sigma_B^2/\sigma_A^2 > 1$ vs. $H_A: \sigma_B^2/\sigma_A^2 \leq 1$.
3. The test statistic will be $F = S_B^2 / S_A^2$ with $v_1 = 7 - 1 = 6$ and $v_2 = 5 - 1 = 4$.
4. Set $\alpha = 0.05$.
5. The critical region is $F \leq F_{6,4,0.05} = 0.220572$.
6. From the data, $S_B^2 = 3.571428$, $S_A^2 = 1.075$ and $F = 3.3223$.
7. The sketch illustrates the *CDF* of *F* with DoF 6 and 4 and the *CDF* = α = 0.05 critical region.

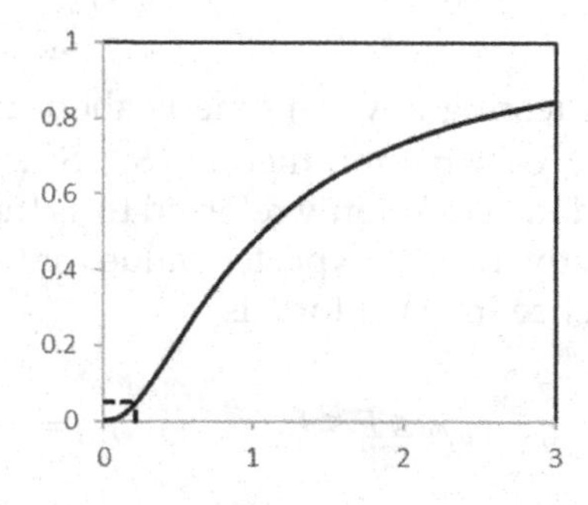

8. As $F = 3.3223 \nleq 0.220572$, we do not reject the *H* at the 95% confidence level. We therefore conclude with 95% confidence:

The variance at Conditions B could be worse than at Conditions A.

The same result is obtained if $F = S_A^2 / S_B^2$ (necessitates the reversal of both inequalities in step 2 above).

Example 6.15: Going back to Example 6.8, let's test the equality of variances. From the data, $F = S_1^2 / S_2^2 = 0.005427 / 0.045689 = 0.1180$. The critical values of F are $F_{9,9,0.005} = 0.15288$ and $F_{9,9,0.995} = 6.54109$. The F calculated from sample data is in the 99% reject region, and does not support $H_0 : \sigma_1^2 = \sigma_2^2$. Therefore, our use of the Case 3 t-test in Example 6.8 was correct.

If the hypothesis is that σ_1^2 is different from σ_2^2 by the ratio k, then the presumption is $\sigma_1^2 = k\sigma_2^2$. then $F = \frac{S_1^2 / \sigma_1^2}{S_2^2 / \sigma_2^2} = \frac{S_1^2}{S_2^2}\frac{\sigma_2^2}{\sigma_1^2} = \frac{S_1^2}{S_2^2}\frac{1}{k}$ and if true then $F = \frac{S_1^2}{S_2^2}\frac{1}{k} = \frac{k}{1}\frac{1}{k} = 1$. But this test requires a presumption of k.

6.9 Characterizing Experimental Distributions

Another frequent use of the χ^2 test is in comparing observed counts (frequencies) and the corresponding expected values if the population is distributed according to a particular theoretical distribution (normal, exponential, etc.). See Section 3.4.7. In this case, we use χ^2 as defined by

$$\chi^2 = \sum_{i=1}^{k} \frac{(O_i - E_i)^2}{E_i} \tag{6.23}$$

where the O_i are the observed frequencies (the count in each category), the E_i are the expected frequencies (counts), and k is the number of classes (categories) into which the data have been divided. Under certain conditions, this statistic is approximately distributed as χ^2 as defined by Equation (3.62). The corresponding confidence limit on χ^2 is

$$P\left(\chi^2_{v,\alpha/2} \le \chi^2 \le \chi^2_{v,1-\alpha/2}\right) = 1 - \alpha \tag{6.24}$$

If the value of χ^2 calculated from sample data falls within the interval described by Equation (6.24), we will accept the null hypothesis that the population probably is distributed as we expected. The use of this confidence interval is termed a *goodness-of-fit test*. One caution: Percentages can only be used as counts in this test if the sample size is exactly 100.

To reject the null hypothesis, H_0: The distribution is as expected, the χ^2 value calculated from sample data must be unusually large. (If the experimental distribution perfectly fit the theoretical, then the perfect value of χ^2 would be zero.) Accordingly, this approach uses a one-sided test described by

$$P\left(\chi^2 \le \chi^2_{v,1-\alpha}\right) = 1 - \alpha \tag{6.25}$$

If $\chi^2 > \chi^2_{v,1-\alpha}$, this null hypothesis should be rejected. If $\chi^2 = 0$, the distribution is exactly as expected (but this also might raise the flag of suspicion).

Note that the degrees of freedom are $k - 1 - p$, where k is the number of classes and p is the number of population parameters which have been calculated from the sample data. If you are testing whether a distribution is normal, $\bar{X}$ and S^2 might be estimated from sample data. As a result, $v = k - 1 - 2 = k - 3$.

If some of the expected counts you calculate are "small", the results of the goodness-of-fit test may be incorrect. As a conservative rule of thumb, if all of the $E_i \geq 5$, the test is valid. If $E_i < 5$, that ith class should be combined with a neighboring class. A less conservative, but also accepted restriction is that each E_i must be >1 and no more than 20% of the E_i may be <5. In addition, if $v = 1$, the absolute value of each nonzero $(O_i - E_i)$ difference must be reduced by 0.5 before calculating χ^2.

Example 6.16: An electronic component is under study. A sample of 100 components yields the following empirical distribution of life length (in 100-hour increments), where x denotes a typical life length. The hypothesized distribution is exponential $pdf(x) = \alpha e^{-\alpha x}$, for which the parameter $\alpha = 1/\mu$ will be calculated with the hypothesized $\mu = 4$ (400 hrs).

Class	Observed frequency O_i	Expected frequency E_i	Contribution to χ^2
$0 \leq x \leq 2$	45	39.347	0.81217
$2 < x \leq 4$	16	23.865	2.59201
$4 < x \leq 6$	10	14.475	1.38346
$6 < x \leq 8$	12	8.779	1.18178
$8 < x \leq 10$	7	5.325	0.52688
$10 < x \leq 12$	4		
$x > 12$	6		
$x > 10$	10	8.298	0.34910

Note: There are 100 test samples. Since the expectation for the category $10 < x \leq 12$ only contains <5 items, it will be combined with the $x > 12$ category to make a $x > 10$ class. The $10 < x \leq 12$ and the $x > 12$ classifications are eliminated.

Does the life expectancy of these components follow an exponential distribution with mean $\mu = 4$?

The cumulative exponential distribution is modeled by Equation (3.43) and is $F(X) = CDF(X) = 1 - e^{-\alpha x}$, where $\alpha = 1/\mu$. This α is the population parameter and has nothing to do with a level of significance. For this proposed distribution: Here $\alpha = 0.25$, and the units are per 100 hrs. As with any cumulative distribution function, $P(x_1 < X \leq x_2) = P(x_2) - P(x_1)$. To obtain the expected frequency E_2 for the second class, we need $P(2 < x \leq 4) = (1 - e^{-0.25(4)}) - (1 - e^{-0.25(2)}) = 0.63212 - 0.393469 = 0.23865$ for a single value. For a sample of 100 items, the corresponding value of E_2 is 23.865.

We have tabulated the expected values of the classes under the assumption that the population does follow the anticipated exponential distribution. Also tabulated are the corresponding contributions of each class to χ^2, the test statistic. If we select $\alpha = 0.05$, the critical region is $\chi^2 \geq \chi^2_{v,1-\alpha} = \chi^2_{5,0.95} = 11.0705$. As the calculated χ^2 contributions sum to 6.8454, we accept that the exponential distribution with $\mu = 4$ (100 hrs) may correctly model the life distribution using a χ^2 goodness-of-fit test at the 95% level of confidence.

6.10 Contingency Tests

A third use of the chi-squared test is the so-called *contingency test*. In it, sample members are classified into mutually exclusive classes such as yes/no, success/failure, and on-grade/off-grade. This test is commonly used in quality control and materials testing and for attribute comparisons. Data for a contingency test are organized into rows and columns. The test is used to determine whether the two attributes (represented by the rows and columns) are independent or not. In the contingency test, if $\chi^2 \ge \chi^2_{v,1-\alpha}$, where $v = (r-1)(c-1)$ and r and c are the number of rows and columns, respectively, then the samples are probably not from the same population. For this test, χ^2 is defined as

$$\chi^2 = \sum_i \sum_j \frac{\left(O_{ij} - E_{ij}\right)^2}{E_{ij}} \tag{6.26}$$

Again, the expected values for each ij category must be >1 with no more than 20% of the $E_{ij} < 5$ in order for this statistic to be distributed as χ^2. A more conservative restriction is that each E_{ij} must be >5. In addition, if $v = 1$, the absolute value of each $(O_{ij} - E_{ij})$ pair must be decreased by 0.5 before squaring.

Another use of the chi-squared test is for evaluation of the experimental or sampling procedure itself. It is possible, though improbable, that χ^2 could be very small. If that situation occurs, it suggests that the categories are unusually similar as a result of poor experimental technique, an improper experimental design, contrived data, etc. In this case, the null hypothesis could be H_0: The experiment appears reasonable vs H_A: The data have too little scatter and therefore are suspicious. For these hypotheses, the critical region is $\chi^2 \le \chi^2_{v,\alpha}$, where χ^2 is as defined by Equation (6.26).

Example 6.17: Melt spinning is a method of producing a man-made fiber (also termed a synthetic fiber). In this manufacturing procedure, molten polymer is extruded through tiny holes (on the order of 0.002 in diameter) in a spinneret. Flow instabilities cause breaks in the filaments, leading to the production of bobbins containing insufficient fiber (short bobbins). The shape profile of the spinneret holes can affect the frequency of short bobbins, an important factor in plant productivity. The data below compare two new spinneret designs with the standard design now in use.

	Type A	**Type B**	**Standard**
Full bobbins	300	400	300
Short bobbins	70	40	60
Total	370	440	360

Is there a statistically significant difference in spinneret performance?

1. We are sure that the data are from dichotomous populations: The bobbins can be classified according to yarn weight as full or short.
2. H_0: The samples are from the same population vs H_A: The samples are probably from different populations.
3. The test statistic is χ^2.
4. Select $\alpha = 0.05$.
5. For this data, $v = (c-1)(r-1) = (2-1)(3-1) = 2$. The critical region is $\chi^2 \ge \chi^2_{2,0.95} = 5.9915$.
6. Of the total of 1,170 bobbins, there are a total of 170 short bobbins. Accordingly, if all spinnerets have similar performance, we expect 170/1,170 = 0.145299 or about 15% of bobbins in each category to be short and about 85% to be full. As there are 370 bobbins from type A, 440 from type B, and 360 from the standard design, the expected values, E_{ij} are

	Type A	Type B	Standard
Full bobbins	316.2393	376.0684	307.6923
Short bobbins	53.7607	63.9316	52.3076

as calculated from Equation (6.26), $\chi^2 = 10.282$. The sketch illustrates the *CDF* of $\frac{\chi^2}{\upsilon}$.
With $\upsilon = 2$, the critical value is $\frac{\chi^2}{\upsilon} = \frac{5.9915}{2} = 2.9957$ and the data $\frac{\chi^2}{\upsilon} = \frac{10.282}{2} = 5.141$.

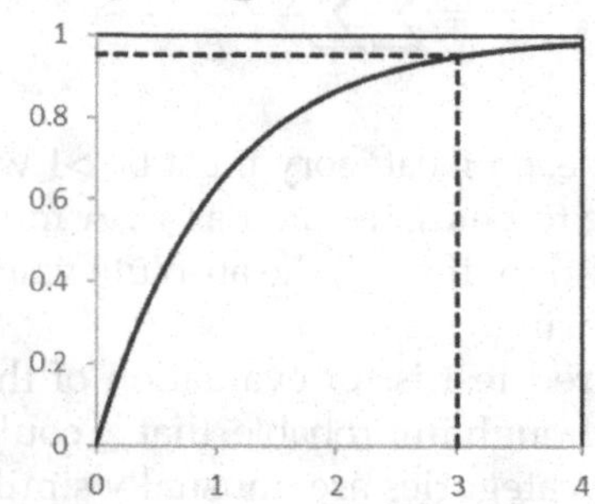

As $\chi^2 > \chi^2_{\text{critical}}$, we must reject H_0 and conclude with 95% confidence that the samples are from different populations, i.e., that:

The spinnerets are probably different.

A quick look at the original data supports this conclusion: The standard and Type A spinnerets produce about 17% short bobbins, but the Type B spinneret produces only 10% short bobbins. But, that quick look does not consider the variability, which the test does. However, the test only indicates that the three treatments are not identical. It does not indicate which is different, nor which is best. (It may be that the cost of Type B spinnerets exceeds the short/full benefit, or that Type B has other not tested deficiencies.)

Example 6.18: Reconsider the comparison of three spinneret designs in Example 6.17 but with different experimental results. The alternate test result is shown below. The question is, "Are the results unexpectedly identical?"

	Type A	Type B	Standard
Full bobbins	300	400	300
Short bobbins	58	81	60

In this new situation, $\chi^2 = 0.0623$ As $\chi^2_{2,0.05} = 0.1026$, we should reject the test results with 95% confidence as the data do not appear to exhibit random scatter:

The test results may be biased.

6.11 Testing Proportions

A proportion, *P*, would be the number of successes, *s*, (or number of classifications) out of *n* trials. From experimental data, $P = s / n$, the binomial distribution usually would model the distribution of the experimental P, given n and p, the probability of an individual trial giving a success. If n is large and P is not near an extreme of 0 or 1 then the z- or t-distribution is valid for hypothesis tests. But, testing proportions should really use the binomial distribution.

If the value of p_0 is known, (and n is large and p_0 is not near the extremes of 0 or 1) then the Z-statistic, is

$$Z = \frac{P - p_0}{\sqrt{p_0 q_0 / n}} = \frac{s - n p_0}{\sqrt{n p_0 q_0}} \tag{6.27}$$

The hypotheses involving single proportions are analogous to those involving a single mean. If you use the definition of Z in Equation (6.27) and replace μ and μ_0 by p and p_0, you will have the table you need for testing hypotheses about single proportions (Table 6.3).

When faced with comparing the equality of two proportions, where P is estimated from n data, use

$$T = \frac{\left[(P_1 - P_2) - (p_2 - p_2)\right]}{\sqrt{\frac{P_1(1-P_1)}{n_1} + \frac{P_2(1-P_2)}{n_2}}} \tag{6.28}$$

The test of this hypothesis is carried out in exactly the same way as the Case 2 t-test. If the inequality of proportions must be tested, use Equation (6.28) after replacing that pooled error with Satterthwaite's estimate.

Alternately, the χ^2contingency test may be used.

Example 6.19: After $n = 25$ coin flip trials, the number of Heads is 10. The experimental proportion is 0.4. The hypothesized proportion is 0.5. Can we reject the fair coin hypothesis?

1. $H_0 : p = 0.5$. This is a test with a two-sided rejection region.
2. $\alpha = 0.05$. Splitting the two rejection regions equally the rejection *CDF* values are $CDF \leq 0.025$ and $CDF \geq 0.975$.
3. The data $P = \frac{10}{25} = 0.4$.
4. With $n = 25$ and the hypothesized $p = 0.5$, the $CDF = 0.025$ and $CDF = 0.975$ limits are $s \leq 8$ and $s \geq 17$. The sketch illustrates the binomial distribution and the critical values.

TABLE 6.3
Testing a Single Proportion p

Statistic: $Z = \frac{P - p_0}{\sqrt{p_0(1-p_0)/n}}$

Null hypothesis	Alternative hypothesis	Rejection region
H_0: $p = p_0$	H_A: $p \neq p_0$	$Z > z_{1-\alpha/2}$ or $Z \leq -z_{1-\alpha/2}$
H_0: $p \leq p_0$	H_A: $p > p_0$	$Z > z_{1-\alpha}$
H_0: $p \geq p_0$	H_A: $p < p_0$	$Z < -z_{1-\alpha}$

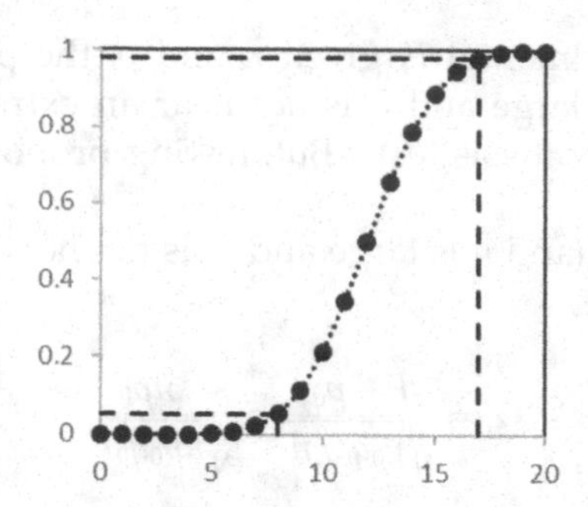

5. Since the experimental $s = 10$ is within the critical value limits we cannot reject the fair coin hypothesis.

Data indicates that the coin might be fair (95% confidence level using the binomial distribution).

The p-value for such an extreme experimental finding, should the hypothesis be true is 0.21, not very close to the rejection value, only raising moderate suspicion.

Repeating the analysis using the standard normal approximation: Z from Equation (6.27) is $Z = \frac{P - p_0}{\sqrt{p_0 q_0 / n}} = \frac{0.4 - 0.5}{\sqrt{0.5(1-0.5)/25}} = -1$. The rejection regions are $Z \le -1.95996$ and $Z \ge +1.95996$. Since the experimental value is not within the rejection region, we accept the null hypothesis. The next sketch illustrates the standard normal distribution with the critical values. Notice the similarity with $Z = -1$ in the sketch below to the binomial sketch above with $s = 10$.

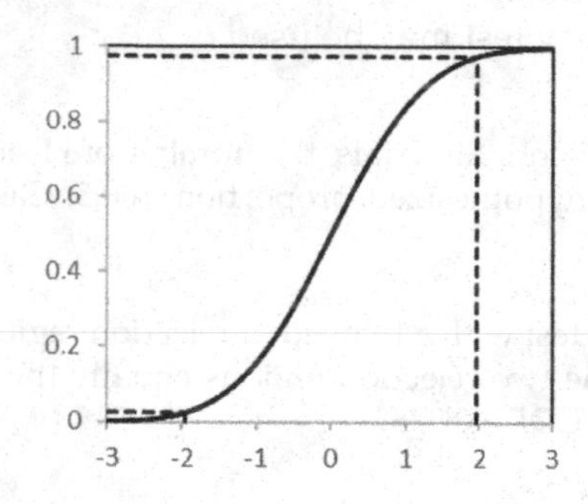

Here the p-value is 0.158, compared to a p-value of 0.21 using the binomial distribution. Although the results are similar with the large n and p far from the extremes, the standard normal approximation is a convenience, not the true approach.

Example 6.20: The manufacturer claims a failure probability of 0.0001 on a particular product, based on quality control testing under normal use. 1,000,000 have been sold and there have been 437 failures reported by customers. Does the manufacturer speak the truth?

1. $H_0 : p = 0.0001$. This is a test with a two-sided rejection region.
2. $\alpha = 0.05$. Splitting the two rejection regions equally the rejection CDF values are $CDF \le 0.025$ and $CDF \ge 0.975$.
3. The data $P = \frac{437}{1000000} = 0.000437$.

4. With $n = 1{,}000{,}000$ and the hypothesized $p = 0.0001$, the $CDF = 0.025$ and $CDF = 0.975$ limits for the binomial model are $x \le 81$ and $x \ge 120$. The sketch illustrates the binomial distribution and the critical values.

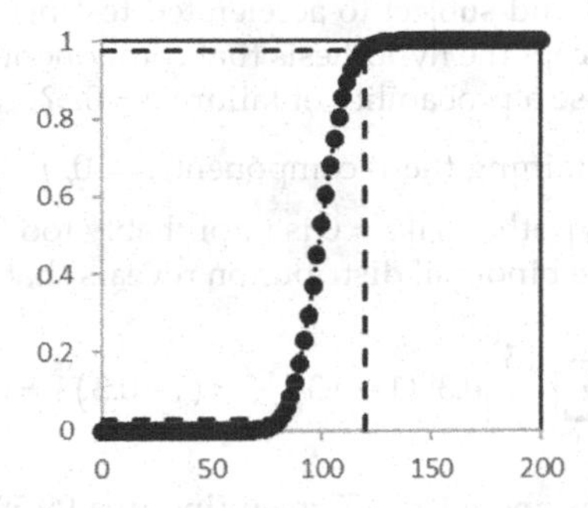

5. Since the experimental $x = 437$ is beyond the critical value limits, we reject the hypothesis that the customer experience is compatible to the manufacturer's claim of a 0.0001 failure probability.

However, this does not mean that the manufacturer did not tell the truth. Soccer balls are designed to be kicked on the grass, but if the dog plays catch with it, and bites a hole in it, the ball fails. If a mechanic leaves the channel-lock wrench out in the rain, the tool will rust, not work, and fail. If a parent lets their kids use the cell phone as a music blaster all day at the side of the pool it will fail. Who knows what abuse a customer might create for a product! If products are not used as the manufacturer designed their use, the customer-reported failure rate, may not be the manufacturer's issue.

Here is a case of a legitimate statistical test, which can be misinterpreted as providing evidence to support a false claim about a supposed cause-and-effect mechanism.

6.12 Testing Probable Outcomes

The probability of an event happening is a proportion, where p is the number of events (successes, s) per number of trials, n, as the number of trials approaches infinity. If you expect a probability to be 50%, and do one trial, you either get a success or a not-success, but with one trial you have no confidence to reject the hypothesis. In $n = 12$ flips of a fair coin, you expect $s = 6$ successes. If $s = 5$, or if $s = 7$ you would not reject the fair coin hypothesis, but if $s = 1$, you might very well claim foul. There is a probability that a fair contest would result in only 1 (or fewer) successes out of $n = 12$ trials when $p = 0.5$, given by the binomial distribution as Equation (3.11).

$$F(x_i \mid n) = P(X \le x_i \mid n) = \sum_{k=0}^{x_i} \binom{n}{k} P^k (1-P)^{n-k} \tag{6.29}$$

With $s = x_i = 1$, $n = 12$, and $p = 0.5$, $F(1 \mid 10) = 0.00317\ldots$. There is only a 0.32% chance that a fair contest would return so few successes.

If you are considering rejecting the fair coin hypothesis if either too few or too many successes, then there is a 0.64% of either getting $s \le 1$, or $s \ge 11$ in $n = 12$ flips. You could claim the hypothesis is not true with a 99.56% confidence.

Example 6.21: There is a claim that a component from Supplier B frequently causes our manufactured product to fail (in 3 out of 10 instances), but components from Supplier A are nearly faultless. Fifteen samples each of product with components from A and from B were randomly sampled and subject to accelerated testing. None of the 30 products failed. Can you reject or accept the hypothesis that components from B make a less reliable product (that they cause a probability of failure $p = 0.3$?

Here, with products containing the B component, $s = 0$, $p = \frac{3}{10} = 0.3$, and $n = 15$. We will reject the one-sided hypothesis if $s = 0$ is improbably too few, given the hypothesis that $p = 0.3$. The cumulative binomial distribution reveals that

$$F(0\,|\,15) = \sum_{k=0}^{0} \binom{15}{k} 0.3^k (1-0.3)^{15-k} = (1-0.3)^{15} = 0.004747\ldots$$

The p-value of the results is about 0.00475, resulting in a 99.5% confidence in rejecting the claim that B leads to product failures.

Example 6.22: A medical test has both false positives and false negatives. If a person has the disease, the test will correctly indicate so in 95% of the cases, but this means there are 5% false negatives. Five percent of the tests on patients with the disease will not detect the disease. Alternately, if a person does not have the disease, the test will correctly identify that in 85% of the cases. This means that it will report 15% false positives. Fifteen percent of the tests on patients without the disease will indicate that the patient does have the disease. The doctor does not know whether the patient has the disease or not, and one test would not be definitive. Two tests could either affirm or contradict each other. So, the doctor prescribes 4 tests. The results are 1 positive indication and 3 negative indications. Can either hypothesis be rejected?

If the patient has the disease, then $s = 1$, $p = 0.95$, and $n = 4$. We will reject the one-sided hypothesis if $s = 1$ is improbably too few. The cumulative binomial distribution reveals that

$$F(1\,|\,4) = \sum_{k=0}^{1} \binom{4}{k} 0.95^k (1-0.95)^{4-k} = 0.004812\ldots$$

The p-value of the results is about 0.0048, resulting in a 99.52% confidence in rejecting the hypothesis that the patient has the disease.

If the patient does not have the disease, then $s = 3$, $p = 0.85$, and $n = 4$. We will reject the one-sided hypothesis if $s = 3$ is improbably too few. The cumulative binomial distribution reveals that

$$F(3\,|\,4) = \sum_{k=0}^{3} \binom{4}{k} 0.85^k (1-0.85)^{4-k} = 0.47899\ldots$$

The p-value of about 0.48, does not provide adequate justification to rejecting the hypothesis that the patient does not have the disease.

6.13 Takeaway

The hypothesis is about the population aspect not about the sample. You have the sample data and can make definitive statements about the sample, but you do not have all the population data; you cannot make definitive statements about the population.

We presume that the sample average, sample standard deviation, or sample proportion are best estimates of the population values. Be sure that this is reasonable.

The precise statement of a statistical conclusion is often unintelligible to a non-statistician, and we usually restate the conclusions to generate the appropriate business engineering action. In doing so, we must take responsibility not to misrepresent the limits of the test.

The hypothesis is about some data-based attribute of the population that would be expected to be manifest if the supposition is true. The hypothesized feature must be legitimate for the supposition.

Statistics is not the decision. Use statistical analysis as one of many criteria to decide action.

The statements will be dependent on the tests used, the level of significance for rejection, the hypothesis, the choice of the number of data, and number of bins in a histogram. All of these are user-choices. It is relatively easy to structure the test to get an outcome that you might want. Be careful to structure the test to be a legitimate indicator of the reality, not a contrived appearance of an affirmation of someone's bias.

We must test all features that are of importance. For example, both mean and variance of seatbelt breaking strength are important. Certainly, a treatment that gives greater strength is a benefit, but if the treatment also causes greater variability, the fraction of belts with breaking strength below some minimum acceptable value may make the effect of that treatment worse. Be sure that all important features are tested before making recommendations. Don't test only one aspect of the treatment.

The important decision is not the "accept" or "reject" of the hypothesis. What is important is the action, and that must be based on the context and the statistical outcome. Be aware of and sensitive to the context.

6.14 Exercises

1. Define these terms:
    a. critical region
    b. hypothesis
    c. primary hypothesis
    d. null hypothesis
    e. alternate hypothesis
    f. level of significance
    g. critical values
    h. Type 1 error
    i. Type 2 error
    j. Reject
    k. Rejection region
    l. Accept
    m. Acceptance region
    n. Two-sided test

o. One-sided test

p. *p*-value

2. Repeat Example 6.14 with two alternate hypotheses: 1) The variances are equal, and 2) A is more variable than B. If you choose one of these, and another person chooses another, how would you settle the conflicting "accept" claims?
3. Choose any example from this chapter and change the value of alpha. Do the outcomes make sense? How would you choose an appropriate value of alpha?
4. Repeat the analysis of Example 6.21 or of Example 6.22 using either the normal *z*- or *t*-statistic. Comment on the new results and validity of the z- or *t*-statistic.
5. Use Example 6.8 to test other possible hypotheses.
6. Show that the discussion at the end of Section 6.6 is true.

7

Nonparametric Hypothesis Tests

7.1 Introduction

Probability distributions are described by specific mathematical functions (Chapter 3) with a few parameters that adjust the location and scale of the curve. For the normal distribution, those parameters are μ and σ. For the Poisson distribution, the single parameter is λ. Parametric tests are hypothesis tests concerning the values of those parameters. Parametric hypothesis tests, such as in Chapter 6, are valid only if the distribution of the sampled data is similar to that of the population.

Parametric tests are to test the values of the distribution coefficients (parameters) such as mean, standard deviation, or probability. However, there are many other data attributes that can be meaningful. These include the number of runs in the data, the median value, the largest deviation, or the count of the value of data that is above or below a hypothesized value. These are not coefficient values in a mathematical model of the distribution.

Further, hypothesis tests commonly involve the assumptions that the random variable is normally distributed and that it has a uniform variance throughout the test region. However, many situations are nonnormal (particle size distributions, queue incidence distributions, failure rate distributions) or have a nonuniform variance throughout the variable range (pH, composition). For such situations, hypothesis tests such as those in Chapter 6 are inappropriate.

Nonparametric tests are used when the population distribution is either not known or cannot be assumed.

Because nonparametric tests are not based on the assumption of population distribution, they are more widely applicable than are parametric tests. However, because knowledge of the population distribution is not utilized, nonparametric tests are less powerful than parametric tests. Nonparametric tests require more data than parametric tests to arrive at the same statistical conclusion with the same confidence level.

The bottom line is as follows: If the population distribution is known, use the appropriate parametric tests because of their greater efficiency. Use nonparametric tests in either of these cases:

1. The population distribution cannot be assumed.
2. The population distribution type is being tested.
3. The within-treatment samples are known to come from different populations (perhaps the variance is not the same).
4. The data are ranked values.

DOI: 10.1201/9781003222330-7

5. The data are nominal or ordinal classifications (chair/table, on/off, preferred/not preferred, above/below, greater than/less than).

7.2 The Sign Test

The sign test is used to compare the median (not mean) values of either paired data or data and their corresponding estimates. We wish to emphasize that the parametric paired tests of Chapter 6 required treatments to be homogeneous and tested for differences of the means. This nonparametric test does not require homogeneous experimental units and compares median (not mean) differences.

Consider a crop experiment in which two seed preservation treatments are being compared to "determine" their effect on crop yield. Seeds from treatments A and B are planted on side-by-side 1-acre plots throughout the geographic region in which the crop is grown. In this way, both A and B will experience a variety of similar soil conditions and vagaries of climate. Consequently, median crop yields from A and B will be directly comparable over a wide range of conditions. There is no justification for assuming that the yield data will be normally distributed. In fact, the soil and climate treatment that seeds A and B experience in each planting are different. The data pairs do not come from the same population of uniform growing conditions. One treatment will be rated as better if it shows a larger median crop yield in a significant number of the adjacent plots.

It is common practice to use paired observations to exclude the effect of uncontrolled variables. In general, there will be a set of paired data, the measured and/or predicted test results of treatments A and B, $\{(A_1\ B_1), (A_2, B_2), \ldots, (A_i, B_i), \ldots, (A_n, B_n)\}$. If treatments A and B are equivalent, then we expect A_i to be greater than B_i in half of the pairings. If the number of pairings in which A_i is greater than B_i is significantly more or less frequent than half, then the A and B treatments are probably not equivalent.

For this crop example, the data might be as shown in Table 7.1. The number of pairings, 10, is low but was chosen to keep the example simple. The third row in the table is the sign of the $A_i - B_i$ data. If the treatments are equivalent, we expect 5 + and 5 – signs; however, there is only one + sign. If the treatments are equivalent, 1 + sign is an unexpectedly low number and suggests that the treatments are different. However, if A and B are equivalent, it is indeed possible to have only 1 + sign out of 10 pairings, just as it is possible to have only 1 Head out of 10 coin flips.

What is the probability of one event occurring out of 10 possible events when the individual event probability is 50%? Using the binomial distribution, the probability for each composite event is listed in Table 7.2 as calculated from Equation (3.9).

TABLE 7.1

Crop Yield (Tons/Acre) for Treatments A and B

Location	1	2	3	4	5	6	7	8	9	10
Treatment A	2.1	1.8	2.5	2.6	1.1	1.7	2.2	1.5	1.2	2.1
Treatment B	2.6	1.7	2.7	3.0	1.5	1.8	3.3	2.0	1.3	2.3
Sign of $A_i - B_i$	–	+	–	–	–	–	–	–	–	–

TABLE 7.2

Binomial Probabilities for Composite Events When Individual Event Probability is 1/2

P(0 | 10)	0.0009765625
P(1 | 10)	0.0097656250
P(2 | 10)	0.0439453125
P(3 | 10)	0.117187500
P(4 | 10)	0.205078125
P(5 | 10)	0.24609375
P(6 | 10)	0.205078125
P(7 | 10)	0.117187500
P(8 | 10)	0.0439453125
P(9 | 10)	0.0097656250
P(10 | 10)	0.0009765625

From Table 7.2, one can read that the probability of 0 events or 10 events out of 10 possible events (all + or all – signs) is $P(0 \mid 10) + P(10 \mid 10) = 0.001953125$. Further, the probability of 1 event or 9 events out of 10 possible events (one + or one – sign) is $P(1 \mid 10) + P(9 \mid 10) = 0.01953125$. Consequently, the cumulative probability of 1 or 0 occurrences of either sign is 0.00195… + 0.01953… = 0.021484375. If treatments *A* and *B* are equivalent, there is only a 0.0214… chance that such a lopsided or even more distorted distribution would have occurred. We can reject the hypothesis that medians (not means) of the treatments *A* and *B* are equivalent at the $(1 - 0.0214\ldots) \cdot 100 = 97.85\ldots\%$ confidence level. Normally, this statement would become, "we are 98% confident that *B* is better than *A*." But be aware of the personal selection of *B* and of the untested evaluation "better", when only one of many attributes are measured.

To clarify this caution, observe that the hypothesis this illustration tested was that the yield of treatments *A* and *B* are equivalent. The statistic generated was the number of + signs, and the hypothesis would be rejected if there were *either* improbably too few *or* too many + signs. Had the hypothesis been that treatment *B* is better than treatment *A*, the test would be quite different and would require more information. We would still choose the number of + signs as the statistic, and we would expect all of the signs to be +. If there are improbably too few + signs, we would reject the hypothesis. To do so, however, we would need the probability that the B_i would be better than the A_i. This situation requires knowledge of the distribution of A_i and B_i, information that we don't have. This nonparametric test illustration hypothesized that *A* and *B* are equivalent, hence the random chance that A_i is better than B_i is 50% regardless of the distribution. You must take care to follow the hypothesis test procedure precisely and to state the conclusion carefully.

As a helpful hint, you do not have to use Equation (3.9) to generate either the point binomial probabilities or the cumulative binomial probabilities. The Excel Function is $CDF = \text{BINOM.DIST}(s,n,p,1)$.

Alternately, Table A.1 provides critical values for the sign test.

As a final observation, the discrimination ability of the test increases with the number of paired observations. For example, to be 98% confident that *A* and *B* are different in a 10-pair sample, one must have 1/10 or fewer pairs with one type of sign. For a 40-pair sample, only 30% or fewer of the data must have the same sign.

To apply the sign test:

1. Obtain paired observations.
2. Hypothesize that the median values of the treatments are equivalent.
3. Choose α.
4. Determine the number of + signs, m, and – signs, n. Disregard tied data. Nominally $m + n = N$ = the total number of data, with tied data $m + n < N$.
5. Set $R = \min\{n,m\}$.
6. Use $F = CDF = 1 - \alpha$ and N to determine r_F from the binomial distribution. In Excel $r_F = \text{BINOM.INV}\left(N, p = 0.5, F\right)$. Alternately use Table A.1 to obtain the critical value of R.
7. If $R \leq r_F$, reject the hypothesis at the α level of significance.

Note that the median, not the mean, is tested. The median is the data value for which half of the observations are larger (better) and half are smaller (worse). The sign test does not require numerical values. One could use any classification system to obtain observed differences (like/don't like, window/door) that could also be classified as +/–.

Example 7.1: Automobile tire rubber compound formulations A and B have been laboratory-tested and appear equivalent. If in-use tests confirm their equivalence, your company will switch to formulation B. As a first in-use test, 6 cars in your company fleet were fitted with formulation A tires on the left side and formulation B tires on the right side. The tires were switched sides every month so that there was no preferential roadside treatment. The tread lifetime (miles) of each of the 24 tires was measured and recorded as follows:

	Car 1	Car 2	Car 3	Car 4	Car 5	Car 6
Front tire A	73,080	69,117	84,307	78,692	75,453	71,439
Front tire B	75,325	71,970	83,076	84,417	73,208	75,003
Rear tire A	77,215	74,744	66,090	82,422	78,161	78,214
Rear tire B	79,039	79,556	68,446	80,913	87,912	76,500

Are the medians of treatments A and B equivalent?

The A and B treatments are paired by car and by location so that there are 12 pairings. The hypothesis is that the median tread lifetime (miles) of formulation A is equivalent to that of formulation B. Choose $\alpha = 0.05$. Using $A_i - B_i$ to determine the paired mileage difference, the number of "+" differences is $n = 4$. The number of "–" differences is $m = 8$, making the total number of nonzero differences $N = 12$ and $R = 4$. Assuming p = 0.5 (the tires are equivalent), At $F = 1 - \alpha = 0.95$ and $N = 12$, $r_F = 3$. We find $R = 4 \nleq r_F = 3$.

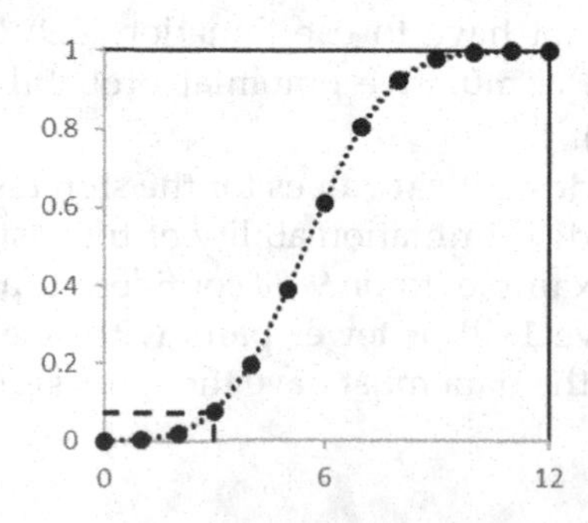

Accordingly, using a sign test based on tread life (miles) of 12 tire pairs, we cannot reject the hypothesis that the median tread life of formulations A and B are equivalent at the 95% confidence level. However, the count is very near to the reject criterion.

One would normally simply state that based on tire wear, the two formulations are not confidently different, but that there is suspicion that B is better.

When observations have relative numerical values, the sign test is primitive and inefficient when compared to a chi-squared contingency or ANOVA test. However, the sign test is applicable to any dichotomous classification and forms the conceptual basis of the signed-rank test that follows.

If formulations A and B gave different performances in Example 7.1, the differences were not great enough to be shown by the sign test. However, note that formulation B was better more often than A: when B was better, it was markedly better; and when B was worse, it was slightly worse. The data suggest that tire B is better, but the sign test does not utilize enough information to make such a statement. Signed-rank tests, however, are more powerful.

7.3 Wilcoxon Signed-Rank Test

The Wilcoxon signed-rank test uses the relative magnitudes of the "+" and "−" deviations as additional information to include in a sign type of test. Consequently, it is a more efficient test (requires fewer points at the same level of significance or is more discriminating with the same number of points) than the sign test. However, while the Wilcoxon signed-rank test is a nonparametric test, it carries the additional restrictions that the treatment differences can be ranked and that the two populations are symmetric: The chance of a sample being z units greater than the median is identical to the chance of a sample being z units less than the median.

To apply the Wilcoxon signed-rank test:

1. Obtain N paired observations.
2. Check that the data are symmetrically distributed.
3. Hypothesize that the treatment medians are equivalent.
4. Choose α.
5. Rank the *absolute values* of the paired observation differences in ascending order. Ranks go from 1 to N and include pairs with zero difference. In case of pairs with identical differences, assign each the average rank value.
6. Sum the ranks of the paired differences that had positive differences, S.
7. Use $F = 1 - \alpha$ and N to determine $s_{\alpha/2,N}$ and $s_{1-\alpha/2,N}$ from Table A.2.
8. If $S \le s_{\alpha/2,N}$ or if $S \ge s_{1-\alpha/2,N}$, reject the hypothesis.

Example 7.2: Use the Wilcoxon signed-rank test to analyze the data in Example 7.1.

The following table lists the tread lifetimes of the 12 pairs of tires, the difference, the ranks based on absolute values, the ranks of positive differences, and the sum of the ranks of positive differences. The hypothesis (A and B have equivalent median tread life) and level of significance ($\alpha = 0.05$) are the same as those in Example 7.1. Visually, the mileage data of each treatment appears symmetric about its respective median.

Pair	Formulation A tread life (miles)	Formulation B tread life (miles)	Difference (miles)	Rank of absolute value	Rank of + difference
1	73,080	75,325	−2,245	5.5	
2	69,117	71,970	−2,853	8	
3	84,307	83,076	+1,231	1	1
4	78,692	84,417	−5,725	11	
5	75,453	73,208	+2,245	5.5	5.5
6	71,439	75,003	−3,564	9	
7	77,215	79,039	−1,824	4	
8	74,744	79,556	−4,812	10	
9	66,090	68,446	−2,356	7	
10	82,422	80,913	+1,509	2	2
11	78,161	87,912	−9,751	12	
12	78,214	76,500	+1,714	3	3
					$\Sigma = 11.5 = S$

Note that the tied absolute values of differences for pairs 1 and 5 share the 5th and 6th rank values of 5.5. From Table A.2, using $F = 1 - \alpha = 0.95$ and $N = 12$, $s_{0.025,12} = 14$ and $s_{0.975,12} = 64$. We find that $S = 11.5 < s_{0.025,12} = 14$. Accordingly, using the Wilcoxon signed-rank test based on the tread life (miles) of 12 tire pairs, we reject the hypothesis that formulations A and B have equivalent median performance at the 95% confidence level.

One would normally simply state: Based on tire wear, the medians of formulations A and B are probably different.

Note that the test does presume that the mileage data of each treatment appear symmetric about its respective median. This is just visually supported. Even though each of the two tread life columns seem symmetric about their respective medians, the difference column is not symmetric.

In this example, the Wilcoxon signed-rank test recognized that there was probably a difference in formulations A and B. By contrast, the sign test ignored the relative magnitude of the paired differences and could not confidently report any formulation effect. The Wilcoxon signed-rank test is preferred but is valid only if the data are symmetrically distributed. You can use the technique of Sections 7.6 or 7.7 to test for distribution symmetry.

7.4 Modification to the Sign and Signed-Rank Tests

Treatment A may be better than treatment B. If the differences can be numerically quantified, you can hypothesize a median difference and adjust the $A_i - B_i$ difference by C (units or %). Then use either the sign or Wilcoxon signed-rank test to test the hypothesis that the adjusted medians are equivalent. You could carry out this procedure for a variety of C values and find the hypothesis rejection region and thereby the confidence interval on C.

Example 7.3: Is the median tread life (miles) of formulation B in Example 7.1 better than that of formulation A by 2,200 miles?

Recalculating the differences in Example 7.1 as $A_i - B_i + 2200$ gives the following data:

Pair	Adjusted difference (miles)	Rank of absolute value	Rank of + differences
1	−45	1	
2	−653	4	
3	+3,431	7	7
4	−3,525	8	
5	+4,445	11	11
6	−1,364	5	
7	−376	3	
8	−2612	6	
9	−156	2	
10	+3,709	9	9
11	−7,551	12	
12	+3,914	10	10
			$\Sigma = 37 = S$

There is no evidence that the tire mileage data are not symmetric; therefore, the Wilcoxon signed-rank test can be used. The hypothesis is that the median value of the 2,200-mile-adjusted data is 0. Choose $\alpha = 0.05$. There are $N = 12$ data pairs and the sum of the positive ranks is $S = 37$. From Table A.2, we obtain $s_{0.025,12} = 14$ and $s_{0.975,12} = 64$. We find $S = 37 \nleq s_{0.025,12} = 14$ and $S = 37 \ngeq s_{0.975,12} = 64$. Accordingly, using the Wilcoxon signed-rank test based on the tread life (miles) of 12 tire pairs, we cannot reject the hypothesis that the median tread life of formulation B is 2,200 miles greater than that for formulation A at the 95% confidence level.

Or more simply: the median life of formulation B could be 2,200 miles greater than that of A.

Note: This does not mean that B is 2,200 miles greater than A. If you hypothesized that the difference was 1,500 miles you would conclude that might be true. Also 2,500 miles might be true. Someone will make a decision on the findings. Be sure to present the results in a manner that does not mislead a reader.

7.5 Runs Test

If there is a dichotomous population with $p = 0.5$, then one expects half of the events to be of one kind and half of the other kind, such as A/B, heads/tails, +/−, ... In addition, if each event is independent of previous events, then each kind of event should be randomly scattered throughout the sequence of trials. For example, sequentially measured process data may be observed to increase, I, or decrease, D, with respect to the previous measurement. If a process is operating steadily, then measured data should reflect random and independent events, and one would expect Is and Ds to occur randomly throughout a sequence with the probability of each $p = 0.5$. If, however, the sequence were IIIIIDDDIIIIDDDDIII DDDIII, you would suspect that the system was oscillating under the influence of some driving force (perhaps an improperly tuned controller) and not at steady conditions. In addition, if the sequence were DDIIIIDIDIIDIIIIIIII, you would suspect that some external event is causing the process variable to increase regularly and that the process is not at steady conditions.

A run is a sequence of identical dichotomous events. The eight runs in the following series are underscored.

HH T H TT HHHHH T H TT

If there are either too few or too many runs in a dichotomous population with $p = 0.5$, you would suspect that the events are not independent of all previous events. Too few runs indicate long periods where some influence persists. Too many runs, if one value is high making the next low, and vice versa, suggest that some "force" is shaping sequential data.

The runs test is useful in statistical process control (see Chapter 21). It is also useful as a test for independence of sequential data (Example 7.4), and for identification of systematic nonlinear differences between data and models (Chapter 19). The runs test will require at least 12 observations for you to be able to detect or "see" improbably too-few runs, and 20 or more to "see" improbably too-many runs.

To apply the runs test to detect improbably too-few number of runs (persistence in the data):

1. Collect N sequential data from a dichotomous population with an expected $p = 0.5$.
2. Hypothesize that the events are not clustered, but that one type of event is randomly distributed throughout the sequence.
3. Choose $F = 1 - \alpha$.
4. Count m and n, the number of each type of events, m is the number of fewer-occurring events. Count U, the number of runs.
5. Use α, m, and n to interpolate $u_{\alpha,m,n}$ from Table A.3a. If $N > 20$ observations, assume $m = n$ and use Table A.3b.
6. If $U \leq u_{\alpha,m,n}$, reject the hypothesis.

You may occasionally encounter a third type of event: A tie or neutral rating. For example, there are also 8 runs in this sequence:

H0 T H TT HH0HH T H T0

A tie extends the prior run. Occasionally, one may test for too many runs, or oscillating behavior; then if $U \geq u_{1-\alpha,m,n}$, reject the null hypothesis. Or if one is testing for either too few or too many runs, use the two-sided test and reject if either $U \leq u_{\alpha/2,m,n}$ or $U \geq u_{1-\alpha/2,m,n}$.

Example 7.4: The country of Zard has two political parties, Nationalists and Patriots, which supply most of the elected officials. The following sequence represents the controlling party over the past 15 years.

NNNNPNNPPPNPPPP

Is the controlling party an independent random variable? Or does control by one party extend its control?

The population is dichotomous. Since the numbers of Ns, m, and Ps, n, are 7 and 8, respectively, there is no reason to suspect that event probability is other than 0.5. Consequently, we can apply the runs test to test for independence. There are $6 = U$ runs in the sequence. We will choose a 0.05 level of significance. From Table A.3a, we find that $u_{0.05,7,8}$ is 4, and find that $U = 6 \nleq u_{0.05,7,8} = 4$. Consequently, using the runs test on a 15-year sequence of controlling party, we cannot reject the hypothesis that the party in power has no influence that extends its control at the 95% confidence level.

More simply, one would state, there is insufficient information to conclude that once a Zard political party gets in power, it stays there for a while.

7.6 Chi-Squared Goodness-of-Fit Test

In Section 6.10, we showed how the chi-squared distribution can be used to compare measured frequency (or number of data of a particular classification) with the expected frequency (or number). The same test can be used to compare a measured probability distribution to a hypothesized distribution, perhaps to test whether the measured data are normally distributed.

To apply the chi-squared goodness-of-fit test:

1. Collect N data points.
2. Partition the range of the data into n cells, with no fewer than one observation in any cell and no fewer than 20% of the cells with less than five observations. Preferentially with at least 5 observations in each cell.
3. Count O_i, the number of observations in each ith cell.
4. Hypothesize that deviations of the data histogram from a particular probability density function are small. You could use the data $\bar{X}$ and S, for instance, to parameterize a particular normal probability density function.
5. Choose α.
6. Calculate the degrees of freedom, $v = n - 1 - k$, where k is the number of parameters in the hypothesized distribution that were determined from the data. If the data had been used to parameterize a Gaussian distribution, k would be 2.
7. Compute the chi-squared statistic

$$\chi^2 = \sum_{i=1}^{N} \frac{(E_i - O_i)^2}{E_i}$$

where E_i is the expected number of observations in the ith cell. For the special case where $v = 1$, some sources advise a correction for continuity by reducing the absolute value of each nonzero value of $(E_i - O_i)$ by 0.5.

8. If $\chi^2 \geq \chi^2_{v,1-\alpha}$, reject the hypothesis that the data fits the model.

Our guide for the number of bins, N, is that it should be about the square root of the number of observations, n. Use equal bin intervals to be able to "see" the *pdf*. Alternately, choose bin intervals so that each bin has the same number of expected observations, to have the strongest test results.

The restriction on the number of observations per cell stated in Step 2 is standard accepted practice but is a subjective rule based on experience with the chi-squared test. The chi-squared distribution is approximately equal to the statistic as calculated in Step 7 when E_i is not too small. Some statisticians suggest choosing partitions such that no cells have fewer than five observations in order to ensure the valid use of the χ^2 distribution.

If the observed distribution is different from the hypothesized distribution, then $(E_i - O_i)$ will be large and the calculated χ^2 statistic will be large. If it is improbably too large, it indicates that the two distributions are probably different. However, an improbably too-small χ^2 is also cause for concern. One expects some deviations from the observed and hypothesized distributions. Too small a χ^2 could indicate contrived data or poor experimental technique. However, because most χ^2 tests only look for unexpectedly large differences, a single-tailed test based on $\chi^2_{v,1-\alpha}$ is indicated in Step 8.

Example 7.5: Liquid is continuously flowing at 1 m³/min through a 4 m³ volume in a well-stirred tank. The liquid volume remains constant. One thousand tracer particles are instantaneously dumped into the tank, rapidly become uniformly dispersed, and begin flowing out of the tank with the liquid. As some of the particles leave, the rest become uniformly dispersed, and the probability of a particle being near the exit and caught by the exit fluid decreases. As time progresses, the rate at which particles leave diminishes. Initially, many particles exit rapidly and have a low residence time in the tank. Later, some particles still remain in the tank and have a longer residence time. The particle residence time distribution is an important mixing and chemical reaction design criterion. The distribution has been measured for this tank by screening the tank effluent for successive 1-minute periods and counting the number of particles. The data follow. Does the measured distribution match the expected exponential distribution?

Period:	**1**	**2**	**3**	**4**	**5**	**6**	**7**	**8**
Measured number:	203	185	132	110	85	65	57	40

The expected distribution is exponential with $\tau = F/V = (1\ \text{m}^3/\text{min})/(4\ \text{m}^3) = 0.25\ \text{min}^{-1}$. No distribution parameter was calculated from the measured frequency data, consequently $k = 0$, and $v = 8 - 1 - 0 = 7$. As measured, the data are partitioned into eight cells, none of which have less than five particles. Choose $\alpha = 0.05$. With $\tau = 0.25\ \text{min}^{-1}$, the expected number of particles leaving in each minute of operation are obtained from Equation (3.43).

Period:	**1**	**2**	**3**	**4**	**5**	**6**	**7**	**8**
Expected number:	221	172	135	104	81	64	49	39
$E_i - O_i$	18	−13	3	−6	−4	−1	−8	−1

$$\chi^2 = \sum_{i=1}^{8} \frac{(E_i - O_i)^2}{E_i} = \frac{18^2}{221} + \frac{(-13)^2}{172} + \frac{3^2}{135}$$

$$+ \frac{(-6)^2}{104} + \frac{(-4)^2}{81} + \frac{(-1)^2}{64}$$

$$+ \frac{(-8)^2}{49} + \frac{(-1)^2}{39} = 4.4063613\ldots$$

The expected chi-squared value could be obtained from many sources. In Excel, $\chi^2_{7,0.95} = \text{CHISQ.INV}(0.95,7) = 14.0671$. We find $\chi^2 = 4.4\ldots \ngtr \chi^2_{7,0.95} = 14.0671$. Accordingly, using the χ^2 goodness-of-fit test, on the 0- to 8-minute tank data, we cannot reject the hypothesis that the tank has a residence time distribution equivalent to the expected exponential distribution with $\tau = F/V$, at the 95% confidence level. There is insufficient evidence to say that the mixer design is bad.

Simply, the mixer appears to work properly.

7.7 Kolmogorov-Smirnov Goodness-of-Fit Test

Although the χ^2 test is a parametric test based on normally distributed data, in practice its efficacy is not undermined by other distributions and its use as a goodness-of-fit test has become standard practice. However, the chi-squared test results are sometimes questioned because the experimenter can bias the results with the selection of the sample partition locations. In addition, ten or more cells (of five or more observations each) are preferred for the chi-squared goodness-of-fit test in order to discriminate between distributions. One may not always have the luxury of $10 \times 5 = 50$ observations. The Kolmogorov-Smirnov test avoids experimenter bias and can be effectively applied to five, but preferably ten or more, observations. The Kolmogorov-Smirnov test is applied to the cumulative distribution of the data to determine whether it is different from a specified cumulative distribution function, $F(X)$, where X is a random variable.

To apply the Kolmogorov-Smirnov test:

1. Specify the cumulative distribution function, $F(X)$.
2. Hypothesize that the observed distribution is equivalent to the specified distribution.
3. Choose $F = 1 - \alpha$.
4. Sample n times and list the observed values of X in ascending order: $x_1, x_2, \ldots, x_i, x_{i+1}, \ldots$.
5. Compute the empirical cumulative function $F_n(X) = k/n$ where k is the number of samples with a value less than or equal to x. $F_n(X)$ is a step function that changes value at each x_i and holds that value for $x_i \le X \le x_{i+1}$.
6. Compute D, the maximum absolute value of $F_n(X) - F(X)$ over the entire range of x, not simply at observed x_i values.
7. Obtain $d_{F,n}$ from Table A.4.
8. If $D \ge d_{F,n}$, reject the hypothesis.

Example 7.6: Computer chips are made by repeatedly applying films and then removing unwanted micron-sized areas by a chemical etch. This process leaves "wires" and "insulators" and eventually builds transistor-type components. In the 1980s, about 300 computer chips were simultaneously built on a 4 inch disk of single-crystal silicon.

In an effort to understand the manufacturing process for making computer chips, an engineer has developed a phenomenological model that attempts to predict the probability distribution of the number of bad chips made on a 325-chip silicon wafer during an etch step. If the prediction is equivalent to the measured distribution, then a phenomenological understanding of the process can be claimed, and the engineer will know how to improve the process.

Empirical data are obtained by randomly sampling and inspecting the half-built computer chips for evidence of bad etching. The number of etch-damaged chips found on 30 wafers follows.

Wafer	1	2	3	4	5	6	7	8	9	10
No. of bad chips	6	11	5	7	6	13	10	15	13	14
Wafer	11	12	13	14	15	16	17	18	19	20
No. of bad chips	19	6	15	15	18	13	14	14	13	5
Wafer	21	22	23	24	25	26	27	28	29	30
No. of bad chips	16	14	12	6	13	13	7	20	12	9

The cumulative probability function that was predicted is

X = No. of bad chips	0	2	4	6	8	10
$F(X)$ =	0	0.02	0.05	0.11	0.20	0.30
X = No. of bad chips	12	14	16	18	20	
$F(X)$ =	0.43	0.76	0.90	0.94	0.96	

Apply the Kolmogorov-Smirnov test to determine whether the theoretical and empirical distributions are equivalent.

The random variable X is the number of bad chips. Values of x_i are those measured. In ascending order, the x_i values are 5, 6, 7, 9, 10, 11, 12, 13, 14, 15, 16, 18, 19, and 20, representing x_1 through x_{14}. For x_4 (9 bad chips), there are $k = 8$ wafers with 9 or fewer bad chips. Consequently, $F_n(x_i)$ are listed below:

x_i	5	6	7	9	10	11	12
k	2	6	8	9	10	11	13
$F_n(x_i)$	0.067	0.200	0.267	0.300	0.333	0.367	0.433
x_i	13	14	15	16	18	19	20
k	19	23	26	27	28	29	30
$F_n(x_i)$	0.633	0.767	0.867	0.900	0.93	0.967	1

Although x_i had only 14 values from 5 to 20, X may have 326 values from 0 to 325. The absolute value of the difference in $F_n(X)$ and $F(X)$ must be checked at all possible values of X. They are listed below, where $F(X)$ at odd values of X has been obtained by linear interpolation.

X	$F(X)$	$F_n(X)$	$\lvert F_n(X) - F(X)\rvert$
0	0	0	0
1	0.01	0	0.01
2	0.02	0	0.02
3	0.035	0	0.035
4	0.05	0	0.05
5	0.08	0.067	0.013

(Continued)

X	F(X)	$F_n(X)$	$\lvert F_n(X) - F(X) \rvert$
6	0.11	0.200	0.090
7	0.155	0.267	0.112
8	0.20	0.267	0.067
9	0.25	0.300	0.050
10	0.30	0.333	0.033
11	0.365	0.367	0.002
12	0.43	0.433	0.003
13	0.595	0.633	0.038
14	0.76	0.767	0.007
15	0.83	0.867	0.037
16	0.90	0.900	0.00
17	0.92	0.900	0.02
18	0.94	0.933	0.007
19	0.95	0.967	0.017
20	0.96	1.00	0.040
21-326	>0.96	1.00	<0.040

The largest deviation is $D = 0.112$ for $X = 7$. With a choice of $\alpha = 0.05$, $F = 1 - \alpha = 0.95$, and $d_{0.95,30} = 0.241$ from Table A.4. We find $D = 0.112 \ngtr d_{F,n} = 0.241$. Accordingly, using the Kolmogorov-Smirnov test on 30 wafers, we cannot reject the hypothesis that the model-predicted and measured distributions of badly etched chips are equivalent at the 95% confidence level.

Since nobody will understand that statement, we tell our boss, the model may be correct.

Note: In this example the random variable, the number of bad chips on a wafer, was a discrete variable. The Kolmogorov-Smirnov test is also applicable to continuous variables. For continuous variables, the step function $F_n(X)$ would have to be compared to the continuous function $F(X)$ for all values of X. It is likely that maximum deviations would occur at either just to the left or just to the right of each x_i where the steps occur. Consequently, deviations probably need to be calculated only at x_i and at x_i-minus-a-tiny-bit.

The chi-squared goodness-of-fit test could have been used instead of the Kolmogorov–Smirnov test. However, with five observations in each cell, there would only have been six cells, or five degrees of freedom for χ^2.

7.8 Takeaway

The hypothesis is about the population aspect not about the sample. You have the sample data and can make definitive statements about the sample. But you do not have all the population data; you cannot make definitive statements about the population.

The precise statement of a statistical conclusion is often unintelligible to a non-statistician, and we usually restate the conclusions to generate the appropriate action. In doing so, we must take responsibility not to misrepresent the limits of the test.

The statements will be dependent on the tests used, the level of significance for rejection, the hypothesis, the choice of the number of data, and number of bins in a histogram. All of these are user-choices. It is relatively easy to structure the test to get an outcome that you might want. Be careful to structure the test to be a legitimate indicator of the reality.

Test all features of the supposition that are important.

The important decision is not the "accept" or "reject" of the hypothesis. What is important is the action that must be based on the context and that statistical outcome. Be aware of and sensitive to the context.

7.9 Exercises

1. Use the data of Example 7.1 to test if front tire wear is the same as rear tire wear.
2. Use the data of Example 7.1 to determine if A on the front is the same as B on the front for tire wear.
3. Use a paired *t*-test on Example 7.2. Is the data compatible with the t-test criteria?
4. Flip a coin 20 times, and then use the runs test to see if the coin was fair.
5. Use the data from Example 7.5 and the K-S method to test if the theoretical *CDF* model might be exponential.
6. Repeat Example 7.5 but use a χ^2 test. This might be a bit confusing, because the category and the number of events in each category are both counts. How will you group bins that have fewer than 5 counts? Does an alternate grouping give exactly the same results?
7. Since 1921 the presidents of the US have come from either of the two major political parties, Republican and Democrat. There is a new election every four years, and from 1921 to 2021 here has been the pattern: RRRDDRRRDDDDDRRDDRDRRRD DRRDDR. Since the 2021 president is a Democrat, the last R is a run of 1. But since we don't know what happens in the future (whether the D will have a run of 1, 2, 3, etc.) the last D data cannot be included. Does the pattern support the hypothesis that leadership by one party causes the voters to vote out that party and vote in the other? Does the pattern support the hypothesis that leadership by one party causes that party to persist as preferred? The expected average run length (ARL) is 2, if the party voted into the presidency is random. What is the actual average run length? Is it different from the expectation random ARL?

8

Reporting and Propagating Uncertainty in Calculations

8.1 Introduction

Since measurements are not error-free, the results of calculations are not error-free. In fact, other sources of error in engineering calculations include both the models we use (the ideal gas law is an approximation of the truth), coefficients in equations (convective heat transfer coefficients, viscosity, the speed of light, any finite truncation of π, etc.), and any data value representing the future (the cost of utilities, the tax rate).

Commonly, the term for uncertainty in a calculated value is "error", but this does not mean a mistake, such as the wrong equation or procedure was used, or that numbers were transposed. There is natural variation in data, which leads to "error" or uncertainty on a value calculated from the data. "Uncertainty" might better represent the range of possible data values, but historically the term used is "error".

To make appropriate management decisions, engineers must properly reflect the propagation of errors in calculated results so as not to misrepresent the uncertainty in the calculated value.

Here is a simple example of a model parameter error: If the diameter of a circle is exactly 100 inches, then using $\pi = 3.1415926535897932...$, the circumference is calculated to be 314.15926535897932… inches. But it is likely you might use a convenient approximation to π such as $\pi = 3.14$. Then even with perfect knowledge about the diameter, the calculated value will be in error of 0.159265… inches. And an example of a measurement error: In a more likely scenario, you would measure the diameter, but for a variety of reasons you can only obtain an approximate measure. For instance, your ruler was not calibrated exactly to a primary standard, or perhaps your length measurement was on a chord and not exactly on a diameter. Perhaps you can only discriminate the ruler markings to $\frac{1}{8}$ inch, or the shape is not exactly circular. If the true diameter were 100 inches but you measured 99.6 inches, then you would calculate a 312.902628… inch circumference, which would be in error by 1.2566… inches. Since that error is over 1 inch, it would be both pretentious and misleading to report the circumference to 36 digits because you know π that well, or to 10 digits because your calculator displays them.

Properly, one should acknowledge measurement uncertainty and report the calculated value in a manner that appropriately reflects the precision of the calculated value. For this circumference example, one might choose to report explicitly

$$\text{circumference} = 313. \pm 1. \text{ in}$$

DOI: 10.1201/9781003222330-8

or report implicitly

$$\text{circumference} = 3.13 \times 10^2 \text{ in}$$

8.1.1 Applications

If an experiment is replicated (exactly duplicated as far as humanly possible) the results should be exactly the same. But they are not. There is uncontrolled variation in the processing, testing, source materials, environmental conditions, test machines, operator techniques, sensor discrimination, etc., and these lead to variation in the supposed-to-be exactly-the-same outcomes.

This propagates to uncertainty in the models that are fit to the data. Then, model-calculated uncertainty becomes uncertainty in the forecast values in the model. Further, model "givens" are often estimates of the values of properties or future events, which have uncertainty that affects predictions. The questions about model certainty are related to determining uncertainty in experimental data, model coefficients, and model predicted values.

There are diverse applications for analysis of uncertainty in models. These include determining:

1. Uncertainty of calculated values in laboratory analysis – taking test data and models to calculate analysis or properties, then determining the range of uncertainty in analysis or properties.
2. Uncertainty of calculated values in modeling or prediction – using a model to see what the range of outcomes might be due to uncertainty in the basis or givens. Or in inverse of models when sensor data is used to determine the process variable (for example measured differential pressure in an orifice is used to calculate flow rate, or conductivity is used to calculate composition).
3. Uncertainty of models that were fit to empirical data – whether empirical models or phenomenological models, determining uncertainty in:
   a. regression coefficients
   b. model prediction due to vagaries in experimental data.
4. Validation of models by comparing residuals to expected uncertainty of the data – if residual σ matches data replicate σ (or propagated σ on data) then the model might be good.
5. The number of replicates or number of experimental datasets – to obtain an average measurement or a model with adequate uncertainty.
6. When a time series has achieved steady-state – such as measuring sample moisture content and claiming it is dry when three measurements in a row are "equal" within uncertainty.

8.1.2 Objectives/Rationale

There are diverse objectives or reasons for applying uncertainty analysis:

1. One objective in analysis of uncertainty, in either laboratory or process results, is process quality improvement. Analysis of uncertainty will let you see how uncertainty (often termed random error, but it could also be a systematic error)

is propagated through calculations to quantify the uncertainty of the calculated value. Uncertainty could be quantified as the range of the values or the standard deviation of values (when ideally the replicate values are identical). If the calculated uncertainty matches the actual, then you understand your process. Then, this analysis can point you to process or procedural changes that reduce variability, improve quality.

2. As a second objective, uncertainty analysis provides legitimacy to a claim about uncertainty that you report to others.
3. Alternately, if the expected and actual variation do not match, then this would be a trigger to find out why; and thereby learn more about your process. A third objective is to truly understand your process. Then you can be better at trouble shooting and improving quality.
4. Analysis of uncertainty also relates to developing models. Models generated from empirical data reflect the perturbations in the particular data sample. A replicate set of experiments will be subject to different perturbations. In modeling from empirical data, the question is, "How does experimental variability impact the certainty limits of the model?"
5. Further, in developing phenomenological models (mechanistic approach, from first principles), the question relates to the residuals (the deviations between model and data). If residuals are greater than propagation of uncertainty expects, then the model should be rejected. Additionally, if residuals are either unexpectedly large or small, the process understanding should be questioned.
6. We use models to predict, forecast, design, analyze, etc., but "givens" in the models have uncertainty. These include empirically derived coefficient values, but also the estimates of future costs, compositions, fees, tax rates, production/sales/demand/load/duty and the like. The ability to propagate uncertainty of the "givens" in a problem statement to the calculated value will provide an uncertainty range on designs and engineering calculations that are essential to assign safety factors or assess risk.
7. In optimization we seek the best value: Lowest cost, least risk, most reliable, etc. However, if experimental results guide the evolution of the decision variables (DV) in optimization, the uncertainty on the data will lead to uncertainty on the optimum DV values. Further, if the optimization is using models, uncertainty on the coefficients and other "givens" will create uncertainty on the optimum DV values. Analysis of uncertainty is needed to reveal how the model or experimental uncertainty affects the range of the resulting "optimum" DV values.
8. Propagation of uncertainty is useful in experimental design. For example, in determining the number of data samples or measurements to reduce uncertainty to an acceptable level, or to assess when sequential measurements indicate that a process has come to steady-state or equilibrium.

8.1.3 Propagation of Uncertainty

In all of those diverse applications, the concept is to use a model to propagate uncertainty. For instance, the calculation might be to compute stress, S, from tensile load and cross-sectional area, $S = L/A$. If there is uncertainty, possible error, on both load and area, ε_L and ε_A, then propagation of maximum error on S using the analytical approach is $\varepsilon_S = \left|\frac{\partial S}{\partial L}\varepsilon_L\right| + \left|\frac{\partial S}{\partial A}\varepsilon_A\right|$, which reduces to $\varepsilon_S = \left|\frac{1}{A}\varepsilon_L\right| + \left|\frac{L}{A^2}\varepsilon_A\right|$, and still more simply $\varepsilon_S / S = \varepsilon_L / L + \varepsilon_A / A$. Here "L" and "A" would be the called the independent or input variables, the "givens" in the calculation, and "S", the calculated consequence, termed the dependent or response variable.

Whether analytical or numerical propagation, whether reporting probable or maximum error, and whether uncertainty on the independent variables is evaluated by range or standard deviation, to analyze uncertainty on a calculated value (dependent variable) requires 1) a model, and 2) some measure of uncertainty on the elements (independent variables).

In spite of the diverse applications listed above, there are common techniques for propagation of uncertainty.

8.1.4 Nomenclature

Bias – a consistent deviation from true – a systematic error – accuracy – the difference between the population means of the measurement and the truth.

Precision – a measure of repeatability, such as variance.

Error – when applied to data, it commonly means the random error due to individual data variation, but it could mean the bias. When applied to a calculated value, propagation of error does not mean a mistake in calculation or equation. It means the uncertainty of the calculated value due to uncertainty on the values input to the calculation (random error and/or bias). Carefully convey how you use the term "error".

R = range high minus low values.

ε = error, half range, $\pm\varepsilon$ should include 98%, or so, of the values.

x, y, z = variable. It is variously used to represent the calculated value, or the independent values. It is also used to represent the givens or coefficient values.

σ = the true population standard deviation, which is unknowable. To obtain an experimental value for σ, you need to sample the entire population, infinite samples. If you have a theoretical basis for σ, then you have an ideal value, which can only be confirmed with infinite number of samples from the entire population. Either way, the value of σ is unknowable, it is either a theoretical idealization of reality, or an empirical approximation.

s = the estimate of the population standard deviation.

w or $\tilde{w}$ = model calculated response variable, the model output.

x^* = optimum value of a model input, an independent variable, the decision variable.

a, b, c, = model coefficients.

8.2 Fundamentals

8.2.1 What Experimental Variation is and is Not

To experimentally estimate uncertainty, only use replicate trials, independent trials that should provide the exact same result, because they nominally were run at exactly the same conditions. Do not use an s-value (standard deviation) that is calculated from all the experimental observations, use the s-value from replicates only.

Here is an example to clarify the difference: I measure the height of each grandchild when they come to visit. Stand them next to the wall, eye-ball level the pencil from the top of their head to the wall and mark the wall. If the pencil is not perfectly horizontal, or their socks are thick, or I don't start from top-dead-center on their head, then the mark on the wall has some error to it. Perhaps each mark could be ±1/8 inch from true. Each mark for each grandchild for each visit is off by a maximum of about ±1/8 inch; but not every mark is off that much, some are closer to the exact value. Using an approximation that the range of a variable is about 5 standard deviations, a ±0.125-inch maximum error on any one height, means about 0.125/2.5 = 0.05 inches standard deviation. If I were to replicate one kid's measurement 100 times, and look at the distribution of marks, I expect the sigma of the variation for any one measurement would be about 0.05 inches.

At one point in time, Kennedy was 63 inches ("), Jamaeka 62", Parker 61", Conor 45", Ashton 40", and Kain 33". Landon was not old enough to stand up yet. The average of 63, 61, 62, 45, 40, and 33 is 50.67". But none of them is 50.67" tall. The standard deviation of the same set of height data is 13.003". But each of my measurements are not in error by 13 inches.

Propagation of uncertainty is the measure of uncertainty in a particular measurement (the s = 0.05"), not the standard deviation of all the data in your experimental conditions (the 13.003"). To experimentally estimate uncertainty, use only replicates, independent trials that should provide the exact same result.

As another example, roll a standard cubical game die. The average value of many rolls should be 3.5. But you can only see values of 1, 2, 3, 4, 5, or 6. In N rolls the discrete uniform distribution indicates $\sigma_{\text{on the average}} \cong 1.71/\sqrt{N}$. Alternately, on a 10-sided die, the average is 5.5, and with N rolls $\sigma_{\text{on the average}} \cong 2.87/\sqrt{N}$. Replicating would only be rolling one type of dice. Only analyze data that is expected to have the same value. Do not collectively include values that are not expected to be identical.

So, when analyzing uncertainty, don't include Product A data with that of Product B. Don't include test data at one temperature with that of another temperature. Likely, you should not include data from one lab with data from another, or perhaps data prior to and after a device recalibration.

8.2.2 Measures of Random Variation

There are two key measures of deviation from the average (the expected or nominal value): error (ε) and standard deviation of the variable (s). These are related.

There is a range of values that might be expected when something is replicated. Range (R) is the difference between the highest of replicates and the lowest. Ask someone what the range might be, and they will not tell you the highest and lowest all-time possible extreme values. Similarly, they will not tell you the last two measurements. They will tell you the normally expected range, perhaps the range that includes about 98% of all possible values. This would be approximately the 5σ range. If the data is symmetric, then $R = 2\varepsilon \cong 5s$. So, $s \cong \varepsilon / 2.5$. You need many samples, perhaps 100 or more, to see the range. Two measurements, for instance, will provide a range, but probably not a range that is representative of the population.

Standard deviation is calculated from all of the available sample data, $s = \frac{1}{N-1}\sum_{i=1}^{N}(x_i - \bar{x})^2$, which is a portion of the infinite number in the possible population. Range is only calculated from two values, the high and low in the dataset, $R = x_{high} - x_{low}$. So, s, calculated from all N data, is a better, estimate of variability. However, when sigma is estimated from the range or half-range error, $s \cong \varepsilon / 2.5$, it is a reasonable, but remains a coarse, two-point approximation.

But neither are perfect measures of variability. Roll a standard 6-sided die 2 times; theory indicates that you expect $\sigma_{\text{of the values}} \cong 1.71$, but you will probably not get that. For instance, if you roll a 3 and a 5 then $s = 1.41$, and if a 1 and a 4 then $s = 2.12$. These are 1.7-ish, which would indicate that the process is understood. Rolls with a greater number of dice would lead to a sigma that is closer to the expected value of about 1.71, but it may take a million rolls to experimentally see the 1.71 value.

Since you won't have millions of replicates, since you will only have a small sample of the population, any method that you use to estimate the sigma or range will be imperfect.

The chi-squared distribution can be used to indicate the expected range on s, which can indicate whether the experimental value is close enough to the model-calculated expected value. In normal situations the true σ-value might be between half and triple the calculated value (roughly the 97% limits with 10 data). However, formal chi-squared or F-statistic testing is not required to assess uncertainty on the variability. With a little experience an intuitive decision as to whether the model-calculated and experimental s-values are close enough to claim they are the same is often all that is warranted.

So: In uncertainty analysis, there is uncertainty in the outcome. Do not expect to predict exact values. One must be willing to judge whether a calculated value is close enough or not. A one-third to four times ratio is bordering the extremes, but not unexpected.

8.2.3 Sources of Variation

We have many names for the uncertainty of numerical values. These include error, fuzz, noise, bias, fluctuation, and variance. Since the term "error" often connotes a mistake, and "bias" denotes a systematic deviation, the term uncertainty is a better representation. Sources of uncertainty are not necessarily human mistakes. They include the naturally occurring variability on measurements due to either systematic bias or random fluctuation. They also include process-to-model mismatch that results from idealizations, truncation, or the use of tabulated data. This chapter will use the term uncertainty; however, both the terms "error" and "propagation of error" remain as commonly used labels.

A few sources of uncertainty encountered in engineering analysis follow.

Estimation: Often the basis of a calculation is an estimate. "Oh, I expect we'll be able to sell 25 metric tons per year." But a plant that large may be a large capital risk for the company resources. Accordingly, the investment managers will want to know the likelihood of only being able to sell 15 or 20, or of the potential to sell 30 or 35 metric tons per year. Whether termed "givens" or "basis" such estimates of the situation are uncertain. Similarly, economic profitability indices such as Net Present Value or Long-Term Return on Investment, are based on estimates of future sales, tax rates, raw material costs, etc. Who can precisely predict the future? Again, such basis for investment choices have uncertainty.

Discrimination: No measurement device is capable of infinitely small measurement intervals. For instance, a pressure gauge may be marked in increments of 10 psi, and one might be able to estimate a gauge reading to the nearest 2 psi. Discrimination limits the reading to ± 1 psi. Similarly, if a 0 to 12,800 gpm flow rate reading is processed by an 8-bit computer, the discrimination ability is 2^{-8}, about 0.4% of full scale, about 50 gpm. Consequently, as the flow steadily increases from 900 to 1,000 gpm, the computer would continue to report

900 gpm during the period in which the flow was changing from 900 to 910, to 920, etc. At 950 gpm it would report (jump-to and hold) 950 gpm until the flow reached 1,000 gpm. Similarly, discrimination error in reading numerical data from charts and diagrams depends on the thickness of the pencil line or the scale of the axis. From tabulated data, discrimination error is related to the number of digits displayed in the data. As a common example, a digital clock indicates time in hours and minutes. It displays 7:23am until the time becomes 7:24. Even though time progressively increases, the reading remains at 7:23 until it jumps to 7:24. The discrimination interval is 1 minute.

Calibration Drift: A metal ruler lengthens and contracts due to temperature change. A wooden ruler also changes length due to humidity. Temperature affects spring stiffness in pressure gauges or weigh scales. Temperature changes the resistance and capacitance in the electric circuitry of sensors and transmitters, which may be due to either ambient conditions or unit warm-up in use. In general, calibration drift is affected by instrument age, temperature, humidity, pressure, oxidation, actinic degradation, fouling, and catastrophic events such as dropping the device.

Accuracy: Sensors and transmitters are not exactly linear. Calibrations are usually performed by adjusting a transmitter zero and span so that the instrument reports a value "close enough" to the values of two standards. Then one assumes that the instrument response between the standards is linear even though the response is not exactly linear. (In fact, the local calibration standards are not perfect, either.) The *accuracy* of a device, its reading deviation from the true value, is often called *systematic error* or *bias*.

Technique: The measurement procedure may measure the wrong thing. For example, if knowledge of the steam pressure is required to size a reboiler, one must have the pressure downstream of the steam flow control valve and not the steam header pressure. As another example, a thermowell might be measuring the temperature in a hot spot in a furnace, but this would not represent the average temperature over all of the space which drives the heat transfer.

Constants and Data: Most calculations involve fundamental constants, but their values come from previous measurements that are not exactly known. Examples of constants include the gas law constant, the gravitational constant, the speed of light, and the molecular weight of sodium. Typically, these values are known to errors of only 0.001% or 0.01%. Examples of data include a tabulated viscosity or thermal conductivity. Typically, these data are known only to 0.1% to 1% error. Often, data are obtained from correlations or graphs, such as pipe-flow friction factors, thermodynamic properties of mixtures, and convective heat transfer coefficients. Such values may have a 10% to 30% uncertainty.

Noise: Process flow turbulence, changing electromagnetic fields around equipment, equipment mechanical vibrations, etc. may cause the measured value to randomly vibrate or fluctuate about its average value. Due to the fluctuation, any one instantaneous reading may not reflect the time-local average. One has to average many values to temper the uncertainty of random noise. Noise reduces your ability to obtain a precise measurement. Precision, repeatability, is related to such random influences.

Model and Equations: Nonidealities are usually not expressed in models of phenomena, and, accordingly, inaccuracy is reflected in the equations that are used to calculate the results. For instance, the ideal gas law ($PV=nRT$) is a model of particle dynamics that neglects the effects of intermolecular forces and molecular size. Although the law is often used for gases at ambient conditions, it can introduce a 5% error at those conditions and up to an 80% error at conditions near the critical point of real gases. The volume of a tank may be calculated by $V=\pi r^2 h$, an equation that is based on a right-circular cylinder model.

Such a model neglects real surface irregularities such as the effects of dents or ribs, sensor intrusions into the side of the tank, and the curved bottom and the length of drainpipe above the bottom discharge valve. The square root relation between differential pressure and flow rate, derived from the ideal Bernoulli Equation for inviscid streamline flow, is commonly accepted in orifice calibration equations, and may lead to a 5% error. The Beer–Lambert relation for the concentration-dependent light absorption is commonly accepted in spectrophotometric devices. Models are often incorrect because they are incomplete; therefore, the calculated values (whether part of the measurement device or offline) are also imperfect.

Humans: Humans are often part of the data acquisition and transmission process. We might transliterate, reversing the position of adjacent digits. We might decide when a process has achieved steady-state for sampling, but a desire to finish the project might override waiting long enough for the transient to settle. A noisy signal might be biased by using a convention that reports the mental average of the upper (or lower) extreme that was observed, or the most frequent value that was observed. A human might judge that a data point is faulty because it is inconsistent with expected trends in the data, and may discard that point, when in fact the data was good, and the understanding of the process was wrong.

8.2.4 Data and Process Models

Models could be classified as either data models or process models. The distinction between data model and process model is in its use. The distinction is not fundamental to a model.

Data models are used in the experiments to convert sensor data into the composite measurement. For example, consider a bucket and stopwatch method to measure flow rate. Weigh the empty bucket, time interval to collect material, and weigh the full bucket. These are the measurements, then use the data model to calculate volumetric flow rate, $F = \left(W_{\text{full}} - W_{\text{empty}}\right) / \left(t_{\text{end}} - t_{\text{start}}\right) / \rho$. In this case the data model is the definition of flow rate. Here, the estimated uncertainty on the five data values (W, W, t, t, ρ) would be used to evaluate uncertainty on the calculated F.

We often think of the calculated flow rate from the data model as the measurement because it is the property or response that we want to see. But F is truly not the measurement; it is the calculated value from the measurements. If we are interested in how valve position affects flow rate, then we plot the data-model-calculated flow rate "measurement" w.r.t. the experimental control. But, the "measurement", was not measured. Weights and times were the measurement. The "measured response" was calculated by a data model.

Similarly, in getting a yield stress measurement from a tensile test, the actual measurements are breaking load and cross-sectional dimensions, and stress is calculated from a definition, the data model. Again, similarly, when using electrical conductivity as a sensor to infer salt composition, the measurement is conductivity, and the inverse of the calibration model is the data model that is used to report composition. Although the lab might report breaking stress and composition as the measurements, they are truly calculated from a data model and the true measurements.

Contrasting the data model, the process model is the model that describes the response behavior of a process or product or procedure. For example, a process model might be that used to calibrate an orifice flow meter using the F results from above. The model might be $F = a\left(i - i_0\right)^b$, where i (transmitted mA signal) will be used to calculate flow rate for display. (Ideally, the Bernoulli relation would indicate that $b = 0.5$, and you may have expected to see a square root relation in the orifice equation. But numbers in the $b = 0.46$ range often provide a better representation of the nonideal reality.) In calibration, the objective is to adjust the model coefficients (a, b) to make the calibration model best match the (F, i) experimental data. Here, uncertainty on the data-model calculated F, and sensor-transmitted (i, i_0) values will affect certainty on the model coefficients (a, b). In this regression application, the response value is known, not calculated. It is the model coefficients that are the unknowns.

Subsequently, in online process monitoring use, this calibration process model becomes the data model, converting sensor data (i) into "measured" flow rate. So, the process model could be considered the data model in an alternate use.

The process model could also represent how the process responds. For instance, the question might be, "How does flow rate depend on valve position?" Perhaps the model could be an equal-% type, $F = ae^{o/R}$ where o is the controller output and (a, R) are model coefficients. Regression would be used to determine the (a, R) values. Once the model is obtained, the inverse of the process model could be used to determine the controller output needed to achieve a target flow rate: $o = R\ln\left(\frac{F}{a}\right)$.

Again, what is classified as the data model or process model depends on the application of the model, the same model could be viewed as representing either category, depending on the situation. The model could be used to calculate the response, or the inverse of the model used to calculate the influence to achieve a desired response, or the model could be used in regression to adjust coefficients to best fit data.

We'll use several common models as examples in this chapter. The first is the calculation of volumetric flow rate with the bucket and stopwatch method, $F = \left(W_{\text{full}} - W_{\text{empty}}\right) / \left[\rho\left(t_{\text{full}} - t_{\text{start}}\right)\right]$. The second is calculation of sample breaking stress from a tensile test, $S = L / dw$. The third will be surface area and volume from simple geometric shapes. Last is the volumetric flow rate calculated from the milliamp signal from an orifice flow meter, $F = a\left(i - i_0\right)^b$. The examples are all relatively simple; but they reveal modest derivative complexity, which partly motivates the choice of numerical, not classic analytical approaches to uncertainty propagation.

8.2.5 Explicit and Implicit Models

Models could be classified as either explicit or implicit. An explicit model can be rearranged to isolate the response variable on one side of an equation. Then the functionality on the other side directs how to explicitly calculate the value.

However, some models contain functionality that prevents explicit solution of the relation. This might be an algebraic relation such as $\sqrt{x} + 0.1\ln(x) = 17$, which seems impossible to rearrange to isolate the variable x. For such relations, a numerical root finding procedure is an option. Additionally, implicit models might require solving an optimization, or determining the point along a path or time that meets a particular criterion. This could appear as $\min_{\{x\}} J = f(x) = \frac{17}{x} + x^{0.1}$, or $17 = \int_{0.1}^{x} f(x)dx$, or any number of forms with path or transient model formats. But, whether implicit or explicit, uncertainty on the coefficients 0.1 and 17, create uncertainty on the x-value.

In addition, implicit models have a convergence criterion that determines when the solution is close enough to claim convergence. Implicit solutions do not give the exact solution, but iteratively approach it. Be sure that your threshold is chosen to have insignificant impact (two or three orders of magnitude less) on the answer, relative to other sources of uncertainty.

8.2.6 Significant Digits

Digits in a number that are significant are the digits that have values of which we are fairly confident. Whether explicitly stated or implied, you should reveal the uncertainty in reporting numerical values. The following is a reporting convention for integers and real numbers.

An integer has no decimal point. It is used to represent the number of whole events or whole things. A real number has a decimal point and is used to represent the value of a continuum variable that can have fractional values. Some uncertainty is associated with both integers and real numbers. For instance, if you tried to count about 2,000 items, you might lose track, or you might eye-ball and remove groups of ten and might have a count that is off by 10 or so. If your count was 2,013, the number should then be reported as 2,013 ± 10 to explicitly acknowledge the counting precision. Similarly, the speed of light in a vacuum is reported as 2.997925×10^8 m/s ± 0.000003×10^8 m/s to explicitly acknowledge uncertainty.

By custom, the uncertainty in numerical values is usually not explicitly reported but is implied by the number of digits reported.

For continuum-valued numbers, the last (right-most) reported digit is the largest digit with uncertainty. When the precision is implied, we do not explicitly know whether the last digit is accurate to ±1 or ±2 or ±3 or ±4. If the uncertainty were ±5 with a range of 10, the next digit to the left would be uncertain and would have been the last reported digit. If ±0.2 the last reported digit would be known with certainty, and the next digit to the right, the first fuzzy digit should be reported. Therefore, the uncertainty on the last digit reported could range from ±1 to ±4. Without specific guidance, we will assume a mid-value for the uncertainty of ±2.5. For continuum-valued numbers, the last reported digit from the left, even if zero, is the fuzzy number. For example, a length may be reported as 100.0 in., which implies an error of about ±0.25 in. Similarly, the atomic weight of Hydrogen = 1.00794 g/gmole (based on the ^{12}C isotope) implies that the value is known with an error of about ±0.000025 g/gmole.

For integers, the last nonzero digit from the left represents the fuzzy number. For example, the 2010 census reported the population of the United States as 308,700,000. The last (smallest value) of the four nonzero digits, 7, is fuzzy and reflects a counting error of about ±250,000 people. This method is the convention for reporting integers, but it has the unfortunate aspect that significant zeros are not identified. For instance, if there are 2,000 ± 10 items, one reader might interpret a reported 2,000 as 2,000 ± 250, and another as 2,000 ± 2.5.

The ±2.5 is the average error if the last digit is the first (from the left) fuzzy (uncertain) digit. The maximum error would be twice that. Roughly, the sigma is the maximum uncertainty divided by 2.5, using range = 5 sigma. For example, if a table of liquid density indicates a value of 0.624, the "6" and "2" are certain, but the "4" is the first uncertain digit, then the nominal or average uncertainty on the data is expected to be ±0.0025, and the maximum uncertainty on the data value might be ±0.005. But, the deviation is not always the maximum value. Sometimes the deviation might be very small. The standard deviation would be about 0.001.

By contrast, if the last reported digit was the last digit with a certain value, then the uncertainty would be ±0.25. Accordingly, both reporters and readers must use care in reporting and interpreting implied precision. Depending on the convention, the reader may make an order of magnitude error on interpreting an implied precision.

When reporting significant figures with implied precision, the convention is to round off the number to the nearest significant digit (ASTM Standard E29-88). For base 10 digits, the procedure is as follows: select the last (rightmost) reportable digit. If the digit immediately to the right of the last reportable digit is less than 5, truncate to the last reportable digit. If the digit immediately to the right of the last reportable digit is either greater than 5 or is a 5 followed by a nonzero digit, increase the last reportable digit by 1. If the digit immediately to the right of the last reportable digit is either exactly 5 or is a 5 followed by only zeros, then either truncate if the last reportable digit is even or increase the last reportable digit by 1 if it is odd.

As examples, the following numbers are to be rounded to report only significant digits.

Original number	Rounded value
41,276 ± 350	41,300
6,150.02 ± 73.61	6.2×10^3
117.5 ± 1.2	118
116.5 ± 1.2	116
6,149.99 ± 207.13	6.1×10^3

Rounding should be done only on the final reported value. Do not round intermediate results. Each rounding operation changes a numerical value; hence, it introduces error. If several intermediate roundings are performed, this error accumulates through the calculational procedure and could have a significant effect on the results. Of course, the command, "Do not round intermediate results," is impossible when a calculator display has limited digits. In practice, it is fully sufficient to maintain two digits past the fuzzy digit.

8.2.7 Estimating Uncertainty on Input Values

Before we can propagate uncertainty through calculations, or include uncertainty in the x^* values, or report the uncertainty on calculated (or dependent) variables, we need to know the uncertainty on the numerical values that we use within the calculation. There are many legitimate ways to determine uncertainty on independent variables.

We often get data from tables (viscosity, thermal conductivity, density, etc.). If the table does not explicitly report an uncertainty, and you believe that the last reported digit is the largest uncertain digit, use ±2.5 on the right-most reported value as an estimate of the average uncertainty. This means that the maximum uncertainty is ≈±5.0 on the right-most reported value and is the maximum uncertainty, or roughly that $s \cong 1$ on the right-most digit.

If it is a measurement, replicate it enough times to be able to calculate the standard deviation. Usually, 10 replications will be ample to provide an adequately definitive estimate (the range on the true sigma could be within 0.6 to 2 times the 10-sample estimate), but 5 replications are often minimally adequate, and better than 3 which provides an uncertain estimate (the range on the true sigma could be within 0.4 to 14 times the true value). If the distribution is Gaussian (most measurements are close enough to this ideal distribution) then about 99% of the values, the range, fall within about $\pm 2.5\sigma$ of the mean.

If the instrument has a calibration record or manufacturer specifications that indicate precision, use that for repeatability. (If it reports accuracy, use that for systematic error.)

If the numerical value is calculated from a linear model with coefficients determined in a least-squares regression or correlation, use the "standard error of the estimate" as the standard deviation on the model coefficient value.

If the regression is of a nonlinear model to data, use bootstrapping (See Section 19.3) to estimate uncertainty on the model coefficient and prediction values.

If a numerical optimization or a root-finding procedure bounded the optimum, use convergence criteria to estimate the uncertainty in DV* as $\pm\frac{1}{2}\Delta x$ threshold.

Use your judgment and experience to estimate the possible uncertainty on a value. For instance, in using a stopwatch to time collection of material in a bucket, you might estimate the start and stop time to each have a 0.5 second error. For another instance, in reading data from a curve, you can estimate the error that might happen due to pencil line thickness, your lines not being exactly parallel to the axis lines, discrimination ability on the two axes, or curvature of the graph due to photocopying.

To experimentally estimate uncertainty, only use replicate trials, independent trials that should provide the exact same result, because they nominally were run at exactly the same conditions. Do not use a σ that is calculated from all of the data.

Because uncertainty on input values to an equation are often estimates, one cannot expect the model-propagated value to be the true value of uncertainty. So, in uncertainty analysis, there is uncertainty in the model-calculated value as well as the experimental outcome. Again, one must be willing to judge whether a calculated value is close enough or not.

Perhaps, if the sigma experimental and the model-calculated sigma are within a 3:1 or 0.5:1 ratio, you might be justified to claim that σ-modeled is equivalent to σ-experimental. But, a 10:1 ratio would be an indication that the analysis does not match the data. A 4:1 ratio would not lead to confidence in either assessment.

8.2.8 Random and Systematic Error

A systematic error is a consistent bias, a deviation from true that persists for all measurements. In contrast, a random error is an independently changing deviation for each measurement.

For instance, in using a deliver-to graduated cylinder that is marked in 1 ml-increments to measure a volume for a recipe, the user might be able to fill the cylinder to ±0.2 ml. One liquid drop more or less might cause a noticeable deviation and would be corrected. So, the fill point may randomly change by an unnoticeable half-a-drop with each use. However, the glass expands and contracts with temperature (room T, heat from the technician's hand, sample T); and as a result, the marking of the volume might actually include ±0.1 ml more or less than the reading. Then, when the contents are poured out, the residual liquid in the cylinder may vary from 0.1 to 0.2 ml, due to user technique and temperature-induced liquid viscosity, for an additional ±0.1 uncertainty. Primitively, the random error on each use could be the sum of all sources, ±0.4 ml. Not all uses, however, would have that maximum error. In some uses, the three error sources could be counter to each other, and in some uses the individual errors would not be at the maximum values. If the errors could be known, ideally, they would average zero, be symmetric, and have a Gaussian distribution with 95% or so of the error values within the ±0.3 ml range. This would be random error.

However, the graduated cylinder is likely not a primary standard, and the marking of 132 ml, for instance, might actually deliver 0.8 ml less than expected. This error would be the same in each use. Again, if it were possible to assess the error, it would not average zero, but would average −0.8 ml with a random fluctuation between 0.4 and 1.2 ml. This persistent offset of −0.8 ml would be the systematic error or bias.

The systematic bias may have a known value. If, for instance, you know that your bathroom scales read 2 lbs too low, then add 2 lbs to the reading. This would be a systematic error of a known value. If you know the systematic error value, correct for it. However, you may be aware that no home scales are perfect, know that there is a systematic bias, but not know its value. Without knowing the value, you cannot correct the reading. This bias uncertainty on the reading imparts uncertainty to the measurement (observation) value.

In concept, the idealization of segregating random and systematic error is useful, but if the graduated cylinder is progressively heating (or cooling) over sequential uses, then that influence would not be random w.r.t. use. It would appear as a systematic error that is progressively changing with each use, an autocorrelated bias.

The error then, might not be ideally a fixed bias with an independent sample-to-sample fluctuation. The error might be autocorrelated or it might change from day to day. Or, if several devices are used for the measurement, the error might correlate to which device is used.

The device could be marked with the magnitude of the systematic error, for example ±0.8 ml. But, you do not know whether the particular device is reading 0.8 ml high or low or some intermediate value. The value of the systematic error is unknown.

If the value of the systematic error were known, you would use that knowledge to correct the measurement. By contrast, if the systematic error value is unknown, then use uncertainty analysis to determine the impact of the bias on the measurement.

Both random error and unknown bias can be propagated using the same techniques from this chapter, but they should be independently analyzed and independently reported (ASTM Standard E177-86).

8.2.9 Coefficient Error Types

There are coefficients (numerical values) in equations. For some, the value is known with certainty. For some, there is a systematic (always the same) bias, and for others there is a random bias. Finally, for others there is random uncertainty. You need to be able to differentiate the types.

Coefficients with values that are known with certainty would include the 100 when converting a fraction or proportion to a %, or the ½ that converts diameter to radius, or the 4 that multiplies the length of one side of a square to obtain the perimeter. There is no need to propagate uncertainty on such variables.

However, some apparently fundamental coefficients are truncated or rounded, and accordingly have uncertainty. If for instance you substitute 3.14 for the value of π then you are deciding to include a bias of 0.0015926535…. Similarly, if you decide to use 18 for the molecular weight of water, then you are including a bias of 0.01528 (based on ^{12}C isotope). Those errors are systematic bias errors. But the value of the systematic error can be known. Since they are known, you could correct the error in the calculated values that they create, but, why use the wrong value, then post-correct the calculated value? The proper approach is to use the right value to begin with. However, many coefficients are irrational numbers, and it is not possible to use an infinite set. The rule of thumb is to use the number of places that makes the systematic bias negligible in the calculated value. Perhaps use enough digits so that the consequential error on the model prediction is two or three orders of magnitude less than the uncertainty due to other contributors.

Similarly, do not truncate or round values of intermediate calculations to use the shortened version (systematic bias) in subsequent calculations. Keep two or three more digits than would be justified in reporting the value. Using a truncated value would impose a bias to subsequent calculations, but the bias would be random, because it would change in subsequent applications due to the vagaries of the input data. However, you should report the shortened version to reveal your (and your audience's) appreciation for significant digits.

Similarly, when numerical methods of root-finding in implicit relations or optimization are used to determine a value, be sure that the convergence criterion is so small that the uncertainty it imparts to the solved for value is inconsequential to the result.

"Givens" have uncertainty. These are the basis, forecast loads, or assumptions in an application statement, although the givens may not seem uncertain when "given" by authority. For example, the boss may say, "Twenty people are coming to the meeting. Be sure we have 20 seats." Or, the client might direct, "Design the process to produce 100 kg of product per day, when the raw material is 23 wt-% useful." Don't be misled by authority or a problem statement to think that the givens have zero uncertainty.

Of course, coefficients related to material properties (viscosity, molecular weight, thermal expansion, half-life, growth rate), fundamental constants (speed of light, gas law constant), unit conversions (kPa to psia, miles to km), material quantity (lbs per package, product purity), economics (tax rate, labor cost, utility price), probability (reliability, variance), and coefficients in equations ($Nu = 0.023Re^{0.8}Pr^{0.4}$) are all uncertain. Don't accept them as absolute truth.

8.3 Propagation of Uncertainty in Models

There are two common metrics of the uncertainty: maximum error and probable error. For each, there are two methods to propagate uncertainty: analytical and numerical. And, for each there is systematic error and random error which are propagated by the same methods, but independently, and separately reported. What follows is how to propagate uncertainty in general for either systematic or random error.

In general, consider this model:

$$w = f(x, y, z) \tag{8.1}$$

The input or independent variables are x, y, and z. However, they could be listed as x_1, x_2, and x_3. The independent variables, the inputs could also be considered the "givens", the basis for the calculation or the properties of materials. Some of the x, y, z values could represent model coefficients, and there is no requirement that there are three inputs. There could be 1, or 2, or 10, or n, and represented as:

$$w = f(x_1, x_2, x_3, x_4, \ldots) = f(\underline{x}) \tag{8.2}$$

If you have more or fewer, adjust the formulas that follow. The response or dependent variable in this presentation is w. The influence variables are more than just experimental conditions, they include model coefficients and givens – any variable with uncertainty.

8.3.1 Analytical Method for Maximum Uncertainty

A Taylor series approximation to the model about nominal (x, y, z) values is:

$$w = f(x,y,z) = f(x_0,y_0,z_0) + \left.\frac{\partial f}{\partial x}\right|_0 (x-x_0) + \left.\frac{\partial f}{\partial y}\right|_0 (y-y_0) + \left.\frac{\partial f}{\partial z}\right|_0 (z-z_0) \tag{8.3}$$

This is truncated to exclude quadratic and higher order terms, which means that (x, y, z) deviations from the base case values of (x_0, y_0, z_0) are small. Note that the derivatives are evaluated at the base case (x_0, y_0, z_0).

Representing deviations as $\varepsilon_x = (x - x_0)$, and removing the explicit notation on the derivatives:

$$\varepsilon_w = f(x,y,z) - f(x_0,y_0,z_0) = \frac{\partial f}{\partial x}\varepsilon_x + \frac{\partial f}{\partial y}\varepsilon_y + \frac{\partial f}{\partial z}\varepsilon_z \tag{8.4}$$

Maximum uncertainty on the dependent variable would happen when each independent error is at its extreme value and each is "pushing" the independent variable in the same direction. So, use the absolute values and let ε_x represent the maximum expected deviation, not a particular realization.

$$\varepsilon_w = \left|\frac{\partial f}{\partial x}\varepsilon_x\right| + \left|\frac{\partial f}{\partial y}\varepsilon_y\right| + \left|\frac{\partial f}{\partial z}\varepsilon_z\right| \tag{8.5}$$

In general:

$$\varepsilon_w = \sum_{i=1}^{n} \left|\frac{\partial f}{\partial x_i}\varepsilon_{x_i}\right| \tag{8.6}$$

In the special case of w being a product of the factors, $w = xyz$, or in general: $w = \prod x_i$

$$\frac{\varepsilon_w}{w} = \sum_{i=1}^{n} \left|\frac{\varepsilon_{x_i}}{x_i}\right| \tag{8.7}$$

This presumes:

1. The impact of each independent variable has a linear impact on the dependent variable (which is usually acceptable if the deviation is small relative to any nonlinearity curvature effects).
2. That the sensitivity (partial derivative) of one response is independent of the value of other inputs.
3. The (x, y, z) deviations are independent (uncorrelated and unconstrained).
4. The derivatives are all evaluated at the same base case $(x_{10}, x_{20}, x_{30}, \ldots)$.
5. All influences are at their maximum value.
6. All influences are "pushing" the response in the same direction.

A few simple examples reveal some caution issues for those a bit rusty in Calculus I.

Example 8.1: For the yield stress model, $S = L/dw$, use Equation (8.6) to obtain a formula for propagation of maximum error. The answer is:

$$\varepsilon_S = \frac{1}{dw}\varepsilon_L + \frac{L}{d^2 w}\varepsilon_d + \frac{L}{dw^2}\varepsilon_w$$

Note that the negative signs due to the derivatives w.r.t. d and w became positive after the absolute value, and that each term in the equation has the same units of stress.

Example 8.2: For the bucket and stopwatch method, $F = \frac{W_{\text{full}} - W_{\text{empty}}}{\left[\rho\left(t_{\text{full}} - t_{\text{start}}\right)\right]}$, determine the formula for propagation of maximum error. The answer is:

$$\varepsilon_F = \frac{1}{\rho\left(t_{\text{full}} - t_{\text{start}}\right)} 2\varepsilon_W + \frac{\left(W_{\text{full}} - W_{\text{empty}}\right)}{\rho^2\left(t_{\text{full}} - t_{\text{start}}\right)}\varepsilon_\rho + \frac{\left(W_{\text{full}} - W_{\text{empty}}\right)}{\rho\left(t_{\text{full}} - t_{\text{start}}\right)^2} 2\varepsilon_t$$

Note that the factor of 2, is the result of the two weight and of the two time measurements having independent variation with identical uncertainty and each having the same absolute value of the derivatives.

Example 8.3: Determine the maximum error for the orifice flow meter, $F = a\left(i - i_0\right)^b$.
In use there would be noise (uncertainty) on the transmitted mA, but also an uncertainty due to the development of the model. One could consider that the three model coefficients (a, b, i_0) are independently uncertain, but they would be correlated, so we'll lump the model uncertainty into one term such as would be quantified as the 95% limit in a bootstrapping analysis. Here, $F = a\left(i - i_0\right)^b + \pm\varepsilon_{F\text{-model}}$. Then,

$$\varepsilon_F = ab\left(i - i_0\right)^{b-1}\varepsilon_i + \varepsilon_{F\text{-model}}$$

Note the exponent decrement $b - 1$ in the relation. Also note, if one were to consider uncertainty on the coefficient b, then there would be a term $a\left[\ln\left(i - i_0\right)\right]\left(i - i_0\right)^b \varepsilon_b$ in the expansion for ε_F.

Again, don't be misled by the nomenclature to limit x to experimental conditions. In the above, x represents any variable with uncertainty (givens, coefficient values, basis, experimental measurements, etc.).

This method presumes that you can take the analytical derivative. Many common models, like the simple ones above, stretch the calculus ability that you were supposed to have learned in the Calculus 1 class, many years ago. Further, many models are so complex that even with skill, there is a good chance of making an error in many lines of the calculus and algebra associated with getting the analytical equation of a derivative. If so, the analytical derivatives can be replaced by a finite difference estimate. If using the central difference formula:

$$\frac{\partial f}{\partial x} \cong \frac{f(x + \Delta x, y, z) - f(x - \Delta x, y, z)}{2\Delta x} \tag{8.8}$$

You choose the Δx value. It should be small relative to the curvature in the function. We often use $\Delta x = 0.1\varepsilon_x$, but, it needs to be not too small, so that numerical truncation error does not compromise the derivative estimate. Accordingly, we like to use double precision variables in the calculations.

Except in trivial cases, our default is to use the numerical approximation for the derivative to avoid making calculus/algebra errors.

Here are some examples using numerical values:

Example 8.4: What is the volume of a cube in which each side is 0.81 m?

The implied precision of the side length is about ±0.025 m. Since $V = S_1 \cdot S_2 \cdot S_3$, $V =$ 0.531441 m^3. Using Equation (8.7),

$$\frac{\varepsilon_V}{V} = \pm\left[\frac{\varepsilon_1}{x_1} + \frac{\varepsilon_2}{x_2} + \frac{\varepsilon_3}{x_3}\right]$$

$$= \pm\left[\frac{0.025}{0.81} + \frac{0.025}{0.81} + \frac{0.025}{0.81}\right] = \pm 0.0925925\ldots$$

Rearranging,

$$\varepsilon_V = 0.531441(0.0925925\ldots) = 0.0492075\ldots\text{m}^3$$

Explicitly $V = 0.53 \pm 0.05(\text{max error})\text{m}^3$

Implicitly $V = 0.5\ \text{m}^3$

Note that the error propagation type (max error) is explicitly stated.

Example 8.5: Report the value of y if $y = \ln x$ and $x = 7.389 \pm 0.005$.

Using Equation (8.6)

$$y = \ln(7.389) \pm \frac{0.005}{7.389}$$

$$y = 1.999992408\ldots \pm 0.00067668\ldots$$

Explicitly $y = 2.0000 \pm 0.0007$ (max error)

Implicitly $y = 2.000$

Example 8.6: Repeat Example 8.5 but use the central difference numerical method to estimate the derivative value. Report the value of y if $y = \ln x$ and $x = 7.389 \pm 0.005$.

Using Equation (8.8) with $\Delta x = 0.0001$, small relative to other values,

$$\frac{dy}{dx} \cong \frac{\ln(x+\Delta x) - \ln(x-\Delta x)}{2\Delta x} = \frac{2.000005941 - 1.999978874}{0.0002} = 0.1353346$$

Using Equation (8.6)

$$y = \ln(7.389) \pm (0.13533)(0.005)$$

$$y = 1.999992408\ldots \pm 0.000676673\ldots$$

Explicitly $y = 2.0000 \pm 0.0007$ (max error)

Implicitly $y = 2.000$

Example 8.7: Calculate y if $y = \cos(x)$ and $x = 1.571$ radians.
From Equation (8.6)

$$|\varepsilon_y| = \left|\frac{dy}{dx}\right| \cdot |\varepsilon_x| = |\sin(x)||\varepsilon_x|$$

then

$$y \pm \varepsilon_y = \cos(x) \pm |(\sin x)\varepsilon_x|$$

The implied uncertainty on the value of x is $\varepsilon_x = \pm 0.0025$
Substituting numerical values,

$$y = \cos(1.571) \pm |(\sin 1.571)(0.0025)|$$

$$y = -0.0002036732 \pm (0.9999\ldots)(0.0025)$$

Explicitly $y = -0.0002 \pm 0.0025$ (max error)
Implicitly $y = -0.000$

Note: This analysis has been for propagation of the maximum error. We have always taken the worst case and added errors although we acknowledge that errors are random (some "+" and some "–") and that the errors are not always the largest possible value. The probable (or likely) cumulative error on the answer will be less than the maximum error.

8.3.2 Analytical Method for Probable Uncertainty

It is unlikely that each variable will be at its extreme and also that all sources of perturbation are pushing in the same direction. More likely, the perturbation values are normally (Gaussian, bell-shaped) distributed, and independent of each other. In this case, propagation of variance will provide an estimate of the probable error.

We've seen about 4 ways to derive this. Here is our favorite. Start with the definition of variance:

$$s_w^{\,2} = \frac{1}{N-1}\sum_{i=1}^{N}(w_i - \overline{w})^2 \tag{8.9}$$

If the average is the true value at the base case, $\overline{w} = f(x_0, y_0, z_0)$, and if the linear Taylor series approximation is valid,

$$s_w^{\,2} = \frac{1}{N-1}\sum_{i=1}^{N}\left(\frac{\partial f}{\partial x}\varepsilon_{x_i} + \frac{\partial f}{\partial y}\varepsilon_{y_i} + \frac{\partial f}{\partial z}\varepsilon_{z_i}\right)^2 \tag{8.10}$$

Recall that derivatives are evaluated at the base case, not at some i-th value.

Expanding the squared argument of the sum and explicitly showing some of the terms:

$$s_w^{\,2} = \frac{1}{N-1}\sum_{i=1}^{N}\left[\left(\frac{\partial f}{\partial x}\varepsilon_{x_i}\right)^2 + \left(\frac{\partial f}{\partial y}\varepsilon_{y_i}\right)^2 + \ldots + \frac{\partial f}{\partial x}\varepsilon_{x_i}\frac{\partial f}{\partial z}\varepsilon_{z_i} + \ldots\right] \tag{8.11}$$

And regrouping the sum:

$$s_w^2 = \frac{1}{N-1}\sum_{i=1}^{N}\left(\frac{\partial f}{\partial x}\varepsilon_{x_i}\right)^2 + \frac{1}{N-1}\sum_{i=1}^{N}\left(\frac{\partial f}{\partial y}\varepsilon_{y_i}\right)^2 + \ldots + \frac{1}{N-1}\sum_{i=1}^{N}\frac{\partial f}{\partial x}\varepsilon_{x_i}\frac{\partial f}{\partial z}\varepsilon_{z_i} + \ldots \quad (8.12)$$

The squared terms should be recognized as the variance estimate from each input. The cross-product terms are called co-variance and measure the correlation between independent variable values. If the variation in each input variable is independent, then the co-variance terms will have as many + values as – values and the sum of them will tend to remain about a value of zero. The squared terms, however, will always be positive and the sum will be relatively large. If there is little or no correlation, then the co-variance terms can be ignored, and conventionally representing sample standard deviation with the population value, propagation of variance is:

$$\sigma_w = \sqrt{\left(\frac{\partial f}{\partial x}\sigma_x\right)^2 + \left(\frac{\partial f}{\partial y}\sigma_y\right)^2 + \left(\frac{\partial f}{\partial z}\sigma_z\right)^2} \quad (8.13)$$

In general

$$\sigma_w = \sqrt{\sum_{i=1}^{n}\left(\frac{\partial f}{\partial x_i}\sigma_{x_i}\right)^2} \quad (8.14)$$

Example 8.8: What is the uncertainty on the average, $\bar{X} = \frac{1}{N}\sum_{i=1}^{N} x_i$?

The derivative of $\bar{X}$ w.r.t. x_i is $\frac{1}{N}$. From Equation (8.14) $\sigma_{\bar{X}} = \sqrt{\sum_{i=1}^{N}\left(\frac{1}{N}\sigma_{x_i}\right)^2}$. If each observation comes from the same population, then $\sigma_{x_i} = \sigma_x$, then

$$\sigma_{\bar{X}} = \sqrt{\sum_{i=1}^{N}\left(\frac{1}{N}\sigma_x\right)^2} = \sqrt{N\left(\frac{1}{N}\sigma_x\right)^2} = \frac{1}{\sqrt{N}}\sigma_x$$

Note: This is a key formula from the Central Limit Theorem that indicates that the uncertainty on the average of observations from the same population scales with the square root of the number of samples.

If you do not know the values of σ_i you could estimate them from $\sigma_i \cong \varepsilon_i / 2.5$. If you cannot analytically take the derivatives, then use the numerical approximation of Equation (8.8).

This also presumes that:

1. The impact of each independent variable has a linear impact on the dependent variable (which is usually acceptable, if the deviation is small relative to any curvature effects).
2. The sensitivity (partial derivative) of one response is independent of the value of other inputs.

3. The (x, y, z) deviations are independent (uncorrelated).
4. N is large enough to make the covariance sums become negligible.
5. The derivatives are all evaluated at the same base case $(x_{10}, x_{20}, x_{30}, \ldots)$.

Contrasting propagation of maximum error, this does not presume:

1. That all influences are at their maximum value.
2. That all influences are pushing the response in the same direction.

Using probable error as meaning the 95% interval

$$\varepsilon_{w,0.95} = 1.95996\ldots\sigma_w \tag{8.15}$$

However, the estimates on the σ_i values are probably ±50% of the true value, which does not justify digit precision on the 1.95996… justifying this relation:

$$\varepsilon_{w,0.95} \cong 2\sigma_w \tag{8.16}$$

If wanting the 99% probable error use:

$$\varepsilon_{w,0.99} \cong 2.5\sigma_w \tag{8.17}$$

Since the 2.5 coefficient is the same in the estimating relation between error and sigma, this reduces to a propagation of errors (as opposed to variance).

$$\varepsilon_{w,0.99} \cong \sqrt{\left(\frac{\partial f}{\partial x}\varepsilon_x\right)^2 + \left(\frac{\partial f}{\partial y}\varepsilon_y\right)^2 + \left(\frac{\partial f}{\partial z}\varepsilon_z\right)^2} \tag{8.18}$$

Or generically:

$$\varepsilon_{w,0.99} \cong \sqrt{\sum_{i=1}^{n}\left(\frac{\partial f}{\partial x_i}\varepsilon_{x_i}\right)^2} \tag{8.19}$$

Again, considering the uncertainty on the σ_i or ε_i values, and variability in measured σ_w, if a model-predicted $\varepsilon_{w,0.99}$ is within about a 3:1 or 1:3 ratio of the measured $\varepsilon_{w,0.99}$ then the analysis cannot be rejected. However, if outside of the 4:1 or 1:4 ratio, question the model or the sources or estimates of variation.

The propagation of variance will provide a more realistic estimate of probable error than the propagation of maximum error. However, often when safety or risk is involved, people want to know what might be the worst-case scenario, a possible worst outcome. Then propagation of maximum error would be preferred.

Example 8.9: What is the probable range on the volume of a right parallelepiped of sides 5. ± 1 cm, 10. ± 1 cm, and 100. ± 3 cm? The maximum measurement errors are as stated.

By definition, the volume $V = S_1S_2S_3$. Substituting numerical values,

$$V = x_1x_2x_3 = (5.)(10.)(100.) = 5000.\text{cm}^2$$

We need to estimate σ_1, σ_2, and σ_3. Using the maximum measurement errors to indicate the $\pm 2.5\sigma$ range,

$$\hat{\sigma}_1 = \frac{1\text{cm}}{2.5} = 0.4\,\text{cm}$$

$$\hat{\sigma}_2 = \frac{1\text{cm}}{2.5} = 0.4\,\text{cm}$$

$$\hat{\sigma}_3 = \frac{3\,\text{cm}}{2.5} = 1.2\,\text{cm}$$

From Equation (8.14),

$$\hat{\sigma}_V = \left[(10\cdot 100\cdot 0.4)^2 + (5\cdot 100\cdot 0.4)^2 + (5\cdot 10\cdot 1.2)^2\right]^{1/2} = 451.2205\ldots\text{cm}^2$$

From Equation (8.17),

$$\varepsilon_{V,0.99} = 2.5\sigma_V = 2.5\left(451.2205\ldots\text{cm}^2\right) = 1128.05\ldots\text{cm}^2$$

Explicitly $V = 5000. \pm 1128.$ (probable error) cm^3, or
$V = 5000. \pm 1650.$ (max error) cm^3
Implicitly $V = 5. \times 10^3\ \text{cm}^3$

The explicit maximum error of $\pm 1650.\ \text{cm}^3$ was obtained from Equation (8.6).

Example 8.10: Calculate the surface area of a right parallelepiped of sides 5. ± 1 cm, 10. ± 1 cm, and 3. ± 1 cm. The indicated measurement errors are maximum errors.

The formula for surface area is $A = 2S_1S_2 + 2S_1S_3 + 2S_2S_3$, and by substituting numerical values, $A = 190.\text{cm}^2$.

To propagate uncertainty, we need the derivative of A w.r.t. each side length. The first of the three partial derivatives is:

$$\frac{\partial A}{\partial S_1} = 2(S_2 + S_3) = 26\ \text{cm}$$

The other two are similar. Estimating sigma on each measurement as the maximum error divided by 2.5 and using Equation (8.14):

$$\sigma_A = \left[2^2(10+3)^2\left(\frac{1}{2.5}\right)^2 + 2^2(5+3)^2\left(\frac{1}{2.5}\right)^2 + 2^2(5+10)^2\left(\frac{1}{2.5}\right)^2\right]^{1/2} = 17.120747\ldots\text{cm}$$

From Equation (8.17):

$$\varepsilon_{V,0.99} = 2.5\sigma_V = 42.8018\ldots$$

Explicitly $A = 190. \pm 34.$ (probable error) cm^2, or
$A = 190. \pm 43.$ (max error) cm^2
Implicitly $A = 2. \times 10^2\ \text{cm}^2$

8.3.3 Numerical Method for Maximum Uncertainty

The analytical method presumes a) that sensitivities of one variable are independent of the value of other variables, b) that influences are linear, c) that you can obtain values for the sensitivities (derivatives), d) that the individual uncertainties are either at their extremes or that they are normally distributed, and e) independent of each other. The numerical procedure does not impose such conditions.

Here is pseudo-code for an approach to numerically propagate maximum uncertainty. It is a simple exhaustive search.

1. Assign base case values to each x_i variable, and ε_i, the maximum expected error, to each input variable.
2. Perturb the base case with each possible combination of $+\varepsilon_i$, 0, and $-\varepsilon_i$. With 3 variables and three possible perturbed values for each, there are $3^3 = 27$ cases. With n variables and three possible perturbations for each, there are 3^n number of combinations. Calculate the w-value for each combination.
3. Search through the 3^n w-values to find the maximum and minimum.

If the function is nonlinear and the ε_i values are large relative to curvature, it is possible that the extreme w-value could be at some intermediate input value, not one of the extremes, or zero. You could use a 9-point discretization of the input error: Zero, $\pm 0.25\varepsilon_i$, $\pm 0.50\varepsilon_i$, $\pm 0.75\varepsilon_i$, and $\pm 1.00\varepsilon_i$, or any such exploration that seems appropriate. The 9-point analysis would require 9^n cases.

8.3.4 Numerical Method for Probable Uncertainty

Again, the analytical method presumes a) that sensitivities of one variable are independent of the value of other variables, b) that influences are linear, c) that you can obtain values for the sensitivities (derivatives), d) that the individual uncertainties are normally distributed, and e) independent of each other. The numerical procedure does not impose such conditions.

Again, propagation of maximum error provides the worst possible outcomes, but it is unlikely that each of the many input errors are at their extreme values and simultaneously pushing the response in the same direction. Intermediate confluences of perturbations are much more likely. When there are several (or more) independent influences, the probable error will be more representative of the outcome uncertainty.

Here is pseudo-code for a Monte Carlo approach to numerically propagate probable uncertainty:

1. Assign base case values to each x_i variable.
2. Perturb each base case x_i to a possible value within the probable range, a realization. Use whatever distribution is appropriate. If uncertain, use the normal distribution.
3. Calculate w from the perturbed x_i values.
4. Record the possible w-value from that realization.
5. Repeat Steps 2, 3, and 4 a high number of times; 100 is a good starting number for the number of realizations, but you might find that the σ_w value is noticeably changing with sequential realizations. You might need 100,000 realizations to

have the σ_w value settle to a steady value (effectively unchanging with subsequent realizations). Observe how σ_w changes with additional realizations and stop when its value seems to have settled.

6. Calculate the desired statistic from the list of many w-values. This could be the maximum and minimum values, the 99% limits, the 95% limits, quartiles, the standard deviation, etc.

In Step 2, if you want to create a uniformly distributed perturbation over the possible range then use

$$x_{i,j} = x_{i\ \text{base}} + 2\left(r_{i,j} - 0.5\right)\varepsilon_{x_i} \tag{8.20}$$

Here, r is a uniformly distributed random number in the 0–1 interval, and independent for each realization. Subtracting 0.5 from r has the perturbation range from −0.5 to +0.5. The coefficient 2 converts the perturbation to $\pm\varepsilon_{x_i}$. The subscript j indicates realization number. Do the same for each influence. Use an independent r-value for each variable for each realization.

However, influences are more likely to be Gaussian-distributed. In this case the Box–Muller formula (Box, G. E. P.; Muller, Mervin E. (1958). "A Note on the Generation of Random Normal Deviates". The Annals of Mathematical Statistics. 29 (2): 610–611) is a good approach.

$$x_{i,j} = x_{i\ \text{base}} + \sigma_{x_i}\sqrt{-2\ln\left(r_{1,i,j}\right)}\sin\left(2\pi r_{2,i,j}\right) \tag{8.21}$$

Here, r_1 and r_2 are each independent uniform random numbers, $0 < r \le 1$, and independent for each variable and each realization. Here σ_x could be estimated as $\varepsilon_x/2.5$. If your random number generator provides r-values in the range $0 \le r < 1$, then it is possible to have the computer attempt to calculate ln(0). So, use $1 - r$ in the calculation rather than r.

If you feel that the distribution is Poisson, or log normal, or other, it is not too difficult to make perturbations represent realizations from those distributions. Use the random number generator to generate a random number, assign that as the *CDF* value of the distribution, then use the inverse of the distribution to get the random variable value.

Also in Step 2, you could add a cross correlation to the independent variables. For instance, if you believe that what makes y high (perhaps humidity or operator technique) also makes z high, then you could calculate the perturbation on z from that on y:

$$z = z_{\text{base}} + \frac{\sigma_z}{\sigma_y}\left(y - y_{\text{base}}\right) \tag{8.22}$$

There are hundreds of options to model forms of correlation, such as including independent variation with the correlated perturbation, autocorrelation in time-series of values, and adjusting sigma values with variable value or other variables.

We like this numerical and propagation of variance approach best and use Equation (8.21). However, it does require some computer programming, but that is not very complex. Although it does not seem as mathematically sophisticated as the calculus-derived formulas of the analytical approaches, it has fewer assumptions about linear and independent relations, and it is much easier to change the underlying distribution of variation and to add correlated variation. It avoids the errors that might arise when attempting to take calculus derivatives.

8.4 Identifying Key Sources of Uncertainty

One of the uses of uncertainty analysis is to determine what variables to consider to reduce variability of the dependent variable, to improve uniformity, to improve quality. Variance propagated is:

$$\sigma_w^{\ 2} = \left(\frac{\partial f}{\partial x}\sigma_x\right)^2 + \left(\frac{\partial f}{\partial y}\sigma_y\right)^2 + \left(\frac{\partial f}{\partial z}\sigma_z\right)^2 \tag{8.23}$$

and the fractional impact of x variation on the total is:

$$\frac{\sigma_{x\text{ on }w}^{\ 2}}{\sigma_{\text{total }w}^{\ 2}} = \frac{\left(\frac{\partial f}{\partial x}\sigma_x\right)^2}{\left(\frac{\partial f}{\partial x}\sigma_x\right)^2 + \left(\frac{\partial f}{\partial y}\sigma_y\right)^2 + \left(\frac{\partial f}{\partial z}\sigma_z\right)^2} \tag{8.24}$$

Also do this for the y and z elements. The values will sum to unity. If something has less than 0.1 relative value, then there is no justification to seek ways to reduce its impact on the dependent variable uncertainty. Focus on reducing the variation of the input variable that has the largest relative impact.

If you want to assess the individual impact in a numerical approach, then make one of the $\varepsilon_{\text{input}}$ values equal to zero and repeat the realizations and analysis with realizations of the others. The reduction in ε_w, or σ_w would indicate the impact if the ith source of variation could be entirely eliminated. You could halve the $\varepsilon_{\text{input}}$ value (or make another choice based on what you think might be a possible reduction in an influence), and look at the reduction in ε_w, or σ_w. Do this for each of the input variables to see where it is most beneficial to invest effort to seek improved quality.

8.5 Bias and Precision

Consider measurement with an instrument that is not calibrated correctly. If 1,000 people use that instrument, they will report a spectrum of values that expresses random events in the measurement process. The average of those measurements will nearly "eliminate" the random error and give the average measurement of the improperly calibrated instrument, but that average will be different from the true value.

Bias, B, is the systematic error due to instrument calibration or measurement technique. According to ASTM Standard E177-86, bias is a generic term that represents a measure of the consistent systematic difference between results and an accepted reference.

By contrast, the random error, ε, is due to random effects on a particular measurement, and precision is a generic term which is a measure of the closeness of repeated measurements. An indicator of precision, such as σ, can be calculated from replicate measurements.

However, an evaluation of bias must be obtained either by calibration to an absolute standard or by estimation and judgment. Once done, estimated bias error is individually propagated as a variance, as was random error in Equation (8.19). If $y = f(x_1, x_2, x_3,\ldots)$, then

$$B_y = \left[\sum_{i=1}^{N} \left(B_{x_i} \frac{\partial y}{\partial x_i} \right)^2 \right]^{1/2} \tag{8.25}$$

There have been several attempts to develop methods to combine B_y and σ_y, as propagated through calculations, into a single probable error measure. However, the method must be explicitly stated, and any single measure reduces the information communicated. In concurrence with the ASTM Standard E177-86 view that no formula combining precision and bias is likely to be useful, we recommend that separate statements of precision and bias should be made.

8.6 Takeaway

All calculations and values have uncertainty. Those taking action on the values need to be aware of the uncertainty to be able to make rational decisions. Report uncertainty.

This technique will be valuable to compare variability of residuals in regression (best fitting models to experimental data) to the expected values, if the model matched the phenomena being modeled.

Carefully convey how you use the term "error".

8.7 Exercises

1. Derive Equation (8.7) from (8.6).
2. Use an equation from your recent experience and define the ε_i values for each input, and each "given", and each model coefficient.
3. Derive Equation (8.19) from (8.14).
4. An elementary model for terminal velocity (the point at which gravitational pull is balanced by fluid drag) of a free-falling sphere is $v_t = \sqrt{\frac{2mg}{C_D A_p \rho_a}}$. v_t is the terminal velocity, m/s; m is the mass of the particle, kg; g is the gravitational acceleration, nominally 9.807 m/s^2; C_D is the drag coefficient, dimensionless, ~0.5 for smooth spheres in air; A_p is the cross-sectional area of the particle, m^2; ρ_a is the density of air, nominally at sea-level 1.225 kg/m^3. Consider a hailstone. The density of pure ice is nominally 934.0 kg/m^3. The diameter of pea-sized hail is about 0.01 m. With these numbers a sphere of ice would have a mass of 4.8904×10^4 kg, and a cross sectional area of 7.854×10^{-5} m^2, resulting in a simplistic terminal velocity calculation of 14.121 m/s. Consider reasonable values for the variation of particle density (which affects both mass and cross-sectional area), the particle shape (which affects cross-sectional area), and drag coefficient, and calculate the associated maximum and probable uncertainty on terminal velocity.
5. Consider a bucket and stopwatch method to measure flow rate. Weigh the empty bucket, time-interval to collect material, and weigh the full bucket. These are

the measurements, then use the data model to calculate volumetric flow rate, $F = \left(W_{\text{full}} - W_{\text{empty}}\right) / \left(t_{\text{end}} - t_{\text{start}}\right) / \rho$. Choose reasonable values for each of the five values, (W, W, t, t, ρ), estimate uncertainty on each from your experience, and calculate both maximum and probable uncertainty on *F*.

6. A simple model for car acceleration is $F = ma/g_c$, rearranging and substituting for acceleration, $\frac{dv}{dt} = \frac{g_c}{m} F = \frac{g_c}{m} \sum F_i$, and $\sum F_i$ represents the sum of all forces. One is the accelerating force of the engine, another is the rolling friction, a third is the air resistance, and fourth is the gravitational pull if the road is not level. If, for a particular car, foot-pedal movement, road angle, and speed, the nominal acceleration is 5 mph/s, what is the uncertainty on acceleration due to the range of mass that might represent the car (full or empty fuel tank, trunk, passengers)?
7. What is the maximum and the probable uncertainty on the composition of a 25/75 by weight mixture? Consider mixing 10 g of salt measured in a paper tray on a bench-top scale displaying to 0.01 g, and 30 ml of water measured in a 100 ml graduated cylinder, with tick marks for each ml. The nominal density of water is 1.0 g/ml. The salt and water are delivered to a beaker where they are mixed. Consider random error on the measurements, random error on the deliver-to quantity (some water and salt does not leave the measurement vessel). Estimate reasonable values. Consider implied uncertainty on the density. Consider reasonable values for calibration error on the two measuring devices, and independently propagate maximum and probable error due to both systematic bias and random error.
8. Use a chi-squared or an *F*-statistic to defend the guide in this chapter that if the experimental standard deviation is between about 1/3 to 3 times that predicted, they should be considered acceptable.
9. Show the transition from Equation (8.12) to (8.13). Recognize that the partial derivative values are determined at the base point, so that $\frac{\partial f}{\partial x}$ is a constant, not a function. Factor it out of the sum. Explain that $\varepsilon_{x_i} = x_i - x_0 = x_i - \bar{x}$.
10. The Poisson distribution reveals the probability of x = 0,1,2,3,... events in a time-space interval when the expected average number of events is λ per time-space interval. If λ is reported as 1 event per interval, what is the implied uncertainty on λ and the consequential uncertainty of *P*(0 events/interval), *P*(1 events/interval), and *P*(3 events/interval)?
11. Use the logistic model approximation to the Gaussian distribution, $CDF(x) = \alpha = \frac{1}{1 + e^{-\frac{\sigma}{1.7}(x-\mu)}}$, and analytically derive the uncertainty on $x_{\alpha=0.25}$ due to uncertainty on μ and σ. First invert the equation to determine x as a function of α, μ, and σ, then propagate uncertainty. If you cannot analytically obtain the derivatives, show how you would obtain values numerically.
12. Use the result from Exercise 8.11 and your choice of values for of α, μ, and σ to compare the propagation of uncertainty value to that of the Gaussian distribution. Use numerical approximations on the Gaussian *CDF* inverse.

9

Stochastic Simulation

9.1 Introduction

Simulation is an essential tool for generating data to be able to visualize and to test data-processing techniques. We encourage readers to use simulation to generate their own data, apply the techniques in this book to the data, and experience the procedure and outcomes. Hopefully, simulation will validate procedures.

This section reveals simulation techniques that are useful for methods of exploration and validation. There are two basic concepts in the following. The first is the generation of data to represent what might have been sampled from a distribution. The second is to use that data in a simulation of a process to produce surrogate data that might represent experimental results.

9.2 Generating Data That Represents a Distribution

The Excel cell function RAND() or the Excel/VBA function Rnd() generates uniformly and independently distributed values on the range of 0 to 1 (excluding 0, including 1). $0 < x \le 1$. The variable x is continuum-valued. The nomenclature is UID(0,1). The *CDF* is illustrated in Figure 9.1, where the vertical axis is the *CDF* and the horizontal axis is the continuum-valued x. To generate x-values use the cell function $x = RAND()$ or the VBA function

$$x = Rnd() \tag{9.1}$$

If you wish to exclude 1 and include 0 then use $x = 1 - Rnd()$.

Note: Figure 9.1 shows that there is a probability of 0.2 that x will have a value of 0.2 or lower. The values for x and $CDF(x)$ are identical, which will be a useful attribute in more complicated distributions.

Most data-processing packages have an equivalent random number generator.

To generate a uniformly distributed continuum variable on any interval from a to b, with $a < x \le b$, use

$$x = a + (b - a) * Rnd() \tag{9.2}$$

DOI: 10.1201/9781003222330-9

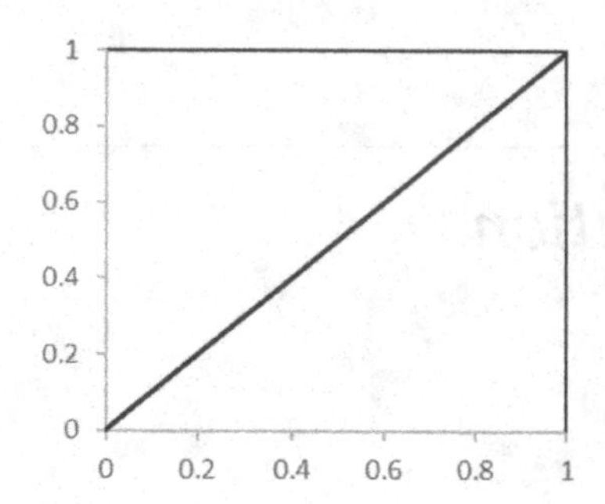

FIGURE 9.1
CDF of a uniformly distributed continuum variable on the range of 0 to 1.

To generate a uniformly distributed continuum variable that is centered on 0 on an interval from $-a$ to $+a$, with $-a < x \leq +a$, and $\bar{x} = 0$ use

$$x = 2a * (Rnd() - 0.5) \tag{9.3}$$

If you want to simulate uniformly distributed discrete events, with n possible outcomes, then multiply the complement of the standard random number by n and take the integer of the result. For example, to simulate the outcomes of rolling a $n = 6$-sided die use

$$x = 1 + \text{INT}\left[n\left(1 - Rnd(\ )\right)\right] \tag{9.4}$$

The complement of *Rnd* is needed to eliminate the extremely rare event of the *Rnd*() = 1 possible value, which would lead to a numerical value of rolling a 7.

Actually, *Rnd* is not truly a random number. It is calculated by a deterministic procedure, which generates a sequence that is seemingly random. It is truly a pseudo-random number, but it is so very nearly true to the concept that it is a very useful function.

The Excel functions are single precision valued, and some people object to the fidelity of the distribution to the uniform ideal. We find them fully adequate for most experiments, however, when the number of samples exceeds a million, the imperfections can be visible. If you need extraordinarily large sample numbers you can use a Linear Congruent Generator to generate double precision, uniformly distributed, "continuum-valued" pseudo-random numbers. The continuum-valued is in quotes because it still is limited to a finite number of values.

Perhaps the second most useful random number generation represents what might be sampled from the Gaussian (normal) distribution. Our favorite formula is the Box-Muller method to generate an independently distributed, continuum-valued random variate with mean of zero, $\mu = 0$, and standard deviation of σ, *NID*(0, σ). (Box, G. E. P.; Muller, Mervin E. (1958). "A Note on the Generation of Random Normal Deviates". The Annals of Mathematical Statistics. 29 (2): 610–611).

$$x_i = \sigma\sqrt{-2\ln\left(r_{1,i}\right)}\ \sin\left(2\pi r_{2,i}\right) \tag{9.5}$$

Here, the variables $r_{1,i}$ and $r_{2,i}$ represent uniformly distributed independent random numbers on the 0 to 1 range, the *Rnd*() function. The subscript, *i*, indicates the sample number, and $r_{1,i}$ and $r_{2,i}$ are two separate random numbers. The complement to $r_{1,i}$, $(1 - r_{1,i})$, in the natural logarithm function may be needed for some random number generators to eliminate the rare possibility of taking the log of zero.

If you wish to generate normally distributed numbers with a specified mean, μ, then translate Equation (9.5). To generate $NID(\mu, \sigma)$

$$x_i = \mu + \sigma\sqrt{-2\ln(r_{1,i})}\,\sin(2\pi r_{2,i}) \tag{9.6}$$

Example 9.1: Generate a set of $n = 10$ normally distributed independent random numbers representing what might be experimentally sampled from a population with $\mu = 15.3$ mg/L and $\sigma = 2.2$ mg/L.

Use Equation (9.6). Here is a core of the code in VBA:

```
Dim x(10)
mu = 15.3
sig = 2.2
Twopi = 2 * 3.1415926536
Randomize
For I = 1 to 10
    r1 = Rnd()
    r2 = Rnd()
    x(i) = mu + sig * Sqr( -2 * Log(r1)) * Sin( twopi * r2)
next i
```

The "Randomize" command initializes the random number generator seed value with the computer clock. This ensures that each random number sequence is different from a former realization (unless of course, you start the code at exactly the same clock time). In VBA, the "Log" function is the natural log.

Several realizations of the random numbers from this code are:

Sample	Realization 1	Realization 2	Realization 3	Realization 4	Realization 5
1	17.80781328	17.66581803	14.92054458	18.3263979	15.43748412
2	15.88920007	14.12270341	15.98023409	16.67980317	16.31679972
3	14.25760708	18.40588201	20.8927006	13.17113698	17.8311549
4	11.83099507	16.80929385	14.56222289	17.52029735	14.12315023
5	14.42720561	15.287185	12.39291446	13.24464237	14.70560673
6	11.76398997	11.48129346	12.74394403	17.57691798	18.5764986
7	15.83189451	14.98749109	13.33220284	16.21385103	13.60738408
8	13.20503636	14.43964068	14.71663892	16.24135465	13.31143105
9	11.31508163	16.16750036	16.99776206	11.00376699	18.10429335
10	14.66471639	10.82236726	14.72878776	16.43574963	12.98184958
Average	14.1	15.0	15.1	15.6	15.5
Std Dev	2.1	2.5	2.5	2.4	2.1

Note: Each realization of ten numbers is independent of the others. The averages and standard deviations are rounded to reflect the variation in their values. The standard deviation of the data uses (N – 1). The average and standard deviation are not exactly the target values, because there are only 10 observations in each sample; but they cannot be rejected as representing the target mean and sigma.

With only $n = 10$ samples, a graph of empirical *CDF* w.r.t. data value does not clearly reveal the underlying distribution. So, here is the empirical *CDF* from $n = 100$ samples, compared to the ideal normal distribution (dashed line) with the generating μ = 15.3 mg/L and σ = 2.2 mg/L.

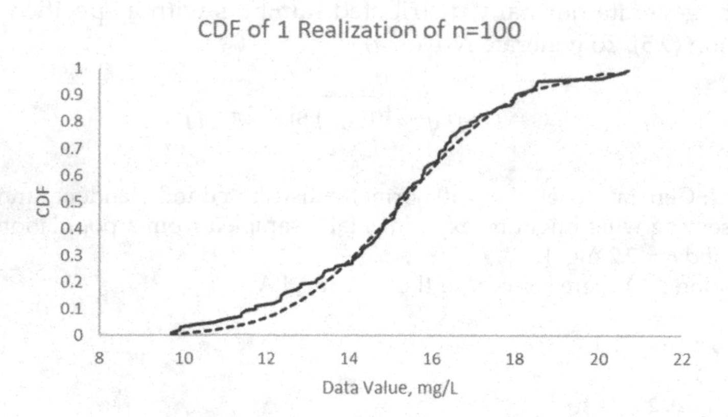

The distribution shape is close to the ideal, but even with n = 100 samples, not exactly ideal. It takes thousands of samples to have experimental results satisfyingly match the ideal distribution.

To generate random numbers with alternate distributions, use the inverse of the distribution *CDF* function. For example, the exponential distribution is $CDF(x)=1-e^{-ax}$, and its inverse is

$$x=-\frac{1}{a}Ln(1-CDF) \tag{9.7}$$

By letting the uniformly distributed random number represent the *CDF*, the exponentially distributed continuum variable is

$$x=-\frac{1}{a}Ln(1-Rnd) \tag{9.8}$$

You can use Excel Cell functions to do this for many functions. For instance, to generate normally distributed variable from r = *UID*(0,1),

$$z=\text{NORM.INV}(r,0,1) \tag{9.9}$$

Or the number of successes from a binomial process

$$s=\text{BINOM.INV}(n,p,r) \tag{9.10}$$

If either 1) you do not have a built-in distribution inverse function, 2) cannot analytically derive the inverse, or 3) the data does not seem to match any conventional distribution model, then a useful approach is to approximate the distribution with a piecewise linear approximation. Then you can use the random number representing the *CDF* and the piecewise linear inverse to determine the random variate value. First, select the break points that, when connected with straight lines, the connect-the-dot lines still adequately match all of the data. The choice for the number of break points is not critical as long as the connect-the-dots model generally matches the inverse data. The more you use, the better the match to the data will be, but the data contains uncertainty, so an exact fit just fits noise and takes more work. Use as low a number of break points as you can and still get a reasonable representation of the data. The points do not need to be equally spaced but should include 0 and 1 *CDF* values. The break points for the piecewise model do not need to exactly match any of the data. Instead, selected break points to represent a local middle of the data.

The piecewise linear model is relatively simple. Label the break points with an index $i = 1, 2, 3, 4\ldots$ from lowest to highest. Between points use a linear model to convert *CDF* to x:

$$x = x_{i-1} + \left(CDF - CDF_{i-1}\right)\frac{\left(x_i - x_{i-1}\right)}{\left(CDF_i - CDF_{i-1}\right)}, \quad CDF_{i-1} \le CDF \le CDF_i \tag{9.11}$$

Example 9.2: In a fairly recent exploration of the grades of 151 students, the empirical cumulative distribution of their grade point average (gpa) in their major is shown in the illustration.

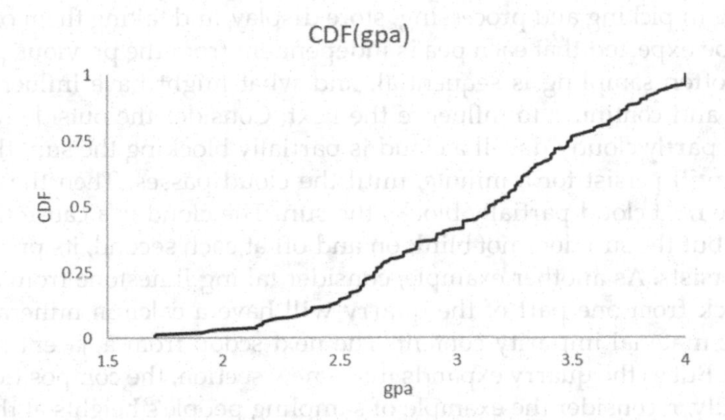

One might expect the grade distribution to be normal, and it does have the long lead-in tail, but the central and right-hand data seems linear (uniformly distributed). The empirical distribution does not seem to match the shape of any conventional distribution, so to create a simulator that mimics the empirical gpa distribution, fit the gpa values as a piecewise linear response to the *CDF*.

The above figure presents the data in the normal view – *CDF* on the vertical axis as a response to the data variable on the horizontal axis. But what is wanted is a model of the gpa that matches the *CDF*. So, reverse the axes as shown in the figure below. Choosing four break points for the linear segments (and also the two end points) the result is

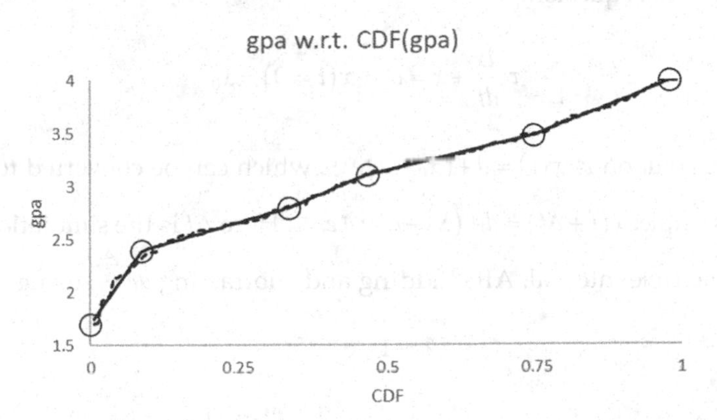

The internal break points are shown as open circles. The experimental data is the dashed line and the solid line is the piecewise linear model. There are 4 internal break points in this example. However, 2 internal break points (3 line segments) also gave a satisfying result.

9.3 Generating Data That Represents Natural Variation over Time

When samples are independent, when the value of the prior sample has no influence on the next, then the methods in Section 9.2 are appropriate. An example might be going to a county fair and sampling the height of every tenth person you meet. You expect the people to be unrelated and the heights independent. As another example, consider buying unshelled peas in the grocery store, then shelling them and counting the number of peas in each pod. Maybe peas from one plant have more peas in a pod than another plant, but with the mixing in picking and processing, store display, and taking them one at a time to shell, it would be expected that each pea is independent from the previous pea.

By contrast, often sampling is sequential, and what might have influenced the prior value, persists, and continues to influence the next. Consider the outside radiant energy of the sun on a partly cloudy day. If a cloud is partially blocking the sun, then it is likely that the shade will persist for a minute, until the cloud passes. Then the clear sky will persist until the next cloud partially blocks the sun. The cloud is a cause that makes the sunshine vary, but the sun does not blink on and off at each second, its presence persists, its influence persists. As another example, consider taking limestone from a quarry. One shovelful of rock from one part of the quarry will have a calcium mineral content and also an organic material impurity content. The next scoop from adjacent rock will have similar content. But as the quarry expands into a new section, the composition is expected to change. Finally, reconsider the example of sampling people's heights at the county fair. Child cousins and schoolmates travel together, and their parents are close by but not randomly mixed with the children. So, if you sample the height of one child, next in line will be another, and next another, and what makes one short (age) persists to the next, and there will be a sequence of short values.

In many instances, there is sequential sampling from a process that has persisting attributes, which tends to make sequential samples not have independent values representing the entire population of possible values.

A classic and very simple model of persistence that approaches a new value, d, is the first-order differential equation:

$$\tau \frac{dx}{dt} + x = d, \quad x(t=0) = x_0 \tag{9.12}$$

Analytically, the solution is $x(t) = d + (x_0 - d)e^{-\frac{t}{\tau}}$, which can be converted to an incrementally new time sample: $x(t+\Delta t) = d + (x_0 - d)e^{-\frac{t}{\tau}}e^{-\frac{\Delta t}{\tau}}$. Here Δt is the simulation time step, or sample-to-sample time interval. After adding and subtracting $de^{-\frac{\Delta t}{\tau}}$, using

$$\lambda = 1 - e^{-\Delta t/\tau} \tag{9.13}$$

and relabeling $x(t+\Delta t)$ as x_{new} and $x(t)$ as x_{prior}

$$x_{new} = \lambda d + (1-\lambda) x_{prior} \tag{9.14}$$

If the influence, d, is not a constant, but continually changes, then the new x-value is influenced by the new d-value $x_{new} = \lambda d_{new} + (1-\lambda) x_{prior}$, which is the implicit numerical representation of the differential equation. If the forcing function for the differential equation, d, is modeled as randomly changing in a Gaussian manner with a mean of 0, then the first-order persistence driven by NID(0,σ) noise is modeled as:

$$x_{new} = \lambda \sigma \sqrt{-2\ln(r_1)}\ \sin(2\pi r_2) + (1-\lambda) x_{prior} \tag{9.15}$$

If $\Delta t \ll \tau$, which is usually the case in simulation of real events, then the simplified approximation $\lambda = \Delta t/\tau$ can be used. In creating a simulation, the user would choose a time-constant for the persistence that is reasonable for the effects considered, and a σ-value that would make the disturbance have a reasonable variability. Since the first-order persistence tempers the response to the random influences, the σ-driver in Equation (9.15) needs to be greater than the desired response σ_x.

How should one choose values for λ and σ in the simulator? First consider the time-constant, τ, in Equation (9.12). It represents the time-constant of the persistence of a particular influence. In an ideal mixer it would be volume divided by flow rate. Roughly $\tau \approx 1/3$ of the lifetime of a persisting event. (Because the solution to the first-order differential equation indicates that after 3 time-constants, x has finished 95% of its change toward d.) So, if you considered that the shadow of a cloud persists for 6 minutes, then the time-constant value is about 2 minutes. Once you choose a τ-value that matches your experience with nature and decided for a time interval for the numerical simulation, Δt, calculate λ from Equation (9.13).

To determine the value for σ in Equation (9.15), propagate variance on Equation (9.14). This indicates that the σ-value depends on the user choices of σ_x and λ (which is dependent on Δt and τ).

$$\sigma = \sigma_x \sqrt{\frac{2-\lambda}{\lambda}} \tag{9.16}$$

Choose a value for σ_x, the resulting variability on the x-variable. To do this, choose a range of fluctuation of the disturbance. You should have a feel for what is reasonable to expect for the situation that you are simulating. For instance, if it is barometric pressure the normal local range of low to high might be 29 to 31 inches of mercury, if outside temperature in the summer it might be from 70 to 95°F, or if catalyst activity coefficient it might be from 0.50 to 0.85. The disturbance value is expected to wander within those extremes. Using the range, R, as

$$R = \text{HIGH} - \text{LOW} \tag{9.17}$$

And the standard deviation, σ_x, as approximately one-fifth of the range, then

$$\sigma = \frac{R}{5} \sqrt{\frac{2-\lambda}{\lambda}} \tag{9.18}$$

The reader is welcome to use other methods for generating NID(0,σ) values, alternate models for disturbance persistence, or other distributions that are deemed appropriate. For example, in particle size results the distribution could be log-normal, or in queuing or radiation intensity it could be Poisson. Section 9.2 indicated how to generate noise and disturbances with alternate distributions.

The standard deviation of the noise could be proportional to the measurement, as in orifice flow-rate measurement, in which fluid turbulence causes the pressure drop perturbations, and turbulence is proportional to the flow rate. The disturbances could have a higher order autoregressive moving average behavior mimicking environmental conditions or blended raw material properties. And, additional disturbances could be impacting model coefficients related to fouling factors, efficiency factors, etc.

Equation (9.14) is variously termed an exponentially weighted moving average, an autoregressive moving average model of orders 1 and 1, ARMA(1,1), a first-order filter, or a first-order lag. Since computer assignment statements update the new variable value from the old, in implementation, the subscripts are not needed.

$$x := \lambda \sigma u + (1-\lambda)x \tag{9.19}$$

Using the symbol ":=" to represent a computer assignment, and u is the NID(0,1) normal random variate. Structured text would be:

```
U := Sqr(–2 * Log(rnd())) * Sin(twopi * rnd())
X := lambda * sigma * u + (1–lambda) * x
```

An illustration of one 200-sample realization with tau = 20 and Range = 10 is shown in Figure 9.2.

Note: The first drift into positive values lasted for about 30 samplings and the next drift into low values lasted for about 90 samplings. The lifetime of any one particular event is approximately 3 time-constants, and should be about 60 samples, but continual changes in the driver mask the action of past events. The period of persistence, high or low, is consistent with the specified time-constant.

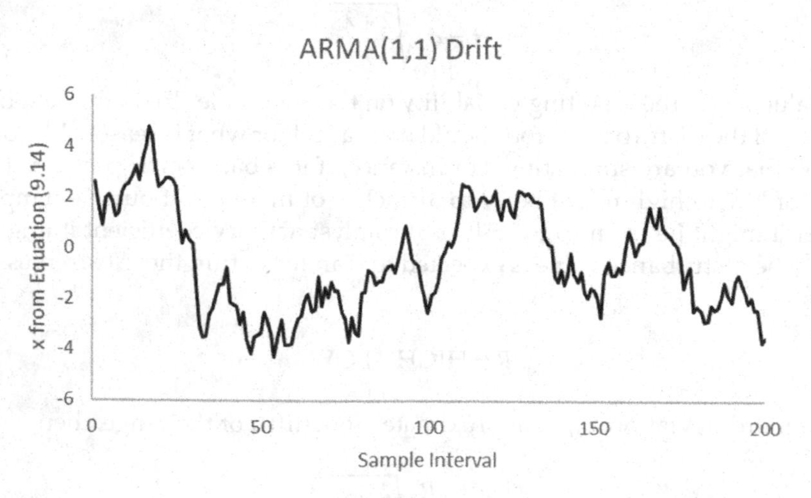

FIGURE 9.2
A realization of an ARMA(1,1) drift.

Note also: The range of the data in these 200 samples is about 10, when the model was designed for a range of 10. In this limited realization of 200 samples neither the full range of the disturbance value nor the full range of its persistence are expected to be revealed.

This approach modeled a drifting influence as a first-order response to a Gaussian distributed driver. One certainly could model drifts as higher-order or driven by alternate forcing function distributions. However, in the authors' experience this adequately mimics how randomly driven drifts in natural disturbances appear to respond and provides a relatively simple method to both execute and parameterize.

The variable x randomly drifts about a value of zero. You can shift it to fluctuate about any given value x_{base}. Also you could add additional random independent perturbations to mimic noise as well as drifts.

```
u := x_base + Sqr(-2 * Log(rnd())) * Sin(twopi * rnd())
x := lambda * sigma * u + (1-lambda) * x
n := sigma_noise * Sqr(-2 * Log(rnd())) * Sin(twopi * rnd())
y := x + n
```

9.4 Generating Stochastic Models

Simulators use models of phenomena (processes, equipment) to indicate how the physical item might respond to influences. Models take inputs and predict responses.

It might be a stationary (steady-state, end of trial, end of batch, or equilibrium) model. For instance, the amount of salt used in a recipe would be the input, and the diner's opinion of flavor would be the response. Or the automobile accelerator position as the influence, and the final speed it achieves as the model output. Or the pressure over the vapor of a liquid mixture as the influence, and the equilibrium concentration in the vapor space as the response.

Alternately, it might be a transient (or dynamic, time-dependent) model. For instance, predicting how the speed of a car responds over time to the car accelerator position.

The influences on a model are usually considered to be the humanly chosen input variables (alternately termed decision variables, independent variables, or controlled variables). And often we think that these are the only variables that will be subject to change. By contrast, there are also uncontrolled environmental influences and time-changing attributes of the process or device (ambient temperature, raw material purity, catalyst activity, tire inflation pressure, heat exchanger fouling, etc.). And also, there are calibration drifts and noise on the measurement.

Further, even the "controlled" variable is not perfectly controlled! The common view is that an experiment is exactly controlled, but inputs to an experiment are not exactly controlled. Inputs are the result of the final control element (valve, knob, dial, resistor, etc.) which is physically set or selected to make an input reading (flow rate, temperature) have a desired value. Either way, the true value of the input is not the desired value, it is the measured value adjusted for the unknowable measurement error.

Four types of variables have variation – Inputs, Environmental Variables, Process Coefficient Values, and Measurements.

To create a simulator, first define the equation or procedure to obtain y_{true}, the truth about nature, from $x_{desired}$, the desired value of the influence.

$$y_{true} = f\left(x_{desired}, p_{ideal}, e_{ideal}\right) \tag{9.20}$$

The p_{ideal} represents the ideal values of the process or equipment characteristics (catalyst reactivity, tire air pressure, heat exchanger fouling, fluid flow pressure loss coefficient, etc.). The e_{ideal} represents the ideal values of the environmental influences (temperature, %RH, raw material composition, etc.). Before using the model, first perturb each variable in each of the three categories by appropriate disturbances (drifts) and noise. Then use the model to calculate the true response. Then finally, perturb the response with measurement drift and noise:

$$x_{actual} = x_{desired} + d_x + n_x \tag{9.21a}$$

$$p_{actual} = p_{ideal} + d_p + n_p \tag{9.21b}$$

$$e_{actual} = e_{ideal} + d_e + n_e \tag{9.21c}$$

$$y_{model} = f\left(x_{actual}, p_{actual}, e_{actual}\right) \tag{9.21d}$$

$$y_{measured} = y_{model} + d_y + n_y \tag{9.21e}$$

Although the disturbances, d, and noise, n, are added here, it might be more appropriate to make them a multiplicative factor when you anticipate that the noise level or perturbation magnitude scales with the input value.

Example 9.3: What is the distribution of terminal velocity of pea-sized hail?

When a free object is falling, the downward accelerating force is gravity, but air resistance provides an opposing force. We will ignore the relative inconsequential buoyancy force in this case and will use common physics/engineering models. The gravitational force is $F_g = mg / g_c$. The drag force is modeled as $F_D = C_D A_p \rho_a \frac{v^2}{2g_c}$. At terminal velocity the two forces are equal, resulting in $v_t = \sqrt{\frac{2mg}{C_D A_p \rho_a}}$.

This is a stationary model. The calculated variable is not time dependent.

v_t is the terminal velocity, m/s
m is the mass of the hail particle, kg
g is the gravitational acceleration, nominally 9.807 m/s^2
C_D is the drag coefficient, dimensionless, ~0.5 for smooth spheres in air
A_p is the cross-sectional area of the particle, m^2
ρ_a is the density of air, nominally at sea-level 1.225 kg/m^3

The density of pure ice is nominally 934.0 kg/m^3. The diameter of pea-sized hail is about 0.01 m. With these numbers a sphere of ice would have a mass of 4.8904×10^{-4} kg, and a cross sectional area of 7.854×10^{-5} m^2, resulting in a simplistic terminal velocity calculation of 14.121 m/s.

This would be a nominal value, and for a solid sphere of ice. Let's consider some possible examples of random perturbations to the nominal values and more reasonable nominal values.

The drag coefficient has a nominal value of 0.5, implying an uncertainty on the 5. Using an average uncertainty of 0.25, and a uniform distribution, the stochastic simulation of a possible drag coefficient values is $C_D = 0.5 + 2(.25)(Rnd(\ ) - 0.5)$.

Air density is affected by temperature and pressure, which change with elevation and time. Since there are many sources of influence, we'll model the variation of density as normally distributed and a sigma of about 3% of the nominal values seems reasonable. So, the stochastic simulation of air density affecting terminal velocity will be modeled as $\rho_a = 1.225 + 0.03(1.225)\sqrt{-2Ln(r_1)}\sin(2\pi r_2)$.

Hailstones are usually not a solid sphere of ice, but an agglomeration of many particles. They have interstitial spaces. The maximum density would be 934.0 kg/m^3. A minimum might be half that. Without any specific knowledge, model the distribution of hailstone density as uniform: $\rho_p = 943.0 - 471.5 * Rnd()$.

Finally, particle size is usually log-normally distributed. Rather than using the inverse of that, use a piecewise linear approximation, with these points:

Diameter, m	*CDF*
0.0050	0.00
0.0075	0.20
0.0100	0.50
0.0120	0.75
0.0200	1

There are other aspects of variation. These include the irregular shape of hailstones, wind up-drafts, etc. This analysis idealized the sources of variation.

However, with this set of considerations, the results of 1,000 realizations are illustrated in the figure.

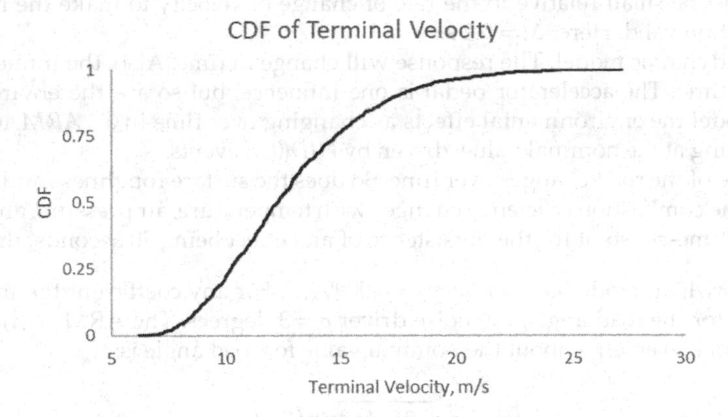

In this realization, the average terminal velocity, the 50th percentile value is about 12.3 m/s. The quartiles are about 10.2 and 14.8 m/s. The range is from about 6 to 25 m/s.

Note: Due to nonlinearities in this example, the distribution is neither symmetric nor centered on the nominal value of 14.121 m/s.

Note: Initially we used 100 realizations but recognized that one set of realizations gave different (although somewhat similar) results to another. With 1,000 realizations the set-to-set variation is small relative to the range of values.

This terminal velocity treatment is simplistic. More rigorous models could include the buoyancy effects, the velocity impact on the drag coefficient, and a more representative exponent on the velocity. Further, more variables have variation (such as the shape of the hailstones) than are considered here. And, the distributions included here are chosen to reveal three separate approaches, not because they are phenomenologically correct.

Example 9.4: A car speed model is $F = ma/g_c$, rearranging and substituting the rate of change of speed for acceleration, $\frac{dv}{dt} = \frac{g_c}{m} F = \frac{g_c}{m} \Sigma F_i$, and ΣF_i represents the sum of all forces. One is the accelerating force of the engine, simplistically represented as proportional to the accelerator pedal position, u, $F_{\text{engine}} = ku$. Another is the rolling friction, due to mechanical rubbing and tire deformation, simplistically represented as proportional to speed, $F_{\text{rolling}} = av$. Another is the aerodynamic drag, simplistically represented as proportional to square of the difference between car and air speed, $F_{\text{aero}} = b(v - v_{\text{air}})^2$. The last to be considered is the gravitational pull when going up or down hills, proportional to the weight of the car and the sine of the climb angle from horizontal, $F_{\text{gravity}} = (mg / g_c)\sin(\theta)$. Combining and merging coefficients

$$\frac{dv}{dt} = \alpha u - \beta v - \gamma (v - v_{\text{air}})^2 - \delta \sin(\theta).$$

Here, $0\% \le u \le 100\%$, the velocity of both the car and the air is in mph, and time is in seconds. The nominal numerical values are $\alpha = 0.2$ mph/%s, $\beta = 0.05$ s^{-1}, $\gamma = 0.001$ s^{-1}-mph^{-1}, and $\delta = 22$ mph/s.

Using the explicit finite difference $\frac{dv}{dt} \cong \frac{\Delta v}{\Delta t} = \frac{v(t+\Delta t) - v(t)}{\Delta t}$, the simulator is

$$v_{t+\Delta t} = v_{t+\Delta t} + \Delta t \left[\alpha u - \beta v - \gamma (v - v_{\text{air}})^2 - \delta \sin(\theta) \right]_t .$$

Given prior values, you can calculate the incrementally next velocity. The time interval, Δt, must be small relative to the rate of change of velocity to make the numerical approximation valid. Here, $\Delta t = 0.2$ s.

This is a dynamic model. The response will change in time. Also, the influences will change in time. The accelerator pedal is one influence, but so are the environmental effects. Model the environmental effects as changing over time in an *ARMA*(1,1) manner, averaging at the nominal value, driven by *NID*(0,σ) events.

The angle of the road changes over time. So does the surface roughness and the wind velocity. The combustion efficiency changes with temperature, air pressure, and humidity. With a time-constant for the persistence of any effect being 20 seconds, the lambda for the *ARMA*(1,1) model is $\lambda = 1 - e^{-\frac{\Delta t}{\text{tau}}} = 0.00995\ldots$. For any coefficient the model will be similar, for the road angle, the noise driver $\sigma = 3$ degrees. The ARMA(1,1) model of Equation (9.14) averaging about the nominal value for road angle is:

$$d_t = \sigma \sqrt{-2Ln(r_1)} \sin(2\pi r_2)$$

$$\theta_t = (1 - \lambda)\theta_{t-\Delta t} + \lambda \theta_{\text{nominal}} (1 + d_t)$$

Drift models for other terms are similar. At each simulated time step, first calculate the new values for the model coefficients that change in time, then use the finite difference model to calculate the incrementally next velocity.

Here is one realization of the car speed over time subject to changes in the four variables – air speed, road friction, engine power, and road angle:

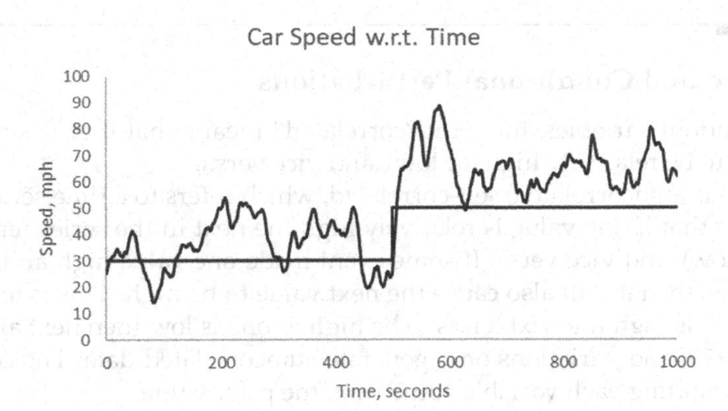

The horizontal line that makes a step change is the pedal position, moving from 30% to 50% at a time of 500 seconds.

The before and after pedal position change durations are 500 seconds, or about 8 minutes. Does the variation in car speed match what you think it should if the accelerator pedal position was fixed and wind and hills affected car speed? Before you claim a simulator is realistic, check that it matches your experience.

During either period of constant accelerator pedal position, the range of the speed is about ±20 mph, and the persistence (an eyeball average of the time above average, and time below average) is about 70 seconds. That persistence value indicates a generating time-constant of about 23 seconds, affirming the models used, but it is barely enough data to make a confident claim that the simulation gives expected results. The ±20 mph variation is on the car speed, a consequence of the several factors changing. So, one would have to propagate uncertainty through the model to see if it is as expected.

9.5 Number of Realizations Needed for Various Statistics

A few stochastic simulations will generate an outcome, but it will not represent the population value. It will have some variation. A new realization of the few stochastic simulations will result in a different outcome value. If the outcome is normally distributed, then the central limit theorem can be used to estimate the number of simulations needed to return an average that is within a tolerable range. The uncertainty on average scales with the square root of the number of individual data. So, if N_1 simulations result in an outcome that has a standard deviation of σ_1, but it is desired to have an outcome that has variation, σ_2 then the number of simulations must be

$$N_{\text{required}} = N_1 \frac{\sigma_1^2}{\sigma_{\text{desired}}^2} \tag{9.22}$$

The uncertainty on a probability is $\sigma_p = \sqrt{pq/n}$, so after a few trials, that provides a reasonable estimate of p, the number of trials that would result in a desired uncertainty on the probability is

$$N_{\text{required}} = \frac{p(1-p)}{\sigma_{\text{desired}}^2} \tag{9.23}$$

9.6 Correlated and Conditional Perturbations

Referring to random variables, the term "correlated" means that if one is relatively high the other tends to be relatively high (or low), and vice versa.

A special case is autocorrelated, self-correlated, which refers to a time series of one variable, and means that if one value is relatively high the next in the series tends to be relatively high (or low), and vice versa. If some event made one value high and the event has some persistence, then it will also cause the next value to be high. This is termed positive correlation, if one is high the next tends to be high, if one is low then next also tends to be low. Equation (9.19) and variations on it generate autocorrelated data. The correlation can be observed by plotting each variable value w.r.t. the prior value.

Negative autocorrelation can happen. If one measurement is high and there is something to "fix" it, such as an overly aggressive controller, then excessive control action might overcompensate, and the next measurement might tend to be low, and vice versa. (This can happen in management of an organization or a country, as well as with automatic control of temperature.) To generate negatively autocorrelated data use a negative lambda value.

Cross correlation means that if one variable is high then another variable will tend to be high or low. For example, if a person has a lot of gray hair (one variable) then they are likely to also have many face wrinkles (another variable). That would be a positive cross correlation, high means high. As another example, the more breakfasts a person eats (one variable) the shorter is their remaining life expectancy (another variable). That is a negative cross correlation. Hopefully, these examples reveal that cross correlation does not prove a cause-and-effect relation, because both variables could be the result of another. The speed of a car does not increase because road noise increases, they are both consequences of the accelerator pedal position.

Conditional probability can be easily expressed in simulation. The mean, or sigma, or other perturbation generator coefficient, can be related to a desired variable. It could be IF-THEN models, Fuzzy Logic models (or other Natural Language Processing models), linear models, exponential models, etc. as appropriate.

9.7 Takeaway

Use simulators to validate your techniques and understanding.

As Example 9.1 indicates, even 100 realizations only created a fair realization of the true distribution. You may need to run thousands of simulations with randomly perturbed variable values to clearly see and quantify the resulting distribution.

Critically evaluate the simulator results to see that it makes sense with your experience.

9.8 Exercises

1. Use Equation (9.1) to generate uniformly distributed variate. Use the $CDF(x)$ to show that it is uniform over the range of 0 to 1. Balance the work of the number of data you generate with acceptable sufficiency that it works as claimed.

2. Use Equation (9.2) to generate uniformly distributed variate over a desired range. Use the *CDF*(*x*) to show that it is uniform over the desired range. Balance the work of the number of data you generate with acceptable sufficiency that it works as claimed.
3. Derive Equation (9.3) from (9.2).
4. Show that Equation (9.4) generates uniformly distributed integer values for n = 1, and n = 6. Balance the work of the number of data you generate with acceptable sufficiency that it works as claimed.
5. Show that Equation (9.5) generates $NID(0,\sigma)$. Balance the work of the number of data you generate with acceptable sufficiency that it works as claimed.
6. Show that Equation (9.6) generates $NID(\mu,\sigma)$. Balance the work of the number of data you generate with acceptable sufficiency that it works as claimed.
7. Show that the inverse of the logistic function, as a *CDF* model, generates reasonably Gaussian distributed random variables with appropriate mean and sigma. The logistic function is $CDF = 1/\left(1+e^{s(x-c)}\right)$. Use $c = \mu$, and $s = -\sigma/1.7$. Balance the work of the number of data you generate with acceptable sufficiency that it works as claimed.
8. Show that Equation (9.8) generates exponentially distributed values with the desired mean and sigma. Balance the work of the number of data you generate with acceptable sufficiency that it works as claimed.
9. Show that Equation (9.11), the piecewise linear approximation to the *CDF* inverse, generates random values from a desired distribution. Balance the work of the number of data you generate with acceptable sufficiency that it works as claimed.
10. Use Equation (9.15) to generate a sequence of autocorrelated numbers. Use Equations (9.13) and (9.16) to determine model coefficients. Generate data and show that the mean is about zero and the standard deviation is as expected. Show that the persistence is as expected. Balance the work of the number of data you generate with acceptable sufficiency that it works as claimed.
11. What is the composition of a 25/75 mixture? Consider that scales are used to weigh 25 g of salt, and that a beaker is used to deliver a volume of 75 ml of water with a density of 1 g/ml. Ideally this results in 25 g of salt mixed with 75 g of water, which is a 25/75 mix. But of course, there is uncertainty in reading the graduated cylinder, a drop or so might not come out, the scales have a discretized display of 0.01 g intervals, and neither are perfectly calibrated. If hundreds of labs were using independent equipment to create a 25/75 mix, what is the distribution of actual compositions?
12. Calculate the probable error of the surface area of a right parallelepiped of sides 5. ± 1 cm, 10. ± 1 cm, and 3. ± 1 cm. The indicated measurement errors are maximum errors. The formula for surface area is $A = 2S_1S_2 + 2S_1S_3 + 2S_2S_3$, and by substituting numerical values, A = 190. cm^2. Use the method of Example 9.3 Compare the results to that indicated in Example 8.9.
13. Derive Equation (9.13) from (9.12).
14. Propagate variance on Equation (9.14) to determine the value for sigma on d to create the desired sigma on x, as revealed in Equation (9.16).
15. Start with Equation (9.13) and show that if $\Delta t << \tau$ then $\lambda \cong \Delta t / \tau$.

Section 2

Choices

10

Choices

10.1 Introduction

Statistics is an exact mathematical science. However, when applied in decision-making, there are many human choices that need to be made. The choices can make the statistical outcome go one way or the other – to reject a treatment or to accept the treatment. The choices could be made to support a person's bias or preference – to justify a particular action or to condemn the action. This chapter is not about the science of statistics, but about the cautions needed when applying statistics to support or direct decisions.

This chapter about nonscience, about nonmathematical truth, may be the most important in this book! If you are using statistics to make claims about reality, you need to be aware of the possible distortion your choices could make. If you are the audience of a statistical report, you need to understand the potential distortions that the person doing statistical analysis could be making (either intentionally to distort, or unintentionally).

10.2 Cases

The choices include formulating the hypothesis, choosing the level of significance, choosing the number of data, ensuring that data are independent, and many other issues related to the context of the decision to be made.

10.2.1 The Hypothesis

The hypothesis could be either equivalent-to or better-than. Either $X = Y$ or $X > Y$. The less-than hypothesis is equivalent, just switch the X and Y classification. With continuum-valued variables, the greater-than-or-equal-to is really no different from the greater-than condition when you consider that when enough digits are revealed the experimental statistic will not have the exact same value as the critical value of the statistic. In mathematics a point has no width, and theoretically the $\geq$ and $>$ conditions are different. Of course, if you truncate or round the experimental statistics and critical values to a few digits, or are using counts, you might need to include the $\geq$ condition.

There are several aspects of choosing a hypothesis. First, you must decide what aspect of the comparison is going to become the test statistic. For instance, consider that X and Y are two raw material suppliers. If they are equivalent, then either could be used. If equivalent means that the yield of manufacturing outcome is the same, then that is what you test. But equivalent might be considered as product uniformity, delivery reliability, energy used to convert the raw material into your product, or any of many valid metrics. If you interpret the X is equivalent to Y in the myopic view that this means product yield is equal, and only test manufacturing yield, your claim of "reject" or "accept" may be true about that single aspect, but it may not characterize the entire reality of the context. Be sure the hypothesis is comprehensive. The accept or reject equivalence might require several tests on each of the context-relevant aspects.

Second, you will choose the equivalent-to or better-than hypothesis. This can shape the statistical conclusion to fit your bias. If you choose equivalent-to, then you have a two-sided test and might reject a treatment that is better-than. If you are currently using, and have an allegiance to treatment X, and you want to reject treatment Y, especially if it is better, then choose the equal-to hypothesis. You could legitimately claim "the hypothesis that Y was equivalent to X has been rejected". However, if you want to increase the chance that Y is accepted then choose the $Y > X$ hypothesis. Then if not rejected, even if they are equivalent, you can claim, "Our testing indicates that we accept $Y > X$". The outcome of the statistics will likely be shaped by your choice of the hypothesis. Don't let your bias or agenda misrepresent the reality within the context. At least, support an "accept/reject" conclusion with the associated p-value. Perhaps, also reveal the results with alternate hypotheses choices.

The hypothesis might be that the treatments X and Y are economically equivalent. As an example, consider that the treatments are raw material suppliers for a process. The economics would include many factors, such as price, receiving packaging, waste, possible processing speed, yield, and many more. X may be better in some attributes and worse in others. The hypothesis would not be on one attribute, but on the holistic economic impact. If the current supplier is X, and Y is being qualified by tests, then there will be an additional cost if Y is accepted. This would represent the management of change (MoC), which would include the revision of operating documents to acknowledge the change, the additional burden of tracking which raw material is in which products, etc. It also considers the cost of the trials. So, the hypothesis should not be simply based on processing economics, but it should include the MoC costs. In order to accept Y, the economic benefit of Y, $\dot{E}_Y$, should exceed X by the MoC. If the MoC is a one-time cost, and the economic benefits of X and Y are annual benefits (rates), then the two need to be normalized. The MoC could be normalized by a target Pay-Back-Time. Then the t-statistic for comparing X and Y would be

$$T = \frac{\dot{E}_Y - \left(\dot{E}_X + \frac{\text{MoC}}{\text{PBT}} \right)}{\sigma_{\text{pooled on } \dot{E}}} \tag{10.1}$$

Be sure that all the numerator and denominator terms have the same units.

Further, it may be difficult to economically quantify some important factors. One such factor could be the peace of mind that having an alternate supplier brings to robustness to a supply chain failure, if there is a single supplier. Another might be the potential leveraging of relations to supplier Y that are related to other initiatives, maybe even a possible merger. If there are such concerns, apply a value to the sum of all of them, and an equivalent value, $\dot{E}_{\text{equivalent}}$, to the nominal $\dot{E}$. Then structure the T statistic as

$$T = \frac{(\dot{E}_Y + \dot{E}_{\text{equivalent}}) - \left(\dot{E}_X + \frac{\text{MoC}}{\text{PBT}} \right)}{\sigma_{\text{pooled on } \dot{E}}} \tag{10.2}$$

Again, be sure that the terms are dimensionally consistent.

10.2.2 Truncating or Rounding

Conventionally, the rule is to carry two more digits than are justified by uncertainty, then to round the answer to only report significant digits. But humans can choose to truncate rather than round or can choose the number of digits that are significant. If the statistic is close to the critical value, then these human choices can make a reject outcome become accept, or vice versa. For instance, if the critical value is 2.48452 and the experimental value is 2.4199, you would not reject it. But if you claim there are only two significant digits and truncate both to 2.4, you could show the audience that the conclusion is to reject.

10.2.3 The Conclusion

The conventional conclusion is to either accept or reject. But this does not reveal how close the situation was to the alternate conclusion. So, be sure to indicate to your audience how close the claim is. Report a p-value or report the level of confidence that would generate the alternate conclusion.

10.2.4 Data Collection

Design the experiment to represent the population, all conditions, not just a subset of environmental factors, equipment condition, operators, etc. Randomize experimental order to minimize correlation.

Consider that you have historical operating data on raw material from supplier *X*, and before the process equipment is to be taken down for maintenance you run tests on material from supplier *Y*. If the equipment needs maintenance, then the equipment condition may have a larger impact on manufacturing statistics than the difference between material from the two suppliers.

Consider that when operators are on the day shift, and in the presence of process engineers, maintenance personnel, and plant management, they may pay closer attention to manufacturing conditions than when they are on the evening and night shifts. If you choose to run tests of a new treatment during the day shifts (so that it can be properly supervised) or on the evening shifts (so that the trials don't interfere with daytime activities), then the tests will not represent the same conditions as all production.

10.2.5 Data Preprocessing

Grouping data into bins is commonly performed to see the histogram, or to use chi-squared tests. But if the bin is too wide, all the data fits into one, and the detail is not visible. Alternately, if there are too many bins, perhaps as many as there are data, the histogram with just 0, 1, or 2 counts in each bin just reveals noise. The user must choose bin centers and bin widths. However, in doing so, the user can make it appear that treatment *X* is different from treatment *Y*.

As a political parallel, gerrymandering is the practice of manipulating political boundaries so that number of delegates from votes will favor the control of a political party.

If you wish, you can pretend that you have a good reason for your choice of the selection of bins, when the real reason is to bias the outcome of a statistical test. But don't.

10.2.6 Data Post Processing

Elimination of outliers should be a permissible activity. If, after the data analysis, it is realized that a data point, or sequence of data, had been corrupted, and does not represent the population, then it should not be included in statistics that seek to describe the population.

Like allocating data to bins in preprocessing, the user can choose to not seek outliers, if the data suits a personal bias, or use any number of justifications to select outliers if doing so makes the remaining data support a bias. Don't be so tempted.

There are several accepted statistical practices for detecting outliers, but these are associated with the requirement that the action be reported, and that some legitimate mechanism be identified to justify the data rejection. See Chapter 15 for one criterion, Chauvenet's.

10.2.7 Choice of Distribution and Test

If the distribution of the data is normal, then a t-test would be appropriate for assessing the difference between means. But if the statistic is a proportion, or if the results are Poisson or Binomially distributed, the normal approximation may not be valid. In nonnormal cases, if the value of the statistic is near to an extreme value or generated from a low N, then the t-test would not be appropriate.

Similarly, F-tests should be used on variances, and nonparametric tests should be used on medians and runs. Choose the test that is appropriate to the distribution of the data.

Of course, if you do not like a particular outcome from a test, you might be tempted to see if an alternate test returns the decision you might favor. But, please don't shape your test to support your desired result.

So, defend your choice of a test as correct for the distribution of the data.

10.2.8 Choice of N

If you want to accept the hypothesis, use a small number of trials to generate the statistic. In such a case, there will be so little certainty about the statistic value, that you will not be able to confidently reject the hypothesis. Alternately, if you want to reject the null hypothesis keep collecting data. Eventually, the statistic will be beyond the critical value. If two treatments are identical, and $\alpha = 0.05$, then in 5% of the testing, the statistic will be beyond the critical value. Keep collecting data until this happens, then you can claim reject with statistical validity.

Classically in statistics, the number of data is defined by choices of α (the Type 1 error, the probability of rejecting the hypothesis when it is true), and β (the Type 2 error, the probability of accepting the hypothesis when it is false, which also depends on a specified difference between the rejectable and true statistic). Of course, even in the idealized mathematics and truth of statistics, N, alpha, and beta are human choices. Be sure you can defend the legitimacy of your choices within the situation context.

10.2.9 Parametric or Nonparametric Test

Both test types are legitimate. If you want to accept a hypothesis, and a parametric test rejects it, then you may be tempted to try a nonparametric test. Parametric tests are grounded in the coefficient values (parameters) of a particular distribution. For example, the parameters of a normal distribution are the mean and variance. Predicated on a particular data distribution, parametric tests are more sensitive than nonparametric tests.

Parametric tests can reject hypotheses with fewer data. If the parametric test is appropriate, don't discard it because you want an alternate outcome.

10.2.10 Level of Confidence, Level of Significance, T-I Error, Alpha

The four terms, Level of Confidence, Level of Significance, T-I Error Probability, and Alpha are equivalent in hypothesis testing. Three of the terms, Level of Significance, T-I Error Probability, and Alpha are identical. They range between 0 and 1, and level of confidence is 100 $(1 - \alpha)$. If you are using the classic dichotomous hypothesis testing, you use α to define the critical value of the statistic, the boundary between the reject and the accept results.

For instance, consider the test is two-sided using a t-statistic to compare Treatments X and Y. The null hypothesis is that $X = Y$, and we will reject if X is confidently either greater or less than Y. If there are $n = 10$ replicates, and the experimental value is $T = 2.2153$, then a choice of $\alpha = 0.05$ (the 95% confidence) has a *t*-critical value of $t_{\nu,\alpha/2} = 2.2622$ would not reject the equivalence hypothesis. The claim would be to accept that X and Y are equivalent. The report might be, "I can state with a 95% confidence that X and Y are equivalent." However, at the 90% confidence, with $\alpha = 0.1$, the t-critical value is $t_{\nu,\alpha/2} = 1.8331$, and the claim would be to reject the hypothesis that X and Y are equivalent. Now the report might be, "I can state with a 90% confidence that X and Y are different." Both 90% and 95% confidences are fairly strong, and they could be used to influence an audience to accept either claim. If you want product Y to be accepted as equivalent to X, then use the 95% confidence. If you want the company to reject product Y and favor X, then use the 90% confidence.

The data processor's choice can distort the situation. So, reporting a *p*-value is strongly recommended to let the audience know how close the accept/reject decision is. In the above case the *p*-value is $\cong 0.054$, which has a corresponding reject confidence of 94.6%. The interpretation is: If X and Y are equivalent, and the experimental results are normal, the T-statistic value should be about zero. There is only a 5.4% chance that the experimental T-value could be as extremely nonzero as the 2.2153 value (or higher value) obtained.

The 5.4% value can be placed in perspective: If X and Y are not equivalent, if their means difference, $\mu_X - \mu_Y = \bar{X}_X - \bar{X}_Y$, is $2.2153\sigma / \sqrt{10}$, then the $T = 2.2153$ result would be expected. (Recall that $T = \left|\bar{X}_X - \bar{X}_Y\right| / \left(\sigma / \sqrt{N}\right)$.) There is a 50% chance that the experiment would generate data with that, or a higher, value. So, if one hypothesis has a 5% chance of such a high or higher value, and the other hypothesis has a 50% chance, the 10:1 ratio of probabilities (odds) favors the not-equivalent case. We should reject the equivalent hypothesis.

Note: The 10:1 odds of that example are like flipping a fair coin 4 times and losing each time, four in a row. It is certainly a possible outcome, but not expected.

Whether the odds are 9:1 or 11:1, it seems the better decision is to reject the null hypothesis. Whether alpha is 7% or 5% or smaller, the odds indicate a reject decision. There is not certainty, just probable cause. Classic statistical testing uses values in the range of 90% to 99% confidence levels, which means alpha is in the 0.01 to 0.1 range. A question is what is the right value?

Suppose, however, the *p*-value was 0.28 (meaning if the two treatments were equivalent there is a 28% chance that the experiment could generate data that extreme). With the same consideration as above: If X and Y are not equivalent, if their means difference, $\mu_X - \mu_Y = \bar{X}_X - \bar{X}_Y = T\sigma / \sqrt{N}$, then the T result (whatever value it is) would be expected. There is a 50% chance that the experiment would generate data with that, or a higher, value. So, if one hypothesis has a 28% chance of such a high or higher value, and the other hypothesis has a 50% chance, the ~2:1 ratio of probabilities slightly favors the not-equivalent case, but it is not unexpected. To put this into perspective, this is the same as flipping a fair coin once and not winning.

But, somewhere between the *p*-values of $\alpha \cong 0.28$ and $\alpha \cong 0.05$ is the "could be" region where no decision can be confidently made. One option is to continue experiments, to collect more data to have greater precision in the comparison. But, that may not be practicable.

The question remains: What are the right values of α and β for testing a hypothesis? We would offer that you consider the consequences of rejecting the hypothesis if it is true, and the consequences of accepting a minimally wrong condition, weighted by the undesirability of the time and cost penalty of excessively large number of trials. This exercise seems to be a nonlinear optimization which is dependent on quantifying several categories of concerns. So, as a general rule, perhaps aim for the conventional $\alpha \cong 0.05$ for economic-only consideration, $\alpha \cong 0.02$ for quality issues, and much lower values where issues of very high concern (health, safety, life, loss prevention, etc.) are involved.

If you deviate from the historical norm, clearly defend why with logic grounded in the application context.

10.2.11 One-Sided or Two-Sided Test

For the same level of confidence, or the complement level of significance, alpha, the two-sided test, places half the rejection area on each side. The one-sided test places all of the rejection area on the same side. This means that the critical value of the statistic is nearer to centrality for the two-sided test and farther for the one-sided test (into the more extreme region).

As an example for degrees of freedom $\upsilon = 10$, and level of significance $\alpha = 0\ .05$, the one-sided t-critical is $t_{\text{critical}} = 1.8125$. The two-sided t-critical is $t_{\text{critical}} = 2.2281$. If you choose the two-sided critical value, a larger difference between two treatments will be required to reject the hypothesis.

Similarly, for degrees of freedom $\upsilon = 10$, and level of significance $\alpha = 0.05$, the one-sided chi-squared-critical to reject a large difference is $\chi^2{}_{\text{critical}} = 18.307$. The two-sided upper value is $\chi^2{}_{\text{critical}} = 20.4832$. Again, if you choose the two-sided critical value, a larger difference between two treatments will be required to reject the hypothesis.

But you don't have a choice about using a two-sided or a one-sided test. If the hypothesis is that two treatments are equal, then if one is either too large or too small you would reject the equal hypothesis. You must use the two-sided test. By contrast, if the hypothesis is that one treatment is better than the other, then you are interested in rejecting the hypothesis if the data indicates it is not better. Here you need to use the one-sided test.

There may be a temptation to use the one-sided or two-sided critical value that supports a decision you might want to have supported. Don't make that choice. Let the hypothesis determine the rejection region.

10.2.12 Choosing the Quantity for Comparison

There are probably multiple criteria for judging whether one treatment is better than another. And the criteria have disparate dimensional units and impact.

Here is one author's example: Should I drive on path A or path B? My son's house was across town, about 20 minutes away. I could drive there, first going east, then north (Treatment A). Or first going north then east (Treatment B). Path A is more direct, but it has more traffic. Path B has more open roads but many turns and hills. So, initially we often took alternate paths and found that the travel time of Treatment A (Choice A, Path A) is less on average, but not always. Treatment B is a mile longer, meaning it costs more in gasoline. Treatment B goes through the country, and the isolation from population, traffic lights, traffic, service facilities is more aesthetically pleasing. Alternately, B is less available to rescue if needed, and more prone to hitting a wild animal. Although the traffic and aesthetics of Treatment A is less desired, the security is higher, but so is the risk of being involved with a "fender-bender".

We could measure only travel time for the two treatments, and then do a t-test to see if they are different. But this discounts the several other attributes that are important. This example discussed a few:

- Travel time, minutes.
- Travel cost, $ associated with gas, tire wear, etc.
- Security, an emotional rating associated with the convenience, or lack, of help if needed. This might be ranked or valuated on a 0 to 10 basis.
- Aesthetics, an emotional preference for open land beauty contrasting the undesirability of traffic congestion and old buildings on crumbling parking lots. Again, this might be ranked or valuated on a 0 to 10 basis.

In manufacturing, the disparate issues might be product yield, customer-perceived aesthetics, political capital, processing cost, quality variability, manufacturing flexibility, environmental risk, or many others. Whatever is your application, there are likely to be many aspects of desirability or undesirability of the treatment.

One treatment might be better for one criterion, but worse for another criterion. Don't make a comparison solely on one of the many criteria that are probably important. If you choose only one criterion, then you might bias the management decision.

One approach is to measure, compare, and report treatment characteristics relative to each of the several criteria. If one treatment is better in one aspect, but worse in another, you could report the p-values for each aspect, and let the audience choose the treatment. This is a multi-objective approach, reporting the metric for each attribute separately, and letting the decision-maker balance the several objectives and make the choice.

Alternately, one could combine the several metrics. However, they will often have different dimensional units and impact relevance, so that cannot simply be added. In the travel example above, the several metrics for the path options were travel time, travel cost, aesthetic appeal, and safety.

The metric values for each criterion, v_i, could be multiplied for a total evaluation value.

$$E_{\text{treatment}} = \prod v_{i,\text{ treatment}} \tag{10.3}$$

Then the units on the combined evaluation for each treatment, $E_{\text{treatment } A}$ and $E_{\text{treatment } B}$, would be the same. However, this makes the importance of one value proportional to the value of the others.

An additive combination assesses each value as remaining independent of the others. However, some sort of scaling is needed to be able to add terms with disparate units. A Lagrange-type multiplier approach is commonly used.

$$E_{\text{treatment}} = \sum \pm \lambda_i v_{i,\text{ treatment}} \tag{10.4}$$

But this means that someone must provide λ_i values that provide both unit scaling and impact scaling. Our favored approach to determine the lambda-weighting is to use equal concern scaling from an ideal.

$$E_{\text{treatment}} = \sum \pm \frac{v_{i,\text{treatment}} - v_{i,\text{ideal}}}{EC_i} \tag{10.5}$$

The ideal value, $v_{i,\text{ideal}}$ represents the ideally desired outcome. In the travel path option, for travel time it would be zero minutes. For travel cost it would be zero \$. In aesthetic beauty it would be a perfect 10. In the equation, $v_{i,\text{ideal}}$ is actually not needed. In the difference between treatments it is added and subtracted.

$$E_{\text{treatment change}} = E_{\text{treatment }B} - E_{\text{treatment }A} = \sum \pm \frac{v_{i,B} - v_{i,A}}{EC_i} \tag{10.6}$$

But often, acknowledging the ideal helps assign the equal concern scaling values. However, just considering the sum of scaled performance values, without the ideal is also fully functional.

$$E_{\text{treatment}} = \sum \pm \frac{v_{i,\text{treatment}}}{EC_i} \tag{10.7}$$

The ± sign is required. The impact of some terms (such as travel time) is undesirable, and some (such as trip aesthetics) are desirable. To make the treatment evaluation, E, represent desirability use a "+" sign for all terms which represent desirability, and a "−" sign for all representing undesirability.

The equal concern scaling factors, EC_i, represent the relative subjective concern and importance for each term. The *EC* values have the same dimensional units as the respective numerators; so, each of the terms in the sum are dimensionless.

To assign these:

1. First, choose a deviation for one item that is cause for concern (or joy). For instance, a 5-minute lengthening of the trip time, or a 3 lb/sec improvement in throughput, or a 0.5% increase in interest. Choose a value that has some meaningful concern but is not excessive. This is one of the *EC* values.
2. Consider all of the concern (and joy) issues that such a deviation might invoke. Perhaps someone is waiting and needs to wait 5 minutes more. Perhaps the round-trip costs you 10 minutes from getting back to watching the game or returning to housework. Some considerations may have little to no impact. If the person waiting deserves it, then adding 5 minutes to their wait has little negative impact. If the person waiting is going to do you a favor, then making them wait has a substantial impact. Consider all aspects of the deviation, the *EC* factor, and feel the level of concern (or joy) associated with that deviation.
3. One-by-one, for each remaining term in the sum, choose a value of the *EC* factor that raises the same level of concern (or Joy) as that chosen in Step 1.

Note: The *EC* factors are context dependent. As an example: During a period of sold-out capacity, getting 1% more production might be very important. On another day, when sales are only 75% of capacity, the 1% throughput increase is of no impact.

Example 10.1: Treatments A and B represent processing temperatures. Treatment A is the current operating temperature. Treatment B, the higher temperature being explored, speeds up production rate by 5 units/day, but cost $1/unit more, and increases product quality variability by 0.8 kg/L. The increase in production is moderately desirable and is used as a basis for assigning equal concern factors. $EC_{\text{productivity}} = 5 \text{ units/day}$. It has been decided that a cost change of $3/unit has the same level of concern. $EC_{\text{cost}} = \$3/\text{unit}$. It has also been decided that a change in variability of 0.2 kg/L has the same level of impact. $EC_{\text{variability}} = 0.2 \text{ kg/L}$. From Equation (10.6):

$$E_{A\text{ to }B} = E_B - E_A = \sum \pm \frac{v_{i,B} - v_{i,A}}{EC_i} = \frac{+\frac{\$5u}{\text{day}}}{\frac{\$5u}{\text{day}}} + \frac{-\frac{\$1}{u}}{\frac{\$3}{u}} - \frac{\frac{\$0.8\text{ kg}}{\text{L}}}{\frac{\$0.2\text{ kg}}{\text{L}}} = -3.\overline{33}$$

Even though the productivity increase is beneficial, the concern over loss in quality makes that quality term dominate the total performance assessment of the treatment difference. The switch to the higher temperature is not desirable.

Note: In another year, the perceived loss of quality might not be of much concern, and if the equal concern over variability is $EC_{\text{variability}} = 2\frac{\text{kg}}{\text{L}}$, then $E_{A\text{ to }B} = +0.2\overline{66}$, then a change would be justified.

Equation (10.7) represents an overall performance, quality, or attribute evaluation for the treatment. After N trials it will have a variability, which can be calculated as the standard deviation of the N $E_{\text{treatment}}$ values.

Alternately, you might only have one trial, one value for each element in Equation (10.7). If you have experience of the variation of individual factors, you can estimate the variation on $E_{\text{treatment}}$ by propagation of variance.

$$\sigma_E = \sqrt{\sum \left(\frac{\sigma_{v_i}}{EC_i} \right)^2} \tag{10.8}$$

Then the standard error of the average treatment evaluation would be

$$\sigma_{\bar{E}} = \sqrt{\sum \left(\frac{\sigma_{v_i}}{EC_i} \right)^2 / N} \tag{10.9}$$

And a t-statistic used to compare averages (assuming the variation is equivalent, Case 2) is

$$T = \frac{\bar{E}_B - \bar{E}_A}{\sqrt{\frac{\sigma^2_{E_B}}{N_B} + \frac{\sigma^2_{E_A}}{N_A}}} \tag{10.10}$$

If you have multiple trials from which to calculate the standard deviation of the $E_{\text{treatment}}$, it would be good to compare that with the estimate from Equation (10.8).

10.2.13 Use the Mean or the Probability of an Extreme Value?

Figure 10.1 illustrates the *CDF* of the distribution of observation values from two treatments. Looking at the 50th percentile values, you can see that the mean of B (about 4 on

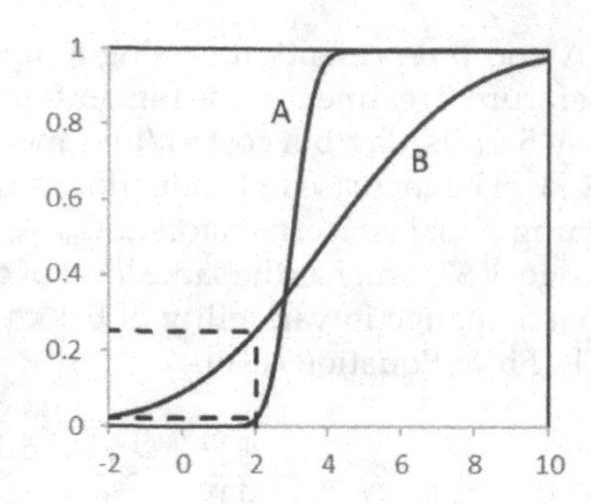

FIGURE 10.1
CDF of sample values from two populations.

the x-axis) is greater than the mean of the other, A (about 3). Since greater is desirable, Treatment B might be chosen.

However, the left-tail of the two distributions indicates that B is more likely to be below the threshold of disaster, a value of 2, as illustrated in this figure. Perhaps this indicates a profitability index, or a safety index. Treatment B has a 25% chance of being a failure, while A only has a 1% chance of being a failure. If the penalty for a failure is equivalent to a 5 on the x-axis then risk difference of 1.2 (= 0.25 * 5–0.01 * 5) is larger than the possible benefit of choosing B over A.

Treatment B may be better on average, but if Treatment A is adequate, the security associated with that choice may be more important than the benefit. Consider this example: The thrill of cliff-jumping into the water far exceeds the thrill of wading into the water from the shore. But for most people, the thrill is not worth the possible risk.

Often, we should be using data to reveal the possibility of an undesired possibility, not just comparing averages to assess treatments.

The undesirability, or penalty, might not be a fixed value, but might increase with deviation into the undesired region. As well, one might look at the glorious possibility of getting a 10 from Treatment B when the best that Treatment A will provide is a 4. A gambler might choose B over A.

This can be subjectively quantified. You need a model of the distributions, and a sense of concern or joy about the *x*-values. Normally risk is associated with a cost of an event. But there are many more undesirable aspects associated with personal and corporate reputation, the concern can be qualitatively mapped to the *x*-value. Concern could be rated on a scale of 0 to 10, from no concern to the ultimate most undesirable set of outcomes that could happen. Similarly, joy over exceeding expectations can be qualitatively mapped w.r.t. the *x*-value. Joy could also be rated on a scale of 0 to 10, from indifference over the outcome to the ultimate most desirable set of outcomes that could happen. Joy should be equivalent to Concern. Figure 10.2 illustrates a possible mapping of Joy and Concern w.r.t. the *x*-value. Here, an *x*-value of 0 creates a concern level of 2. Equivalent to a concern value of 2 is a Joy value of 2, which corresponds to an x-value of about 7.

Note: This illustrates both some level of Joy and some level of Concern at the x-value of 2. What the shapes are do not need to be continuum-valued, and they will be relative to a particular context. Perhaps get several people with a comprehensive view of the situation to collectively judge the trend.

The probability of getting an x-value within a particular range is the *CDF* difference. If within that *x*-range, the subjective estimates from Figure 10.2 can be used to get the

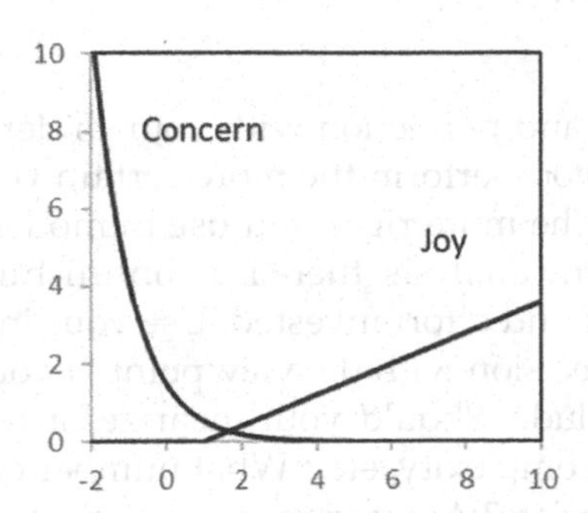

FIGURE 10.2
Level of Joy and Concern w.r.t. X.

corresponding Concern and Joy values, the product of probability times Concern and Joy summed over all of the Δx values will represent the overall rating for each treatment.

$$E_T = \sum \left[CDF(x+\Delta x) - CDF(x) \right] \left[\text{Joy}\left(x + \frac{\Delta x}{2} \right) - \text{Concern}\left(x + \frac{\Delta x}{2} \right) \right] \tag{10.11}$$

10.2.14 Correlation vs Causation

Just because there is a correlation, does not mean there is a cause-and-effect relation. A relatable example is that gray hair on humans is strongly correlated to face wrinkles. This correlation does not mean that gray hair causes wrinkles. There is a strong correlation between sunrise and people awakening, but this does not mean that people awakening causes the sun to rise.

Often however, the presence of a correlation is used to defend a flawed mechanistic viewpoint, to defend an illogical claim, or to justify action that is driven by a hidden agenda (self-serving, or politically or emotionally based). A 99% confidence in the correlation does not mean that a faulty supposed mechanism is confidently true.

In this big-data machine-learning era, we have the computing tools to analyze relations and trends in vast quantities of data. I think that it is very useful in detecting the possibility of fraud or spam, and in suggesting preferences, "If you like that music, you might also like these suggestions." However, these findings are correlations between variables, and although they might discover cause-and-effect mechanisms, they do not represent the mechanism. And often they discover two effects of the same cause. Flowers blooming do not make it become a warm season.

If there is a mechanistic cause-and-effect relation, if there is a phenomenological pathway between one event and the outcome, then you should be able to describe each relation in the sequence and test each internal variable to see if the relation is true. You can measure each internal variable, and also block it to see if the outcome does not happen. For example: Turning the key in the ignition does not start the car, although there might be a 99.9% correlation that it does. Turning the key closes an electrical circuit that energizes the starter motor, engages it with the engine, and closes the choke. It requires gasoline in readiness, energy in the battery, mechanical engagement of the starter motor gears, and many other factors. If you are going to claim causation, be able to defend each cause-and-effect mechanism in the sequence.

There may be very confident statistical evidence for a correlation. Don't use it to claim causation. The statistical hypothesis is not the mechanism.

10.2.15 Intuition

Balance cost, time, certainty, and perfection with your understanding of sufficiency. The greater the number of trials you perform the more certain you can be about the outcome distribution and a decision. The more rigor you use in modeling the more confident you can be in the model. But for any analysis, there is a diminishing return of the benefit with respect to the amount of time and effort invested. Use your intuition to make such a decision, but also validate your decision with the viewpoint of your customers/stakeholders.

Intuitive choices could include: Should you linearize or not? What is the appropriate power series order, network complexity, etc.? What number of trials is appropriate? What level of significance is appropriate? Are variances uniform over the range?

10.2.16 A Possible Method to Determine Values of α, β, and N

As a possible approach to quantifying values of α, β, and N for testing a hypothesis, consider this situation: You are currently using Treatment X, and want to know if Treatment Y could be an alternate. Perhaps X is a raw material, from a single supplier, and you want to qualify Y for security of supply, for leveraging of price, etc. You have decided on a key metric to evaluate the performance of X and Y, which may be a composite of several critical aspects; and from extensive operating experience know values for μ_X and σ_X for this key metric, and that the variation in metrics is normally distributed.

You would accept Y if $\mu_Y \geq \mu_X$. Trials with Y, of course will reveal $\bar{Y}$, not μ_Y. So, if $\bar{Y}$, indicates that μ_Y could be $\geq \mu_X$, accept Y. Assume that $\sigma_Y = \sigma_X$, then if $\mu_Y \geq \mu_X$, the lower limit on $\bar{Y}$ would be calculated as $\bar{Y}_{crit} = \mu_X - z_\alpha \dfrac{\sigma_X}{\sqrt{N}}$, where N is the number of trials. Accept Y if $\bar{Y} \geq \bar{Y}_{crit} = \mu_X - z_\alpha \dfrac{\sigma_X}{\sqrt{N}}$. You need values for two factors here, α and N. In Excel $\bar{Y}_{crit} = \text{NORM.INV}\left(\alpha, \mu_X, \dfrac{\sigma_X}{\sqrt{N}}\right)$.

Another issue is that you would like to reject Y if $\mu_Y \leq \mu_X - \Delta$, where you choose the value of Δ. However, if $\mu_Y = \mu_X - \Delta$, because of experimental vagaries, there is a chance that $\bar{Y} \geq \bar{Y}_{crit}$. You would like enough trials so that there is a small chance of this happening. β is this chance, the probability that $\bar{Y} \geq \bar{Y}_{crit}$ given that $\mu_Y = \mu_X - \Delta$. You need values for two factors here, β and Δ. You choose Δ. In Excel $\beta = 1 - \text{NORM.DIST}\left(\bar{Y}_{crit}, \mu_X - \Delta, \dfrac{\sigma_X}{\sqrt{N}}, 1\right)$.

Increasing the number of trials makes the uncertainty on $\bar{Y}$ smaller. Permitting desirably smaller values for α (the chance of rejecting Y if Y is good) and β (the chance of accepting Y if Y is bad). But, large N creates a concern associated with the cost and duration of the trials. The method offered here seeks to determine values of α, β, and N that minimize all the concerns. Table 10.1 reveals possible issues and associated level of concern (LoC).

Each level of concern stated in Table 10.1 is a qualitative number. It is similar to ratings that humans give to products, movies, restaurants, vacations, bosses, employees, and attractiveness of each other. It represents how important all the diverse aspects are. It is qualitative, and it would change with the context. The values in the table are not a universal truth. These values represent the authors' general experience. Use values relevant to your application.

The concern for an extensive number of trials also needs to be included. In general impatience, and undesirability, rises as the square of the magnitude of something. And

TABLE 10.1

Outcomes and Level of Concern

Truth	Action: Reject Y	Action: Accept Y
$\mu_Y \geq \mu_X$	Wrong Action. Concerns include you still need to find an alternate to X, wasted trials, and you missed a good thing. On a 0–10 basis the level of concern, c_1, might be a 4. The probability of this happening is α.	Correct Action. You are happy to have an alternate but need to still do the work to manage the change of implementing Y in production, and there is always the nagging possibility that something unforeseen will cause an upset. On a 0–10 basis the level of concern, c_2, might be a 1. The probability of this happening is $(1-\alpha)$.
$\mu_Y \leq \mu_X - \Delta$	Correct Action. However, you still do not have an alternate, and have invested in trials. Finding and testing a new alternate, is in the future. On a 0–10 basis the level of concern, c_3, might be a 3. The probability of this happening is $(1-\beta)$.	Wrong Action. You have accepted a bad treatment. Certainly, it will be revealed after substantial use, and everyone associated will be embarrassed. Trial expense was wasted, Y will be rejected, and you still need to find an alternate to X. It will show up on your annual appraisal. On a 0–10 basis the level of concern, c_4, might be an 8. The probability of this happening is β.

assigning the concern of 20 trials as equivalent to a level of concern value of 3, the level of concern of N trials would be $3\left(\frac{N}{20}\right)^2$.

In total the sum of all the concern factors would be $4\alpha + 1(1-\alpha) + 3(1-\beta) + 8\beta + 3\left(\frac{N}{20}\right)^2$. Expressing it generically, with c symbols representing the coefficients:

$$C_{\text{total}} = c_1\alpha + c_2(1-\alpha) + c_3(1-\beta) + c_4\beta + c_5\left(\frac{N}{c_6}\right)^2 \tag{10.12}$$

The objective is to minimize this term. It appears that the decision variables are α, β, and N, but since β depends on $\alpha, \sigma, \Delta, N, \mu_x$, there are really only two adjustable parameters, α and N.

The optimization statement is

$$\min_{\{\alpha, N\}} J = c_1\alpha + c_2(1-\alpha) + c_3(1-\beta) + c_4\beta + c_5\left(\frac{N}{c_6}\right)^2 \tag{10.13}$$

S.T. User-specified LoC values for each aspect

μ_X, σ_X, and Normal Distribution from historical data

Δ as chosen by the user, to represent a deviation that should be detected

$$\bar{Y}_{\text{crit}} = \text{NORM.INV}\left(\alpha, \mu_X, \frac{\sigma_X}{\sqrt{N}}\right)$$

$$\beta = 1 - \text{NORM.DIST}\left(\bar{Y}_{\text{crit}}, \mu_X - \Delta, \frac{\sigma_X}{\sqrt{N}}, 1\right).$$

From the LoC factors in Table 10.1, the solution is $\alpha = 0.13$, $N = 7$, and $\beta = 0.065$. If, as an example, the concern for excessive trials was much lower, with $N = 20 = c_6$ trials equivalent to a c_5 of 1, the solution is $\alpha = 0.066$, $N = 11$, and $\beta = 0.035$. Finally, as another example wherein the concerns for not making a mistake dominate the concern for excessive trials, the solution is $\alpha = 0.004$, $N = 35$, and $\beta = 0.0006$.

The procedure to determine appropriate α, β, and N values is somewhat complex. It requires the user to provide rating for the six LoC factors; and the mixed integer–continuum optimization may not be within the skillset of the person designing the experiment. But satisfyingly, the procedure returns conventional values (with conventional economic LoC values), and it extrapolates to traditional level of significance where life-related concerns dominate.

10.2.17 The Hypothesis is not the Supposition

The statistical hypothesis is not the supposed cause-and-effect mechanism. The hypothesis and associated attribute are expected observable outcomes if the mechanism is as supposed. It is what you are considering as a data comparison to test a supposed mechanism.

Maybe a person wants to claim Supplier A of a component in your product is the reason that customers return your product as being defective, and that person wants to claim that Supplier B is better for your company. So, they count returns and find that 5,317 have the A component and only 684 have the B component. Uncertainty associated with the count is about 10%, implying that $\sigma_A = 0.1 \cdot 5317 / 2.5 = 141$ and $\sigma_B = 0.1 \cdot 684 / 2.5 = 27$. A test on the numbers will clearly reject the hypothesis that A = B.

However, if 97,689 products were made with the component from A, and 13,472 products were made with a component from B, then a more reasonable test would be on the portion of products returned. Here the portions of returned products are $p_A = 5317 / 97689 = 0.05443$, and $p_B = 684 / 13472 = 0.05077$, and the equivalent hypothesis cannot be rejected.

Be sure that the choice of the hypothesis and the attribute (statistic) being tested is a legitimate test of the claim about a mechanism.

10.2.18 Seek to Reject, Not to Support

I hope that you enjoy the fallacious claim in this section, and its message. "The Theory of Positional Invariance" was an article (R. Russell Rhinehart, Develop Your Potential Series in CONTROL magazine, Vol. 33, No. 11, November 2020, p. 41) and reproduced here with kind permission of the editor.

The "Theory of Positional Invariance" states that regardless of the observer's viewpoint the object retains its properties. There are many examples: Whether observed from the north or south poles, the moon has the same mass and craters and rotational speed. Although, the moon does appear upside down to one observer, the viewer orientation does not change the properties of the object.

Whether you look at a person from the top or back or front, it is still that person with the same color eyes and personality. I asked my grandchildren, "What's my name?" Their first thought was, "Oh, no! It is happening to him." Then they said very tentatively, "Pop." "Good," I said, then turned around and asked, "Now, what is my name." One said, "It's still Pop." The other called to their grandmother, "DeeGee, something's wrong. We need help in here."

A theory starts with corroborating observations, acquires the rule, then a sophisticated sounding name to help validate it. Regardless of the observer's viewpoint, an object retains its properties: "Positional Invariance".

Applying the principle, observe that except for the 45° rotation, the $\times$ and the $+$ symbols are the same; so the theory claims that $2 + 2 = 2 \times 2$. There you are! Let's try with some other numbers

$3+1.5=3\times1.5$, and $(-4)+0.8=(-4)\times0.8$, and $1+2+3=1\times2\times3$. But, what about complex numbers!? Here are some demonstrations: $[(-1)+2i]+[0.75-0.25i]=[(-1)+2i]\times[0.75-0.25i]$, and $[(-1)+\sqrt{2}i]+[2/3-\sqrt{2}/6i]=[(-1)+\sqrt{2}i]\times[2/3-\sqrt{2}/6i]$. There are an infinite number of corroborating examples. I chose these examples to show that it works with negative numbers, fractions, and irrational numbers, but I kept the numbers convenient for your affirmation of the truth of the Theory of Positional Invariance.

The theory is intuitively logical, has a sophisticated name, and is confirmed by data which has an infinite number of cases. So, the claim must be true. I use this truth to support my claim that we should not be wasting time and mental effort by having students memorize both addition and multiplication facts. Addition is all that is needed. (One can have fun?)

Just because there is some corroborating evidence and some intuitive basis for a fancily packaged claim, does not mean that the claim is true. Don't blindly accept either the technical folklore of your community or your preferred explanation. Don't seek evidence to support the claim. Seek evidence that could refute it. Data cannot prove. Data can only disprove. So, critically shape trials and examples to see if you can disprove the claim.

10.3 Takeaway

The fundamentals of statistics, the equations and mathematical science underlying the techniques, as presented in Chapters 1 through 9, are somewhat important. But the choices you make are even more important to have the analysis and decision to be rational and legitimate.

Ensure the basis of the test (what quantity is tested, what level of significance is used, what statistic is used), matches the situation. Understand how to explain the "Accept–Reject" dichotomy, p-values, confidence/significance, and the relation of the statistical method to the decision or action.

Differentiate causation and correlation.

"There are two reasons for everything: A good reason, and the real reason." That is one version of a century-old adage.

"Statistics don't lie, but liars use statistics." That is one version of another century-old adage.

Don't give your audience a reason to accuse you of either age-old frauds.

Be transparent with your choices and conclusions, their limits, and their influence on decisions.

10.4 Exercises

1. Practice inventing seemingly proper and justifiable choices for statistical analysis to lead to improper outcomes (not to become proficient at deception, but to more easily detect it). Choose any of the topics presented in this chapter.

2. Listen to what is said or written in the media and critically evaluate how the claim might be false, even if couched as true.
3. Consider a situation that you are involved in and assign equal levels of concern and joy for possible outcomes.
4. Create an exercise that could be an example to show how Equation (10.13) works.

Section 3

Applications of Probability and Statistical Fundamentals

11

Risk

11.1 Introduction

Risk is a long-term average of the undesirable effect of possible events. Related to engineering business decisions based on economics, the concept of risk can be defined as expected frequency of an undesired event times the cost consequence of the event:

$$R = fc \tag{11.1}$$

Example 11.1: The expected frequency is 0.01 events per year (once in a hundred years), and the cost associated with the event is \$10,000 per event. What is the risk?

Using Equation (11.1)

$$R = 0.01\left[\frac{E}{\text{yr}}\right]10000\left[\frac{\$}{E}\right] = 100\left[\frac{\$}{\text{yr}}\right]$$

This \$100/yr would be considered as an average annual cost associated with the enterprise and treated like an expense in the economic profitability analysis. For instance, if the expected life of the enterprise is 20 years, then, primitively, the cumulative risk is $100\left[\frac{\$}{\text{yr}}\right]20\left[\text{yr}\right] = 2{,}000\left[\$\right]$.

Risk is not what will happen. One cannot forecast the future. Also, risk is not the worst that could happen. Risk is the expectation, the average, possibility over many years or uses.

Risk can be used in design considerations. If the risk is \$2,000 over a 20-year horizon, and the one-time design cost to prevent the event is \$1,000, then the \$1,000 investment up front makes sense over the life of the enterprise (in a simple pay-back-time analysis not considering the time-value of money). Perhaps an alternate option for prevention of the event is by training, which is estimated to cost \$500/yr, then over the 20-year period, the prevention costs more than the consequence, and investment in that method of prevention would not be a good business decision.

Risk, with the units of \$/yr, can be considered as a possible annual expense and included in economic profitability indices such as pay-back time (PBT), long-term return on assets (LTROA), net present value (NPV), discounted cash flow rate of return (DCFRR), or whatever metric your organization prefers. In doing so, you will gain an appreciation for the impact of the risk on the viability of the venture and insight on solutions to reduce the risk.

DOI: 10.1201/9781003222330-11

Alternately, risk can be considered as the probable cost associated with an undesired event over a period. It is the probability of an event happening within some interval times the financial magnitude of the penalty.

$$Risk = P(\text{event in an interval})\ \text{Cost}(\text{of the event}) \tag{11.2}$$

Example 11.2: Past data indicates that in a normal year, the probability of a substantial leak from any pump seal is 0.0001, and the probability of it being considered an environmental transgression is 0.2, and the fine for a "spill" and cost of land reclamation and the value of the lost material all sum to $50,000. What is the risk?

Note that the probability of incurring a cost is that there is a leak, and it is caught. The AND conjunction directs that the probability of the joint event is the product of individual probabilities.

Using Equation (11.2)

$$Risk = 0.0001 \times 0.2 \times \$50{,}000 = \$1$$

In the example, the probability is on a per year basis. Often risk is defined by replacing frequency in Equation (11.1) with probability. The probability of 1 event occurring in 100 years is 0.01 events per year. The period does not have to be a year, the frequency could be on a per day basis. Alternately, the probability could be on a per event or state or item basis. For example: 1 bag in 50 will have a leak, 1 sales contact per 20 will lead to a sale, 1 in 300 times that a light is switched to ON the light it will pop and fail. If using frequency as the estimate of the likelihood of an event, then the number of undesired events can easily be scaled to an alternate time period. For example, production of 100 units per year, with an expected defect rate of 1 per 50 units, means an expected frequency of 2 defects per year.

There are alternate definitions of risk, and diverse preferences for nonbusiness applications or catastrophic situations such as in healthcare, national security, or personal finance. Some alternate definitions include: Risk = Likelihood of an event times Cost, Risk = Hazard times Vulnerability, or Risk = Danger divided by Resilience. This chapter will only use Equation (11.1) or (11.2) and its variation.

Note several features in the simple leaking pump seal example, Example 11.2. First, the cost of the event is not just from one aspect of the event, it must consider all associated costs. Here there were three. However, it is easy to imagine that many more aspects should be considered.

Second, the probability was for one pump in one year. Likely, there will be several pumps increasing the probability that more than one will fail. Also, the probability for the pump seal to fail was for a one-year interval, but likely the plant will have a longer expected operating life. So, there is a higher chance of a failure over the plant lifetime.

Third, if there was a prior event, then the second environmental event will have a higher penalty.

Fourth, the fine only happens if you are caught. There is a probability of the leak, and a probability of being caught. The probability of the fine is if you have a leak and get caught. Since $P(A \text{ and } B) = P(A)P(B \mid A)$ the 0.0001 and 0.2 are multiplied to get the compound probability of the event incurring a cost.

Fifth, any forecast of probability of events over a multi-year future and the associated financial impact of such an event will have substantial uncertainty.

Finally, this presented an economic valuation of an undesired event, but the violation could concern either ethics or things we should do. Ethics would direct that you do not let a leak happen. Good manners say that you should respond when someone acknowledges you. If you can evaluate the equivalent economic value of the penalty for such events, then you could include that in the risk calculation.

11.2 Estimating the Financial Penalty

The event must be explicitly identified. It cannot be a vague statement like "something bad might happen", nor can it be a nonspecific statement, like the undesired event is "a pipeline leak". It needs to be identifiable and mechanistically specific enough to be able to assess both the likelihood of occurrence and the costs of the consequence. Expanding on the pipeline leak example, specific statements might be: "a pipeline leak of 1 gal per hour of crude oil lasting for 5 days because a welded seam fails", "a pipeline leak of 100 gal of potable water because of an illegal tap into the line to access material", "a pipeline leak of 50 gal of gasoline because of a pump seal failure". Each of those cases have their individual probability and individual cost of consequences. Of course, risk analysis extends well beyond pipelines and includes legal terms in a contract, driving a car, rules in raising children, operation of a chemical facility during a hurricane, hiking a woodland trail, etc. But, in any case, the event needs to be explicitly and specifically defined.

Three sources of cost associated with the toxic material spill were included in Example 11.2, but there are many more. For instance, if the event happens, then there will be adverse company image issues in the public domain. This may lead to a loss of sales to sensitive customers or to backlash protests by environmentalists that cause insurance premiums on the facility to rise. An event will likely lead to a change in maintenance procedures on pumps, which will have associated costs with increased inspections and the management of change associated with revised training, scheduling, and document updating. When you are considering the cost of an undesirable event, be sure to consider all aspects.

Continuing discussion of the financial impact: Also consider the personal impact of those in the company that get blamed for the event. The corporate folks will blame the Plant Manager who will blame the Maintenance and Operations Supervisors. The event will show up on their annual appraisal and will impact their hopes for promotion, or salary increase, or profit sharing. These managers probably also have families, with schoolchildren who will be harassed by classmates because their parent was responsible for polluting their environment. (A good corporate citizen of the local community should add value, not do harm.) The perceived personal impact to these folks might be much larger than the $50,000 cost to the company. How much more will it cost the company in donations to the local schools, library, and community centers to restore community approval of the employees?

The total financial impact of the undesired event is context-dependent. If for instance, there have been several recent undesired events associated with the company then the magnitude of the fine may be larger, loss of allegiance by the public larger, and the personal impact on employees larger.

It is not easy to forecast all the consequences that might happen, or what the costs of each might be. There is substantial uncertainty on such an estimate. Often the cost associated with an undesired event will be a reasonable "guestimate".

The penalty for an event is context-dependent. People need to be fully aware of the context to be able to estimate the cost.

11.3 Frequency or Probability?

A common formula for risk uses probability of an event, Equation (11.2). $R = pc$, which is equivalent to Equation (11.1) for low probability events. Here is why: The Poisson distribution gives the probability of n events occurring in an interval (of space or time) in which the average expected is λ events per time/space unit considered. (This Poisson model also presumes independence of the event, and stationary λ.) The Poisson point probability is $p(n) = \frac{\lambda^n e^{-\lambda}}{n!}$. If $\lambda <\sim 0.05$, then $p(n=1) \cong \lambda$, and $p(n>1) \cong 0$. So, for infrequently expected events, the probability of a single occurrence in the time interval is effectively the same as the expected frequency, which makes the f and p interchangeable in the risk formula.

Frequency could have a value of 2 events per year. However, if frequency in Equation (11.1) is replaced with probability, the value of 2 is not possible. Probability of an event can only range from 0 to 1. Instead of frequency, one could use the term probable number of events in the period, the expected number, not the probability of an event. Alternately, one could use the expected long-term average, λ, and the Poisson distribution to give the probability of $n = 1$, $n = 2$, $n = 3$, etc. number of events. This is developed in the next section.

11.3.1 Estimating Event Probability – Independent Events

In the introductory Example 11.2, note that the probability of 0.0001 is for one particular pump failing, and failing in a one-year period. Such a value must come from past experience with similar pumps, similar service, and under similar maintenance programs. It also has substantial uncertainty. Further, the plant will likely have more than one pump in the do-not-let-this-leak service, and likely plans to operate for more than one year. So, we need to use probability to project what might happen with k number of pumps over an n-year horizon.

The Poisson distribution models the number of events, x, expected in an interval if the average event rate per interval, λ, is known and events are independent.

$$P(x) = \frac{\lambda^x e^{-\lambda}}{x!} \tag{11.3}$$

This reveals that the probability of one failure of the one pump in a given year is

$$P(1) = \frac{\lambda^1 e^{-\lambda}}{1!} = \frac{0.0001e^{-0.0001}}{1} = 0.000099990\ldots \tag{11.4}$$

which is a bit less than (but nearly equal to) the λ value, but also that there is a chance that the same pump might fail twice in the year. That probability is

$$P(2) = \frac{\lambda^2 e^{-\lambda}}{2!} \cong 5.\times 10^{-9} \tag{11.5}$$

If a pump fails a second time, then what is the second financial penalty? Probably larger than the first. This calculation presumes independence of the second event. If maintenance procedures create a common cause for failure, then given a first event, the second might have a higher probability. Alternately, if maintenance on the failed item replaces worn out parts with new parts, then a second failure might be much less. The Poisson model assuming that events are independent may not be true.

Also, in an n-year consideration there is the probability of a pump failing in the first year, or the second, or the third, …. For simple analysis, consider that the probability of a second fail in the same year, $P(2) \cong 5.\times 10^{-9}$, is negligible, and use the p = 0.0001 value. If there are k pumps and an n-year operation, then there are nk total number of possibilities for a failure. The binomial distribution reveals the probability of x number of events when there are nk trials and the probability of an event is p.

$$P(x \mid nk) = \frac{(nk)!}{x!(nk-x)!} p^x (1-p)^{nk-x} \tag{11.6}$$

So, if there are k = 10 pumps in the critical process lines, and an n = 10 year consideration then nk = 100, and from Equation (11.5) with p = 0.0001:

$$\begin{aligned} P(0 \mid 100) &= .99004933\ldots \\ P(1 \mid 100) &\cong .01 \\ P(2 \mid 100) &\cong 5.\times 10^{-5} \\ P(3 \mid 100) &\cong 1.6\times 10^{-7} \end{aligned} \tag{11.7}$$

The probability of one event is effectively nk = 100 times that of a single expected event per year per pump. And the probability of a second event is about 10,000 times greater than that from a single pump in a single year.

A more proper analysis would be to use the Poisson distribution. If the failure rate for one pump in one year is $\lambda = 0.0001$ then the expected rate for ten pumps in a ten-year interval is $nk\lambda = 10 \cdot 10 \cdot 0.0001 = 0.01$, then the probabilities for the first few failures are

$$\begin{aligned} P(0 \mid 100\ \text{pump\_years}) &= \frac{\lambda^0 e^{-\lambda}}{0!} = .9900498\ldots \\ P(1 \mid 100) &= \frac{\lambda^1 e^{-\lambda}}{1!} = 0.00990049\ldots \\ P(2 \mid 100) &= \frac{\lambda^2 e^{-\lambda}}{2!} = 0.000049502\ldots \\ P(3 \mid 100) &\cong 1.65008\times 10^{-7} \end{aligned} \tag{11.8}$$

Note the equivalence to the binomial approximation of Equation (11.7) when the probability of a second pump failure is small.

After these considerations on both the financial penalty for an event and the probability of an event, returning to Equation (11.2), and a 10-year estimate, with 1) an event probability of 0.01, 2) the probability of it being considered an environmental transgression is 0.2, and 3) all aspects of financial penalty have been re-estimated as \$500,000. Then the risk is $\text{Risk} = 0.01\times 0.2\times \$500{,}000 = \$1{,}000$, which may be large enough to justify taking action to prevent the event, or to contain the material in case there is an event, or it still may be small enough to accept the possible future penalty. Consider: We go outside even though there is the possibility of a mosquito bite.

The probable number of the events is often termed the expected frequency. For a continuously operating process, it is the probable number of the events per a specified time interval, and per a unit of the item. For instance, considering the "per unit" aspect of the item, a pipeline with two welded seams will likely have fewer seam failures than another single pipeline of 5,000 welded seams operating for the same duration. And, considering the time interval, the probability of a car accident is low in any one-minute interval, but may be high when considering a 60-year driving period for one person; or again on a per item basis, a fleet of 5,000 delivery vans in a one-minute interval will have a higher expected number of undesired events than a single van in a one-minute interval.

Values for such data come from past experience with similar processes, equipment, and operating conditions, but past data need to be adjusted for the forecast situation being considered.

If risk is calculated with probability, there is the probability of one event, and the probability of two events, and three events, etc. in any use interval. If there are three events, then the total penalty is the sum of that for the first, and for the second, and for the third. So, the equation needs to be modified to sum the risk associated with all event types:

$$R = p(1)(c_1) + p(2)(c_2) + p(3)(c_3) + \cdots \tag{11.9a}$$

$$R = \sum_{n=i}^{\infty} c_n p(n) \tag{11.9b}$$

Here the costs are explicitly shown as different values. The cost consequence of a second event in a period would likely be greater than the first, and a third event may be greater yet. Consider speeding tickets, the first penalty may just be a fine, but the third arrest in a year may lead to license suspension. Consider insults in public, the first might be dismissed as poor judgment by the other, but the third is evidence of a hostile agenda and needs responsive action. In Equation (11.9a) the term $p(3)$ means the probability of only the third event, not the cumulative probability of the first and the second and the third. To have the third, the first two must have happened.

A problem with Equation (11.1) is that evaluation of risk scales linearly with the number of events. In it, a frequency of 10, an expected occurrence of 10 undesired events in a year, is 10 times more costly than a single event. Consider: A customer may forgive a single transgression of product or service expectations, but 10 transgressions in a year will create a reputation that could lead to loosing many customers. So, the penalty should increase for repeated transgressions, and Equation (11.9) permits that.

Example 11.3: What is the risk over a 10-year period, which is the forecast life of a product? The frequency of an event is 0.001 per year per unit. There are 5 units in use. The cost of the first event is \$20,000, the second event is \$50,000, and the third event is \$150,000. If 4 events, the plant is shut down.

We'll assume that all events are independent and are modeled by the Poisson distribution. With 5 units and a frequency of 0.001 events per year per unit, $\lambda = 5 \cdot 0.001 \cdot 10 = 0.05$ events per 10-year horizon. The probability of a 1st, 2nd, 3rd … events can be obtained from the Excel cell function POISSON.DIST$(x, \lambda, 0)$ and is

$$P(1) = \frac{\lambda^1 e^{-\lambda}}{1!} = 0.047561\ldots$$

$$P(2)=\frac{\lambda^2 e^{-\lambda}}{2!}=0.001189\ldots$$

$$P(3)=\frac{\lambda^3 e^{-\lambda}}{3!}=1.9817\ldots\times 10^{-5}$$

$$P(4)=\frac{\lambda^4 e^{-\lambda}}{4!}=2.477\ldots\times 10^{-7}$$

From Equation (11.8) and using rounded probability values

$$R\cong 0.0476\cdot \$20k+0.00119\cdot(\$50k)+1.9817\ldots\times 10^{-5}\cdot(\$150k)\cong \$1{,}015$$

The chance of a complete shutdown of operations would be the chance of more than 3 events, $2.5\ldots\times 10^{-7}$, less than 1 in a million.

11.3.2 Estimating Event Probability – Common Cause Events

Often events are not independent but are influenced by common cause. As a common cause example, a fleet of cars may be comprised of the same make and model, and each may have the same driver's blind spot. Then, if that attribute causes one driver to have an accident, the same attribute will likely cause many to have a similar accident. If one pipeline weld fails because of the incompatibility of material, welding practices used, or pressure shocks in the line during operation; then many welds will likely fail for the same common cause mechanism. If one employee quits because of the style of a boss, it is likely that others will also quit. Values for event frequency or probability can be obtained from human experience, but they should be adjusted for the specific application, which requires subjective judgment. And, mechanisms for compounding probability for common cause mechanisms also require human judgment. Again, although the equation seems simple, the evaluation of probability of the event is subjective.

Equation (11.2) is simple and often good enough, but it is not applicable if the frequency of the event is greater than about 0.05 events per interval, nor does it admit that a second or third event would have a larger penalty. Equation (11.9) permits greater penalty for repeated instances and conditional probability techniques can be used to estimate the likelihood of a second, third, etc. event due to common cause mechanisms. However, it is a step more complicated to use.

Again, however, either the Poisson or the binomial model is for independent events, and common cause may make the second much more probable if there is a first. Further, over a n-year horizon there may be an interval with a lapse in maintenance increasing the probability of an event, or because of external events an increase in attention that reduces the probability.

Again, there is much uncertainty, and human judgement is needed to determine appropriate probability models and to temper results.

11.3.3 Intermittent or Continuous Use

The use may not be continuous, but discrete, or intermittent. Staplers only run out of staples when they are used. Baseball bats only break when they are used. In such cases of intermittent use, the probability of an undesired event is not on a per time basis but on a per instance-of-use basis. The per instance-of-use basis can be converted to a per year basis (or any time interval) by the number of expected uses per year.

11.3.4 Catastrophic Events

An event may have a 1 in 1,000 (relatively low) chance but it may have a $1,000,000 (very high) penalty. Normalized on a per year risk basis it is $1,000 per year, which may have little impact on the economic analysis. If, however, the business after-tax profit is $500,000 per year, a one-time $1,000,000 event would wholly consume two years of profit and could bankrupt the enterprise. So, the risk of Equations (11.1) or (11.9), representing an expected average loss as a business expense, do not wholly represent the reality of a catastrophic loss.

There are also events that cannot be valued. A value cannot always be placed on the cost consequence of the event. Consider 1) How much is a life worth? Would you simply assign the value as the person's earning potential over their expected remaining life? How does one place an equivalent monetary value on grief to loved ones, impact on children, etc. Consider 2) How much is a reputation worth? If your enterprise goes bankrupt, is the loss only the value of equity in the enterprise? What are the consequences to personal initiative, future credibility of the leader, loss of friends? Consider 3) How much is unethical practice worth? Some people might say that it is unethical to permit an event with a risk value of $100/yr, just because prevention costs $500/yr, and you think it is a poor business decision to lose $400/yr. Their opinion for you might be, "No undesired events should be permitted." Consider 4) environmental damage and that impact on endangered species.

In some situations, the basic equation does not have meaning, because either 1) emotional or ethical issues are controlling, and a cost cannot be assigned to the consequence of the event, or 2) the event has one-time catastrophic implications that cannot be normalized on a per year basis.

On high-consequence events where even one occurrence is catastrophic, traditional risk analysis is not an acceptable tool for analyzing them. Equations (11.1) through (11.9) represent a value-to-value analysis of cost and possible savings for a traditional business economic analysis. For some events, however, there may be no acceptable risk probability, no acceptable frequency.

When human life is at risk, a common practice is to make the next generation of facilities or cars have half the probability of causing a death as the current best state of the industry.

$$p_{\text{new}} = \left(\frac{1}{2}\right) p_{\text{state of the art}} \tag{11.10}$$

Acknowledging the issues related to the risk equations and catastrophic events, in a LinkedIn discussion Steve Cutchen states, "Risk is a function of the combination of probability and consequence, which may or may not be strictly mathematical, and is impacted by various outside factors." I'll add that valuation of probability (or frequency) and of consequence has considerable uncertainty, so that any assessment of risk is somewhat uncertain requiring substantial human judgment.

11.4 Estimating the Penalty from Multiple Possible Events

The above analysis considered only one event, such as a pump seal leak. But there are many possible ways that material can spill – from a leaky valve, from a corrosion-hole in a tank, from a tow-motor hitting and breaking a line, from tank filling, from dropping a barrel during transport, etc. And there are many, many possible undesirable events – a fire, an injury, an overpressure gas release, a squirrel shorting out the transformer, etc.

Determine each risk as above. If all the events are independent, then the combined risk is the sum of all risks.

11.5 Using Risk in Comparing Treatments

Risk can be reduced by prevention, such as greater maintenance attention, which would reduce the probability of an event. Greater maintenance attention would be a different treatment, and it would have its own cost. There are many types of risk and all have preventive approaches. Flu vaccination is a treatment to prevent (reduce the probability) of getting the disease. Proper exercise and diet are a treatment to reduce the probability of unhealthy habits on the body. Training with regard to safety, compliance, ethics, etc. reduces the probability of undesirable events.

Alternately, risk can be reduced by containment. Here one might permit the event to happen but provide some sort of treatment to prevent it from incurring a penalty. Perhaps place catch trays under a seal, or place pumps, pipes, valves, and tanks in a catch basin. If the risk is a fire, have a fire extinguishing system ready. If the risk is embezzlement have an independent auditing system in place to catch its beginning to eliminate the player. If the risk is associated with motorcycle accidents on a highway, impose a helmet law to reduce the severity of an accident.

Risk can also be reduced by alternate designs. Instead of conventional direct-drive pumps that need a seal, use magnetically coupled pumps. Instead of using a raw material or making an intermediate product that is toxic, use an alternative approach to making the final product. Relocate manufacturing to a place that does not penalize the possible spill.

The solution might be building a financial reserve to handle the costs if they occur, or influencing legislation to deflect accountability, or getting insurance, or having contracts with stand-by assistance sources.

Each of those solutions to a problem is a treatment. Usually, the alternate treatment has an economic cost associated with it. When looking at alternate treatments, include risk.

Analysis of risk does not provide a solution as to how to minimize risk, but it reveals mechanisms and quantifies aspects, and thereby permits designers to improve a design. The solution depends on the situation and the creativity of the designer.

11.6 Uncertainty

There is uncertainty associated with each term in the risk equation. The expected frequency will depend on equipment age, operator's physical ability, operator's experience, maintenance practices, supplier of replacement parts, process operating intensity, etc. And, the consequences, have similar uncertainty associated with predicting the future situation and context. Both the frequency and cost will change in time. The evaluator must assign a probable frequency and the cost impact that represent the future for both. One can use past data, but it must be translated to the new application and anticipated future context. These are necessarily subjectively assigned values.

If a person can estimate the frequency and cost, then one should be able to also estimate the possible range of the frequency and cost estimates (or provide a subjective feel for the 95% confidence intervals). Then there are several ways to adapt that. One method would be to use the worst-case scenario for the risk, the maximum of both the estimated frequency and cost. This would be a conservative approach, assigning a worst-case situation, but that would not ground the impact in standard economic decision-making. Alternately, one could use a Monte Carlo approach to simulate 1,000 (or many) combinations of the product and use an average. The average, if the uncertainty on both variables is modeled as uniform and symmetric is simply the product of the nominal values. It is not too difficult to show that the average of Equation (11.1) becomes:

$$\begin{aligned}\bar{R} &= \frac{1}{N}\sum_{i=1}^{N}\left[f_{\text{nominal}} + f_{\text{range}}\left(r_{1,i} - .5\right)\right]\left[c_{\text{nominal}} + c_{\text{range}}\left(r_{2,i} - .5\right)\right] \\ &= f_{\text{nominal}} c_{\text{nominal}}\end{aligned} \tag{11.11}$$

Here $f_{range} = f_{max} - f_{min}$, and $r_{1,i}$ is a uniformly distributed random number on the 0 to 1 interval.

But, if not symmetric or from a uniform distribution it will not be so trivial an equivalent. If many Monte Carlo simulations are performed, one could use the average, median, or 75th percentile worst case, or whatever. But, in any case, the estimate of the uncertainty and the distribution is subjective.

One needs to be careful to not zealously overestimate the frequency or cost if seeking to minimize risk or underestimate the values if seeking to not let risk impact a project or design decision.

The uncertainty associated with choosing frequency or penalty values, or their ranges or distributions, leaves us with the situation in which a best "guestimate" is all that is rationally justified. An estimate of frequency and penalty values that are grounded in past experience must be adjusted to match changes in the new situation.

11.7 Detectability

Sometimes risk is only a liability if the event is detected. For instance, if the product container is supposed to contain 30 lbs of material, but the filler machine occasionally places only 28 lbs in a sack, and a customer detects it, then the customer might return the product. As a minimum this will cost the supplier handling of product, return shipping, and clerical time, and possibly future purchases. Similarly, your product is not supposed to contain a particular impurity that contaminates a customer's product, and if detected you might have to buy all the customer's contaminated product, etc. This could be very expensive. However, the customer may never realize the defect, and then there is no cost consequence to the supplier. So, Equations (11.1) and (11.9) could be modified to contain a detectability probability, d.

$$R = fcd \tag{11.12}$$

$$R = p(1)c_1 d_1 + p(2)(c_2)d_2 + \cdots \tag{11.13}$$

11.8 Achieving Zero Risk

Ideally, we prevent any undesired events from happening, but, the reality is that undesired events cannot be avoided. The only way to not have an accident is to not do anything. Don't manufacture anything, because there might be a forklift accident. Don't prepare food to eat, because it might contain a microbe. Don't go for a walk, because you might get a mosquito bite. Don't shower, because you might slip and fall. But, not doing anything adds no value to human aspirations. We only add value to life by doing; and in doing, one must balance the benefit of activity to the undesired consequences of the activity. Zero risk is neither possible nor desirable. The objective is to balance anticipated benefit to the potential downside.

11.9 Takeaway

The calculation of risk should be grounded in data on probability; but substantial interpretation is needed to estimate the financial penalty of the event, and also to propagate the single event probability to parallel cases and multi-year consideration. Although risk could be presented as using simple probability rules, that would misdirect the reader. It is more engineering craft than mathematical science.

A useful standard is the AS/NZS ISO 31000:2009 (ISO Guide 73:2002).

It is important. Consider risk in design and operational procedures.

11.10 Exercises

1. Person A is driving a rare antique car, and the probability of an accident is the same as that of Person B who is driving a 7-year-old conventional car. Which has greater risk?
2. How would you assess the value of: Losing your wallet with $15 in it? Losing your car keys? Missing a day of work because of illness?
3. How would you assess jamming your finger and needing to wear a cast on your left hand? On your right hand? How would you value the risk for a professional athlete or a surgeon?
4. Derive Equation (11.9b) from (11.9a).
5. Prove Equation (11.11). Expand the sum, argue that the average of the deviation terms is zero.
6. Use the guide in Section 11.3 to derive the equivalence of Equations (11.1) and (11.2).
7. Use Example 11.3 to determine the 99% error on the probable risk, based on the implied uncertainty of each of the four "givens" in the example statement.
8. Draw a timeline of 10 years and consider that there is an average rate of 0.3 events per year. Indicate when the independent and random events might occur on your timeline. Explain your locations.

12

Analysis of Variance

12.1 Introduction

Analysis of variance (AOV or ANOVA) is a powerful technique for comparing the means of treatments. The test statistic is F.

Analysis of variance is often used as a screening technique to determine whether there is any probable qualitative relationship between the treatments before additional effort and resources are spent in an attempt to develop a quantitative relationship. The terms factor and treatment are used interchangeably to refer to the independent variable.

One advantage of ANOVA over other statistical methods to reveal impact is that the treatments do not have to be represented by continuum-valued numbers (those with numerical values representing physical quantities). Although ANOVA can use treatments with such numerical values, ANOVA can also test the effects of treatments that are category or class variables.

12.2 One-Way ANOVA

Consider the data array in Table 12.1. Every entry is subscripted by row and column. By convention, the columns represent treatments, and the rows represent replicate observations. The subscripts i and j refer to the rows and columns, respectively; then $Y_{i,j}$ denotes the value of the ith observation of the jth treatment. There are $j = 1, 2, \ldots, J$ number of treatments. The number of observations for each treatment does not have to be the same. The notation I_j represents the number of replicate observations in the jth treatment. The observations are replicates, meaning that samples in each column represent observations from a population expected to have the same distribution (mean and sigma, if the population is normal).

A treatment can consist of levels of a single variable (e.g., different temperatures) or different types of a discrete variable (e.g., different types of closures, fasteners, respirators). In either case, only the treatment has different values or classifications. All other conditions between columns should be the same. In a one-way ANOVA, the number of rows represents the number of replicate observations per treatment. Although it is convenient for every treatment to have the same number of observations, that situation is not required.

There are a total number of $N = \sum_{j=1}^{J} I_j$ observations in the table. Be aware that the table uses both upper case and lower case I and J.

DOI: 10.1201/9781003222330-12

TABLE 12.1

Data Array for a One-Way ANOVA

Column#→		1	2	...	j	...	J
Row#→	1	$Y_{1,1}$	$Y_{1,2}$	...	$Y_{1,j}$	...	$Y_{1,J}$
	2	$Y_{2,1}$	$Y_{2,2}$	...	$Y_{2,j}$	...	$Y_{2,J}$
	.	.	.		.		.
	.	.	.		.		.
	.	.	.		.		.
	i	$Y_{i,1}$	$Y_{i,2}$	...	$Y_{i,j}$	...	$Y_{i,J}$
	.	.	.	...	$Y_{I_j,j}$	...	.
	.	.	$Y_{I_2,2}$				.
	.	.		...		...	.
		$Y_{I_1,1}$		...		...	$Y_{I_J,J}$

12.2.1 One-Way ANOVA Method

The assumption is that the treatments are all identical, then we expect that the variance associated with all observations should be identical. The supposition is that there is no difference between the means of each treatment. If true, the expected reveal in the data is that the variance, as calculated in any number of ways should be the same, and the variance ratio will be unity $\frac{\sigma_1^2}{\sigma_2^2} = 1$, or alternately, $\frac{\sigma_1^2}{\sigma_2^2} - 1 = 0 = \frac{\sigma_1^2 - \sigma_2^2}{\sigma_2^2}$. The hypothesis is that $F = \frac{\sigma_1^2 - \sigma_2^2}{\sigma_2^2} = 0$. The statistic will be an F-ratio of variances, but not a ratio of the same construct as that of Chapter 6. The ANOVA F-ratio should be near to zero if the hypothesis is true, and the ratio will be large if not true. The rejection will be one sided, only testing if the F-statistic is too large.

If all treatments have the same mean and variance, if there is no difference in the effect of the treatments on the observation, then the variance measured in any number of ways will be the same. All N of the data can be used to calculate estimates of the mean and variance.

$$\bar{\bar{X}} = \frac{1}{\sum_{j=1}^{J} I_j} \sum_{j=1}^{J} \sum_{i=1}^{I_j} Y_{i,j} \tag{12.1}$$

$$\begin{aligned} s^2 &= \frac{1}{\left(\sum_{j=1}^{J} I_j\right) - 1} \sum_{j=1}^{J} \sum_{i=1}^{I_j} \left(Y_{i,j} - \bar{\bar{X}}\right)^2 \\ &= \frac{1}{N-1} SSD_{\text{Total}} = MS_T \end{aligned} \tag{12.2}$$

The term SSD_{Total} means the sum of squared deviations total (for all data) $SSD_{\text{Total}} = \sum_{j=1}^{J} \sum_{i=1}^{I_j} \left(Y_{i,j} - \bar{\bar{X}}\right)^2$, and MS_T means the mean-squared deviation or average (per degree of freedom).

Alternately, we could calculate the mean and variance of each column, associated with each treatment.

$$\bar{X}_j = \frac{1}{I_j}\sum_{i=1}^{I_j} Y_{i,j} \tag{12.3}$$

$$s_j^2 = \frac{1}{(I_j - 1)}\sum_{i=1}^{I_j}\left(Y_{i,j} - \bar{X}_j\right)^2 = \frac{1}{(I_j - 1)} SSD_j \tag{12.4}$$

Here the SSD_j means the sum of squared deviations for the treatment.

The several s_j^2 values could be pooled to have a collective estimate of the variance of all data.

$$s_p^2 = \frac{\sum_{j=1}^{J}(I_j - 1)s_j^2}{\sum_{j=1}^{J}(I_j - 1)} = \frac{SSD_w}{(N - J)} = MS_w \tag{12.5}$$

Here the SSD_w means the sum of squared deviations for all the treatments, within all treatments. And MS_w means the mean squared deviation within, the average (per degree of freedom).

If there is no treatment difference, then the pooled variance should ideally have the same value as the overall variance. And the difference should be zero. Alternately, if there is an effect of the treatments, then MS_T will be larger than MS_w due to the treatment making one or more columns of values lower or higher than the others.

The difference in the sum of squared deviations is one possible measure of the impact of the treatments.

$$SSD_b = SSD_T - SSD_w \tag{12.6}$$

If there is no difference between treatments, then SSD_b should be zero.

The degrees of freedom (DoF) for SSD_b is the difference in DoF for SSD_T and SSD_w.

$$\nu_b = (N - 1) - (N - J) = (J - 1) \tag{12.7}$$

So, the per DoF impact of the treatments is the variance due to the treatments, or between treatments, scaled by the DoF.

$$s_b^2 = \frac{SSD_b}{(J - 1)} = MS_b \tag{12.8}$$

The F-statistic is

$$F = \frac{MS_b}{MS_w} \tag{12.9}$$

with DoF for numerator and denominator of $\nu_b = (J - 1)$ and $\nu_w = (N - J)$.

The F-value is ideally zero. So, reject the hypothesis that all treatments are equivalent if F is too large.

Example 12.1: Data in the following table represent the outcome of four treatments. The experiments were controlled to reasonably exclude other possible influences on the outcome. The observations are replicates. Each column represents samples from the same population.

	Treatment			
Data n	T1	T2	T3	T4
1	30.51	31.66	28.95	29.82
2	29.63	32.55	30.57	29.96
3	31.46	30.71	30.8	31.29
4	30.42	30.87	30.91	30.21
5	30.87	31.31		29.88
6		31.66		28.32
7				29.03
8				30.46
9				29.38

One-way ANOVA can be used to detect if the treatments have differing impact on the observations. The following table is the output from the Excel Data Analysis Add-In "ANOVA: Single Factor" with a 0.05 level of confidence.

ANOVA: Single Factor

Summary

Groups	Count	Sum	Average	Variance
Column 1	5	152.89	30.578	0.44787
Column 2	6	188.76	31.46	0.44024
Column 3	4	121.23	30.3075	0.839092
Column 4	9	268.35	29.81667	0.726675

ANOVA

Source of variation	SS	df	MS	F	P-value	F-crit
Between groups	9.886041	3	3.295347	5.348133	0.007199	3.098391
Within groups	12.32336	20	0.616168			
Total	22.2094	23				

The top section of the analysis table reveals the data count, average and variance values. The lower part of the analysis table indicates the sum of squares, DoF, MS, and F-values as discussed above.

The F-value of the data, 5.348…, exceeds the F-critical value of 3.098…. Accordingly, the null hypothesis (that the treatments have no effect on the observation means) should be rejected. The p-value is 0.007…. There is only a 0.7% chance that treatments that have zero impact could generate data that has such an extreme F-value.

Note: The test does not indicate whether one was worse or better than the other three, or that two were different from two. It just indicates that not all treatments were equivalent.

Example 12.2: Samples of steel from four different batches were analyzed for carbon content. The results are shown below for quadruplicate determinations by the same analyst. Are the carbon contents (given in weight percent) of these batches the same? What are the 99% confidence limits on the average carbon content of each batch?

Percent carbon in steel batches			
A	B	C	D
0.39	0.36	0.32	0.43
0.41	0.35	0.36	0.39
0.36	0.35	0.42	0.38
0.38	0.37	0.40	0.41

1. Assume the data are normally distributed.
2. H_0: $\tau_j = 0$ for all j (there is no treatment effect) vs H_A: $\tau_j \neq 0$ for all j (a treatment effect exists).
3. The test statistic is F.
4. Set $\alpha = 0.05$ for the ANOVA.
5. The critical value is $F > F_{(J-1),J(I-1),1-\alpha}$ or $F > F_{3,12,0.95}$.

ANOVA: Single Factor

Summary

Groups	Count	Sum	Average	Variance
Column 1	4	1.54	0.385	0.000433
Column 2	4	1.43	0.3575	9.17E-05
Column 3	4	1.5	0.375	0.001967
Column 4	4	1.61	0.4025	0.000492

ANOVA

Source of variation	SS	df	MS	F	P-value	F crit
Between groups	0.00425	3	0.001417	1.899441	0.183559	3.490295
Within groups	0.00895	12	0.000746			
Total	0.0132	15				

7. As $F = 1.8994 < F_{3,12,0.95} = 3.49$, we accept H_0: $\tau_j = 0$ for all j and conclude:

The carbon content of the batches is probably not significantly different.

To find the 99% CI on the mean of each batch, we note that $\sqrt{MS_{EE}/I}$ is $Sp = 0.013655$. The intervals are

$$(0.3483 < \mu_1 < 0.4267)$$

$$(0.3158 < \mu_2 < 0.3992)$$

$$(0.332 < \mu_3 < 0.4167)$$

$$(0.3608 < \mu_4 < 0.4442)$$

All four confidence intervals overlap to some degree, providing further evidence (but not a valid statistical condition) that the carbon content of the batches could be the same.

Example 12.3: The life of four types of casing designs was estimated for gas wells in a particular field. Ten wells with the same design were chosen for each of the four types. To eliminate any bias due to variations in depth, only wells of the same relative depth were considered. Operators in this field are interested in determining whether or not these four designs have the same life. The results in projected service life (years) as a result of accelerated life testing follow. Do the designs differ?

Design 1	Design 2	Design 3	Design 4
30	13	45	21
20	15	36	14
32	20	10	30
36	21	15	28
18	18	12	19
40	16	28	25
22	14	20	16
19	30	25	13
28	15	30	12
35	12	12	29

ANOVA: Single Factor

Summary

Groups	Count	Sum	Average	Variance
Column 1	10	280	28	62
Column 2	10	174	17.4	28.04444
Column 3	10	233	23.3	134.9
Column 4	10	207	20.7	48.01111

ANOVA

Source of variation	SS	df	MS	F	P-value	F crit
Between groups	600.5	3	200.1667	2.933322	0.04645	2.866266
Within groups	2,456.6	36	68.23889			
Total	3,057.1	39				

As $F_{DESIGN} = 2.93 > F_{3,36,0.95} = 2.888$, we reject H_0 and conclude:

Casing design probably does affect the service life.

Our conclusion is strengthened by observing the p-value of 0.04645. There is only a 4.64% chance of finding a larger value for F_{DESIGN}.

Note: The test does not indicate whether one was worse or better than the other three, or that two were different from two. It just indicates that not all treatments were equivalent.

Example 12.4: Four vertical elutriators were used to obtain samples of the concentration of cotton dust in the open-end spinning room of a large textile mill. The results, in micrograms per cubic meter, are shown below. Do these samplers give equivalent results?

VE_1	VE_2	VE_3	VE_4
182.6	174.3	182.0	181.7
173.4	178.5	182.1	183.4
190.1	180.0	184.6	180.6
178.6		180.9	
188.2			

The hypothesis is that there is no effect of the sampling device on the analysis.

ANOVA: Single Factor

Summary

Groups	Count	Sum	Average	Variance
Column 1	5	912.9	182.58	47.062
Column 2	3	532.8	177.6	8.73
Column 3	4	729.6	182.4	2.446667
Column 4	3	545.7	181.9	1.99

ANOVA

Source of variation	SS	df	MS	*F*	*P*-value	*F*-crit
Between groups	55.032	3	18.344	0.92976	0.458762	3.587434
Within groups	217.028	11	19.72982			
Total	272.06	14				

As the calculated value $F_{Tr} = 0.93 < F_{3,11,0.95} = 3.59$, we accept the null hypothesis and conclude:

The vertical elutriators gave statistically equivalent results.

Note: The *p*-value of 0.4587... is not near to a small value (such as 0.05) that would represent confidence in rejecting the hypothesis.

Note: Statistically equivalent does not mean that they are the same. They may be different, but the variation masks detecting if they are different.

12.2.2 Alternate Analysis Approaches

There are four treatments in Example 12.1. The ANOVA does not indicate whether all, or just one of the treatments reveals a difference. As an alternate analysis one could do a *t*-test on each pairing, there are six in this case. The six *p*-values are indicated in Table 12.2.

Now however, there are 6 tests. For each there is a chance that a comparison of treatments that are equivalent will generate data that makes it appear to have a difference, a T-I error. If the desire is to have an overall T-I error probability of 0.05, then each comparison needs to have a lower threshold. If there is no treatment difference, then the

TABLE 12.2

p-Value Results of a *t*-Test on the Data of Example 12.1

	T2	T3	T4
T1	0.0565	0.6231	0.1121
T2		0.0486	0.0016
T3			0.3681

T1–T2 comparison should not reveal a difference, AND the T1–T3 should not, …, AND T3–T4 should not. The AND conjunction indicates that probabilities should be multiplied, and if the desired level of significance 0.05, the individual test threshold needs to be $\alpha_{\text{individual}} = 1-\sqrt[6]{(1-\alpha_{\text{overall}})} = 1-\sqrt[6]{(1-0.05)} = 0.0085124\ldots$.

Table 12.2 indicates that the *p*-value for the T2–T4 comparison of 0.0016 exceeds this 0.0085 threshold. That could trigger the null hypothesis rejection, and as well indicate where the improbable difference exists. As it happens T2 is also near to having a rejectable difference between both T1 and T3. Although the ANOVA of Example 12.1 just revealed the probability of a difference, this *t*-test is a more detailed inspection of the individual differences and provides insight as to what treatment causes a difference.

We have compared the one-way ANOVA to the multiple *t*-tests on many simulated data cases, and find that both are effective, but that the one-way ANOVA is a bit more powerful in detecting small differences than the multiple *t*-test approach. The common rule is to do an ANOVA first, and if a difference is detected, then look at the detail to see what the cause is.

There are also alternate ways to process the ANOVA ratio of variances. These have more or- less the same results as the method in Section 12.2.1, which is the accepted standard.

12.2.3 Model for One-Way Analysis of Variance

The classic model for one-way ANOVA is

$$Y_{ij} = \mu + \tau_j + \varepsilon_{ij} \tag{12.10}$$

where μ is the grand population mean, τ_j is the effect of the *j*th treatment (the variable is the *j*th column), and ε_{ij} is the random error associated with the *ij*th observation on *Y*.

As the grand population mean is a measure of location, the τ_j are measures of the displacement of a group mean μ_j from the overall mean.

If we can show that the τ_j are probably zero, then presumably the differences in the column means are all zero. This situation would indicate that the treatment effects are probably nil, i.e., that there is no column effect. The corresponding null hypothesis is H_0: $\tau_j = 0$ vs H_A: $\tau_j \neq 0$ for all *j*. We use the *F*-statistic of Equation (12.9) to evaluate the null hypothesis. The rejection region is $F > F_{(J-1),J(I-1),1-\alpha}$ for constant *I*. Otherwise, the critical value of *F* for H_0: $\tau_j \neq 0$ is $F_{(J-1),\Sigma j(Ij-1),1-\alpha}$ for variable I_j.

Here, we have a one-tailed null hypothesis involving equality of treatment effects. Why do we put the entire rejection region on one side of the distribution? Although the null hypothesis is stated as H_0: $\tau_j = 0$ for all *j*, the basis of the hypothesis is in a different, but equivalent, null hypothesis that the difference between column means is ≤0. This original null hypothesis requires a one-tailed alternative hypothesis. Since the treatment effect τ_j

TABLE 12.3

Analysis of Variance for Completely Randomized Design with Equal Numbers of Observations per Treatment

Source	df	Normalized SS	EMS
Mean	1	$\dfrac{\left(\sum_i \sum_j Y_{ij}\right)^2}{IJ} = SS_M$	—
Between columns (treatments)	$J-1$	$\dfrac{\sum_{j=1}^{J}\left(\sum_{i=1}^{I} Y_{ij}\right)^2}{I} - SS_M = SS_{Tr} = SS_B$	$\left.\begin{array}{l}\text{Model I}: \sigma^2 + I\dfrac{\sum_j \tau_j^2}{J-1} \\ \text{Model II}: \sigma^2 + I\sigma_\tau^2\end{array}\right\} EMS_{Tr}$
Within columns (experimental error)	$J(I-1)$	$\sum_i \sum_j Y_{ij}^2 - SS_M - SS_{Tr} = SS_{EE} = SS_W$	$\sigma^2 = EMS_{EE}$
Total	IJ	$\sum_i \sum_j Y_{ij}^2 = SS_T$	

measures the displacement of the column means μ_j from the overall mean μ, we are only interested in whether the τ_j are probably >0. If they are, then there probably is a measurable effect due to the treatments. As a result, the null hypothesis about the τ_j has as its true alternative H_A: $\tau_j > 0$ for all j, which requires a one-sided critical region. If the calculated value of F falls in that region, the null hypothesis is rejected, and we conclude that the treatment effects were probably nonzero.

Table 12.3 is the standard format for testing H_0: $\tau_j = 0$ for all j. The total sum of squares, SS_T, is important in subsequent analyses, as the sum of squares for experimental error, SS_{EE}, is obtained by difference.

The term EMS in Table 12.3 stands for "expected mean square," i.e., the contribution of the corresponding source of variation to the total population variance. We will not derive any of the EMSs for you; they are derived in many statistical theory texts. We include the EMS column in Table 12.3 and subsequent ANOVA tables only so that you can quickly identify the appropriate F test for the variance components.

In a one-way ANOVA, the treatments can be either different classes of a discrete variable or different levels of the same continuous variable. This dichotomy in variable type requires different mathematical expressions for each type. Model I requires $\sum_{j=1}^{J} \tau_j^2 = 0$ and is only concerned with the J treatments (discrete variables) present in the experiment. Model I is thus a "fixed-effects" model, and the null hypothesis is that no treatment effects are present, i.e., $\tau_1 = \tau_2 \ldots = \tau_j$ or H_0: $\tau_j = 0$ for all j. Examples of fixed treatments often involve equipment or processes: Types of spinning frames, soldering irons, heat exchangers, pumps, etc. Model II is concerned with random variables, which are assumed to be normally and independently distributed with mean 0 and variance σ_τ^2. This concept is usually abbreviated as NID $\left(0, \sigma_\tau^2\right)$. By their very nature, such treatments are part of a continuous distribution: temperature, flow rate, concentration, etc. The fact that we have selected certain levels of such a variable as a treatment is immaterial: The variable is still part of a continuous, infinite distribution of possible values.

12.2.4 Subsampling in One-Way Analysis of Variance

This is often termed sampling with replicates.

Subsampling occurs when the samples (material collected for observation) are divided into subsamples before testing or analysis. For example, in making morning coffee for the family the treatment might be the number of scoops of ground coffee beans. The sample might be the pot of coffee. If the coffee in the pot is divided into cups, one expects each cup to have identical properties. The cup of coffee would be a subsample. From pot to pot, even with the same recipe, the beans have variation, the measurement of bean volume has variation, the metering of water has variation, and the ambient temperature and humidity have variation. So, even with the same treatment, one expects pot-to-pot variation. If the coffee in each pot is perfectly mixed the subsample in each cup should be identical. Then, differences in evaluations of the coffee subsample would be due to measurement device error and sampling handling procedures.

As another example, a scoop of fertilizer might be sampled from a manufacturing line, and the scoop divided into 3 replicate subsamples for lab analysis. The scoop represents the material produced by the process operating conditions, the treatment. All of the material within the scoop is expected to be homogeneous, so any variation in subsample-to-subsample analysis should represent lab chemical analysis variation. For the same treatment, scoop-to-scoop variation would represent the process variation of influence on the material.

A *sample* is a part or all of an experimental unit to which a *treatment* (a set of predetermined values of the independent variables) has been applied. A *subsample*, or replicate measurement, is the analysis of separate portions of the sample.

Certain types of observations (temperature, pressure, etc.) cannot be divided as those variables represent properties of the system or material. Other types can be readily subdivided. Samples of a pharmaceutical can be obtained, divided, and then analyzed. Many test specimens for the determination of tensile strength can be prepared from a single formulation of a polymer. On the other hand, resistors, integrated circuits, etc. cannot be subdivided for testing after manufacture.

For this analysis each sample will be divided into the same number of subsamples.

Note that in a subsampling situation, n samples are taken, and each is divided into m subsamples. The *subsamples are not independent observations* but only provide an estimate of the sampling (sample preparation and analysis) error, σ_η^2.

Differences in the results of subsamples within a sample provide an estimate of experimental measurement error. Differences in the results samples from the same treatment reveal the combined effect of natural variability of the process and the measurement. And differences between the treatments yield an error component associated with all three: The treatments, natural process variation, and measurement error.

The model for one-way ANOVA with subsampling is

$$Y_{ijk} = \mu + \tau_j + \varepsilon_{ij} + \eta_{ijk} \tag{12.11}$$

where , $j = 1, 2, \ldots, J$ are the treatments (levels), $i = 1, 2, \ldots, I$ are observations per treatment, and $k = 1, 2, \ldots, m$ are samples per observation (replicates or subsamples). In this model, τ_j represents the effect of the treatment. We assume that the experimental errors, ε_{ij}, are normally and independently distributed with mean 0 and variance σ^2. In notational form, we write ε_{ij} are $\text{NID}\left(0, \sigma^2\right)$. We also assume that the sampling errors η_{ijk} are NID $\left(0, \sigma_\eta^2\right)$. In our notation, ε_{ij} is the effect of the ith sample subjected to the jth treatment, and η_{ijk} is the effect of the kth subsample from the ith sample subjected to the jth treatment.

TABLE 12.4

Analysis of Variance for Subsampling in a Complete Randomized Design (Equal Subclass Numbers)

Source	df	Normalized SS	EMS
Mean	1	$\frac{\left(\sum_i \sum_j \sum_k Y_{ijk}\right)^2}{JIm} = SS_M$	—
Treatments (columns)	$J-1$	$\frac{\sum_j \left(\sum_i \sum_k Y_{ijk}\right)^2}{mI} - SS_M = SS_{Tr}$	$\begin{cases} EMS_I \\ EMS_{II} \end{cases}$
Experimental error	$J(I-1)$	$\frac{\sum_i \sum_j \left(\sum_k Y_{ijk}\right)^2}{m} - SS_M - SS_{Tr} = SS_{EE}$	$\sigma_\eta^2 + m\sigma^2$
Sampling error	$JI(m-1)$	$\sum_i \sum_j \sum_k Y_{ijk}^2 - SS_M - SS_{Tr} - SS_{EE} = SS_{SE}$	σ_η^2
Total	IJm	$\sum_i \sum_j \sum_k Y_{ijk}^2 = SS_T$	

The procedure for one-way ANOVA with subsampling is shown in Table 12.4. "Equal subclass numbers" indicates that each of the n samples is composed of exactly m subsamples.

Two expected mean squares, EMS_I and EMS_{II}, are included in Table 12.4, corresponding to Model I (discrete variable) and Model II (continuous variable) treatment effects.

The shorthand notation in the table uses a bold dot to indicate a summation. The location of the dot subscript tells you whether the summation is over the rows or the columns. The sum of observations in the ith row is thus $Y_{i\cdot}$, indicating that all values in the ith row have been added, or

$$Y_{i\cdot} = \sum_{j=1}^{J} Y_{ij} \tag{12.12}$$

Similarly, to sum all the rows (entries) in the jth column, we write

$$Y_{\cdot j} = \sum_{i=1}^{I} Y_{ij} \tag{12.13}$$

The individual row and column means are represented as

$$\bar{Y}_{i\cdot} = \frac{Y_{i\cdot}}{J} \text{ and } \bar{Y}_{j\cdot} = \frac{Y_{\cdot j}}{I} \tag{12.14}$$

Note: Each sum has been divided by the number of entries in the column or row, respectively, that contributed to the sum. The overall sum of the matrix or data array is found by summing in one direction and then adding those sums to obtain the overall or total sum, which is

$$Y_{\cdot\cdot} = \sum_{i=1}^{I}\left(\sum_{j=1}^{J} Y_{ij}\right) = \sum_{i=1}^{I} Y_{i\cdot} = \sum_{j=1}^{J} Y_{\cdot j} \tag{12.15}$$

The overall, or grand, mean is the sum of the array divided by the number of members, or

$$\bar{Y}.. = \frac{Y..}{IJ} \tag{12.16}$$

We will further simplify the notation by dropping the explicit statement of the range for all summations so that

$$\sum_i = \sum_{i=1}^{I} \quad \text{and} \quad \sum_j = \sum_{j=1}^{J} \tag{12.17}$$

We will continue to use uppercase letters for random variables and lowercase letters for individual values of the corresponding variable.

The expected mean squares for these models are

$$\text{EMS}_\text{I} = \sigma_\eta^2 + m\sigma^2 + Im\sum_{j=1}^{J} \frac{\tau_j^2}{J-1} \tag{12.18}$$

and

$$\text{EMS}_\text{II} = \sigma_\eta^2 + m\sigma^2 + Im\sigma_\tau^2 \tag{12.19}$$

The ANOVA procedure to follow when subsampling is chosen, is to compare F_{EE} = $\text{MS}_\text{EE}/\text{MS}_\text{SE}$ to the critical region for experimental error. If the ratio $\text{MS}_\text{EE}/\text{MS}_\text{SE}$ is significantly greater than one, the null hypothesis of no experimental error is rejected. In that case, treatments are tested against experimental error by $F_\text{Tr} = \text{MS}_\text{Tr}/\text{MS}_\text{EE}$. If experimental error is not significant (H_0: $\sigma^2 = 0$ is accepted), sampling error σ_η^2 is the primary error source. In such a case, the best approach is to pool the experimental error and sampling error variances to obtain a more accurate variance estimate. If you adopt this approach, you'll also have to pool the degrees of freedom. The result is also an increase in the precision with which you can estimate the effect of the treatments by $F_\text{Tr} = \text{MS}_\text{Tr}/\text{MS}_\text{PE}$ where

$$\text{MS}_\text{PE} = \frac{\text{SS}_\text{EE} + \text{SS}_\text{SE}}{\text{df}_\text{EE} + \text{df}_\text{SE}} \tag{12.20}$$

Another use of the expected mean squares is in providing methods for estimating individual variance components. If you need an estimate (S^2) of σ^2, the true experimental error variance, it can be obtained from the error mean squares as $S^2 = (\text{MS}_\text{EE} - \text{MS}_\text{SE})/m$.

Example 12.5: A packed tower is used to absorb ammonia from a gas into a countercurrent flowing liquid. In the evaluation of a new tower packing, three samples of the "cleaned" gas were taken for ammonia analysis at each of five feed-gas concentrations. Four subsamples were taken from each sample bag. The results are shown below in ppm ammonia. Did the inlet ammonia concentration affect the performance of this pilot-scale absorber?

		Ammonia Concentration (ppm) in Absorber Outlet Gas						
		% ammonia in entering gas phase						
		4	8	12	16	20		
		$j=1$	$j=2$	$j=3$	$j=4$	$j=5$	$Y_{i..}$	$Y_{i..}^2$
	$k=1$	12	21	37	48	60		
$I=1$	$k=2$	13	24	37	42	62	725	525,625
	$k=3$	18	26	31	50	57		
	$k=4$	17	20	33	46	71		
	$k=1$	14	25	36	43	66		
$I=2$	$k=2$	15	27	34	49	64	728	529,984
	$k=3$	13	22	31	51	59		
	$k=4$	10	24	39	41	65		
	$k=1$	11	19	36	49	63		
$I=3$	$k=2$	14	23	37	52	61	734	538,756
	$k=3$	12	21	34	48	58		
	$k=4$	19	28	32	47	70		
$Y_{.j.}$		168	280	417	566	756		
$Y_{.j.}^2$		28,224	78,400	173,889	320,356	571,536		
$\sum_j Y_{.j}^2 = 1{,}172{,}405$				$\sum Y_{i..}^2 = 1{,}594{,}365$				
$Y_{...} = 2187$				$SS_M = 79{,}716.15$				
$\sum_i \sum_j Y_{ij.}^2 = 390{,}959$				$SS_T = 98{,}307$				

This example involves sampling (three bags at each inlet gas concentration treatment) and subsampling (four aliquots taken from each bag). We show all the pertinent arithmetic below and to the right of the original data so that you can more easily follow the calculations. The following additional calculations are needed:

$$SS_{Tr} = \frac{\sum_j Y_{.j.}^2}{12} - SS_M = 17{,}984.2\bar{6}$$

$$SS_{EE} = \frac{\sum_i \sum_j Y_{ij.}^2}{4} - SS_M - SS_{Tr} = 39.3\bar{3}$$

$$SS_{SE} = SS_T - SS_M - SS_{Tr} - SS_{EE} = 567.25$$

The analysis of variance table for this random effects (Model II) experiment is shown below. The treatment, inlet ammonia concentration, is a continuous variable.

Source	df	SS	MS	EMS
Mean	1	79,716.15	—	—
Treatment	4	17,984.26667	4,496.06666	$\sigma_\eta^2 + m\sigma_\tau^2 + mI\sigma^2$
Experimental error	10	39.33333	3.93333	$\sigma_\eta^2 + m\sigma^2$
Sampling error	45	567.25	12.60555	σ_η^2
	60			

The first null hypothesis is $H_{0_1}: \sigma^2 = 0$, or that there is no experimental error contribution to overall variance. As

$$F_1 = \frac{MS_{EE}}{MS_{SE}} = 0.312$$

is less than $F_{10,45,0.95} \simeq 2.055$, we accept H_{0_1} as probably true. Thus, we conclude:

The major error source was due to subsampling (creating the subsamples and laboratory analysis).

As we already have 45 DoF for the sampling term, there is no need to pool the error sources to improve the quality of our test on treatment effects. We already have enough DoF for a highly precise test. Our second null hypothesis is

$H_{0_2}: \sigma_\tau^2 = 0$ vs. $H_{A_2}: \sigma_\tau^2 \neq 0$. To test H_{0_2}, we calculate

$$F_2 = \frac{MS_{Tr}}{MS_{SE}} = 356.66$$

which is greater than $F_{4,45,0.95} \simeq 2.585$. We reject H_{0_2} (it is probably false) and conclude:

Absorber performance as measured by outlet ammonia concentration is probably affected by inlet gas concentration.

12.3 Two-Way Analysis of Variance

If two independent variables affect the dependent response, we will have to use two-way analysis of variance. In this situation, the experimental data are tabulated as in Table 12.1 so that the rows represent values (levels) of one of the independent variables and the columns represent values (levels) of the other independent variable. Each Y_{ij} entry in the table is the value of the dependent variable resulting from the corresponding *treatment combination*, i.e., the particular combination of the two independent variables that caused the response. In our discussion, we'll use α_i for the row variable and β_j for the column variable. This analysis will presume no missing data, the number of entries in each column is the same, $I_1 = I_2 = I_3 \ldots$.

Also, this first analysis is for the case in which there is only one observation per treatment effect – without replicates.

12.3.1 Model for Two-Way Analysis of Variance

The first of three assumptions for the two-way model is that the Y_{ij} are normally and independently distributed with mean μ and variance σ^2 (NID(μ,σ^2)). We then write the two-way model as

$$Y_{ij} = \mu + \alpha_i + \beta_j + \varepsilon_{ij}, \quad i = 1 \cdots a, \; j = 1 \cdots b \tag{12.21}$$

where μ is the contribution of the grand mean $\mu_{..}$ to Y_{ij}, $\alpha_i = \bar{\mu}_{i.} - \bar{\mu}_{..}$ is the contribution of the ith level of the row variable, $\beta_j = \mu_{.j} - \mu_{..}$ is the contribution of the jth level of the column variable, and ε_{ij} is the random experimental error. In this simple two-way ANOVA model, the second assumption is that the row and column variables are simply additive, i.e., that no interaction terms exist.

There are two versions of the third assumption One version is that $\sum_i \alpha_i^2 = 0$ and $\sum_j \beta_j^2 = 0$. The two resulting hypotheses for this fixed-effects case (Model I) are

$$H_0 : \alpha_i = 0 \text{ for all } i \text{ vs. } H_A : \alpha_i \neq 0 \text{ for any } i$$

And

$$H_0 : \beta_j = 0 \text{ for all } j \text{ vs. } H_A : \beta_j \neq 0 \text{ for any } j$$

The other way the third assumption for the two-way model can be stated is $\sigma_{Tr}^2 = 0$. The resulting hypotheses for the Model II case are

$$H_0 : \sigma_\alpha^2 = 0 \text{ vs. } H_A : \sigma_\alpha^2 \neq 0$$

and

$$H_0 : \sigma_\beta^2 = 0 \text{ vs. } H_A : \sigma_\beta^2 \neq 0$$

12.3.2 Two-Way Analysis of Variance Without Replicates

Table 12.5 shows the ANOVA calculations for the simple two-way case with only one observation per (*i*, *j*) combination.

You should observe that in this simple two-way ANOVA with only one observation per treatment combination, the *F*-test compares the treatments to experimental error by $F = MS_{Tr}/MS_{EE}$, just as in the one-way case. To determine the proper *F*-ratio for each test, look at the expected mean squares and remember the null hypothesis is that there is no treatment effect. The value of the ratio EMS_A/EMS_{EE} will be 1 if the null hypothesis is true because the ratio will be σ^2/σ^2. Therefore, $F_A = MS_A/MS_{EE}$ is the proper statistic to test $H_0 : \Sigma\alpha_i^2 = 0$. If the null hypothesis is false, the ratio of expected mean squares will not be 1 and the corresponding value of F_A will be significantly different from 1, leading us to accept $H_A : \sigma_A^2 \neq 0$.

TABLE 12.5

Two-Way Analysis of Variance (One Observation per Treatment Combination)

Source	df	Normalized SS	EMS	
Mean	1	$\frac{Y_{..}^2}{ab} = SS_M$	—	
A	$a-1$	$\frac{\sum_i Y_{i.}^2}{b} - SS_M = SS_A$	$\sigma^2 + \frac{b\sum_i \alpha_i^2}{a-1}$	Model I
			$\sigma^2 + b\sigma_\alpha^2$	Model II
B	$b-1$	$\frac{\sum_j Y_{.j}^2}{a} - SS_M = SS_B$	$\sigma^2 + \frac{a\sum_j \beta_j^2}{b-1}$	Model I
			$\sigma^2 + a\sigma_\alpha^2$	Model II
Error	$(a-1)(b-1)$	$SS_T - SS_A - SS_B - SS_M = SS_E$	σ^2	
Total	ab	$\Sigma\Sigma Y_{ij}^2 = SS_T$		

Example 12.6: The torque outputs for several pneumatic actuators at several different supply pressures are given in the table below.

	Torque (in. lb_f)		
Actuator type	**60 psi**	**80 psi**	**100 psi**
A	205	270	340
B	515	700	880
C	1,775	2,450	3,100
D	7,200	9,600	12,100

(a) Do the different supply pressures significantly affect the output?
(b) Does the output vary for the different actuator types?

Using our standard solution format, we obtain the results below.

1. Assume that both populations (actuator type and pressure) are normally distributed.
2. As this example involves a mixed model (actuator type = rows are fixed, pressure = columns are random), the hypotheses are

$$H_{0_1}: \alpha_i = 0 \text{ for all } i \text{ vs. } H_{A_1}: \alpha_i \neq 0 \text{ for any } i$$

$$H_{0_2}: \sigma_\beta^2 = 0 \text{ vs. } H_{A_2}: \sigma_\beta^2 \neq 0$$

3. The test statistic for both null hypotheses is F.
4. Set $a = 0.05$.
5. The critical region for H_{0_1} is $F_R > F_{3,6,0.95} = 4.76$ and the critical region for H_{0_2} is $F_c > F_{2,6,0.95} = 5.14$.
6. The ANOVA table gives the calculated values for the row variable (actuator or type) and the column variable (pressure) – Excel, Data Analysis Add-in, ANOVA: Two-Factor Without Replication:

ANOVA: Two-Factor Without Replication

Summary	Count	Sum	Average	Variance
A	3	815	271.6667	4,558.333
B	3	2,095	698.3333	33,308.33
C	3	7,325	2,441.667	438,958.3
D	3	28,900	9,633.333	6,003,333
60 psi	4	9,695	2,423.75	10,599,873
80 psi	4	13,020	3,255	18,781,767
100 psi	4	16,420	4,105	29,835,300

ANOVA

Source of variation	SS	df	MS	F	P-value	F crit
Rows	1.7E+08	3	56,781,313	46.62563	0.00015	4.757063
Columns	5,653,438	2	2,826,719	2.321143	0.179205	5.143253
Error	7,306,879	6	1,217,813			
Total	1.83E+08	11				

7. As $F_R = 46.63 > F_{3,6,0.95} = 4.76$, we reject H_{0_1} and conclude:

Pneumatic actuator type probably affects the torque output.

Notably, the *p*-value of 0.00015 permits a very strong claim.

As $F_c = 2.32 < F_{2,6,0.95} = 5.14$, we do not reject H_{0_2}. The result is somewhat surprising. If we look at the original data, the effect of pressure on output appears obvious and consistent for each of the actuators. However, just as love and hate are opposites, if you do not hate someone, it does not mean that you love that person. You could feel ambivalent, or you could love them but not be sure. The reason that we cannot reject H_{0_2} from the data is that the variability in the torque due to the pressure change is small compared to the total variability in the torque due to all effects. Had pressure spanned a greater range, perhaps 10 to 100 psi, or had actuator type D not been included, there could have been enough evidence to reject H_{0_2}. Although it is conventional to equate "not reject" to "accept", it can lead to the same erroneous conclusions as equating "not hate" and "love". Thus, our answer to part (b) is:

There is insufficient evidence to reject H_{0_2}: Pressure has a relatively inconsequential effect on torque.

We introduce a new concept to aid in reconciling the results of this ANOVA. The *variance of a treatment mean* in general is the mean square for experimental error divided by the number of observations per treatment. Here,

$$\mathrm{VAR}(\bar{y}_{i\cdot}) = S_{\bar{y}_i}^2 = \frac{\mathrm{MS_{EE}}}{4}$$

$$= \frac{1{,}217{,}813.195}{4} = 304{,}453$$

which is small compared to $\mathrm{MS_{Actuator}}$ but large compared to $\mathrm{MS_{Pressure}}$. The results thus tell us that the variability in response (torque output) induced by pressure is not sufficiently larger than experimental error to cause a situation in which $\sigma^2_{\text{Pressure}}$ is so large ($\gg 0$) that $\left(\sigma^2 + \sigma^2_{\text{Pressure}}\right)/\sigma^2$ is clearly greater than 1 as determined by the *F*-test.

12.3.3 Interaction in Two-Way ANOVA

The elementary situation described by Equation (12.21) needs to be extended when you are interested in evaluating *interactions* between treatments rather than simply looking at the effects of the treatments alone. Two-way (and higher) analyses of variance can identify interactions *provided that the treatment combinations are replicated* (repeated independently as exactly as experimental conditions will allow). Interactions are of two types: Antagonistic and synergistic. You are probably familiar with the combined and almost immediate effects of particulate air pollution and the presence of SO_2 (or smog) on morbidity. When this combination is present, the incidence of respiratory problems has increased dramatically. This event is an example of a *synergistic interaction*: Together, the two types of air pollutants are worse than the sum of their individual effects. An *antagonistic interaction* occurs when the effect of one treatment tends to diminish the effect of the other.

For the situation in which T treatments are composed of independently repeated combinations of a levels of variable A and b levels of variable B, the two-way analysis of variance model, Equation (12.21), must be rewritten to include the presence of an interaction term. The revised model is

$$Y_{ijk} = \mu + \alpha_i + \beta_j + (\alpha\beta)_{ij} + \varepsilon_{ijk} \tag{12.22}$$

where μ is the true mean effect, α_i is the true effect of the ith level of factor A, β_j is the true effect of the jth level of factor B, $(\alpha\beta)_{ij}$ is the true effect of the interaction of the ith level of factor A with the jth level of factor B, and ε_{ijk} is the true effect of the kth experimental unit subjected to the (ij)th treatment combination, $i = 1,2\ldots,a, j = 1,2,\ldots,b, k = 1,2\ldots,n$. The usual assumption is made that the ε_{ijk} are NID(0, σ^2).

Note: The true interaction might not be the simple product $\alpha\beta$ as this model presumes. Nature expresses many types of nonlinear mechanisms. So, if this method detects interaction, it does not necessarily mean it is the simple product type. And, if this model does not detect interaction, it might just be in another form.

Four possible sets of assumptions can be made regarding the treatments. The first set is that we are only concerned with fixed effects α_i and β_j of factors A and B. This Model I assumption says that we are interested only in the α levels of factor A and the β levels of factor B actually present in the experiment. These assumptions are stated as

$$\sum_i \alpha_i^2 = \sum_j \beta_j^2 = \sum_i (\alpha\beta)_{ij}^2 = \sum_j (\alpha\beta)_{ij}^2 = 0 \tag{12.23}$$

Table 12.6 summarizes calculations for two-way analysis of variance with n replications. (Note that the treatments and the interaction are tested against error.)

The second set of assumptions is that α_i and β_j are random effects, or that a Model II situation exists. In this case, we are concerned with both populations of all possible values of factors A and B of which only a random sample is present. We summarize the pertinent assumptions about the populations as follows:

$$\alpha_i \text{ are NID}\left(0, \sigma_\alpha^2\right) \tag{12.24}$$

$$\beta_j \text{ are NID}\left(0, \sigma_\beta^2\right) \tag{12.25}$$

$$(\alpha\beta)_{ij} \text{ are NID}\left(0, \sigma_{\alpha\beta}^2\right) \tag{12.26}$$

The corresponding analysis of variance for this model is given in Table 12.7.

From the entries in the EMS column of Table 12.7, you notice that the first F-test is interaction vs error to evaluate $H_0 : \sigma_{\alpha\beta}^2 = 0$. If that hypothesis is rejected as probably false, the interaction term is the most significant (greatest) source of error, and the treatments are tested against the interaction.

TABLE 12.6

Two-Way Analysis of Variance with Interaction (Model I)

Source	df	Normalized SS	EMS
Mean	1	$\frac{\left(\sum_i \sum_j \sum_k Y_{ijk}\right)^2}{abn} = SS_M$	—
A	$a-1$	$\frac{\sum_i \left(\sum_j \sum_k Y_{ijk}\right)^2}{bn} - SS_M = SS_A$	$\sigma^2 + \frac{nb \sum_i \alpha_i^2}{a-1}$
B	$b-1$	$\frac{\sum_j \left(\sum_i \sum_k Y_{ijk}\right)^2}{an} - SS_M = SS_B$	$\sigma^2 + \frac{na \sum_j \beta_j^2}{b-1}$
AB	$(a-1)(b-1)$	$\frac{\sum_i \sum_j \left(\sum_k Y_{ijk}\right)^2}{n} - SS_M - SS_A - SS_B = SS_{AB}$	$\sigma^2 + \frac{n \sum_i \sum_j (\alpha\beta)_{ij}^2}{(a-1)(b-1)}$
Error	$ab(n-1)$	$\sum_i \sum_j \sum_k Y_{ijk}^2 - SS_A - SS_B - SS_M = SS_E$	σ^2
Total	abn	$\sum_i \sum_j \sum_k Y_{ijk}^2 = SS_T$	

TABLE 12.7

Two-Way Analysis of Variance with Interaction (Model II)

Source	df	Normalized SS	EMS
Mean	1	$\frac{\left(\sum_i \sum_j \sum_k Y_{ijk}\right)^2}{abn} = SS_M$	
A	$a-1$	$\frac{\sum_i \left(\sum_j \sum_k Y_{ijk}\right)^2}{bn} - SS_M = SS_A$	$\sigma^2 + n\sigma_{\alpha\beta}^2 + nb\sigma_{\alpha}^2$
B	$b-1$	$\frac{\sum_j \left(\sum_i \sum_k Y_{ijk}\right)^2}{an} - SS_M = SS_B$	$\sigma^2 + n\sigma_{\alpha\beta}^2 + na\sigma_{\beta}^2$
AB	$(a-1)(b-1)$	$\frac{\sum_i \sum_j \left(\sum_k Y_{ijk}\right)^2}{n} - SS_M - SS_A - SS_B = SS_{AB}$	$\sigma^2 + n\sigma_{\alpha\beta}^2$
Error	$ab(n-1)$	$\sum_i \sum_j \sum_k Y_{ijk}^2 - SS_A - SS_B - SS_{AB} - SS_M = SS_E$	σ^2
Total	abn	$\sum_i \sum_j \sum_k Y_{ijk}^2 = SS_T$	

If the first F-test shows that the interaction is probably insignificant when compared to error, the treatments are tested against the error.

The other two sets of assumptions yield mixed models: one variable is of the fixed type, the other is randomly distributed. By convention, in the third model (III), α is considered fixed and β is random. In the fourth model (IV), α is considered random and β is fixed. The assumptions for these two models are as follows:

TABLE 12.8

F-Ratios for Hypothesis Testing in Completely Randomized Design with Factorial Treatment Combinations

Source	Model I	Model II	Model III	Model IV
Effects	a, b: fixed	a, b: random	a fixed, b random	a random, b fixed
Mean	—	—	—	—
A	$\frac{MS_A}{MS_E}$	$\frac{MS_A}{MS_{AB}}$	$\frac{MS_A}{MS_{AB}}$	$\frac{MS_A}{MS_E}$
B	$\frac{MS_B}{MS_E}$	$\frac{MS_B}{MS_{AB}}$	$\frac{MS_B}{MS_E}$	$\frac{MS_B}{MS_{AB}}$
AB	$\frac{MS_{AB}}{MS_E}$	$\frac{MS_{AB}}{MS_E}$	$\frac{MS_{AB}}{MS_E}$	$\frac{MS_{AB}}{MS_E}$
E	—	—	—	—

Model III:

$$\Sigma\alpha_i = \sum_i (\alpha\beta)_{ij} = 0, \quad \beta_j \text{ are NID}\left(0, \sigma_\beta^2\right) \tag{12.27}$$

Model IV:

$$\sum_j \beta_j = \sum_j (\alpha\beta)_{ij} = 0, \quad \alpha_i \text{ are NID}\left(0, \sigma_\alpha^2\right) \tag{12.28}$$

The assumptions of the four ANOVA models are summarized in Table 12.8.

The *F*-tests of treatment effects for mixed models are accomplished in a simple fashion:

Always test the interaction vs error.

Test the fixed term vs the interaction term.

Test the random term vs error.

If the interaction is not significantly different from error, all treatments are tested vs error. If, however, the hypothesis $H_0(\alpha\beta)_{ij}$, for each *i*, *j* pair is rejected, then the acceptance of H_0: $\alpha_i = 0$ (or H_0: $\beta_i = 0$) over its range should be interpreted to mean that there is probably no significant difference in the levels of *A* (or *B*) when averaged over the levels of *B* (or *A*). The *F*-ratios used for hypothesis testing in two-way ANOVA are summarized in Table 12.8.

Example 12.7: The effect of temperature and steam/hydrocarbon ratio (S/HC) on ethylene production in a cracking furnace gave the yield data below. One sample was obtained from each replicate pilot-plant run. What conclusions should be drawn from this data?

	T_1	T_2	T_3	T_4
$(S/HC)_1$	38	42	43	42
	40	41	45	40
$(S/HC)_2$	36	40	46	44
	37	44	45	42
$(S/HC)_3$	39	43	44	43
	37	42	44	42

The results follow.

1. Assume that the steam/HC and the temperature populations are normally distributed for this Model II situation.
2. The hypotheses are $H_{0_1}:\sigma^2_{\text{RATIO}}=0$ vs. $H_{A_1}:\sigma^2_{\text{RATIO}}\neq 0$, $H_{0_2}:\sigma^2_{\text{TEMP}}=0$ vs. $H_{A_2}:\sigma^2_{\text{TEMP}}\neq 0$, and $H_{0_3}:\sigma^2_{\text{TEMP*RATIO}}=0$ vs. $H_{A_3}:\sigma^2_{\text{TEMP*RATIO}}\neq 0$.
3. The test statistic in all cases is F.
4. Set $\alpha=0.05$.
5. & 6. $\text{SS}_\text{T}=\sum_i\sum_j\sum_k Y^2_{ijk}=41{,}757$,

$$\text{SS}_\text{M}=\frac{Y^2_{\dots}}{abn}=41{,}583.375$$

$$\text{SS}_{\text{RATIO}}=\sum_i\frac{Y^2_{i..}}{bn}-\text{SS}_\text{M}=0.75$$

$$\text{SS}_{\text{TEMP}}=\sum_j\frac{Y^2_{.j.}}{an}-\text{SS}_\text{M}=138.4583$$

$$\text{SS}_{\text{TEMP*RATIO}}=\sum_i\sum_j\frac{Y^2_{ijk}}{n}-\text{SS}_\text{M}-\text{SS}_{\text{TEMP}}-\text{SS}_{\text{RATIO}}=13.916$$

$$\text{SS}_{\text{ERROR}}=\text{SS}_\text{T}-\text{SS}_{\text{RATIO}}-\text{SS}_{\text{TEMP}}-\text{SS}_{\text{TEMP*RATIO}}-\text{SS}_\text{M}=20.5$$

$$F_{\text{TEMP*RATIO}}=\frac{\text{MS}_{\text{TEMP*RATIO}}}{\text{MS}_{\text{ERROR}}}=1.358<F_{6,12,0.95}=3.00$$

7. From the test of H_{0_3}, we conclude that the interaction is probably not significant. Testing the treatment terms vs the interaction term is futile, given the result of the H_{0_3} test. We elect to pool the sums of squares of the interaction and error terms (don't forget to pool the degrees of freedom also) to increase the precision of the treatment tests:

$$\text{MS}_{\text{PE}}=\frac{\text{SS}_{\text{TEMP*RATIO}}+\text{SS}_{\text{ERROR}}}{6+12}=1.912$$

Proceeding to test the treatments against the pooled error, we have

$$F_{\text{RATIO}}=\frac{\text{MS}_{\text{RATIO}}}{\text{MS}_{\text{PE}}}=0.1961<F_{2,18,0.95}=3.566$$

$$F_{\text{TEMP}}=\frac{\text{MS}_{\text{TEMP}}}{\text{MS}_{\text{PE}}}=24.138>F_{3,18,0.95}=3.176$$

We conclude:

1. The interaction is not significant.
2. The range of S/HC ratios studied does not reveal a significant effect.
3. Temperature probably affects yield.

The Excel Add-in, Data Analysis, procedure "ANOVA: Two Factor With Replicates" returns nearly identical results. (It does not pool items after rejecting the interaction term.)

ANOVA: Two-Factor with Replication

Summary	T1	T2	T3	T4	Total
(S/HC)₁					
Count	2	2	2	2	8
Sum	78	83	88	82	331
Average	39	41.5	44	41	41.375
Variance	2	0.5	2	2	4.553571
(S/HC)₂					
Count	2	2	2	2	8
Sum	73	84	91	86	334
Average	36.5	42	45.5	43	41.75
Variance	0.5	8	0.5	2	13.92857
(S/HC)₃					
Count	2	2	2	2	8
Sum	76	85	88	85	334
Average	38	42.5	44	42.5	41.75
Variance	2	0.5	0	0.5	6.214286
Total					
Count	6	6	6	6	
Sum	227	252	267	253	
Average	37.83333	42	44.5	42.16667	
Variance	2.166667	2	1.1	1.766667	

ANOVA

Source of variation	SS	df	MS	*F*	*P*-value	*F*-crit
Sample	0.75	2	0.375	0.219512	0.806064	3.885294
Columns	138.4583	3	46.15278	27.01626	1.27E-05	3.490295
Interaction	13.91667	6	2.319444	1.357724	0.306351	2.99612
Within	20.5	12	1.708333			
Total	173.625	23				

Example 12.8: The masses of airborne cotton dust on sampling filters at a yarn mill are listed below for all four shifts in the warehouse and the bale-opening areas. The same five OSHA-approved samplers were used throughout this annual compliance test. Is there a difference ($\alpha = 0.05$) in the results between shifts? Is one of these mill locations dustier than the other?

	Cotton dust sample weight (μg)			
Site	Shift A	Shift B	Shift C	Shift D
Warehouse	610	350	515	635
	830	130	635	355
	630	380	485	855
	660	460	465	655
	490	400	405	685

(*Continued*)

	Cotton dust sample weight (μg)			
Site	**Shift A**	**Shift B**	**Shift C**	**Shift D**
Bale opening	430	170	845	870
	380	250	765	815
	690	270	605	495
	500	360	610	530
	330	370	670	460

The Excel Add-in, Data Analysis, procedure "ANOVA: Two Factor With Replicates" returns results.

ANOVA: Two-Factor with Replication

Summary	**Shift A**	**Shift B**	**Shift C**	**Shift D**	**Total**
Warehouse					
Count	5	5	5	5	20
Sum	3,220	1,720	2,505	3,185	10,630
Average	644	344	501	637	531.5
Variance	14,980	15,930	7,230	32,420	30,610.79
Bale opening					
Count	5	5	5	5	20
Sum	2,330	1,420	3,495	3,170	10,415
Average	466	284	699	634	520.75
Variance	19,630	6,880	10,817.5	37,217.5	42,969.14
Total					
Count	10	10	10	10	
Sum	5,550	3,140	6,000	6,355	
Average	555	314	600	635.5	
Variance	24,183.33	11,137.78	18,911.11	30,952.5	

ANOVA

Source of variation	**SS**	**df**	**MS**	***F***	***P*-value**	***F*-crit**
Sample	1,155.625	1	1,155.625	0.063712	0.802336	4.149097
Columns	632,511.9	3	210,837.3	11.62398	2.59E-05	2.90112
Interaction	185,086.9	3	61,695.63	3.401433	0.029416	2.90112
Within	580,420	32	18,138.13			
Total	1,399,174	39				

From these results, we reject $H_{01}: \Sigma\beta_j = 0$ and $H_{03}: \Sigma\Sigma(\alpha\beta)_{ij} = 0$ and do not reject $H_{02}: \Sigma\alpha_i = 0$. Our conclusions are:

1. An interaction between shift and work area probably exists.
2. There probably is a difference between shifts.
3. There is probably no significant difference due to work areas.

Statistics are just numbers unless you use them to help you improve safety, quality, production rate, etc. Use the statistics to point to justified actions. How might we interpret the significance of these results? Should the people on Shift B be praised and those on Shift D be reprimanded?

Find the cause for the difference before taking action. In this case, most routine and preventive maintenance is done on the day (A) shift. Only breakdowns are repaired during the night (B) and graveyard (C) shifts. As the swing shift (D) rotates, almost any situation may occur. What does maintenance have to do with dust level? More maintenance creates both more nuisance dust and more of the respirable dust sampled by the samplers. Was the crew in one of those areas trying to "go over the top" when the compliance test was run and so put the squeeze on the people in the warehouse or the opening/cleaning area for a higher throughput? Use the statistical results to help analyze cause-and-effect mechanisms.

12.3.4 Two-Way Analysis of Variance with Replicates

The analysis is similar to that above. The Excel Data Analysis procedure ANOVA: Two-Factor with Replication can provide the analysis.

12.4 Takeaway

ANOVA just seeks statistical evidence of linear correlation. It does not assign causation. It is very useful as a screening tool to see if there is a relation between treatments and data, which can guide model/relation/cause-and-effect development efforts.

Excel Data Analysis has limited power, but it is very convenient for basic cases. If important, use professional statistical software.

If the treatments are values of continuum or discrete variables, then correlation analysis of Chapter 13 may be as or more effective than two-way ANOVA.

12.5 Exercises

1. An oil-cracking unit processed three kinds of heavy oil (30% hydro-treated, 50% hydro-treated, and 70% hydro-treated) to reduce their viscosities. Samples of the output were taken each day for 4 days to analyze the boiling point of the product. The boiling point test was carried out four times for each sample.

Boiling Point (°F) for Processed Oils

Sample day	Feed 30%	Feed 50%	Feed 70%
1	425	457	510
	431	462	507
	436	460	500
	433	455	505
2	431	460	500

(Continued)

Boiling Point (°F) for Processed Oils

Sample day	Feed 30%	Feed 50%	Feed 70%
	423	456	510
	427	463	495
	429	465	498
3	428	482	505
	437	476	511
	436	480	506
	431	475	513
4	433	470	513
	435	476	505
	425	467	507
	430	465	510

Prepare the analysis of variance for these data and interpret the results.

2. The following data give the yields of a product that resulted from trying catalysts from four different suppliers.
   (a) Are yields influenced by catalysts?
   (b) What are your recommendations in the selection of a catalyst to obtain the greatest yield? Assume that economics dictate 95% probability of being right on a decision.

Catalyst			
I	II	III	IV
36	35	35	34
33	37	39	31
35	36	37	35
34	35	38	32
32	37	39	34
34	36	38	33

3. A solution of hot potassium carbonate (HPC) is used to scrub CO_2 from the gas feed stock in the production of an intermediate in the manufacture of nylon. It has been proposed that the addition of an amine solution to the HPC will increase the purity of the scrubbed gas. The pilot-plant data follow. Does the additive have the desired effect?

	HPC flow rate (gal/hr)			
	0.5	0.7	0.9	1.1
Amine conc. (Wt. %)				
0.5	1.64	1.32	0.76	0.62
1.0	1.31	1.09	0.47	0.38
3.0	0.99	0.81	0.15	0.08
6.0	0.61	0.45	0.04	0.01

4. The removal efficiency of a pilot-scale gas absorber is presumed to be a function of liquid rate A and the presence or absence of a buffer B. For the following data collected as a 2^2 factorial with r = three replications, evaluate the probable significance of the variance components. Replicates are used as blocks, and the form of data presentation is conventionally rotated 90° counterclockwise for convenience of display.

	Without buffer		**With buffer**	
Replication	**6 GPH**	**12 GPH**	**6 GPH**	**12 GPH**
1	21	37	48	60
2	24	37	42	62
3	26	31	50	57

5. In Example 12.2 the samples from four batches were all analyzed by the same technician. Suppose, alternately, that each manufacturing source analyzed their own samples in their own labs. Could the data be used to claim that the manufacturer's products are equivalent?

13

Correlation

13.1 Introduction

Correlation means that values of two variables rise or fall with each other. Your personal experience might agree with these examples: 1) The number of apples a tree yields rises with the size of the tree. 2) The noise on a school playground rises with the number of students on the playground. 3) Road noise in a car rises with the car speed. 4) Road noise in a car falls with the how much the windows are wound up.

Correlations can be either positive or negative. If a positive correlation, when one variable increases the other increases, and a decrease in one means a decrease in the other. A negative correlation means that they move in opposite directions. If there is no relation between the variables, the correlation is zero.

An example of a positive correlation is daylight intensity (perhaps lumens) and sun "height" over the local Earth location (as measured by sine of the angle from the eastern horizon). At dawn, the sun is just rising, the angle is zero degrees, and the sine is zero. The dawn light is low. As the sun rises the angle increases, the height (sine of the angle) increases, and day light intensity increases. At midday the sun is at 90 degrees, the sine is 1, and light intensity is at maximum. Then as the sun path continues its arc, the angle from the eastern horizon increases toward 180 degrees, the sun "height" decreases (the sine of the angle decreases) and the light intensity falls. Even though over time the light rises then falls, a plot of light intensity w.r.t. sun height (sine of the angle) only shows a positive trend.

Note: The measurement of light intensity reaching the ground will be affected by the vagaries of cloud movements, changes in local humidity and atmospheric pressure, and sunspot and sun flare activity. So, although ideally, there might be a classroom geometry model of a deterministic trend between light intensity and sun "height" natural vagaries would confound the trend (add noise) as revealed by actual measurements.

Correlation does not mean there is a cause-and-effect relation. The two correlated variables could both be effects of a common cause. Here are some examples 1) Leaves falling from the trees do not make the cold weather happen. 2) Gray hair does not cause face wrinkles, nor do face wrinkles cause gray hair. 3) Because of the driver's aggressive behavior (or due to driving on a dirt/gravel path), cars that have unexpectedly low gas mileage may also have tires that wear out rapidly. This does not mean that low gas mileage causes tires to wear rapidly.

Correlation does not affirm a hypothesized mechanism. Another example: The flowers turn during the day to point to the sun. This correlation does not affirm the hypothesis that flowers have tractor beams and pull and push against the sun to make the world go around.

The mantra is "Correlation is not Causation".

DOI: 10.1201/9781003222330-13

If there was no uncertainty on variables (no noise, no random perturbations, no natural variation, no uncontrolled factors) then either correlation or a lack of correlation would be easy to detect. However, because of variation on the data values, we need statistical tests to reveal the extent of, and confidence in a correlation.

Correlation does not have to represent a linear trend. The trend could be quadratic, exponential, or any other. In this analysis we will only consider that the trend is consistently positive or negative, that it does not reverse, that it is monotonic. Returning to the sun position and light intensity, if light intensity is plotted w.r.t. time, initially, in the morning, the correlation is positive, then changes to negative. However, if light intensity is plotted w.r.t. sun "height" it is always a positive trend.

Correlation can also exist within a single variable, for instance when it is considered over time. Many variables are measured over time: In manufacturing products, quality metrics are observed over time in quality charts (perhaps a month-long window of daily quality values). In process control, the process might be sampled on 100 ms interval (10 Hz frequency) and the trend over the past minute displayed in the operator's panel. If there is no trend in value over time, if the variable is constant over time, then the charts would ideally show a flatline horizontal trend. However, sample-to-sample vagaries will make the ideal trend noisy. If the influences on each of the samples are independent, then there will not be a relation between samples. However, if an influence on the prior sample has some persistence and continues its influence on the next sample, then there will be a positive autocorrelation. This can also be termed serial correlation.

As an example, reconsider sunlight intensity. If a cloud is blocking the sun, the light intensity is low. The clouds do not blink on and off, randomly, each millisecond. If a cloud is there, it remains there for a while before being blown away, the shadow persists prior to permitting the high intensity light. Here the influence on sun intensity is the cloud. If sunlight intensity is sampled and is relatively low, the presence of the influence will persist for a while and the observation on the next second will also tend to be low. When the sky clears, sequential light intensity values will be high.

If the noisy trend is actually changing in time (increasing or decreasing) this background change in the variable will also cause positive autocorrelation between samples.

In Chapter 19, on model validation, we will observe the residuals (the difference between model and data) w.r.t. a modeled variable. If the model is true to the data-generating mechanism, then the residuals should have no autocorrelation. By contrast, if the model does not properly capture the mechanism, then because of process-model mismatch, one residual will be positive (or negative), and an adjacent residual will tend to have the same sign because the model is locally in error.

In positive autocorrelation, if one sample has a high value, the next will tend to be high; and if one has a low value, the next will trend to be low. The first sample value is not the cause for the second value. The persistence of the influence on the first is the cause for the second.

13.2 Correlation Between Variables

13.2.1 Method

Equation (4.12), the Pearson product-moment correlation, provides a classic measure of correlation between two variables, x and y.

$$r = \frac{\sum_{i=1}^{n}(x_i - \bar{x})(y_i - \bar{y})}{\sqrt{\sum_{i=1}^{n}(x_i - \bar{x})^2}\sqrt{\sum_{i=1}^{n}(y_i - \bar{y})^2}} = \frac{\frac{1}{(n-1)}\sum_{i=1}^{n}(x_i - \bar{x})(y_i - \bar{y})}{s_x s_y} \tag{13.1}$$

The values are paired, and the subscript i is a number index for the pairing. The data can be represented as in Table 13.1.

It does not matter whether variable x is to the left or right of y in the table. It does not matter whether the index is a sequence of numbers, or letters, or category labels. It does not matter whether the rows are reorganized (such as listing in reverse order), of if two sets of pairs are switched. What matters is that 1) the variable values are rationally paired, 2) there is no missing data in only one column when using Equation (13.1), and 3) the variance of x and y are nearly the same at all of their values. (If there are missing data values in one column, eliminate the extra value on the other variable and decrement n.)

If there is zero correlation, then when x is above its average y will be below its average as frequently as y is high. Then the numerator sum in Equation (13.1) will have an approximately equal number of "+" and "−" values and will tend to remain about a value of zero. If there is positive autocorrelation, then when x is above its average y will tend to be above its average, also; and terms in the numerator will tend to be positive. Alternately, if there is negative autocorrelation, then when x is above its average y will tend to be below its average, and terms in the numerator will tend to be negative. When scaled by the standard deviations, the range of the correlation statistic is $-1 \le r \le +1$.

If r is about zero there is very little correlation. The closer the value of r gets to ± 1, the stronger is the evidence for correlation.

13.2.2 An Illustration

As a data-based illustration, Table 13.2 shows part of the data from a study of undergraduate student grades in chemical engineering. The column labeled "Major GPA" is a particular student's grade point average in their upper-level major classes, and the column labeled "STEM GPA" is the grade point average for the same student in all of their first- and

TABLE 13.1
Variables for Correlation Testing

Index	Variable y	Variable x
1	y_1	x_1
2	y_2	x_2
3	y_3	x_3
…	…	…
i	y_i	x_i
…	…	…
n	y_n	x_n

TABLE 13.2

A Portion of Student Performance Assessment Data

Student ID	STEM GPA	Major GPA
1	3.804348	4
2	3.058824	1.971429
3	2.926829	2.285714
4	2.686275	2.914286
5	3.780488	3.714286
6	2.5	2.5
7	3.764706	3.8
8	2.705882	2.771429
9	3.313725	3.142857
10	3.28125	3.085714
11	3.431373	3.714286
12	3.680851	3.6
13	2.942308	3.029412
14	3.512195	2.6
15	3.511628	3.852941
16	3.06383	3.142857
17	3.361702	3.314286
18	3.392157	2.542857
19	3.392157	2.142857
20	2.365385	1.714286

second-year STEM classes (STEM is the acronym for Science, Technology, Engineering, and Mathematics). The hope from the study was to provide indicators to the students transitioning from the second to third year about what they might expect, and to encourage those who might not have adequate ability or preparation to either self-select an alternate major or improve preparation prior to starting the upper-level major classes.

Whether the students are listed in alphabetical order, age, number of letters in their name, or distance of their home from the university is irrelevant. The number representing student ID was actually randomly assigned to prevent any traceability to a particular student.

Using Equation (13.1) the correlation coefficient value is $r = 0.694$. With 20 data pairs, this is a strong indicator of a relation. But also, it is not a perfect correlation.

In our study we actually explored many metrics of academic performance in the first two college years to see what metrics would best forecast performance in the upper-level major courses. Table 13.3 shows more of the variables for the same 20 students. For convenience in this presentation, the data are rounded.

The second-to-last column is the same Major GPA of Table 13.2, and the last column was another key factor, the integer number of D or F grades (unsatisfactory grades) the student earned in the major classes. The other columns are other metrics. "Average ENSC" is the student's grade in three lower-level engineering science courses. "CHE 2033" is the lower-level introduction to chemical engineering (material and energy balances). "Avg PhysII CalcIII" is the average of the student's grades in Physics II (light and magnetism) and Calculus III (multivariable calculus). "Adv Chem Lab" is the student's grade in either organic chemistry or biochemistry lab. "Adv Chem" is the student's grade in their

TABLE 13.3

Student Performance Assessment Data

ID	STEM GPA	Avg ENSC	CHE 2033	Avg PhysII CalcIII	Adv Chem Lab	Adv Chem	# repeats	Programming	Major GPA	Major #D&F
1	3.80	4.00	4	4.00	4	2	1	4	4.00	0
2	3.06	2.33	2	3.67	3	1	2	2	1.97	7
3	2.93	3.00	2	3.00	2	3	0	4	2.29	2
4	2.69	3.33	2	2.33	3	2	5	3	2.91	2
5	3.78	3.67	4	4.00	4	3	0	4	3.71	0
6	2.50	2.67	3	2.00	2	2	1	3	2.50	0
7	3.76	4.00	4	3.67	4	3	0	3	3.80	0
8	2.71	2.67	2	2.33	3	2	2	3	2.77	0
9	3.31	3.00	4	3.67	4	3	0	3	3.14	0
10	3.28	3.33	4	4.00	3	2	1	4	3.09	0
11	3.43	4.00	4	3.33	3	2	0	4	3.71	0
12	3.68	3.33	3	4.00	4	3	0	4	3.60	0
13	2.94	2.33	3	3.00	3	2	3	4	3.03	0
14	3.51	3.33	3	4.00	3	2	0	4	2.60	1
15	3.51	4.00	4	3.50	4	3	2	4	3.85	0
16	3.06	3.00	3	3.33	3	2	0	4	3.14	0
17	3.36	3.00	4	3.33	3	3	0	4	3.31	0
18	3.39	3.67	3	3.67	3	2	0	3	2.54	0
19	3.39	3.33	3	3.67	3	3	1	3	2.14	2
20	2.37	2.33	3	2.33	2	2	9	2	1.71	5

advanced chemistry elective. "# repeats" is the number of times a student repeated lower-level STEM courses to get a passing grade. And "Programming" is the computer programming course. In the initial study we actually considered many other possible metrics such as grades in English, or social science courses, or general chemistry, or calculus II. but these had very little correlation. Further, several courses had similar correlation impact on the outcome and were also strongly inter-correlated, so they are presented here as average.

There are 10 data columns in Table 13.3, permitting 45 comparison combinations, or 55 if a column is compared to itself. The Excel Data Analysis Add-In "Correlation" analyzes all 55. Table 13.4 presents the results, with values rounded to 2 decimal digits.

The values of 1 along the main diagonal indicate that the variable is compared to itself. Expectedly, the correlation is the perfect +1. The upper right of the table would have exactly the same values as the lower left. Equation (13.1) reveals that reversing the x and y variables does not change the r-value. So, only half of the full table is presented. Some r-values are negative indicating negative correlation.

Like ANOVA, the correlation study does not necessarily indicate cause and effect, but it does indicate strong correlations that should become a clue to closer mechanistic analysis. The classifications of *STEM GPA, Avg ENSC, CHE 2033, Adv Chem Lab,* and *Programming* each have strong correlation to *Major GPA* (r-values are 0.69, 0.71, 0.69, 0.78, and 0.66), and are the ones that should be viewed as strong indicators of student success. Further, the students' performance in *Programming* has a strong correlation to *Major#D&F* as indicated by $r = -0.71$. This is a negative correlation, meaning that a high programming grade correlates to few D and F grades in the Major. These observations could provide clues as to the innate student attributes that are key to the desired success.

TABLE 13.4

Correlation Results of Data in Table 13.3, Rounded Values

	STEM GPA	Avg ENSC	CHE 2033	Avg PhysII CalcIII	Adv Chem Lab	Adv Chem	# repeats	Programming	Major GPA	Major #D&F
STEM GPA	1									
Avg ENSC	0.77	1								
CHE2033	0.64	0.58	1							
Avg PhysII CalcIII	0.90	0.54	0.51	1						
Adv Chem Lab	0.80	0.60	0.55	0.67	1					
Adv Chem	0.44	0.39	0.42	0.25	0.39	1				
# repeats	−0.66	−0.45	−0.32	−0.59	−0.39	−0.34	1			
Program ming	0.51	0.46	0.42	0.41	0.30	0.37	−0.56	1		
Major GPA	0.69	0.71	0.69	0.43	0.78	0.39	−0.45	0.66	1	
Major #D&F	−0.42	−0.51	−0.53	−0.18	−0.41	−0.45	0.56	−0.71	−0.71	1

13.2.3 Determining Confidence in a Correlation

There seem to be several approaches to approximating a confidence interval, critical values, or a p-value on the Pearson correlation statistic of Equation (13.1). In one method, calculate a T-statistic from the data results:

$$T = r\sqrt{\frac{(n-2)}{1-r^2}} \tag{13.2}$$

The hypothesis is that there is no correlation, so that the ideal r-value and calculated T is zero. Then correlation will be accepted if r is either large positive or large negative. Then use the two-sided t-distribution with $\upsilon = (n-2)$ degrees of freedom.

Alternately, in Fischer's method calculate a modified z-statistic:

$$z' = \frac{1}{2}\text{Ln}\frac{1+r}{1-r} \tag{13.3}$$

Which is approximately normally distributed with a variance of $\sigma^2 = 1/(n-3)$.

Neither method is perfect, but both seem to approach the respective distributions rapidly with increasing n and are in reasonable agreement with each other.

Example 13.1: Compare Equations (13.2) and (13.3) in determining a p-value for the correlation of the Adv Chem Lab grade to the Major GPA from Table 13.3. The correlation statistic is $r = 0.78$, and $n = 20$ sample pairs.

Using Equation (13.2) the data T-value is $T = 0.78\sqrt{\frac{(20-2)}{1-0.78^2}} = 5.288\ldots$ and with $\upsilon = (20-2) = 18$ degrees of freedom, the p-value is 0.00005.

Using Equation (13.3) the modified z-statistic is $z' = \frac{1}{2}\text{Ln}\frac{1+r}{1-r} = 1.045\ldots$ and with $\sigma = \sqrt{\frac{1}{20-3}} = 0.2425\ldots$, the p-value is 0.000016.

The two approximations do not return exactly the same value, but both agree that if there were no correlation it is very improbable (about one out of 20,000 or one out of 61,000 chance) that the data could have generated an r-value so large. We reject the null hypothesis of no relation. We claim that the Adv Chem Lab grade is a strong indicator of the Major GPA, with a confidence greater than 99.99%.

Example 13.2: Compare Equations (13.2) and (13.3) in determining a p-value for the correlation of the Avg PhysII CalcIII grade to the #D&F in the Major from Table 13.3. The correlation statistic is $r = -0.18$, and $n = 20$ sample pairs.

Using Equation (13.2) the data T-value is $T = (-0.18)\sqrt{\frac{(20-2)}{1-(-0.18)^2}} = 0.776\ldots$ and with $\upsilon = (20-2) = 18$ degrees of freedom, the p-value is 0.4476….

Using Equation (13.3) the modified z-statistic is $z' = \frac{1}{2}\text{Ln}\frac{1+r}{1-r} = -0.1819\ldots$ and with $\sigma = \sqrt{\frac{1}{20-3}} = 0.2425\ldots$, the p-value is 0.4530….

Again, the two approximations do not return exactly the same value, but both agree that if there were no correlation it is quite possible that the data could have generated such an r-value. We conclude that there is little evidence to reject the null hypothesis. We accept that there is inadequate evidence to be able to claim a significant correlation between the Avg PhysII CalcIII grade to the #D&F in the Major.

13.3 Autocorrelation

13.3.1 Method

Equation (4.13), reproduced here as (13.4), indicates how to calculate the autocorrelation statistic of lag-1, between sequential variables. And Equation (13.5) indicates the general case of a lag-k, between values that are k-intervals apart.

$$r_1 = \frac{\sum_{i=2}^{n} r_i r_{i-1}}{\sum_{i=1}^{n} r_i^2} = \frac{\frac{1}{(n-1)}\sum_{i=2}^{n}(x_i - \bar{x})(x_{i-1} - \bar{x})}{s_x^{\,2}} \tag{13.4}$$

$$r_k = \frac{\sum_{i=k+1}^{n} r_i r_{i-k}}{\sum_{i=1}^{n} r_i^2} = \frac{\frac{1}{(n-1)}\sum_{i=k+1}^{n}(x_i - \bar{x})(x_{i-k} - \bar{x})}{s_x^{\,2}} \tag{13.5}$$

The variable r_i in the sums is termed a residual, it could be the difference between model and data, but here it is indicated as the difference between data and average. Note that there are $(n - k)$ terms in the numerator sum, and n (a greater number of) terms in the denominator sum. The $(n - 1)$ coefficient is for the denominator translation from the sum of squared deviations to the variance. The method assumes that data variance is uniform throughout the n values. The range of the autocorrelation statistic is approximately $-1 < r_k < +1$. The statistic is normally distributed, but its mean is not zero. The mean is $\approx \frac{-1}{n-k}$. If r_k is near to $\frac{-1}{n-k}$ there is no evidence of autocorrelation.

Note: Unfortunately, in statistics there are too many variables with the same symbol, r. Take care.

The data can be represented as in Table 13.5.

Contrasting Table 13.1 to Table 13.5: In Table 13.5 1) There is only one variable. 2) The index reveals sequential order in time, or spatial position, or the value of another variable. 3) You must preserve the order (you cannot interchange some of the rows, but you can list the data in reverse order from n to 1). Similar to Table 13.1, in Table 13.5 the variance on the x-values needs to be nearly the same at all of their values.

To visualize autocorrelation, you can plot x_i w.r.t. x_{i-1} (See Figure 13.1).

13.3.2 An Autocorrelation Illustration

Table 13.6 presents a window of time series data representing orifice-measured flow rate sampled at 10 Hz (ten times per second).

TABLE 13.5

A Variable Sequence for Autocorrelation Testing

Index	Variable x
1	x_1
2	x_2
3	x_3
…	…
i	x_i
…	…
n	x_n

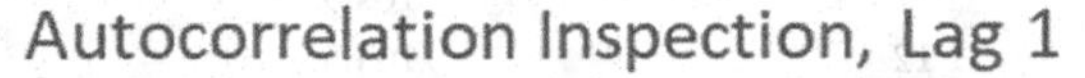

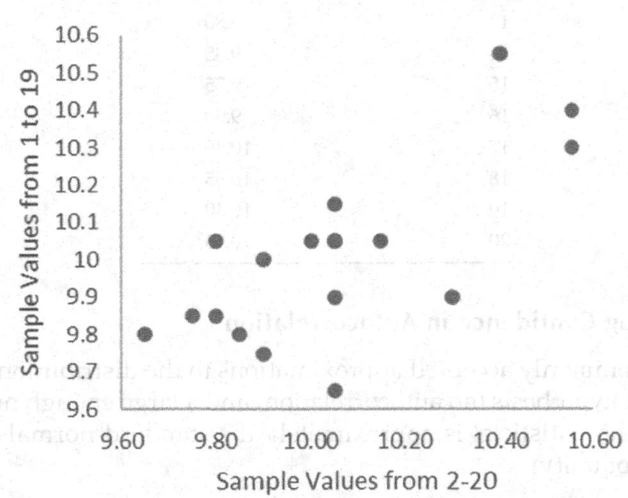

FIGURE 13.1
A visual reveal of autocorrelation with data from Table 13.6.

There is a filter on the transmitter to temper noise (fluctuations due to flow turbulence). Even though turbulence in the flowing fluid should provide random fluctuations at this time interval, the "averaging" by a first-order filter retains some of the prior values, creating autocorrelation.

There is also a rounding of the values to increments of 0.5 for digital presentation, which can contribute to autocorrelation if the variation range is not much larger than the discretization interval.

Figure 13.1 is a plot of the immediate prior data value w.r.t. the current value. Although there are 20 data values, there are only 19 comparisons on the graph. The general diagonal trend reveals autocorrelation. If one value is high, the next tends to be high. If one is low, the next tends to be low.

From Equation (13.4) the r-lag-1 value is $r_1 = 0.6262\ldots$. From Equation (13.5) the r-lag-2 and r-lag-3 values are $r_2 = 0.2886\ldots$ and $r_3 = -0.05319\ldots$.

TABLE 13.6

A Sample of Time-Series Data

Sample number	Flow rate cuft/min
1	9.85
2	9.80
3	9.65
4	10.05
5	10.15
6	10.05
7	10.05
8	10.00
9	9.90
10	10.05
11	10.05
12	10.05
13	9.80
14	9.85
15	9.75
16	9.90
17	10.30
18	10.55
19	10.40
20	10.55

13.3.3 Determining Confidence in Autocorrelation

There seem to be commonly accepted approximations to the distribution of the *r*-lag-*k* statistic. With the null hypothesis (no autocorrelation) and a large enough number of samples ($n-k>\sim 10$) then the statistic z' is approximately the standard normal statistic (mean of zero and variance of unity).

$$z' = \frac{1+r_k\left(n-k\right)}{\sqrt{n-k-1)}} \tag{13.6}$$

Example 13.3: Using the *r*-lag-*k* values from Table 13.6, calculate the *p*-values for lags 1, 2, and 3.
From Equation (13.6) the z' values are 3.040…, 1.502…, and 0.0239…. And from the standard normal distribution, the corresponding *p*-values are 0.0023…, 0.1329…, and 0.9809….

There is strong evidence that the autocorrelation exists for lag-1 (99.7%), modest support to reject the zero autocorrelation hypothesis for lag-2 (86.7%). But there is not enough evidence to confidently claim that autocorrelation does exist through the third following sample.

As an alternate, for large $(n - k)$ Equation (13.6) reduces to

$$z' = |r|\sqrt{n} \tag{13.7}$$

13.4 Takeaway

Correlation is not causation. If there is a strong correlation between two variables it might mean that they are both effects of a common cause. Correlation analysis is a simple way to identify strong relations, as a clue about postulating or affirming or rejecting mechanistic cause and effect conjectures.

Correlation does not necessarily mean that there is a linear relation between variables. However, the value of the square of the correlation r is the same value of the linear regression correlation coefficient r-squared, of a linear relation between the variables.

The methods of this chapter require uniform variance throughout the range of any one variable. If not uniform, then the data in the large variance region will dominate and diminish the effects of the other data.

13.5 Exercises

1. Test a few of the columns in Table 13.3 to see if you get the same values of the correlation r in Table 13.4.
2. Test a few rearrangements of the data order to see if the correlation coefficient value changes.
3. Plot the data in Table 13.6 for Lag-2 and Lag-3 to visually reveal the autocorrelation and the lack of autocorrelation with one and with two samples between data.

14

Steady State and Transient State Identification in Noisy Processes

14.1 Introduction

Identification of both steady-state (SS) and transient state (TS) in noisy time- or sequence-based signals is important. SS models are widely used in process control, online process analysis, and process optimization; and, since manufacturing and chemical processes are inherently nonstationary, selected model coefficient values need to be adjusted frequently to keep the models true to the process and functionally useful. Additionally, detection of SS triggers the collection of data for process fault detection, data reconciliation, neural network training, the end of an experimental trial (when you collect data and implement the next set of conditions), etc. But, either the use of SS models for process analysis or their data-based adjustment should only be triggered when the process is at SS.

In contrast, transient, time-dependent, or dynamic models are also used in control, forecasting, and scheduling applications. Dynamic models have coefficients representing time-constants and delays, which should only be adjusted to fit data from transient conditions. Detection of TS triggers the collection of data for dynamic modeling. Additionally, detection of TS provides recognition of points of change, wake-up data recording, the beginning of a process response to an event, interruptions to the norm, etc.

Characteristic of these online real-time applications is that we only have past and current data. We do not have the future points. In many filtering applications such as image and recording enhancement and detection, filtering is done offline after all the data has been collected. We have access to both the before and after data and can use kernel-type filtering approaches to use both sides to better estimate the in-between value. By contrast, in real-time applications we need to make a decision now, without seeing the future.

Our experience has been applying SSID and TSID to chemical processes, which are characterized by time-constants on the order of 1 second to 1 hour, multivariable (coupled and nonlinear), noise of several types, with mild and short-lived autocorrelation, variance changes with operating conditions, controllers and final element dead-band can cause oscillation, and flatlining measurements are not uncommon (for any number of aspects such as maintenance, sensor failure, data discretization). Additionally, control computers are inexpensive, and the operators have education typically at the associate degree level; both aspects require simplicity in algorithms. The approaches presented here might not be right if mission criticality can afford either powerful computers or highly educated operators, or if rotating machinery creates a cyclic response as a background oscillation to the steady signal.

DOI: 10.1201/9781003222330-14

If a process signal was noiseless, then SS or TS identification would be trivial. At SS there is no change in data value. Alternately, if there is a change in data value, the process is in a TS.

However, since process variables are usually noisy, the identification needs to "see" through the noise and should announce probable SS or probable TS situations, as opposed to definitive SS or definitive TS situations. The method also needs to consider more than the most recent pair of samples to confidently make any statement.

Since the noise could be a consequence of autocorrelated trends (of infinite types), varying noise amplitude (including zero), individual spikes, non-Gaussian noise distributions, or spurious events, a useful technique also needs to be robust to such aspects.

A process might not need to be exactly "flat-lined" to be considered at SS. For practical purposes, a very small trend or oscillation might have a negligible impact on SS data uses.

Finally, in observing data, a process might appear to be at SS due to measurement discrimination intervals, when in fact it is changing but the change has not exceeded the data discretization interval. Time, as an example, continually progresses; but a digital watch only updates the screen on one-minute intervals. Time is not at a SS between the numerical display change events.

14.1.1 Approaches and Issues to SSID and TSID

A conceptually simple approach to identify SS would be to look at data values in a recent time-window and if the range between high and low is acceptably small, declare SS. This approach, however, requires the human to decide the acceptable range for each variable, and the time window duration; and, if the process noise level changes, the threshold should change. Although the approach is simple to understand, easy to implement, and often works acceptably, it is not a universal approach.

Another straightforward implementation of a fully automated method for SSID would be a statistical test of the slope of a linear trend in the time series of a moving window of data. This technique is a natural outcome of traditional statistical regression training. Here, at each sampling, use linear regression to determine the best linear trend line for the past N data points. If the t-statistic for the slope exceeds the critical value, then there is sufficient evidence to confidently reject the SS hypothesis and claim it is probably in a TS. A nice feature of this approach is that the determination is independent of the noise amplitude. However, at the crest or trough of an oscillation centered in the N-data window, this slope would be nearly zero, and SS will be claimed during the TS. The approach is also somewhat of a computational burden.

Another straightforward approach is to evaluate the average value in successive data windows. Compute the average and standard deviation of the data in successive datasets of N samples then compare the two averages with a t-test. If the process is at SS, ideally, the averages are equal, but noise will cause the sequential averages to fluctuate. If the fluctuation is excessive relative to the inherent data variability, then the t-statistic (difference in average divided by standard error of the average) will exceed the critical t-value, and the null hypothesis (process is at SS) can be confidently rejected to claim it is probably in a TS. Again, however, when windows are centered on either side of the crest or trough in an oscillation, the averages will be similar, and SS will be falsely accepted. A solution could be to use three or four data windows, each with a unique N, to prevent possible matching to a periodic oscillation.

Many other approaches have been used, including Statistical Process Control techniques, dual filters, runs tests, wavelets, polynomial interpolation, thresholds on variance or slope, closure of SS material and energy balances, and variance ratios in the data.

Note, these methods reject the null hypothesis, which is a useful indicator that the process is confidently in a TS. But, not rejecting SS, does not permit a confident statement that the process is at SS. A legal judgment of "Not Guilty" is not the same as a declaration of "Innocent". "Not Guilty" means that there was not sufficient evidence to confidently claim "Guilty" without a doubt. Accordingly, there needs to be a dual approach that can confidently reject TS to claim probably in a SS, as well as rejecting SS to claim probable TS. This chapter presents several dual approaches.

Further, such conventional tests have a computational burden that does not make them practicable online, in real time, within most process control computers.

A practicable method needs to be computationally simple, robust to the vagaries of process events, easily implemented, and easily interpreted.

14.2 A Ratio of Variances Methods

Von Neumann (von Neumann, J., "Distribution of the ratio of the mean square successive difference to the variance", The Annals of Mathematical Statistics, 1941, 12, 367–395) and Crowe et al. (Crowe, E. L., F. A. Davis, and M. W. Maxfield, *Statistics Manual,* Dover Publications, New York, NY, 1955) proposed an approach that calculates the variance on a dataset by two approaches – the mean-square deviation from the average and the mean-square deviation between successive data. The ratio of variances is an *F*-like statistic, and assuming no autocorrelation in the data, it has an expected value of unity when the process is at SS.

The filter approach of Cao, S., and R. R. Rhinehart (An Efficient Method for On-Line Identification of Steady-State, *Journal of Process Control,* Vol. 5, No 6, 1995, pp. 363–374) is similar, but computationally simpler.

Begin with this conceptual model of the phenomena: The true process variable (PV) is at a constant value (at SS) and fluctuations on the measurement and signal transmission process create uncorrelated "noise", independently distributed fluctuations on the measurement. Such random measurement perturbations could be attributed to mechanical vibration, stray electromagnetic interference in signal transmission, thermal electronic noise, flow turbulence, etc. Alternately, the "noise" could represent process fluctuations resulting from nonideal fluid mixing, multiphase mixtures in a boiling situation, crystal size, or molecular weight that create temporal changes to the local measurement.

If the noise distribution (mean and variance) were uniform in time, then statistics would classify this time series as stationary. However, for a process, the true value, nominal value, or average may be constant in time, but the noise distribution may change. So, SS does not necessarily mean stationary in a statistical sense of the term.

The first hypothesis of this analysis is the conventional null hypothesis that the process is at SS, H_0: SS. The statistic, a ratio of variances, will ideally have a value of unity, but due to the vagaries of noise, it will have a distribution of values at SS. As long as the Ratio-statistic value is within the normal range of SS distribution of values, the null hypothesis cannot be rejected. When the R-statistic has an extreme value, then the null hypothesis can be rejected with a certain level of confidence, and probable TS claimed.

By contrast, there is no single conceptual model of a TS. A transient condition could be due to a ramp change in the true value, or an oscillation, or a first-order transient to a new value, or a step change, etc. Each is a unique type of transient. Further, each single transient event type has unique characteristics such as ramp rate, cycle amplitude and frequency, and time-constant and magnitude of change. Further, a transient could be comprised of any combination or sequence of the not-at-SS events. Since there is no unique model for TS, there can be no null hypothesis, or corresponding unique statistic that can be used to reject the TS hypothesis and claim probable SS. Accordingly, an alternate approach needs to be used to claim probable SS.

The alternate approach used here is to take a transient condition which is barely detectable or decidedly inconsequential (per human judgment) and set the probable SS threshold for the R-statistic as an improbably low value, but not so low as to be improbably encountered when the process is truly at SS.

So, there are two one-sided tests, one to reject SS, and one to reject TS, and two critical values as illustrated in Figure 14.1, where the vertical axis is the *CDF*, and the horizontal axis is the SS Ratio-statistic. The solid curve is the distribution of R-values when at SS and is centered on a value of 1. The dashed curve is the distribution when not at SS, but nearly so, and is centered on about 1.7.

The vertical dashed lines are the critical values. The right-most line is the trigger to reject SS. It intersects the at-SS *CDF* at a value of about 0.95. There is a 5% chance that an at-SS process will generate an R-value greater than 2. However, there is about a 35% chance (1 – 0.65) that the nearly-but-not-at-SS process will generate an R-value greater than 2.

The left-most line is the trigger to reject TS. It intersects the at-SS *CDF* at a value of about 0.25. There is a 25% chance that an at-SS process will generate an R-value less than 0.8. However, there is only about a 1% chance that the nearly-but-not-at-SS process will generate an R-value less than 0.8.

In this illustration, the odds of correctly rejecting SS are 0.35/0.05 = 7:1, and the odds of correctly accepting SS are 0.25/0.01 = 25:1. However, in most transient conditions the dashed not-at-SS curve is further to the right, and both odds are better than illustrated.

14.2.1 Filter Method

Figure 14.2 illustrates the filter method concept to create an R-statistic. The markers represent process measurements over the 100 sequential samples indicated on the horizontal axis. The process value starts at about 10, ramps to a value of about 15, and then holds steady. The true trend is unknowable, only the measurements can be known, and they are

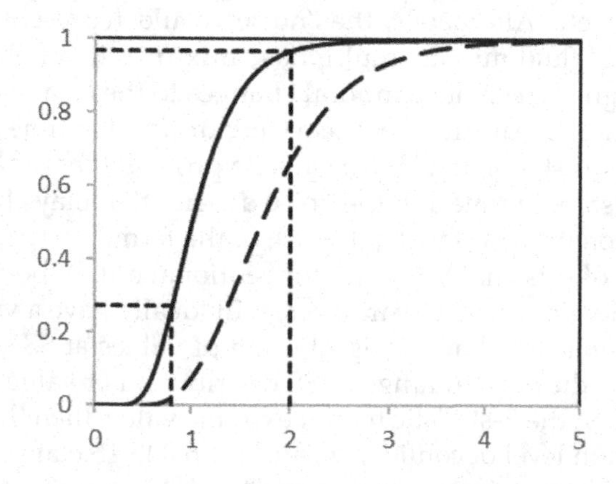

FIGURE 14.1
Illustration of dual rejection regions. The vertical axis is the *CDF*, and the horizontal axis is the SS Ratio-statistic. The solid curve is the *F*-like distribution of R-values when at SS. The dashed curve is the *F*-like distribution when not at SS.

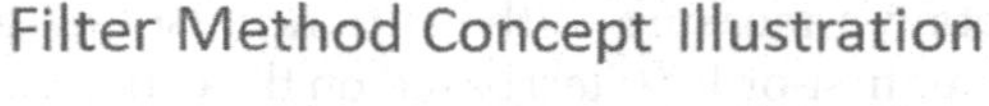

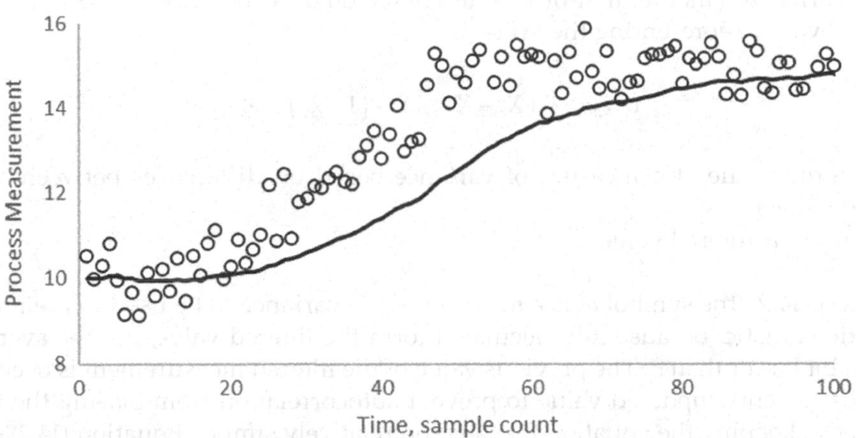

FIGURE 14.2
Filter method concepts.

infected with noise-like fluctuations. The solid line is a first-order filtered value of the data. It starts at about 10 then lags behind the process, and at the end of the chart finally settles to a value representing the process level of 15. The filtered value is not smooth but reveals wiggles due to the high and low vagaries of the data.

The method first calculates a filtered value of the process measurements, then the variance in the data is measured by two methods. One is based on the difference between the measurement and the filtered trend. The other is based on deviations between sequential data measurements.

If the process is at SS, as illustrated in the 0–10 to and 90–100 time periods, the filtered value, X_f, remains almost in the middle of the data. Then a process variance, s^2_1, estimated by differences between data and filtered value will ideally be equal to the true value of σ^2. The variance can also be estimated by the data-to-data (not average-to-data) differences, s^2_2. Then the ratio of the variances, $r = \frac{s^2_1}{s^2_2}$ will be approximately equal to unity. Alternately, if the process is in a TS, such as in the 20–60 time period, then X_f is not the middle of the data, the filtered value lags behind the process, and the variance as measured by the data-to-filter difference will be much larger than the variance as estimated by sequential data differences, $s^2_1 \gg s^2_2$, and ratio will be much greater than unity.

To minimize computational burden, in this method a filtered value (not an average) provides an estimate of the data mean:

$$X_{f,i} = \lambda_1 X_i + (1-\lambda_1) X_{f,i-1} \tag{14.1}$$

X = the process variable
X_f = Filtered value of X
λ_1 = Filter factor
i = Time sampling index

Note: Alternate terms for the filtered value are first-order filter, first-order lag, and exponentially weighted moving average.

The first method to obtain a measure of the variance uses an exponentially weighted moving "variance" (another first-order filter) based on the difference between the data and the filtered value, representing the average:

$$\upsilon^2_{f,i} = \lambda_2\left(X_i - X_{f,i-1}\right)^2 + \left(1-\lambda_2\right)\upsilon^2_{f,i-1} \tag{14.2}$$

$\upsilon^2_{f,\ i}$ = Filtered value of a measure of variance based on differences between data and filtered values
$\upsilon^2_{f,\ i-1}$ = Previous filtered value

In Equation (14.2), the symbol ν^2 is a measure of the variance to be used in the numerator of the ratio statistic. Because it is calculated form the filtered value, not the average, ν is actually a bit larger than σ. The previous value of the filtered measurement is used instead of the most recently updated value to prevent autocorrelation from biasing the variance estimate, $\upsilon^2_{f,i}$, keeping the equation for the ratio relatively simple. Equation (14.2) does not provide the variance, even at SS, because using the filtered X_f, rather than the true average, adds a bit of variability to the difference $\left(X_i - X_{f,i-1}\right)$.

The second method to obtain a measure of variance is an exponentially weighted moving "variance" (another filter) based on sequential data differences:

$$\delta^2_{f,i} = \lambda_3\left(X_i - X_{i-1}\right)^2 + \left(1-\lambda_3\right)\delta^2_{f,i-1} \tag{14.3}$$

$\delta^2_{f,\ i}$ = Filtered value of a measure of variance
$\delta^2_{f,\ i-1}$ = Previous filtered value

This will be the denominator measure of the variance. Also, it is not the variance, but is effectively twice the variance when the process is at SS. So, it uses the symbol δ instead of σ.

The ratio of variances, the R-statistic, may now be computed by the following simple equation:

$$R = \frac{\left(2-\lambda_1\right)\upsilon^2_{f,i}}{\delta^2_{f,i}} \tag{14.4}$$

Since Equations (14.2) and (14.3) compute a measure of the variance, not the true variance, the $\left(2-\lambda_1\right)$ coefficient in Equation (14.4) is required to scale the ratio, to represent the classic variance ratio. At SS it will fluctuate about a value of unity, during a TS, it will fluctuate about larger values. The calculated R-value is to be compared to its two critical values to determine SS or TS. Complete executable code, including initializations, is presented in VBA in the software on the site www.r3eda.com. The essential assignment statements for Equations (14.1) to (14.4) are:

```
nu2f = l2 * (x - xf) ^ 2 + cl2 * nu2f             Equation (14.2)
xf = l1 * x + cl1 * xf                            Equation (14.1)
delta2f = l3 * (x - x_old) ^ 2 + cl3 * delta2f    Equation (14.3)
x_old = x                                         Update prior value
R_Filter = (2 - l1) * nu2f/delta2f                Equation (14.4)
```

The coefficients *l*1, *l*2, and *l*3 represent the filter lambda values, and the coefficients *cl*1, *cl*2, and *cl*3 represent the complementary values. *cl*1 = 1 − *l*1. Equations (14.2) and (14.1) are calculated in reverse order so that the prior *xf* value does not need to be stored.

The five computational lines of code of this method require direct, no-logic, low storage, and low computational operation calculations. In total there are four variables and seven coefficients to be stored, ten multiplication or divisions, five additions, and two logical comparisons per observed variable.

Without prior knowledge of the value for σ, initialize filtered values with 0. This is more convenient than initializing them with a more representative value from recent past data. This leads to an initial not-at-SS value of the initial R-statistic, but after about 35 samples the initial wrong values are incrementally updated with representative values.

Being a ratio of variances, the statistic is scaled by the inherent noise level in the data. It is also independent of the dimensions chosen for the variable.

Critical values for the R-statistic, based on the process being at SS with independent and identically distributed variation (white noise), were also developed by Cao and Rhinehart (Cao, S., and R. R. Rhinehart, "Critical Values for a Steady-State Identifier," *Journal of Process Control*, Vol. 7, No. 2, 1997, pp. 149–152), who suggest that filter values of $\lambda_1 = 0.2$ and $\lambda_2 = \lambda_3 = 0.1$ produce the best balance of Type-I and Type-II errors. The null hypothesis is that the process is at SS. If the computed R-statistic is greater than R-critical (a value of about 2.5) then we are confident that the process is not at SS.

However, if the R-value is a bit less than the upper critical value, the process may be in a mild TS. Consequently, a value of R less than or equal to a lower R-critical value, ~0.9, means the process may be at SS (Shrowti, N., K. Vilankar, and R. R. Rhinehart, "Type-II Critical Values for a Steady-State Identifier", Journal of Process Control, Vol. 20, No. 7, pp. 885–890, 2010).

Often we assign values of either "0" or "1" to a variable, SS, which represents the state of the process. If R-calculated > R-critical1 ~2.5, "reject" SS, assign SS = 0. Alternately, if R-calculated < R-critical2 ~.9, "accept" that the process may be at SS and assign SS = 1. If in-between values happen for R-calculated, hold the prior 0 or 1 (reject or accept) state, because there is no confidence in changing the most recent declaration.

The method presumes no autocorrelation in the time series of the process measurement data at SS. This is ensured by selection of the sampling time interval, which may be longer than the control interval. Running in real time, the identifier does not have to sample at the same rate as the controller.

14.2.2 Choice of Filter Factor Values

The filter factors in Equations (14.1), (14.2), and (14.3) can be related to the number of data (the length of the time window) effectively influencing the average or variance calculation. Simplistically, the effective number of data in the window $N = 1/\lambda$. If $\lambda = 0.1$ then effectively the method is observing $N = 10$ most recent data points. However, based on a first-order decay, long past data retain some influence, and roughly, the number of data effectively influencing the window of observation is about $3.5/\lambda$. If $\lambda = 0.1$ then effectively the method is remembering the impact of $N \approx 35$ data points. However, not all data points have equal weighting. The filter is a first-order decay, an exponentially declining weighting N past data. Truly, the long-past data retain some influence, but the collective fractional influence is $1 - (1 - \lambda)^N$. With $\lambda = 0.1$ and $N = 35$ the old data only have a 0.025 fractional influence.

Larger λ values mean that fewer data are involved in the analysis, which has a benefit of reducing the time for the identifier to catch up to a process change, reducing the average run length (ARL) to a decision. But, larger λ values have an undesired impact of increasing the variability on the statistic, confounding interpretation. The reverse is true: Lower λ values undesirably increase the ARL to detection but increase precision (minimizing statistical errors).

Originally, Cao and Rhinehart recommended values were $\lambda_1 = 0.2$, $\lambda_2 = \lambda_3 = 0.1$, effectively meaning that the most recent 35 data points are used to calculate the R-statistic value. Since then, we usually recommend $\lambda_1 = \lambda_2 = \lambda_3 = 0.1$ for convenience and effectiveness. However, $\lambda_1 = \lambda_2 = \lambda_3 = 0.05$ will have less uncertainty, but a bit longer ARL. These are not a critical choice.

14.2.3 Critical Values

If the supposition is that a process is at SS, then a Type-I (T-I) error is a claim of not-at-SS when the process is actually at SS. The concept is best understood by considering the distribution of the R-statistic of a SS process. The left curve in Figure 14.3 represents the statistical distribution of the R-statistic values at SS. The average value is R = 1. Note that the distribution is not the symmetric, bell-shaped, normal distribution. It is an *F*-type of distribution, skewed, and with no values below zero. The right most curve represents the R-distribution when a process in in the TS, here the average ratio is 3. In either case, the R-statistic will have some variability because of the random fluctuations in the sequential measured data.

If the value of R is larger than upper 99% confidence value of about 2.5, illustrated by the vertical dashed line, there is about a 1% chance that the process could be at SS, but about a 70% chance that value would have come from a process in a TS. Bet on it being TS.

If the R-value is a bit less than 2.5 there is still a substantial probability (about 30%) that the process in the TS could have generated it. So, don't use a single critical value to both reject and accept SS or TS.

There is also a lower critical value, about 0.9, indicated on the figure. If the process is in a TS, Figure 14.3 shows that the TS process has less than about a 1% chance that it will generate an R-value less than the lower critical value. However, if the process is at SS, then as illustrated in Figure 14.3 there is a about a 40% likelihood that it will generate such an R-value or lower. So, if $R < R_{lower\ critical}$ the odds are that process is at SS. Claim SS. This alternate Type-I error is accepting SS if the process is in a TS.

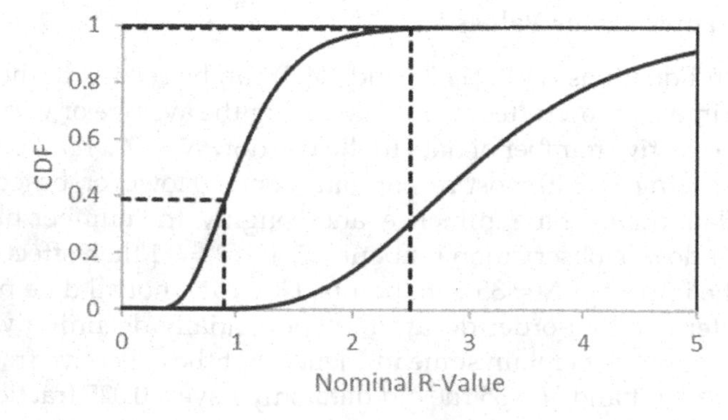

FIGURE 14.3
R-statistic distributions – at SS (left curve), at TS (right curve).

However, if the R-value is in between then there is a high likelihood of the process being either at SS or TS. There is no adequate justification to make either claim. So, retain the last claim.

Both T-I and alternate T-I errors are important. Critical values can be obtained from Cao and Rhinehart (1997), and from Shrowti, et al. (2010). However, it is more convenient and less dependent on idealizations to visually select data from periods that represent a transient or steady period, and to find the R-critical values that make the algorithm agree with the user interpretation.

Experience recommends R_{upper}~3 to 4 and R_{lower}~0.85 to 1.0, chosen by visual inspection as definitive demarcations for a transient and steady process.

14.2.4 Illustration

Figure 14.4 illustrates the method. The process variable, PV, is connected to the left-hand vertical axis (log10-scale) and is graphed with respect to sample interval. Initially it is at a SS with a value of about 5. At a sample number 200, the PV begins a first-order rise to a value of about 36. At sample number 700, the PV makes a step rise to a value of about 40. The R-statistic is attached to the same left-hand axis and shows an initial kick to a high value as variables are initialized, then relaxes to a value that wanders about the unity SS value. When the PV changes at sample 200, the R-statistic value jumps up to values ranging between 4 and 11, which relaxes back to the unity value as the trend hits a steady value at a time of 500. Then when the small PV step occurs at sample 700, the R-value jumps up to about 4, then decays back to its nominal unity range. The SS value is connected to the right-hand vertical axis and has values of either 0 to 1 that change when the R-value exceeds the two limits, of $R_{\beta,TS}$ and $R_{1-\alpha,SS}$.

14.2.5 Discussion of Important Attributes

14.2.5.1 Distribution Separation

A good statistic will provide a large separation of the SS and TS distributions relative to the range of the distribution. The distribution shift is the signal that results of the process

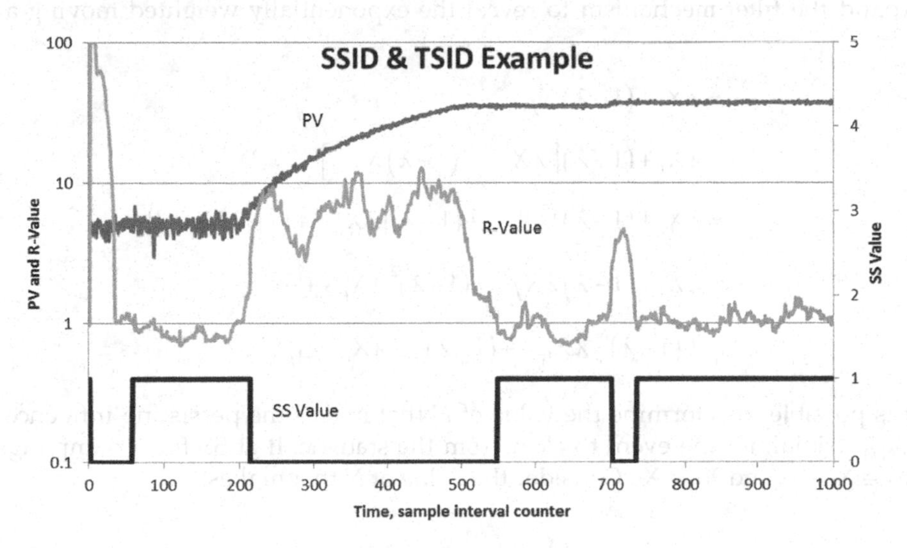

FIGURE 14.4
Illustration of the filter method to identify SS and TS.

shift from SS to TS, and the width of the distribution is the uncertainty, or noise associated with the statistic. A good method to detect SS and TS will have a large signal-to-noise aspect, a large shift in the not-at-SS distribution relative to the width of the distribution.

Figure 14.4 reveals the issue. In several SS periods (sample 100–200, 600–700, and 800–1000) the R-statistic has values in the 0.8 to 2 range of values. In the TS instances (sample 200–500, and 700) the R-statistic has values in the 5–10 range, which are definitely different from the values in the SS periods. There is good separation in the two ranges of values.

Lower filter lambda values, for instance $\lambda = 0.05$, make the variation in the R-statistic smaller, meaning that the separation between SS and TS R-values is more definitive. However, larger lambda values, for instance $\lambda = 0.2$, make time-to-detection, ARL faster. Again, refer to Figure 14.4. Note that after the step change in the PV at sample 700, the PV holds at its new value, but the method did not return to a SS claim until about sample 740.

14.2.5.2 Average Run Length

Another aspect to be considered as a quality performance metric of a statistic is ARL, the number of samples after an event occurs to be confident in declaring a change. In the moving window methods with N data being considered, the last not-at-SS data event must be out of the window for the analysis of the window to be able to declare "at SS". This would appear to be N data, or the average run length of ARL = N. However, when the process is at SS, the statistic will not always be at the extreme value. There is a probability, β, that it is beyond the left value. When at SS, if there is no autocorrelation in the R-statistic, the number of data required to randomly have a value that is less than the β-probable extreme is $1/\beta$. Then the ARL would be the number of samples to clear the window plus the expected number to likely create an extreme value. So, simplistically, ARL = $N + 1/\beta$.

The filter factors in Equations (14.1–14.3) can be related to the number of data (the length of the time window) in the average or variance calculation. Roughly, the number of data with a residual influence on the statistic is about $3/\lambda$ to $5/\lambda$, depending on your choice of what constitutes an inconsequential residual influence of past data. To determine an ARL, first expand the filter mechanism to reveal the exponentially weighted moving average form:

$$\begin{aligned} X_{f_i} &= \lambda X_i + (1-\lambda) X_{fi-1} \\ &= \lambda X_i + (1-\lambda)\left[\lambda X_{fi-1} + (1-\lambda) X_{fi-2}\right] \\ &= \lambda X_i + (1-\lambda)\{\lambda X_{fi-1} + (1-\lambda)[\lambda X_{fi-2} + (1-\lambda) X_{fi-3}]\} \\ &= \lambda X_i + (1-\lambda)\lambda X_{fi-1} + (1-\lambda)^2 \lambda X_{i-2} + \cdots \\ &\quad + (1-\lambda)^N \lambda X_{i-N} + (1-\lambda)^{N+1} \lambda X_{i-(N+1)} \end{aligned} \tag{14.5}$$

Now it is possible to determine the value of N that makes the persisting influence of the old $X_{i-(N+1)}$ trivial, for the event to clear from the statistic. If at SS for N samplings, then $X_i \cong X_{i-1} \cong X_{i-2} \cdots$ and $X_{f_i} \sim X_i$. Consider the value of N that makes

$$(1-\lambda)^{N+1} X_{fold} \ll X_{SS} \tag{14.6}$$

As an estimate assume "<<" means 2%, and perhaps $X_{f_{old}} \sim 5X_{SS}$. If $\lambda = 0.1$, then the condition is

$$(1-0.1)^{N+1}(5X_{SS}) = 0.02X_{SS} \tag{14.7}$$

Rearranging to solve for N,

$$N = \frac{\text{Ln}\left(\dfrac{0.02}{5}\right)}{\text{Ln}(1-0.1)} - 1 = \frac{\text{Ln}(0.004)}{\text{Ln}(0.9)} - 1 \approx 50. \tag{14.8}$$

Assuming "<<" means 5%, then $N = \dfrac{\text{Ln}\left(\dfrac{0.05}{5}\right)}{\text{Ln}(1-0.1)} - 1 \approx 43$. Assuming "<<" means 5% and $X_{f_{old}} \sim 3X_{SS}$, then $N = \dfrac{\text{Ln}\left(\dfrac{0.05}{3}\right)}{\text{Ln}(1-0.1)} - 1 \approx 38$.

So, N for the influence to be gone, to permit X_f, ν_f^2, and δ_f^2 to relax from an upset to be close to their SS values is between 38 and 50 depending on magnitude of event and what decay fraction makes it inconsequential.

With $\lambda_1 = 0.2$ and $\lambda_2 = \lambda_3 = 0.1$, effectively meaning that the most recent 38 to 50 data points are used to calculate the R-statistic value, effectively, $N \approx 4.5/\lambda$. However, alternate lambda values have been recommended to ensure greater confidence in the SS claim, or faster identification.

The ARL for a transition from SS to TS is just a few samples, which is not a problem. However, the transition from TS to SS requires the filter statistics to relax, which takes time and results in an ARL $\approx 4.5/\lambda + 0.7/\beta$.

14.2.5.3 Balancing ARL, T-I and T-II Errors

Larger λ values mean that fewer data are involved in the analysis, which has a benefit of reducing the time for the identifier to catch up to a process change (ARL) but has an undesired impact of increasing the variability on the statistic, broadening the distribution and confounding interpretation. Lower λ values undesirably increase the ARL to detection but increase precision (minimizing T-I and alternate T-I statistical errors) by reducing the variability of the distributions increasing the signal-to-noise separation of a TS to SS situation.

This work recommends filter values of $\lambda_1 = \lambda_2 = \lambda_3 = 0.1$, balancing precision with ARL and convenience of use. At these choices, we observe that ARL for automatic detection roughly corresponds to the time an operator might confidently declare that the process has achieved a SS.

However, rather than choose critical values to reject either SS or TS from ideal theoretical distributions and Equation (14.4), it is more convenient and less dependent on idealizations to visually select data from periods that represent a transient or steady period, and to find the R-critical values that make the algorithm agree with the user interpretation. This work uses $R_{\text{crit to reject } SS} \approx 3$ to 4 and $R_{\text{crit to reject } TS} \approx 0.85$ to 1, chosen by visual inspection as definitive demarcations for a transient and steady process.

14.2.5.4 Distribution Robustness

Cao and Rhinehart (1995, 1997) and Shrowti, et al. (2010) investigated the robustness of the R-statistic distributions to a variety of noise distributions in the process variable. They found that the R-statistic distributions and the critical values are relatively insensitive to the noise distribution. This finding is not unexpected, recognizing the well-known robustness of the *F*- and chi-square statistics (also based on a ratio of variances) for distributions with modest deviations from normal.

14.2.5.5 Autocorrelation

The basis for the R-statistic method presumes that there is no autocorrelation in the time-series process data when at SS. Autocorrelation means that if a measurement is high (or low) the subsequent measurement will be related to it. For example, if a real process event causes a temperature measurement to be a bit high, and the event has persistence, then the next measurement will also be influenced by the persisting event and will also be a bit high. Autocorrelation could be related to control action, thermal inertia, data filters in sensors, signal discretization, etc. Autocorrelation would make the denominator estimate of the variance smaller, which would tend to make all R-statistic distributions shift to the right, requiring a reinterpretation of critical values for each process variable.

As common in Statistical Process Control techniques, it is more convenient to choose a sampling interval that eliminates autocorrelation, than to model and compensate for autocorrelation in the test statistic. This can be done if the source of the autocorrelation is from random perturbations. Although they have some persistence, the influence fades, and new random perturbation values cause measurement fluctuation. This short-lived persistence from random perturbations is different from the regular cycling from a feedback mechanism, or the flat-line constant value that arises when a signal is having a coarse discretization.

A plot of the current process measurement versus the previous sampling of the process measurement over a sufficiently long period of time (equaling several time-constants) at SS is a simple visual method to establish the presence/absence of autocorrelation. To detect autocorrelation, visually choose a segment of data that is at SS and plot the PV value w.r.t. its prior value. Alternately, you could use a data-based statistic from Chapter 13.

Figure 14.5 plots data with a lag of 1 sample (a measurement w.r.t. the prior measurement) and shows autocorrelation. [In statistics jargon, lag means delay. In system dynamics and control jargon, lag means a time-constant. Here, we are using the statistics jargon.] Figure 14.6 plots the same data but with a lag of 5 samples and shows zero autocorrelation.

What this means is that the SS or TS monitor should be only looking at every 5th data value from the process data stream to calculate δ_f^2.

14.2.5.6 Signal Discretization

There is a smallest increment that can be displayed or reported. In a digital image it is pixel size. On a digital clock it is a 1-second interval, and on a stopwatch perhaps 1/100 of a second. On a digital speedometer display, 1 mph, and on a digital temperature display 1°F or 0.5°C. As time, speed, temperature, etc. progresses, the displayed value remains unchanged until it crosses over the discretization threshold to the next interval. If a data transmission system is a 10-bit system, then the binary number can have $2^{10} = 1024$ values. If the calibration uses the middle 80% of values (paralleling a standard 4–20 mA or

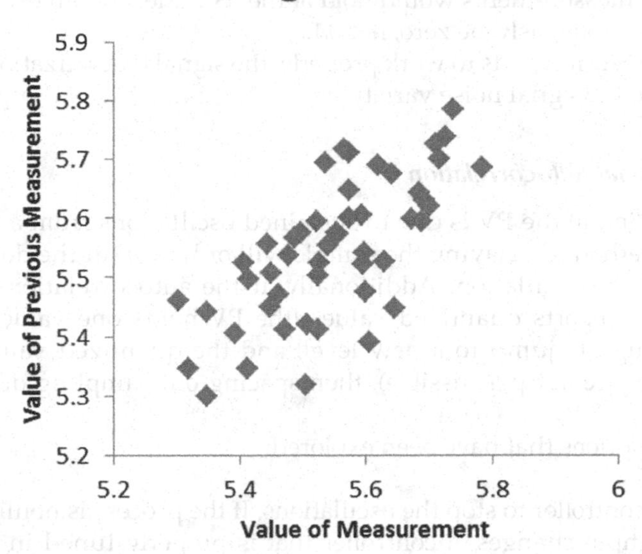

FIGURE 14.5
Data showing autocorrelation.

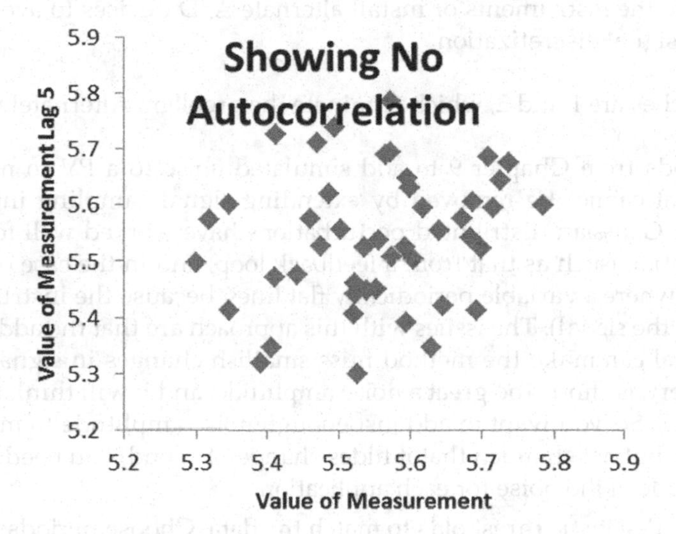

FIGURE 14.6
Data showing no autocorrelation when the interval is 5 samplings.

3–15 psig analog signal), then the PV range is matched with about 820 binary counts. There are only 820 possible values, with a discretization interval of about 0.12% of full scale.

Discretization is visible as a signal that repeatedly jumps between only a few values.

Because many process measurements are nonlinear, such as orifice flow measurements based on orifice pressure drop, the discretization may not be visible in one data range, but clearly distinguished in another.

Discretization confounds an estimate of data variance, or noise amplitude. If, for instance a signal was at SS, with a value of 53.5 and a noise sigma of 0.1, then the noisy measurement will be within 53.5 +/− 3 sigma or about 53.2 to 53.8. But, if the digital discrimination interval was 1, then all measurements would hold at the 53 value, and an estimate of the noise amplitude would, erroneously, be zero, not 0.1.

For most SSID/TSID methods to work properly, the signal discretization interval should be small relative to the signal noise variation.

14.2.5.7 Aberrational Autocorrelation

If the autocorrelation in the PV is due to sustained oscillations from a feedback mechanism, then this method of delaying the sample will only work if the delay (lag) matches the cycle time of the oscillation. Additionally, if the autocorrelation is due to signal discretization that reports quantized values (the PV holds one value until the signal has changed enough to jump to a new level, and the quantized values are repeated, in between values are not permissible), then spacing out sampling does not eliminate autocorrelation.

Here are seven options that have been explored:

1. Re-tune the controller to stop the oscillations. If the process is nonlinear or makes large throughput changes, a controller that is properly tuned in one operating region might be too aggressive in another. Perhaps "gain schedule" the controller coefficient values with operational region.
2. Re-calibrate the instruments or install alternate A/D devices to avoid the flatlining due to signal discretization.

Favored approaches are 1 and 2, which eliminate the problem. Alternately mask it:

3. Use methods from Chapter 9 to add simulated noise to a PV to mask autocorrelation that cannot be removed by extending signal sampling interval. Either uniform or Gaussian distributed perturbations have worked well for both cyclic autocorrelation (such as that from a feedback loop) and in the case of discrimination error (where a variable periodically flat lines, because the instrument system discretizes the signal). The issues with this approach are that the addition of noise to the signal can make the method miss smallish changes in signal level that it would otherwise find. Too great a noise amplitude, and it will think the process is always at SS. So, you want to add just enough noise amplitude to mask the autocorrelation, but not so much that it hides changes. Second, you need to determine the amplitude of the noise for each application.
4. Modify the R-statistic thresholds to match the data. Choose periods that you consider to be steady and look at the range of the R-statistic values in those periods. Use the distribution to select values for the T-I and alternate T-I triggers. This approach requires unique values for each process variable, but common thresholds for all applications would be preferred.
5. Only use the numerator variance measure – the squared deviation of the signal from the filtered value. Do not normalize it by the square of the sequential deviations. If the process is in a TS, the value is high. If at SS, it is low. So, observe the numerator variance in time, and choose threshold values that you observe definitively reveal SS and TS. This simplifies the algorithm by removing a line of code. The ratio approach is more universal because the variables are scaled. If only

testing the numerator variance, each PV needs to have a threshold set by the user, and if the process variance changes with operating conditions, then thresholds are required for each region.

6. Alternately, use one of the simpler approaches from Section 14.1.1 to look at the range of data in a window, and if larger than normal claim TS, and if within a normal range accept SS. But this also requires the user to set individual thresholds for each PV, and possibly in each operating range.

Alternatively, use an alternate technique:

7. Apply alternate techniques that are not confounded by such autocorrelation. A favorite of several is the 4Points method described in Section 14.3, but it is a bit more complicated than the R-statistic filter method.

14.2.5.8 Multivariable Extension

In a multivariable analysis, there is a greater chance that data vagaries will cause a false TS claim in any one monitored variable. Also, there could be an excessive wait for all PV monitors to return to SS.

One approach is to change the threshold to be more cautious about rejecting SS and more accepting in accepting SS. This analysis is predicated on independence of all PVs.

In expanding the technique for a multivariable analysis, we choose to claim that a process is not at SS if any process variable is not at SS, and that the PV might be at SS if all variables might be. This can be easily computed with a single statistic:

$$SS_{\text{process}} = \prod_{j=1}^{N} SS_j \tag{14.9}$$

N = Total number of variables in process
j = variable index
$SS_j = 1$ if the jth variable is at SS and $SS_j = 0$ if the individual process variable is in a TS
$SS_{\text{process}} = 1$ if all variables are at SS and $SS_{\text{process}} = 0$ if any one variable is in a TS

Assign a value of either 0 or 1 to a dichotomous variable, SSID(i), if the ith process variable is at probable transient or probable steady conditions. At incidences of intermediate values of the R-statistic, the dichotomous SSID(i) variable retains its latest definitive value. Therefore, an SS_{process} value of 1 means that the last definitive state of each variable was SS – there has not been a definite transient since the last definite SS. Similarly, an SS_{process} value of 0 means that the last definitive state of at least one variable was TS – there has not been a definite SS since the last definite transient.

If the process is at SS and variations and trends on each variable are independent of other variables then:

$$P\left(SS_{\text{process}} = 1\right) = \prod_{i=1}^{N} P\left(SS_i = 1\right) = \prod_{i=1}^{N} (1-\alpha_i) \tag{14.10}$$

$$\left(1-\alpha_{\text{process}}\right) = \prod_{i=1}^{N} (1-\alpha_i) \tag{14.11}$$

For example, if we want $\alpha_{process} = 0.05$, then to determine the required level of significance for the SSID test on each variable.

$$\alpha_i = 1 - \sqrt[N]{(1 - \alpha_{process})} \tag{14.12}$$

Where
N = number of variables observed in the analysis
$\alpha_{process}$ = desired overall level of significance for the process statement
α_i = required level of significance for each individual test

If, for example, there are $N = 10$ variables with independent behaviors, and the desired level probability of a T-I error is 0.05 (95% confidence), then from Equation (14.12) the level of significance for each individual test would need to be about 0.005 (99.5% confidence).

In a multivariable analysis, there is a greater chance that data vagaries will cause a false TS claim, and there could be an excessive wait for all PV monitors to return to SS.

Further, the signals will probably not be independent. If an event makes one PV change or appear to change, it likely affects several PVs. So, as an alternate solution, the implementer might want to temper the monitors: Claim TS only if 10% or more of the PVs are at TS, and claim SS if 90% or more are at SS. See Section 14.2.5.10.

14.2.5.9 Cross Correlation

In addition to the R-statistic requirement that there is no autocorrelation between data points (during SS), Equation (14.12) also requires that there is no cross correlation between variables (at SS). When the process is at SS, this means that the noise on one variable is not correlated to the noise on another. Variables that are being observed should not have correlated noise at SS conditions. The investigator should find the minimum set of independent process variables which would present a comprehensive view of the process state and which have uncorrelated noise or responses.

For example, consider a tank that has inlet flow, outlet flow, and level measurements. Only two of the three are necessary for observation. The third would be correlated to the other two.

As another example, consider that temperature is measured on both Tray 7 and Tray 8 in a distillation column. If some event changes the composition on Tray 7 it will very soon similarly affect Tray 8, so the temperatures will be cross-correlated. Only one of the two temperatures is needed for observation of that section of the column.

14.2.5.10 Selection of Variables

A process will likely have many variables that can be observed (for instance tray temperatures on many trays in a distillation column), but if one is in a TS, whatever causes it will also be affecting other measurements, so not all variables need to be observed. The user should consider which subset of all possible variables are the minimum number that are key to detecting whether TS or SS, and only set up the procedure on that minimal set. This minimizes computational burden, and as well, reduces the number of false alarms generated by normal vagaries in the data. The selection needs user understanding of the process for recognition of redundancy in the variables. The user should also select variables that have the least autocorrelation. For instance, if there are two temperature measurements that could be used, and one is in a thermowell and the other is not, the thermowell will create greater autocorrelation. Use the other. Also select variables that show the largest change-to-noise ratio during a transient.

Some variables respond rapidly, such as flow rates, liquid inventory, or pressure. Other variables respond much more slowly, such as composition or temperature in a distillation column. (But, in a heat exchanger, temperature could respond rapidly, and in a reactor composition could change rapidly.) Some variables may be grouped into a subcategory for rapidly responding phenomena such as "hydraulics". And other variables might be grouped into a subcategory such as "composition". This would reveal that one part of the process has settled, a comforting thing to know; but we need to wait a bit for the residual changes in compositions to come to SS.

14.2.5.11 Noiseless Data

As a final aspect of the method development, recognize that this statistical method only accommodates noisy data. If the data are noiseless and at SS, the data series will have exactly the same values, which would drive the denominator measure of variance to zero, eventually leading to an execution error in Equation (14.4). But, the numerator variance also approaches zero. When either is very small, digital truncation could lead to a large R-statistic, and a not-at-SS claim, even though the PV is at SS. This truncation effect could be device specific.

If a PV approaches a noiseless SS, then a deterministic method is easier to implement and understand – if the ABS(PV-PVprior) < threshold then at SS. One might choose the threshold to be 1/100 of the PV range used to display the PV.

A noiseless signal could arise from several situations: The flow rate is off, the device is broken, the signal is being overridden, the signal is not changing as much as the discrimination error and the reported value is constant, the process is not being used, etc. Consider the periodic flat-lining due to digital discrimination. This has happened with online analyzers and with temperature sensors. In some operating regions the sample-to-sample fluctuation exceeds the digital discrimination, and the measurement seems normally noisy (but it actually just hops between about 5 or so values). But, in other cases the process variation is less than the digital threshold, and the signal flatlines.

To avoid this condition, one could add a normally and independently distributed noise signal of zero mean and small variance NID(0,σ) to the original data. Alternately, one could limit the minimum measure of the estimate of the data standard deviation. In either case, a guide would be to choose a noise sigma, or limit the sigma to 1/100 of the range used for graphical display.

If adding artificial noise, we recommend the Box–Muller method for adding Gaussian distributed independent noise to a simulated signal.

$$x_{\text{measured},i} = x_{\text{true},i} + w_i \tag{14.13}$$

$$w_i = \sigma\left(\sqrt{-2\ln(r_{1,i})}\,\sin(2\pi r_{2,i})\right) \tag{14.14}$$

Here w_i is a normally and independently distributed "white noise" perturbation of zero mean and standard deviation σ variation that is added to the true value to mimic the measurement. The variables $r_{1,i}$ and $r_{2,i}$ represent independent random numbers uniformly distributed in the interval above 0 and up to 1, $0 < r \leq 1$.

A simpler noise-adding approach is to add uniformly distributed, not Gaussian, perturbations. Here $w_i = 2a(r_i - 0.5)$. Where a is the desired amplitude of the perturbation range.

In some control computers, there is no random number generator function. We've had to use our own pseudo-random number generation code as a subroutine!

In any of these cases, the noise amplitude (or sigma) needs to be appropriate to the signal. It should not be too large to override detecting changes when the signal is making real changes, but large enough to mask the flatline periods. A knowledgeable implementer can select an appropriate value, but this means an additional variable needs to be selected.

However, it might be best to reject such a signal and observe another that represents the same category of process behavior and does not have noiseless periods.

Andy Robinson, Principal, Phase 2 Automation. Raleigh, NC, suggested a simple solution: Set a minimal threshold on the denominator variance. (Andy wishes to share credit with his colleague David Goodman.) If the flatline periods (truly a noiseless SS, or changes masked by signal discrimination) would want the denominator variance to approach zero, this would reset it at the threshold. The threshold value would need to be scaled appropriate to the process. The same variation of 1 inch would have a magnitude of 1.6E-5 if the units changed to miles. A process implementer would know what an appropriate minimum value would be for the process. Square that value to represent variance. The simplicity and effectiveness of this approach is desirable. It only has an impact on the calculation when the signal is noiseless, and the denominator value approaches zero. However, it adds a tuning factor that the user must choose, and which is application specific.

14.2.6 Alternate R-Statistic Structure-Array

The rationale for the three first-order filters of Equations (14.1)–(14.3) is to reduce computational burden. But, they have the disadvantage of permitting a past event to linger with an exponentially weighted diminishing factor. Further, the filter structure is not as comfortable for many who are familiar with conventional sums in calculating variances. Further, the window of fixed length N is easier to grasp than the exponential weighted infinite window, or the interpretation of the three lambda values. Alternately, in using a window of N data, the event is totally out of the window and out of the analysis after N samples, which provides a clearer user understanding of data window length. Finally, the use of a single coefficient, N, is more convenient than specifying three separate λ values.

Returning to the base concept with the R-statistic as a ratio of variances, first calculate the average from the N most recent observations.

$$\bar{X} = \frac{1}{N}\sum_{i=1}^{N} X_i \tag{14.15}$$

Where i is the sample counter starting at the most recent sample and is counting back in time N data points in the window. Then the conventional variance can be expanded:

$$s_1^2 = \frac{1}{N-1}\sum_{i=1}^{N}(X_i - \bar{X})^2 = \frac{1}{N-1}\left[\left(\sum_{i=1}^{N} X_i^2\right) - N\bar{X}^2\right] \tag{14.16}$$

and substituting for the average

$$s_1^2 = \frac{1}{N-1}\left[\left(\sum_{i=1}^{N} X_i^2\right) - \frac{1}{N}\left(\sum_{i=1}^{N} X_i\right)^2\right] \tag{14.17}$$

Assuming no autocorrelation (independent deviations) and at SS, the variance can also be estimated as the differences between successive data.

$$s^2 = \frac{1}{2(N-2)} \sum_{i=1}^{N-1} (X_i - X_{i-1})^2 \tag{14.18}$$

There are N–1 terms in a set of N data because there are N–1 differences in the data window, and hence N–2 in the divisor. Then the ratio of variances is:

$$\begin{aligned} R = \frac{s_1^2}{s_2^2} &= 2\frac{(N-2)\left[\sum_{i=1}^{N} X_i^2 - \left(\sum_{i=1}^{N} X_i\right)^2 / N\right]}{(N-1)\sum_{i=1}^{N-1} (X_i - X_{i-1})^2} \\ &= 2\frac{(N-2)}{(N-1)}\frac{\left[\text{Sum1} - \text{Sum2}^2 / N\right]}{\text{Sum3}} \end{aligned} \tag{14.19}$$

This is essentially the von Neumann (1941) approach, but what appears to be a computationally expensive online calculation can be substantially simplified by using an array structure to the window of data and a pointer that indicates the array element to be replaced with the most recent measurement. With this concept, the sums are incrementally updated at each sampling.

The method is: First increment the counter to the next data storage location. Then decrement the sums with the data point to be removed. Then replace the data array element with the new value. Then increment the sums. Then recalculate the R-statistic. In comparison to the filter method, this requires storage of the N data, the N-squared data differences, and the three sums. The pointer updating adds a bit of extra computation. Our investigations indicate that the array approach takes about twice the execution time as the filter approach. Also, since it stores $2N$ data values, it has about that much more RAM requirement. The increased computer burden over the filter method may not be excessive in today's on-line computing devices.

This approach calculates a true data variance from the average in the data window of N samples and has the additional benefit that the sensitivity of the R-statistic to changes is greater than with the filter method (about twice as sensitive from experience). The array approach can respond to step changes in the PV of about 1.5σ. Accordingly, this defined window, array approach, is somewhat better in detecting small deviations from SS.

Further, the value of window length, N, explicitly defines the number of data after an event clears to fully pass through the data window and influence on the R-value. By contrast, the filter method exponentially weights the past data, and the persisting influence ranges between about $4/\lambda$ to $5/\lambda$.

Additionally, the array method seems less conflicted by autocorrelation and discretization flat spots in the time series of data.

However, the array method is computationally more burdensome than the filter method. Nominally, the filter method needs to store and manipulate 7 values at each sampling. With $\lambda = 0.1$, this characterizes a window of about 45 samples. For the same characterization, the array method needs to store and manipulate over 180 variables at each sampling.

However, there is only one adjustable variable representing the data window, N. An advantage in comprehension.

A preference is to use the filter method for its computational simplicity. But, if data vagaries (changes in autocorrelation, flat spots, coarse discretization, small shifts relative to noise) make it dysfunctional, then the array method is a bit better.

14.3 4Point Method

This is another relatively simple approach and has roots in a statistical process control approach. See Chapter 21. Here, four points in the moving window are looked at, and a filtered value is calculated for each. The first filtered value represents the average of the most recent data value (the first in the window), the second is that of a data point that is (1-gamma)**N* samples old, the third is gamma**N* samples old, and the fourth is that of data that is *N* samples old (the last in the moving window). Gamma is the Golden Ratio, $\gamma = \frac{\sqrt{5}-1}{2} = 0.618\ldots$. The variance is calculated by the sequential data differences from the leading data. The *t*-statistic is composed of the maximum difference between filtered values scaled by the filtered sigma.

This seems very simple and very effective.

A classic approach is to segregate the data into the older and recent halves of the window, then do a *t*-test on the differences of averages in the window. But, in an oscillation, when the split is at the peak or valley, the averages will be the same and SS will be erroneously declared.

A solution could be to divide the entire data window into three sub-windows, and the *t*-test is on the two windows of maximum average difference. Perhaps equal partitioning of the number of data in each sub-window will be fine, but maybe unequal spacing, with the longest the most past (perhaps to best establish a historical base line) and the shortest the most recent (perhaps to make the faster recognition). Unequal spacing might be best to not let symmetry in a data pattern confound the test. Or perhaps, the middle window should be the longest, making the difference between the 1st and 3rd windows largest in a ramp or first-order change.

We chose four points and the golden ratio spacing to eliminate missing a periodic signal.

If multiple windows and a classic *t*-test some questions still remain: What to use for the pooled sigma or *N* in the *t*-test? One could use a filtered data-to-data difference as the simplest to implement, but this would be confounded by autocorrelation in the noise. With autocorrelation, sigma would be small, making normal average differences seem large. One could estimate sigma by the classic deviation from average; but in a ramp, sigma would be large. Filter on all data is simplest.

What to use for DoF in the *t*-statistic if the sub-windows with greatest difference change? A solution would be to just use the maximum differences scaled by sigma estimate (eliminating the degree of freedom scaling) and let data at SS define the two thresholds.

If using averages in windows, the code needs to transfer data from window #1 to #2 to #3 to #4. Alternately, it could have 1 array with *N* data and a pointer for each sub-window to ID the in and out data. Then the multi-window code and calculation would be similar to the array method of the R-statistic.

But here, 4 points in the past *N* data are used to represent 4 windows, and the "averages" of the windows are calculated by the simpler first-order filters. The 4 points are the most recent and Nth past data and two intermediate points spaced by the golden ratio. (This is the same as the spacing in the Golden Section optimization approach.) It seems unlikely that a periodic oscillation would match that pattern and selected that spacing for that reason.

Based on experience with the other techniques, we chose 0.1 for each data filter lambda and 0.05 for the variance filter. Larger filter values create variation that can mask smallish changes, and smaller filter values can lead to a delay in detecting a TS or a return to a SS.

In a t-test, the denominator is a pooled sigma, from variance calculated by the number of items in the two averages in the numerator. Here, with filter factors of 0.1 for the data averages the effective N for averaging is $10 = 1/0.1$. However, the sigma is calculated with a filter value of 0.05, making N effectively 20. It is simpler to not normalize the t-statistic by the DoF.

The statistic is

$$R_i = \frac{x_{f\text{largest},i} - x_{f\text{smallest},i}}{\sqrt{\delta_{f,i}^2}} \tag{14.20}$$

Where i is the sample index, x_f would be calculated by Equation (14.1) and δ_f^2 from Equation (14.3). From experience $\lambda_1 = 0.1$ and $\lambda_2 = 0.05$, and the two critical values for R are 2 to reject SS, and 1 to reject TS give good results. If autocorrelation in the data is significant, just extend the sampling interval for δ_f^2.

In our investigation of diverse confounding effects, the 4Points approach has been best overall. But when the noise is uncorrelated the filter method is effective and remains the simplest.

- Overall 4Points is best, and Array is second best.
- 4Points and Array are tied for not worst.
- Filter is simplest, and fully adequate to recognize SS when uncorrelated noise.
- For autocorrelation 4Points was best.
- For simplicity, Filter is best, with 4Points a close second.

14.4 Using SSID as Regression Convergence Criterion

Nonlinear, least-squares optimization is commonly used to determine model coefficient values that best fit empirical data by minimizing the sum of squared deviations (SSD) of data-to-model. Nonlinear optimization proceeds in successive iterations as the search progressively seeks the optimum values for model coefficients. As the optimum is approached, the optimization procedure needs a criterion to stop the iterations. However, the current stop-optimization criteria of thresholds on either the objective function, changes in objective function, change in decision variable, or number of iterations require *a priori* knowledge of the appropriate values. They are scale dependent, application dependent, starting-point dependent, and optimization algorithm dependent; right choices require human supervision.

Use this filter method of SSID to stop optimizer iterations when there is no statistical evidence of SSD improvement relative to the variation in the data: An observer of an

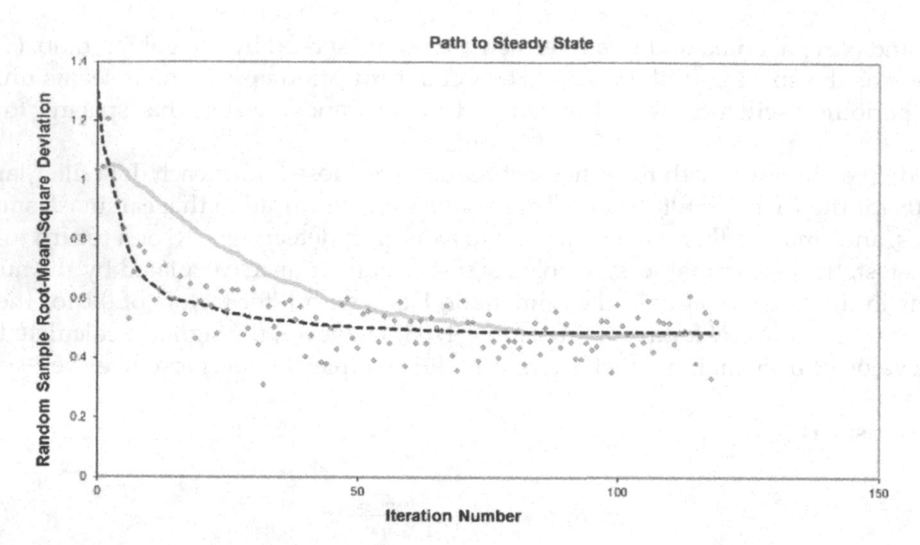

FIGURE 14.7
SSID application to nonlinear regression.

optimization procedure for empirical data will note that the SSD between the data and the model drops to an asymptotic minimum with progressive optimization iterations. The novelty is to calculate the SSD of a random subset of data (a unique, randomly selected subset at each iteration). The random subset SSD (RS SSD) will appear as a noisy signal relaxing to its noisy "steady-state" value as iterations progress. At "steady-state", when convergence is achieved, and there is no further improvement in the SSD, the RS SSD value is an independently distributed variable. When RS SSD is confidently at SS, optimization should be stopped. Since the test looks at signal-to-noise ratio, it is scale independent and "right" for any particular application.

A root-mean-square, rms, value could be used as an alternate to SSD.

Although the stopping criterion is based on the random and independent subset of data, the optimization procedure still uses the entire dataset to direct changes in the Decision Variable (model coefficient) values.

Figure 14.7 illustrates the procedure using rms as the objective. The dashed line represents the rms value of the least squares objective function as it approaches its ultimate minimum with optimizer iterations. The dots represent the rms value of randomly selected fraction of data (15 out of 30 datasets). And the thicker curved line that approaches the dashed line, and lags behind the data, represents the filtered random-subset rms value. The SSID procedure identified SS at iteration number 118 and stopped the optimization procedure. The graph indicates that in iterations beyond 100, the dashed line is effectively unchanging w.r.t. the variability in the data. The variability indicates both data variability and the process-model mismatch. In either case, the optimizer was not making substantive improvement in the model when considered relative to data variability.

14.5 Using SSID as Stochastic Optimization Convergence Criterion

Monte Carlo type simulations are often used to model the outcome of events with probabilistic influences such as rolling the dice, device reliability, molecular behavior, or human interactions. The models support the design of best response rules or system features. The simulations return a stochastic value, a value that changes with each replication, with each duplication of nominally identical inputs. The distribution of outcomes represents what one might expect in the real world, but this means that each simulation provides a different answer. One could average many simulations, but the Central Limit Theorem indicates that the variability would be attenuated by $\sqrt{N}$, not eliminated.

Figure 14.8 represents a search for coefficients in the best lead-lag rules to guide the motion of a defensive robot soccer player attempting to intercept the ball-carrying offensive opponent. The game pieces start in randomized places and the evasive moves of the offensive opponent are also randomized. The surface appears as a curved valley with steep walls and an irregular floor. This is one realization of the surface.

At any realization, the surface retains the same general shape, but the irregularities shift. The numbered axes represent scaled value of the lead and lag coefficients. The vertical axis represents the objective function value, a combination of fraction of defensive failures (goals scored) and playing-time to intercept the ball. The objective is to minimize the objective function (OF) value, to find the lowest place in the valley floor. Unfortunately, at each sampling, the floor has a different value, as if you were standing on a trampoline bed and kids were jumping on it all around you or trying to measure the lake height when it is continually perturbed by waves and ripples. Every re-sampling of the exact same spot yields a different value.

To seek the best set of decision variables (DV) (design parameters of hardware, procedures, or rules), optimizers use the OF value as a search guide, and often also as a stopping criterion. However, the stochastic nature of the OF value is a confounding aspect that will guide the solution toward an improbable, fortuitous, best value. One solution (Rhinehart, R. R., "Convergence Criterion in Optimization of Stochastic Processes", *Computers & Chemical Engineering*, Vol. 68, 4 Sept 2014, pp 1–6) is to use a multiplayer approach and reevaluate the apparently best solution and assign a worst-of-*N* realizations as the OF value.

Any number of stopping criteria can be used, but the common ones related to the OF value, or to the change in OF value, or to the change in DV values are scale dependent and require *a priori* knowledge. However, SSID can also be applied to determine convergence

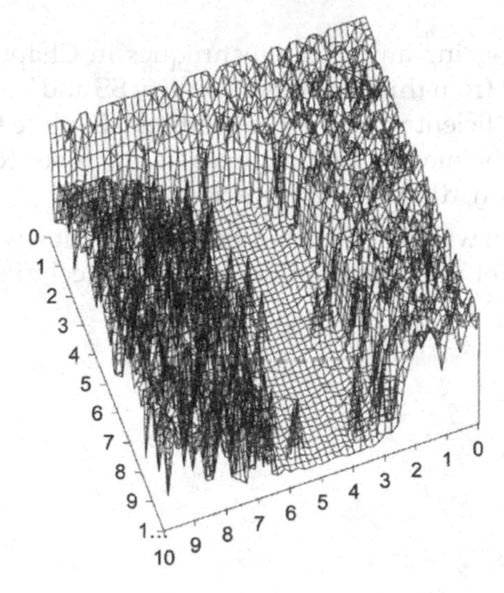

FIGURE 14.8
A stochastic optimization problem.

by observing the worst OF value at each iteration. During the phase of the search in which the optimizer is making improvements, the noisy OF value will progressively change. However, at convergence, the OF value will average at the same value, but sequential trial solutions will return a noisy response. A graph of the OF w.r.t. iteration looks very similar to Figure 14.7. (The iteration sequence value of the best OF has autocorrelation; however, at SS, the worst player OF value has none. So, apply the SS convergence criterion to the worst player.)

14.6 Takeaway

The filter technique is simplest to implement, but autocorrelation and signal discretization can confound it in process monitoring. However, for regression and optimization, the noise is uncorrelated and only the approach to SS needs to be detected. Here the filter approach is fully adequate and simple to implement and well used by the author.

In a multivariable process the user needs to select variables to be monitored, and criteria should be that observed variables need to be the minimum number that observes all aspects of the process. In a multivariable analysis, there is a greater chance that data vagaries will cause a false TS claim. Also, an excessive wait for all PV monitors to return to SS. So, the implementer might want to temper the monitors: Claim TS only if 10% of the PVs are at TS, and claim SS if 90% or more are at SS.

Unlike typical hypothesis-rejecting procedures which have only one critical value, these tests have two for the single statistic. One critical value is used to reject the SS hypothesis, and the other to reject the TS hypothesis.

14.7 Exercises

1. Generate noisy data using any of the techniques in Chapter 9 and apply one or more of the methods from this chapter to identify SS and TS. Explore the impact of variations in the coefficient values of the methods. Explore the impact of a variety of data features (in the mean – step changes, ramp, first-order, oscillation) (in the noise – autocorrelation, discretization, variance).
2. Alternately, instead of writing your own simulator, visit www.r3.eda.com to access a SSID and TSID Excel VBA program to explore the techniques and data features.

15

Linear Regression – Steady-State Models

15.1 Introduction

Regression (curve fitting) is a technique for developing a possible mathematical relationship between the dependent variable (Y_i) and one or more independent variables (X_{1i}, X_{2i}, …, X_{ni}). You have the $(y, \underline{x})$ data and want to find a model that best fits the data. We use the term "possible" deliberately, as many possible models exist, some supported by theory, others strictly empirical in nature.

There are several ways to categorize or differentiate models, one being steady-state in contrast to dynamic (or transient). Dynamic (transient) models have a time-dependent response. Consider that your automobile is at rest and you press the accelerator to 30%. Initially the car speed was zero, then second-by-second it increases to 10 then 20 then 28 then 35 then 40 then 42 and asymptotically approaches the final speed of 45 (km/h, or mph). In this case the independent variable, x, was the accelerator pedal position; and the response, the dependent variable, y, was the speed. In the transient, time-dependent, situation there is not one unique y-value for any x-value. By contrast, steady-state models have a one-to-one relation. The steady-state data pairs, $(y, \underline{x})$, represent the final, settled, equilibrium, end-of-batch value. This chapter is about the use of *steady-state* models.

Linear regression does not necessarily mean that the model is linear. A linear model is a straight line fit of y to x. What linear regression does mean is that the coefficients appear linearly in the model. This quadratic model (describing the nonlinear relation between y and x) is linear in the coefficients.

$$y = a + bx + cx^2 \tag{15.1}$$

The coefficients, a, b, and c each appear to the first power and they do not interact. A plot of how y changes with the value of any coefficient is linear and independent of the values of any other coefficients. A mathematical definition of linearity is that the second derivative of the model w.r.t. any pairing of coefficients is zero.

$$\frac{\partial^2 y}{\partial c_i \partial c_j} = 0 \tag{15.2}$$

Here c_i and c_j represent any two model coefficients.

DOI: 10.1201/9781003222330-15

Example 15.1: Show that model (15.1) is linear in coefficient c.

The model is $y = a + bx + cx^2$. The first derivative w.r.t. coefficient c is $\frac{\partial y}{\partial c} = x^2$, and since the value of the independent variable x does not depend on the value of any of the coefficients a, b, or c, the subsequent derivative of $\frac{\partial y}{\partial c} = x^2$ w.r.t. any coefficient is zero. For instance, $\frac{\partial}{\partial a}\frac{\partial y}{\partial c} = \frac{\partial}{\partial a}x^2 = 0$.

Example 15.2: Show that the seemingly linear model $y = \frac{1}{a+bx}$ is nonlinear. Although the $a + bx$ functionality seems linear in coefficients a and b, the coefficients are in the denominator, and effectively appear to the −1 power.

The first derivative of the model w.r.t. coefficient a is $\frac{\partial y}{\partial a} = -(a+bx)^{-2}$. The subsequent derivative of w.r.t. coefficient a is $\frac{\partial}{\partial a}\frac{\partial y}{\partial a} = \frac{\partial}{\partial a}\left[-(a+bx)^{-2}\right] = 2(a+bx)^{-3} \neq 0$.

Example 15.3: Show that model $y = a + bx_1 + cx_2 + dx_1x_2$ is linear. Although the x_1x_2 functionality makes it nonlinear in the independent variables, it remains linear in each of the four model coefficients.

The first derivative of the model w.r.t. coefficient d is $\frac{\partial y}{\partial d} = x_1x_2$. Since the independent variables are not a function of any of the model coefficient values, the subsequent derivative of w.r.t. any coefficient is zero.

There are several fundamental assumptions behind the conventional least squares regression analyses: 1) The x-values are perfectly known. 2) Only the y-values have uncertainty (error). 3) The errors are normally distributed with mean 0 and variance σ^2. 4) The error variance is uniform, homoscedastic, the same throughout the range of x and y values. (The errors will be heterogeneous if their magnitude depends on the value of the independent variable(s). This situation occurs whenever the precision of a process measurement instrument varies throughout the range. Lack of normality occurs whenever the errors are skewed because an instrument or controller has different tolerances in different parts of the overall range.) 5) The errors on y are independent from each other. 6) The model represents the phenomena that generated the data. (The model is true to the process.)

Assumptions 1) and 2) above lead to a concept of vertical least squares. A conventional plot of y w.r.t. x has the response on the vertical axis (ordinate) and the influence on the horizontal axis (abscissa). Since x is presumed perfectly known, and only the y-value is credited with experimental error; the difference between model value at any particular x-value, $\tilde{y}(x_i)$, and the experimentally obtained y-value, y_i, is $\left[\tilde{y}(x_i) - y_i\right]$, representing the vertical distance between data and model on the graph. In conventional least squares regression, the objective is to adjust the model coefficients to minimize the sum of the squared vertical distances.

$$\min_{\{\underline{c}\}} J = \sum_{i=1}^{n\ \text{data}} \left[\tilde{y}(x_i) - y_i\right]^2 \tag{15.3}$$

Certainly, one can best fit model-to-data to accommodate experimental uncertainty in both the x and y data, using techniques such as Maximum Likelihood or variants of it such as

Akaho's method and perpendicular least squares. Such approaches are a step more complicated, and do not have the extent of analysis associated with vertical least squares. The simpler vertical least squares usually generates fully functional models for engineering, and vertical least squares also seems to be widely accepted as "the way". Although model perfection would be improved with maximum likelihood techniques, this chapter will focus on traditional vertical least squares approaches. (For maximum likelihood and other methods, see Rhinehart, R. R., *Nonlinear Regression Modeling for Engineering Applications: Modeling, Model Validation, and Enabling Design of Experiments*, Wiley, New York, NY, September, 2016. ISBN 9781118597965.)

The squared vertical deviation is a direct outcome of using the Gaussian or normal distribution, the presumed distribution of experimental errors. It is a limit of the Maximum Likelihood objective with the several assumptions. Alternately, an intuitive approach, to minimize the absolute value of the deviations, also works and gets reasonable models.

The models obtained are not "the Truth". There are many reasons: 1) Every realization of the experimental data will vary from another experimental set. The model from any one experimental realization will not have exactly the same coefficient values as the model from another. 2) The rules that nature uses to generate data are not the same as human-chosen models. 3) The six assumptions listed earlier are not necessarily true.

15.2 Simple Linear Regression

Simple linear regression involves the following as the proposed functional relationship.

$$Y_i = \beta_0 + \beta_1 X_i + \varepsilon_i \tag{15.4}$$

The model is linear in both coefficients and the y response to x. The coefficients (often termed parameters) β_0 and β_1 are the intercept and slope, respectively. The subscript i represents the dataset number, the index number for the (y, x) pair. In the model, ε_i represents the total effect of all errors affecting each Y_i. The independent variable X is *not* a random variable but assumes fixed values set by us, the experimenters. We assume that there is no error on the x-values and that the y-errors are independent and have a normal distribution with mean 0 and variance σ^2, regardless of the value of X.

Using $\hat{\beta}_0$ and $\hat{\beta}_1$ as the estimates of β_0 and β_1, then the estimated or predicted value of Y can be expressed as $\hat{Y}_i = \hat{\beta}_0 + \hat{\beta}_1 X_i$, and $e_i = Y_i - \hat{Y}_i = Y_i - \left(\hat{\beta}_0 + \hat{\beta}_1 X_i\right)$ represents the error between the data and the model. The individual error terms $e_i = Y_i - \hat{Y}_i$ are called *residuals* and represent the process-to-model mismatch. They are measures of the failure of the regression equation to predict the Y_i values without error.

The objective now is to obtain the estimates $\hat{\beta}_0$ and $\hat{\beta}_1$ from the sample data. The accepted "best" way to obtain these estimates is the *method of least squares.* This method is based on the principle that the best estimators minimize the sum of squares due to error, SS_E, defined as:

$$SS_E = \sum_{i=1}^{n} e_i^2 = \sum_{i=1}^{n} \left(Y_i - \hat{Y}_i\right)^2 = \sum_{i=1}^{n} \left(Y_i - \hat{\beta}_0 - \hat{\beta}_1 X_i\right)^2 \tag{15.5}$$

To find the values of $\hat{\beta}_0$ and $\hat{\beta}_1$ that minimize SS_E, use a classic analytical optimization procedure. Take the first partial derivatives with respect to $\hat{\beta}_0$ and $\hat{\beta}_1$ of the error sum of squares in Equation (15.5), set those derivatives equal to zero, and solve the resulting equations for the estimators $\hat{\beta}_0$ and $\hat{\beta}_1$. The result of this procedure is:

$$\hat{\beta}_1 = \frac{\Sigma_i X_i Y_i - n\bar{X}\bar{Y}}{\Sigma_i X_i^2 - n\bar{X}^2} \tag{15.6}$$

and

$$\hat{\beta}_0 = \bar{Y} - \hat{\beta}_1 \bar{X} \tag{15.7}$$

This procedure for estimating β_0 and β_1 is called the least squares solution. The estimates are those needed by the regression model of Equation (15.3).

The total sum of squares, SS_T, in regression analysis may be partitioned just as we did in analysis of variance to give

$$SS_T = SS_M + SS_R + SS_E \tag{15.8}$$

where SS_R is the sum of squares (sum of residuals squared) due to regression and SS_M is the sum of squares due to the mean as defined in Table 15.1. The sum of squares due to error is SS_E. In like manner, the total degrees of freedom, n, may be partitioned into one for the mean, one for the regression, and $(n - 2)$ for error. Calculational routines for the terms in Equation (15.8) are presented in Table 15.1.

15.2.1 Hypotheses in Simple Linear Regression

There are several standard hypotheses concerning the simple linear regression model. The first is H_0: $\beta_1 = 0$ vs H_A: $\beta_1 \neq 0$. If the null hypothesis is accepted as probably true, then there is probably no linear relationship between X and Y. First, we need an estimate for σ^2, the true experimental error variance. If we assume that the failure of the model to fit the data is due solely to error, the mean square for error can be used to estimate σ^2:

$$\hat{\sigma}^2 = MS_E = \frac{SS_E}{n-2} \tag{15.9}$$

TABLE 15.1

Analysis of Variance for Simple Linear Regression

Source	df	SS	MS
Due to mean	1	$n\bar{Y}^2 = SS_M$	—
Due to regression	1	$\hat{\beta}_1\Sigma(X_i - \bar{X})(Y_i - \bar{Y}) = SS_R$	$MS_R = SS_R$
Error or residual	$n - 2$	$\sum_i (Y_i - \hat{Y}_i)^2 = SS_E$	$MS_E = \frac{SS_E}{n-2}$
Total	n	$\sum_i Y_i^2 = SS_T$	

In determining the degrees of freedom for error, it may help you to remember to use $v = n-(p+1) = n-p-1$ where n is the number of observations in the experimental dataset and $(p + 1)$ is the number of coefficients in the model.

In order to evaluate H_0: $\beta_1 = 0$, we now need an unbiased estimate of the variance of β_1. This estimate is calculated from sample data and is

$$\sigma^2_{\hat{\beta}_1} = \frac{\hat{\sigma}^2}{\sum_i (X_i - \bar{X})^2} = S^2_{\hat{\beta}_1} \tag{15.10}$$

The random variable

$$T_{\beta_1} = \frac{\hat{\beta}_1 - 0}{\sigma_{\hat{\beta}_1}} = \frac{\hat{\beta}_1 - 0}{S_{\hat{\beta}_1}} = \frac{\hat{\beta}_1 - 0}{\left[\mathrm{MS_E} / \sum_i (X_i - \bar{X})^2\right]^{1/2}} \tag{15.11}$$

has a t distribution with $(n - 2)$ degrees of freedom and $S_{\hat{\beta}1}$ is the standard error associated with $\hat{\beta}_1$. In order to evaluate the more general case $H_0 : \beta_1 = \beta_1'$ vs. $H_0 : \beta_1 \neq \beta_1'$ where β_1' is a particular nonzero value of β_1', replace the 0 with β_1' in Equation (15.11). The hypothesis H_0: $\beta_1 = 0$ can also be tested by use of $F = \mathrm{MS_R}/\mathrm{MS_E}$.

In similar fashion, estimates of the variances associated with $\hat{\beta}_0$ and with Y', the predicted value of Y corresponding to a particular value of X, can be found from

$$\sigma^2_{\hat{\beta}_0} = \left(\frac{1}{n} + \frac{\bar{X}^2}{\sum_i (X_i - \bar{X})^2}\right)\hat{\sigma}^2 = S^2_{\hat{\beta}_0} \tag{15.12}$$

and

$$\sigma^2_{\hat{Y}} = \left(1 + \frac{1}{n} + \frac{(X - \bar{X})^2}{\sum_i (X_i - \bar{X})^2}\right)\hat{\sigma}^2 = S^2_{\hat{Y}} \tag{15.13}$$

Use Equation (15.9) to determine the value of $\hat{\sigma}^2$.

Note: The value of $\sigma^2_{\hat{Y}}$ in Equation (15.13) will depend on the value of X.

The corresponding random variables to be used for test purposes for hypotheses concerning β_0 and Y' are t-distributed.

$$T_{\beta_0} = \frac{\hat{\beta}_0 - \beta_0}{S_{\hat{\beta}_0}} \tag{15.14}$$

and

$$T_{Y'} = \frac{\hat{Y}' - Y'}{S_{\hat{Y}'}} \tag{15.15}$$

If you are interested in evaluating the entire model, i.e., the simultaneous estimate of β_0 and β_1 then

$$F = \frac{n\left(\hat{\beta}_0 - \beta_0\right)^2 + 2n\bar{X}\left(\hat{\beta}_0 - \beta_0\right)\left(\hat{\beta}_1 - \beta_1\right) + \left(\hat{\beta}_1 - \beta_1\right)^2 \Sigma X_i^2}{2\hat{\sigma}^2} \tag{15.16}$$

has an *F*-distribution with $v_1 = 2$ and $v_2 = (n - 2)$ degrees of freedom. Values of *F* are calculated to test the hypothesis by replacing β_0 with $\hat{\beta}_0$ and β_1 with β_1'. All of the test procedures for simple linear regression and their corresponding rejection regions are summarized in Table 15.2.

15.2.2 Interval Estimation in Simple Linear Regression

The $(1 - \alpha)$ 100% confidence interval on β_0 can be determined from

$$\left.\begin{matrix} U \\ L \end{matrix}\right\} = \hat{\beta}_0 \pm S_{\hat{\beta}_0} t_{n-2,1-\alpha/2} \tag{15.17}$$

where $S_{\hat{\beta}_0}$ is obtained from Equation (15.12) and *U* and *L* are the upper and lower confidence limits, respectively.

If you want the $(1 - \alpha)$ 100% confidence interval for β_1, it can be calculated from Equation (15.18).

$$\left.\begin{matrix} U \\ L \end{matrix}\right\} = \hat{\beta}_1 \pm S_{\hat{\beta}_1} t_{n-2,1-\alpha/2} \tag{15.18}$$

where $S_{\hat{\beta}1}$ is obtained from Equation (15.10). If you need the $(1 - \alpha)$100% confidence interval for an individual *Y* value *Y′* associated with a particular *X* value *X*, calculate

$$\left.\begin{matrix} U \\ L \end{matrix}\right\} = \hat{Y} \pm S_{\hat{Y}} t_{n-2,1-\alpha/2} \tag{15.19}$$

TABLE 15.2

Test Procedures in Simple Linear Regression

H_0	Statistic	Rejection region
$\beta_0 = \beta_0'$	$T = \dfrac{\hat{\beta}_0 - \beta_0'}{S_{\hat{\beta}_0}}$	$T \geq t_{n-2,1-\alpha/2}$ or $T \leq -t_{n-2,1-\alpha/2}$
$\beta_1 = \beta_1'$	$T = \dfrac{\hat{\beta}_1 - \beta_1'}{S_{\hat{\beta}_1}}$	
$\mu_{Y\mid X=X_0} = \mu_0$	$T = \dfrac{\hat{\beta}_0 + \hat{\beta}_1 X_0 - \mu_0}{S_{\hat{Y}}}$	
$\beta_0 = \beta_0'$ and $\beta_1 = \beta_1'$	$F = \dfrac{\left[n\left(\hat{\beta}_0 - \beta_0'\right)^2 + 2n\bar{X}\left(\hat{\beta}_0 - \beta_0'\right)\left(\hat{\beta}_1 - \beta_1'\right) + \left(\left(\hat{\beta}_1 - \beta_1'\right)^2\right)\Sigma X_i^2\right]}{2\hat{\sigma}^2}$	$F \geq F_{2,n-2,1-\alpha}$
$\beta_1 = 0$	$F = \dfrac{MS_R}{MS_E}$	$F \geq F_{1,n-2,1-\alpha}$

where $S_{\hat{Y}}$ is obtained from Equation (15.13). Now that you have seen the two-sided confidence intervals, you should be able to write out the one-sided confidence intervals for β_0, β_1, and Y'.

Example 15.4: With the following *x-y* pairs, determine the best linear model for *Y* w.r.t. *X*, and the 99% confidence interval on the modeled *Y*-values.

X	Y
3	12.03
3.3	11.01
3.6	12.36
4	17.88
5	12.35
6	16.04
8	23.73
8.5	23.82
10	21.19
14	36.86

There are $n = 10$ data pairs. The best linear model is $y = 4.8943 + 2.1151x$. (The coefficient values are obtained from Equations (15.6 and 15.7) are rounded here.) This is represented by the central line in the graph. The dashed lines represent the upper and lower 99% confidence limits on the modeled value using Equation (15.19).

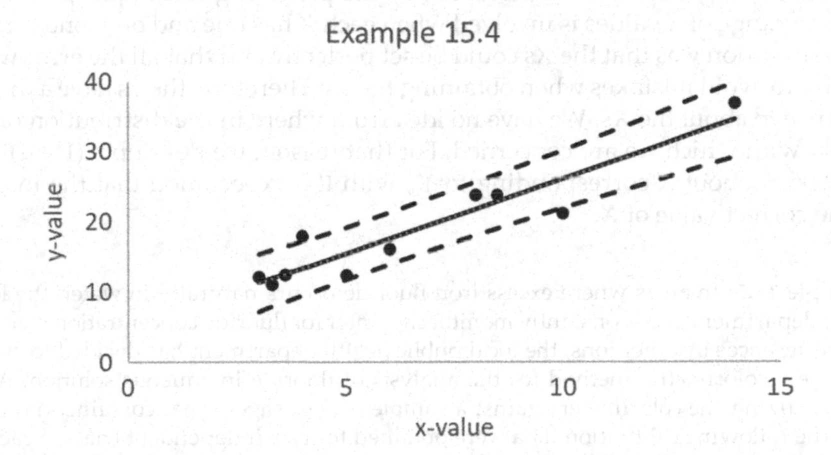

Note that the limits are curved lines, even though the model is linear. They are closer to the model near the central region, where they are surrounded by data, and farther near the ends where there is only data on one side. Also note that the confidence interval is farther from the model on the right side (where there are fewer local data points) than it is on the left side (where there are more local data points).

The Excel Data Analysis routine "Regression" also provides the model coefficient values, and a confidence interval and *p*-value about the significance that the slope is not zero, paralleling the use of Equation (15.11). The standard error or the slope (X_Variable) is reported as 0.2719… and with the slope of 2.115… the *p*-value is 5.3×10^{-5} indicating that it would be exceedingly rare to obtain such a slope if the slope was in fact zero.

15.2.3 Inverse Prediction in Simple Linear Regression

Occasionally, you may question a data point. Could the Y value in your notebook really have come from the experimental test program? You may be able to answer that question by calculating the range of X values that could have generated the Y value in question. This situation is inverse prediction. From the given value of Y_0, calculate

$$\hat{X} = \frac{Y_0 - \hat{\beta}_0}{\hat{\beta}_1} \tag{15.20}$$

The confidence interval for the real but unknown corresponding X value is found from

$$\left.\begin{matrix} U \\ L \end{matrix}\right\} = \hat{X} + \frac{\hat{\beta}_1\left(Y_0 - \bar{Y}\right)}{\lambda} \pm \frac{t\hat{\sigma}}{\lambda}\left[\frac{\left(Y_0 - \bar{Y}\right)^2}{\Sigma\left(X_i - \bar{X}\right)^2} + \lambda\left(\frac{n+1}{n}\right)\right]^{1/2} \tag{15.21}$$

where

$$\lambda = \hat{\beta}_1^2 - t^2 S_{\hat{\beta}_1^2} \tag{15.22}$$

and $t = t_{n-2,\alpha/2}$.

This situation occurs more frequently than you may think. In one boiler trial, we noticed a definite increase in efficiency. The only observed difference on that run was the generated steam pressure, Y_0. Although the corresponding controller setting was not shown in the log, that missing value can be predicted by the preceding technique. If you are wondering why a range of X values is involved when each X has one and only one Y, remember that our assumption was that the Xs could be set perfectly and that all the error was due to our inability to avoid mistakes when obtaining the Ys. Therefore, the Ys have a distribution with variance σ^2 about the Xs. We have no idea from where in the distribution of which X came the Y_0 with which we are concerned. For that reason, we develop a $(1 - \alpha)100\%$ confidence interval about X corresponding to Y_0, with the expectation that the interval will contain the correct value of X.

Example 15.5: In areas where excess iron fluoride occurs naturally in water, the local water department must constantly monitor the water for fluoride concentration. Because of interferences by other ions, the local public health department has decided to evaluate a new colorimetric method for the analysis of fluoride in aqueous solution. After standardizing the colorimeter against a sample of the reagent that contained no fluoride, the following calibration data were obtained for two independent trials at each of ten concentrations. The calibration equation and plot will be used to determine the fluoride (F^-) concentration when transmittance (%T) is measured. Although the data were generated with concentration as the manipulated (independent) variable, our model for the analytical calibration will use %T as the independent variable.

	Percent transmittance	
Concentration (mg F^-/ml)	**Trial 1**	**Trial 2**
0.000608	80.3	80.3
0.001216	80.5	80.4

(*Continued*)

	Percent transmittance	
Concentration (mg F⁻/ml)	Trial 1	Trial 2
0.001824	80.9	81.0
0.002432	81.2	81.8
0.003040	81.6	82.0
0.003648	82.9	82.5
0.004256	83.0	83.1
0.004864	83.9	84.0
0.005472	84.0	84.0
0.006080	85.0	84.8

It is usually a good thing to see what the data looks like! As the figure reveals, the plot appears linear, supporting the model chosen. Note: With duplicate readings for single sample preparations, and the calibration model switching the experimental (X,Y) to (Y,X), the duplicate x-readings make horizontal pairs on the graph. The dotted line is the linear regression, best linear model.

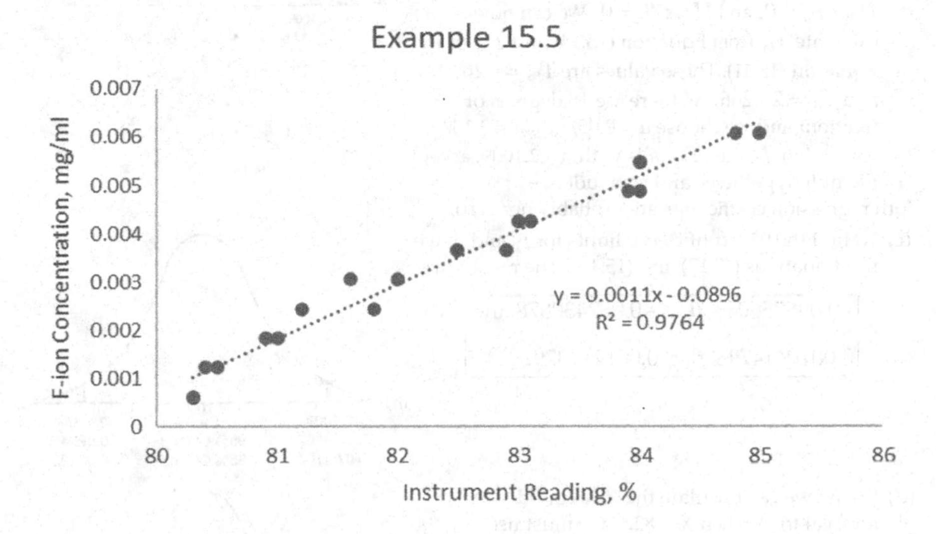

Using the least squares technique,

(a) Find the relationship between concentration (Y) and percent transmittance (X).
(b) Determine the validity of the regression constants by showing that each is probably nonzero.
(c) Give the 95% confidence interval for each regression coefficient.
(d) Find the range of fluoride concentration expected for an unknown sample having 82.1% T.
(e) Estimate the percent transmittance corresponding to a calibration check sample containing 0.00500 mg F^- /ml.

We begin by doing the necessary arithmetic to obtain the following values. (Remember that our standard notation uses lowercase letters for specific values.)

$$\Sigma_i x_i = 1647.2, \qquad (\Sigma_i y_i)^2 = 0.004472934,$$

$$\Sigma_i\left(y_i-\bar{y}\right)^2=6.0995\times10^{-5},\quad \Sigma_i\left(x_i-\bar{x}\right)^2=46.768$$

$$\left(\Sigma_i x_i\right)^2=2{,}713{,}264.84,\quad \Sigma_i x_i \Sigma_i y_i=110.164736,$$

$$\Sigma_i\left(x_i-\bar{x}\right)\left(y_i-\bar{y}\right)=0.0527744$$

$$\bar{x}=82.36,\quad \bar{y}=0.003344$$

(a) Using Equations (15.6) and (15.7), we have

$$\boxed{\hat{\beta}_0=-0.0895934697 \text{ and } \hat{\beta}_1=0.0011284297}$$

(b) From Equations (15.5) and (15.9), the error mean square is $MS_E=\hat{\sigma}^2=8.010407\times10^{-8}$. We next calculate $S_{\hat{\beta}0}=3.409137\times10^{-3}$ and $S_{\hat{\beta}1}=4.138595\times10^{-5}$. The null hypotheses are $H_{0_1}:\beta_0=0$. and $H_{0_2}:\beta_1=0$. We can now calculate T_{β_1} from Equation (15.14) and T_{β_0} from Equation (15.11). These values are $T_{\beta_0}=-26.280$ and $T_{\beta_1}=27.266$. As there are 18 degrees of freedom and we choose $\alpha=0.05$, $t_{18,0.975}=2.1009$. As neither T_{β_0} nor T_{β_1} falls within ±2.1009, accept the null hypotheses and conclude:

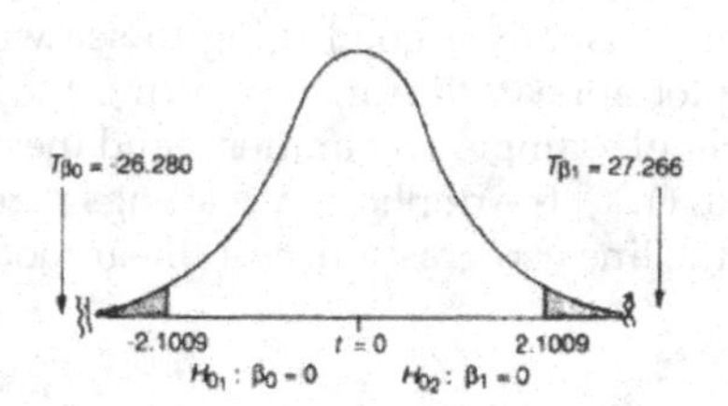

Both regression coefficients are probably nonzero.

(c) To find the 95% confidence limits for β_0 and β_1 we use Equations (15.17) and (15.18). The results are

$$\boxed{\begin{array}{l}-0.09675606\le\beta_0\le-0.082430878 \text{ and}\\ 0.001041479\le\beta_1\le0.001215379.\end{array}}$$

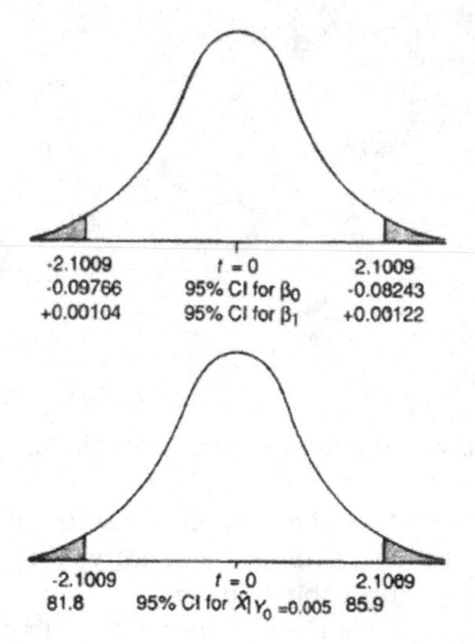

(d) Before we can calculate the 95% confidence interval for $\hat{Y}$ when $X=82.1$, we must use Equation (15.13) to calculate $S_{\hat{Y}}=2.902155\times10^{-4}$. Next, we use the regression equation $\hat{Y}=-0.0895934697+0.0011284297X$ to calculate $\hat{Y}=0.0030506$ when $X=82.1$. From Equation (15.19) we have

$$\boxed{\left.\begin{array}{l}U\\L\end{array}\right\}=0.003660,\ 0.002441}$$

as the required concentration range in mg F⁻/ml.

(e) To estimate $\hat{X}$ when $Y_0=0.005$, we use Equation (15.20) to find $\hat{X}$ and Equation (15.21) for the interval. The results are $\hat{X}=83.82752572$ and

$$\boxed{\left.\begin{array}{l}U\\L\end{array}\right\}=85.9,\ 81.8}$$

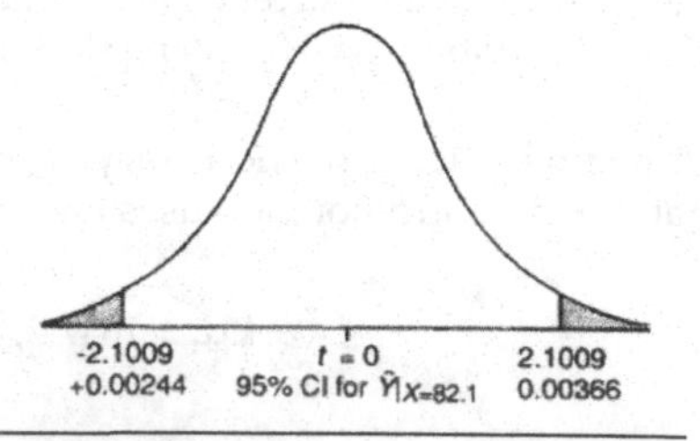

Probably, it will be more convenient to use a regression package to obtain the results. In Excel, the Data Analysis Add-In, Regression, provides the following results. Values are truncated for presentation convenience.

Summary Output

Regression statistics	
Multiple R	0.988106
R Square	0.976353
Adjusted R square	0.975039
Standard error	0.000283
Observations	20

ANOVA

	df	*SS*	*MS*	*F*	**Significance** *F*
Regression	1	5.96E-05	5.96E-05	743.1846	4.34E-16
Residual	18	1.44E-06	8.01E-08		
Total	19	6.1E-05			

	Coefficients	**Standard error**	**t Stat**	**P-value**	**Lower 95%**	**Upper 95%**
Intercept	−0.08959	0.00341	−26.276	8.28E-16	−0.09676	−0.08243
X Variable 1	0.001128	4.14E-05	27.26141	4.34E-16	0.001041	0.001215

15.2.4 Evaluation of Outliers

Although some scatter is expected in all experimental results, *outliers*, or points that do not appear to belong to the dataset, are not as rare as you would hope. For instance, the average of 86., 83., 79., 85., 80., 68., 90., and 83. is 81.75. If the data points represent replicate trials, then they should each have similar values. All are 80-ish except the 68. Was it truly an 86. that was inadvertently transposed? Can one claim that the 68., is invalid, discard it, and average the remaining seven values to obtain 83.71…? The same question arises when data are compared to a model. Can the starred point in Figure 15.1 be discarded, or does it reflect normal process variability?

If as a result of reviewing your data in tabular or graphical form, you spot a point, Y', that you feel is obviously incorrect, you may be able to discard it if sufficient justification exists. First, review all your calculations and data transcriptions to be sure that the "odd-ball" point is not the result of a mistake in arithmetic, number transposition, etc. If you find no obvious error, your second review should include all the original process, production, research, etc. data. Look for indications of incorrect procedure, an upset in performance due to some uncontrolled event, etc. The results of these two reviews may provide evidence that will allow you to discard the outlier so that you can confidently repeat the regression analysis with the remaining data.

If you still feel the data point Y' is incorrect, your last resort is to construct a $(1-\alpha)100\%$ confidence interval around $\hat{Y}$ given X, or $\hat{Y}|_X$. For linear regression, the data point in question, Y', may probably be discarded safely if it is not in the confidence interval defined by Equations (15.13) and (15.19). Chauvenet's Criterion suggests that an appropriate value for

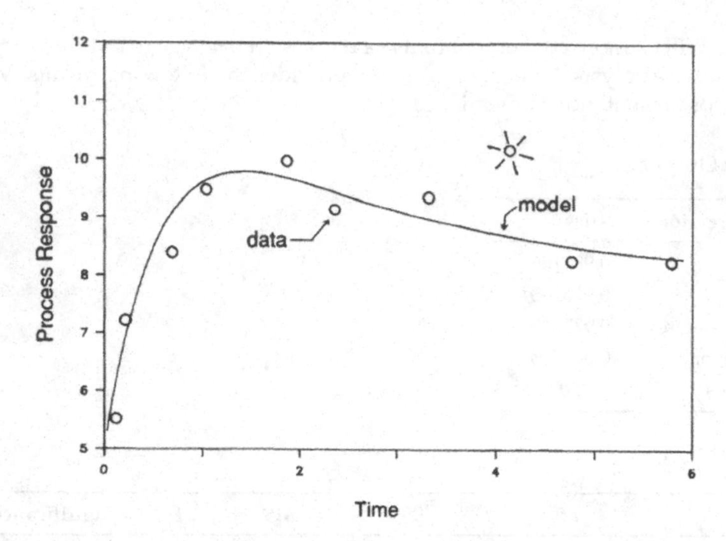

FIGURE 15.1
Comparison of process and model.

α is $1/(2n)$. For $n = 10$ data values, alpha would be $\alpha = \dfrac{1}{2 \cdot 10} = 0.05$ (a conventional value).

For $n = 100$ data values, alpha would be $\alpha = \dfrac{1}{2 \cdot 100} = 0.005$.

This method presumes that the residuals are normally distributed. You should check before using Chauvenet's Criterion.

The procedure just described follows that of ASTM Standard E178-80. Unfortunately, no guidance is given there for the selection of the significance level, α. If in doubt, we recommend the use of Chauvenet's method as previously described.

15.2.5 Testing Equality of Slopes

Have you ever wondered whether the responses of two machines, systems, etc. to an independent variable are the same? A common case involves instruments that don't "read" exactly the same because of calibration differences. Consider the comparison of two groups, (X_{1i}, Y_{1i}) and (X_{2i}, Y_{2i}). The null hypothesis is that the slopes of the corresponding regression lines are the same, or H_0: $\beta_{11} = \beta_{12}$. The two regression lines are $Y_1 = \beta_{01} + \beta_{11}X_1 + \varepsilon_1$, and $Y_2 = \beta_{02} + \beta_{12}X_2 + \varepsilon_2$. As the variances may be presumed equal, their estimates $\hat{\sigma}_1^2$ and $\hat{\sigma}_2^2$ may be pooled (see Section 6.5) directly to obtain $\hat{\sigma}^2$. As the variances of β_{11} and β_{12} are $S_{\hat{\beta}_{11}}^2 = \dfrac{\hat{\sigma}^2}{\sum_j \left(X_{1j} - \bar{X}_1\right)^2}$ and $S_{\hat{\beta}_{12}}^2 = \dfrac{\hat{\sigma}^2}{\sum_j \left(X_{2j} - \bar{X}_2\right)^2}$, the variance of $\left(\hat{\beta}_{11} - \hat{\beta}_{12}\right)$ is $S_{\hat{\beta}_{11}}^2 + S_{\hat{\beta}_{12}}^2$.

Assuming that $\left(\hat{\beta}_{11} - \hat{\beta}_{12}\right)$ has a normal distribution with mean $\left(\overline{\hat{\beta}_{11} - \hat{\beta}_{12}}\right)$ and variance $\sigma_{\hat{\beta}_{11}}^2 + \sigma_{\hat{\beta}_{12}}^2$, then the statistic

$$T = \frac{\hat{\beta}_{11} - \hat{\beta}_{12} - (\beta_{11} - \beta_{12})}{\hat{\sigma}\left[1/\sum_j \left(X_{1j} - \bar{X}_1\right)^2 + 1/\sum_j \left(X_{2j} - \bar{X}_2\right)^2\right]^{1/2}} \tag{15.23}$$

has a t distribution with $v = n_1 - 2 + n_2 - 2 = n_1 + n_2 - 4$ degrees of freedom, where $\hat{\sigma}^2 = \dfrac{(SS_E)_1 + (SS_E)_2}{n_1 + n_2 - 4}$.

The terms $(SS_E)_1$ and $(SS_E)_2$ denote the error sums of squares corresponding to the two sets of data. To test the null hypothesis H_0: $\beta_{11} = \beta_{12}$, calculate T using Equation (15.23). The hypothesis is rejected if $T > t_{v,1-\alpha/2}$ or $T < -t_{v,1-\alpha/2}$.

Example 15.6: The following data were obtained 2 months apart for the relationship between refractive index and composition of mixtures of benzene and 1-propanol. Independent aliquots of the standards were used to prepare the calibration samples, which were retained and refrigerated for periodic calibration checks. Compare the calibration curves at $\alpha = 0.01$. Support your answer statistically.

		Refractive index	
Benzene (ml)	**1-Propanol (ml)**	**Month 1**	**Month 3**
10	1	1.4837	1.4842
10	2	1.4724	1.4716
10	4	1.4650	1.4651
10	6	1.4571	1.4555
10	8	1.4430	1.4420
10	10	1.4385	1.4368
10	14	1.4263	1.4244
10	20	1.4108	1.4100
10	30	1.3941	1.3940
10	50	1.3892	1.3888
10	100	1.3816	1.3819

The calibration data were obtained in terms of the volumes of the two chemicals that were mixed to obtain each calibration sample. We need the calibration data in volume fraction of 1-propanol. In the following, you'll notice that a new variable, Vol fraction of 1-Prop, has been created from the original volumetric data. Both regression programs for these data are the Regression analysis of the Excel Data Analysis Add-In. The normal model concept would be to prepare the mixture and see how refractive index (RI) changes. However, in calibration, we measure RI and use the inverse relation to forecast the composition. So, in this case R. is the x-variable, and vol fraction is the y-variable.

After executing the two programs, we have the following slopes (coefficients) and their standard errors and estimates of experimental error (residual SS):

$$\hat{\beta}_{11} = -7.53042693 \qquad \hat{\beta}_{12} = -7.55369491$$

$$S_{\hat{\beta}_{11}} = 0.30864588 \qquad S_{\hat{\beta}_{12}} = 0.28061160$$

$$SS_{E_1} = 0.01069051 \qquad SS_{E_2} = 0.00880573$$

Pooling the SS_E data, we have

$$\hat{\sigma} = \left(\frac{SS_{E_1} + SS_{E_2}}{n_1 + n_2 - 4}\right)^{1/2}$$

$$= \left(\frac{0.01069051 + 0.00880573}{11 + 11 - 4}\right)^{1/2}$$

$$= 0.032910856$$

We obtain the reciprocal of the sum of squares of deviations in the Xs as

$$\frac{1}{\sum_j \left(X_{1j}-\bar{X}_1\right)^2}=\frac{S^2_{\hat{\beta}11}}{MS_{E_1}}=80.19857997$$

$$\frac{1}{\sum_j \left(X_{2j}-\bar{X}_2\right)^2}=\frac{S^2_{\hat{\beta}12}}{MS_{E_2}}=80.48044281$$

From Equation (15.23), we have, assuming H_0: $\beta_{11}=\beta_{12}$ is true, T = 0.055775. As $t_{18,0.995}$ = 2.8784, we accept the null hypothesis and conclude:

The slopes are probably equivalent.

15.2.6 Regression Through a Point

The method of least squares, as we have seen it so far, is not designed to produce a regression equation that will pass through a predetermined point. Such an event is necessary in many physical situations: At rest, the speed of any object is zero; if no fluid is flowing in a pipe, the flow rate readout should be zero; before an assembly line is started, its production rate is zero; etc. The regression model of the line that must pass through (X_p, Y_p), instead of $\left(\bar{X},\bar{Y}\right)$ is

$$Y_i - Y_p = \beta\left(X_i - X_p\right)+\varepsilon_i, \qquad i=1,2,\cdots,n \tag{15.24}$$

For this particular case, the slope of the regression line is

$$\hat{\beta}=\frac{\sum_i (X_i - X_P)(Y_i - Y_P)}{\sum_i (X_i - X_P)^2} \tag{15.25}$$

Example 15.7: Recent calibration data for the cold-water rotameter on a pilot-scale heat exchanger are listed below. What is the equation of the calibration line for this linear-float unit constrained to pass through (0,0)?

x = Reading (zero to 100% of full scale)	0	10	20	30	40	50	60	70	80	90	100
y = Flow rate (lb/sec)	0	2.0	4.0	6.6	8.3	10.3	12.3	15.7	17.7	21.0	23.0

$$\Sigma x_i y_i = 8588, \qquad \Sigma x_i = 550, \qquad \bar{x}=50$$

$$\Sigma y_i = 120.9, \qquad \Sigma y_i^2 = 1919.61, \qquad \bar{y}=10.990909$$

$$x_P = 0, y_P = 0, \qquad \Sigma x_i^2 = 38{,}500.$$

When the equation is constrained to pass through the origin, there is only one regression coefficient, the slope $\hat{\beta}=0.223064935$. Considering appropriate precision in the coefficient:

$$Y = 0.223X$$

Note: The slope is truncated to reveal approximate precision at $x = 100$ but exaggerates precision by $x = 10$.

The Excel Data Analysis routine Regression provides the same result when the box "Constant is Zero" is checked.

The corresponding standard least squares fit of these data yields $\hat{\beta}_0 = -0.56818182$ and $\hat{\beta}_1 = 0.23118182$. The difference, of course, is because the line was forced through the origin in the first case.

You should notice, again, that we have reversed the roles of scale reading and flow rate in this example. In rotameter calibration, scale reading is the dependent variable and flow rate is the independent variable. However, to develop a calibration equation, scale reading is used as the independent variable X and flow rate is used as the dependent response Y.

15.2.7 Measures of Goodness-of-Fit

After fitting model to data, *residuals* are the remaining vertical deviations between model and data. The sum of the squared residuals, $\sum_{i=1}^{n\,\text{data}} \left[\tilde{y}(x_i) - y_i\right]^2$, represents the unmodeled character of the data. This is usually termed Sum of Squared Deviations (SSD). If this were to be zero, the model would exactly pass through each data point. The variance of the residuals would be the SSD divided by the degrees of freedom. Since there are 2 adjustable coefficients in the simple linear regression, $\nu = n - 2$, and variance of the residuals would be SSD divided by $n - 2$. Here SSD could be subscripted with the number 2 to represent that it is based on a model with two coefficients, or subscripted with m to represent it is the residuals from a model with m number of coefficients, SSD_m.

SSD_m would be one measure of model goodness. Lower is better. However, the value depends on the number of data. Double the number of data, and you double the SSD_m. And it depends on the units you use. Change temperature units from °C to °F and you increase SSD_m by 1.8^2, by more than a factor of 3. By itself, SSD_m is not an exclusive measure of model goodness. So, let's scale it to be independent of both number of data and dimensional units.

There is an original variation in the y data which is due to the influence of the independent variable(s), and the magnitude of this can be evaluated as the sum of squared deviations, also $\sum_{i=1}^{n\,\text{data}} \left[\bar{y} - y_i\right]^2$. The variance in the original y-data would be that SSD divided by $\nu = n - 1$. Here, the "model" has one coefficient, the average, so the SSD could be subscripted with 1, SSD_1. However, it is usually subscripted with "O" for original data, SSD_o.

The objective of the model is to fit the natural phenomena, and not fit the noise or experimental error on the data, to remove all variation due to the independent variable. So, the amount of variation removed, scaled by the original variation, would represent the fraction of variation removed by the model. The ratio is termed regression r-squared.

$$r^2 = \frac{SSD_o - SSD_m}{SSD_o} = \frac{SSD_1 - SSD_m}{SSD_1} \tag{15.26}$$

The numerator and denominator have the same units, and the same number of terms in the sum, so regression r-square is a scaled ratio, independent of those aspects. Conveniently, r-square ranges between 0 (the model does not remove any variation) to 1 (the model perfectly removes all variation).

Although we seek to have values of r-square near unity, this does not necessarily mean the model is closer to the truth about nature. If you only have two data points, a simple linear regression will perfectly fit, and r-square will be unity. But there would be little confidence that nature is linear, and scant chance that replicate trials would return exactly the same data values, hence model.

15.3 Multiple Linear Regression

Multiple linear regression involves, as the name implies, more than one independent variable. The model is still linear with respect to all the coefficients β_j and is expressed as

$$Y = \beta_0 + \beta_1 X_1 + \beta_2 X_2 + \cdots + \beta_P X_P + \varepsilon = \beta_0 + \sum \beta_j X_j + \varepsilon \tag{15.27}$$

All the assumptions inherent in simple linear regression apply to multiple linear regression except that 1) we now have p number of independent variables, and 2) the X_j are not linearly correlated to each other.

The independent variables might truly be separate variables (like temperature, composition, age, size, etc.), or they could be nonlinear transforms of one or a combination of variables. In a power series, or polynomial model of only one independent variable, the Equation (15.27) variable labeled X_1 could represent x, and $X_2 = x^2$, $X_3 = x^3$, etc. The transformations do not have to be in a power series; they could be functionalities chosen by the experimenter that appear to be present in the data. For instance, $X_1 = x$, $X_2 = 1/x$, and $X_3 = \ln(x)$. Alternately, there could be several independent variables (e.g., w and z), and the X_i variables in Equation (15.27) could be chosen as $X_1 = w$, $X_2 = z$, $X_3 = wz$, and $X_4 = w/\sqrt{z}$, etc. Even though the models indicated in this paragraph are nonlinear in the dependent variable response to the independent variable, the models remain linear in the coefficients.

The use of powers or functional transformations could make the magnitudes of terms very different. For example if x ranges between 5 and 10, and you want to use powers of up to x^4 in your model, then x^4 will range between 625 and 10,000. If the x and then x^4 terms have similar impact on the y then the x^4 coefficient, $\hat{\beta}_4$, will have a value that is about 1,000 times smaller than the x coefficient, $\hat{\beta}_1$. The display might have four decimal digits presenting one as 0.2354, and the other as 0.0005 (when it should be 0.0004562). For convenience, and to minimize truncation errors in processing, and rounding misrepresentation of the results, scale the X_j values in Equation (15.31) to have similar numerical ranges.

To obtain the conventional least squares estimators, $\hat{\beta}_j$, we must minimize the error sum of squares with respect to each $\hat{\beta}_j$, set each of the resulting partial derivatives equal to 0, and solve these *normal equations* for the $\hat{\beta}_j$. For p variables with n observations on (Y_i, X_{1i}, X_{2i}, … , X_{pi}), (for $I = 1$ to n) the resulting equations are shown below:

$$\begin{aligned}
&\hat{\beta}_0 n + \hat{\beta}_1 \Sigma X_{1i} + \hat{\beta}_2 \Sigma X_{2i} + \cdots + \hat{\beta}_p \Sigma X_{pi} = \Sigma Y_i \\
&\hat{\beta}_0 \Sigma X_{1i} + \hat{\beta}_1 \Sigma X_{1i}^2 + \hat{\beta}_2 \Sigma X_{1i} X_{2i} + \cdots + \hat{\beta}_p \Sigma X_{1i} X_{pi} = \Sigma X_{1i} Y_i \\
&\hat{\beta}_0 \Sigma X_{2i} + \hat{\beta}_1 \Sigma X_{1i} X_{2i} + \hat{\beta}_2 \Sigma X_{2i}^2 + \cdots + \hat{\beta}_p \Sigma X_{2i} X_{pi} = \Sigma X_{2i} Y_i \\
&\cdots \\
&\hat{\beta}_0 \Sigma X_{pi} + \hat{\beta}_1 \Sigma X_{1i} X_{pi} + \hat{\beta}_2 \Sigma X_{2i} X_{pi} + \cdots + \hat{\beta}_p \Sigma X_{pi}^2 = \Sigma X_{pi} Y_i
\end{aligned} \tag{15.28a}$$

The summation is for $i = 1$ to n. For a three-coefficient model, the matrix notation is

$$\begin{vmatrix} n & \Sigma X_{1i} & \Sigma X_{2i} \\ \Sigma X_{1i} & \Sigma X_{1i}^2 & \Sigma X_{1i}X_{2i} \\ \Sigma X_{2i} & \Sigma X_{1i}X_{2i} & \Sigma X_{2i}^2 \end{vmatrix} \begin{vmatrix} \hat{\beta}_0 \\ \hat{\beta}_1 \\ \hat{\beta}_2 \end{vmatrix} = \begin{vmatrix} \Sigma Y_i \\ \Sigma X_{1i}Y_i \\ \Sigma X_{2i}Y_i \end{vmatrix} \tag{15.28b}$$

Of course, this can be extended to a p-coefficient model, and the solution by linear equation solvers for the unknown $\hat{\beta}_i$ values is a common data processing function.

As with the simple linear case, the sum of squares due to error is

$$\begin{aligned} SS_E &= \sum_i \left(Y_i - \tilde{Y}_i\right)^2 \\ &= \sum_i \left(Y_i - \hat{\beta}_0 - \hat{\beta}_1 X_{1,i} - \hat{\beta}_2 X_{2,i} - \cdots - \hat{\beta}_p X_{p,i}\right)^2 \end{aligned} \tag{15.29a}$$

which can be partitioned into

$$SS_E = SS_{TC} - SS_R = SS_T - SS_M - SS_R \tag{15.29b}$$

which you have already seen as Equation (15.8). The n degrees of freedom are similarly partitioned into p for regression, 1 for the mean of the Y_i, and $(n - p - 1)$ for error. The sum of squares due to regression contains a term for the contribution of each independent variable to the reduction in the sum of squares due to error:

$$SS_R = \hat{\beta}_1 \sum_i \left(X_{1i} - \bar{X}_1\right)\left(Y_i - \bar{Y}\right) + \cdots + \hat{\beta}_p \sum_i \left(X_{pi} - \bar{X}_p\right)\left(Y_i - \bar{Y}\right) \tag{15.30}$$

The most common null hypothesis is H_0: $\beta_1 = \beta_2 = \beta_3 = \ldots = \beta_p = 0$, or in other words that Y is not related to any of the Xs. The alternate hypothesis is that at least one of the $\beta_j \neq 0$. Note that the intercept β_0 is not included in these hypotheses. The null hypothesis can be evaluated by an F test. If $F = MS_R/MS_E > F_{p,(n-p-1),1-\alpha}$, the null hypothesis is rejected. The value of $MS_E = \sum_i \left(Y_i - \hat{Y}_i\right)^2 / (n-p-1)$ is an unbiased estimate of σ^2. The analysis of variance for multiple linear regression is shown in Table 15.3.

TABLE 15.3

Analysis of Variance for Multiple Linear Regression

Source	df	SS	MS
Mean	1	SS_M	—
Regression	P	SS_R	$MS_R = \dfrac{SS_R}{p}$
Error	$n - p - 1$	SS_E	$MS_E = \dfrac{SS_E}{n-p-1}$
Total	n	SS_T	

The mean square for regression, MS_R, can be partitioned and each of the $\hat{\beta}_j$ tested by a t-test to determine whether it is probably some particular value, i.e., zero. Many software packages provide the necessary standard errors to enable you to make such tests. The rejection region for such null hypotheses involving $\beta_j = \beta_j'$ is $T \geq t_{n-p-1,1-\alpha/2}$ or $T < -t_{n-p-1,1-\alpha/2}$, where T is calculated as in Equation (15.11).

Example 15.8: The data below were obtained from the performance test of a degassing tower for which X_1 = liquor rate in hundreds of gallons per hour, X_2 = air velocity in ft/sec, and Y = percent of dissolved CO removed.

x_1	x_2	y
14	1.00	46
15	1.25	51
16	3.00	69
17	3.25	74
18	4.00	80
19	5.25	82
20	5.50	97

Note: In this example, X_1 and X_2 appear to be highly correlated. The data are listed in order of $x1$ to make it easily visible. This event could be the result of poor experimental design or such process factors as allowable ratio of experimental flow rates, concentration of the inlet gas, operating range of the liquid analyzer, etc. Perform a correlation analysis, and you'll see that the correlation coefficient between x_1 and x_2 ($r = 0.982917$) is higher than the r-value between either x and y.

In spite of the correlation, calculate the regression equation for these variables and test its significance at the $\alpha = 0.05$ level.

The values below are first calculated from the raw data.

Σx_{1i}	= 119.0	Σx_{2i}	= 23.25	$\Sigma x_{1i} x_{2i}$	= 417.75
Σx_{1i}^2	= 2,051.0	Σx_{2i}^2	= 95.9375	$(\Sigma x_{1i})(\Sigma x_{2i})$	= 2,766.75
$(\Sigma x_{1i})^2$	= 14,161.0	$(\Sigma x_{2i})^2$	= 540.5625	Σy_i	= 499.0
$\Sigma x_{1i} y_i$	= 8,709.0	$\Sigma x_{2i} y_i$	= 1841.25	Σy_i^2	= 37,487.0
$\Sigma x_{1i} \Sigma y_i$	= 59,381.0	$\Sigma x_{2i} \Sigma y_i$	= 11,601.75	$(\Sigma y_i)^2$	= 249,001.0
$\overline{x}_1$	= 17.0	$\overline{x}_2$	= 3.321428	$\overline{y}$	= 71.2857

$$n = 7$$

The corresponding sums of squares of deviations are:

$$\sum_i (x_1 - \overline{x}_i)^2 = 28.0$$

$$\sum_i (x_{2i} - \overline{x}_2)^2 = 18.71429$$

$$\sum_i (x_{1i} - \bar{x}_1)(x_{2i} - \bar{x}_2) = 22.50$$

$$\sum_i (x_{1i} - \bar{x}_1)(y_i - \bar{y}) = 226.0$$

$$\sum_i (x_{2i} - \bar{x}_2)(y_i - \bar{y}) = 183.8571$$

$$\sum_i (y_i - \bar{y})^2 = 1915.429$$

The regression coefficients are found by writing equation set (15.28a) for $p = 2$:

$$7\hat{\beta}_0 + 119\hat{\beta}_1 + 23.25\hat{\beta}_2 = 499.0$$

$$119\hat{\beta}_0 + 2051\hat{\beta}_1 + 417.75\hat{\beta}_2 = 8709.0$$

$$23.25\,\hat{\beta}_0 + 417.75\,\hat{\beta}_1 + 95.9375\,\hat{\beta}_2 = 1841.25$$

The solution is $\hat{\beta}_0 = -29.2316$, $\hat{\beta}_1 = 5.21934$, and $\hat{\beta}_2 = 3.54925$.

The resulting regression equation is

$$\boxed{Y = -29.2316 + 5.21934X_1 + 3.54925X_2}$$

In this example, the sum of squares due to regression is

$$SS_R = \hat{\beta}_1 \Sigma(x_{1i} - \bar{x}_1)(y_i - \bar{y}) + \hat{\beta}_2 \sum_i (x_{2i} - \bar{x}_2)(y_i - \bar{y})$$

$$= 5.21934(226) + 3.54925(183.8571)$$

$$= 1832.125652$$

The sum of squares due to error is

$$SS_E = \sum_i (y_i - \hat{y})^2$$

$$= \sum_i (y_i - \bar{y})^2 - SS_R$$

$$= 1915.429 - 1832.125652$$

$$= 83.303348$$

For these data, $SS_T = 37{,}487$ and $SS_M = 35{,}571.57143$.

The hypothesis that Y is not related to either x_1 or x_2 is tested by $F = MS_R/MS_E = 916.062826/20.825837 = 43.9868$. As $F > F_{2,4,0.95} = 6.94$, and the p-value for such is 0.001891, the null hypothesis is rejected, and we conclude:

The model reveals a strong relation. Either β_1 or β_1 or both are probably not zero.

The resulting regression analysis table:

Summary Output

Regression statistics	
Multiple R	0.978014
R square	0.956511
Adjusted R square	0.934767
Standard error	4.563436
Observations	7

ANOVA

	df	*SS*	*MS*	*F*	Significance F
Regression	2	1,832.129	916.0644	43.98879	0.001891
Residual	4	83.2998	20.82495		
Total	6	1915.429			

	Coefficients	Standard error	*t* Stat	*P*-value	Lower 95%	Upper 95%
Intercept	−29.2314	61.07108	−0.47865	0.657179	−198.792	140.3291
x_1	5.219316	4.685753	1.113869	0.327751	−7.79042	18.22905
x_2	3.549296	5.731546	0.619256	0.569276	−12.364	19.46262

Although the *t*-statistic and *p*-value for the test for each coefficient indicates we cannot reject that hypothesis; each of the coefficients could be zero, yet the model is strong. There is no confidence in the model coefficient values.

Since there is very strong correlation between x_1 and x_2, perhaps only x_1 is needed. Removing x_2 from the correlation:

Summary Output

Regression Statistics	
Multiple R	0.97588
R Square	0.952342
Adjusted R square	0.94281
Standard error	4.272838
Observations	7

ANOVA

	df	*SS*	*MS*	*F*	Significance F
Regression	1	1,824.143	1,824.143	99.91393	0.000171
Residual	5	91.28571	18.25714		
Total	6	1915.429			

	Coefficients	Standard error	t Stat	P-value	Lower 95%	Upper 95%
Intercept	−65.9286	13.82201	−4.76983	0.005016	−101.459	−30.398
x_1	8.071429	0.80749	9.995695	0.000171	5.995708	10.14715

Now we see that with fewer DoF, the model is a bit stronger (F is higher and its p-value is smaller) and that both model coefficients, β_0 and β_1, are significantly non-zero (p-values are 0.005016 and 0.000171).

The result is the same if you regress Y to X_2 only. And with either, model the other input variable shows no relation to the residuals. These are end-of-chapter exercises.

This data indicates there is no need to use both x_1 and x_2 in the model.

If you believe that both x_1 and x_2 are independent and should be in the model, then redesign the experiment to eliminate, or at least reduce the correlation.

15.4 Polynomial Regression

For polynomial regression, also called curvilinear or power series regression, the model is

$$Y = \beta_0 + \beta_1 X + \beta_2 X^2 + \cdots + \beta_P X^P + \varepsilon \tag{15.31}$$

The basis of validity of this model comes from a Taylor series approximation to a function. The Taylor series approximation has an infinite number of terms, but it is known to be able to have a specified goodness-of-fit with a finite number of terms. Expanding the polynomials of the truncated Taylor series, and combining coefficients of terms of like powers, results in Equation (15.31). The user must choose the number of terms, p.

Although this model involves only integer powers of one independent variable, other functional forms can be included if defined in terms of functions of X. Such an example might be $W_1 = X$, $W_2 = X^2$, $W_3 = \log(X),\ldots$, $W_p = \sin(X)$. With such a transformation, the multiple linear regression model of Equation (15.27) is obtained with independent variables $W_1, W_2, \ldots, W_p$. In order to estimate the β_j of the original polynomial model, solve the corresponding normal equations as discussed in Section 15.3.

As an example, consider the quadratic model, $Y = \beta_0 + \beta_1 X + \beta_2 X^2 + \varepsilon$, for which $W_1 = X$ and $W_2 = X^2$. The necessary normal equations are, for the n observations on X and Y,

$$\begin{aligned}
\hat{\beta}_0 n + \hat{\beta}_1 \sum_i W_{1i} + \hat{\beta}_2 \sum_i W_{2i} &= \sum Y_i \\
\hat{\beta}_0 \sum_i W_{1i} + \hat{\beta}_1 \sum_i W_{1i}^2 + \hat{\beta}_2 \sum_i W_{1i} W_{2i} &= \sum_i W_{1i} Y_i \\
\hat{\beta}_0 \sum_i W_{2i} + \hat{\beta}_1 \sum_i W_{1i} W_{2i} + \hat{\beta}_2 \sum_i W_{2i}^2 &= \sum_i W_{2i} Y_i
\end{aligned} \tag{15.32}$$

where $W_{1i} = X_i$ and $W_{2i} = X_i^2$. With these transformations, the equations become

$$\begin{aligned}
\hat{\beta}_0 n + \hat{\beta}_1 \sum_i X_i + \hat{\beta}_2 \sum_i X_i^2 &= \sum_i Y_i \\
\hat{\beta}_0 \sum_i X_i + \hat{\beta}_1 \sum_i X_i^2 + \hat{\beta}_2 \sum_i X_i^3 &= \sum_i X_i Y_i \\
\hat{\beta}_0 \sum_i X_i^2 + \hat{\beta}_1 \sum_i X_i^3 + \hat{\beta}_2 \sum_i X_i^4 &= \sum_i X_i^2 Y_i
\end{aligned} \tag{15.33}$$

Equation set (15.32) can be solved directly for $\hat{\beta}_0$, $\hat{\beta}_1$, and $\hat{\beta}_2$. Extensions to polynomials of higher degree are straightforward.

Other models that are linear in the model coefficients are

$$\text{(A)} \quad Y = \beta_0 + \beta_1 \log X + \beta_2 X^2 + \varepsilon$$

$$\text{(B)} \quad Y = \beta_0 + \beta_1 e^{-X} + \beta_2 X^{1/2} + \varepsilon$$

$$\text{(C)} \quad Y = \beta_0 + \beta_1 e^{-X1} + \beta_2 X_2^2 + \beta_3 X_3 + \varepsilon$$

The log X term in model A can be equated to W_1 and the X^2 term equated to W_2. Model B may be reduced to a multiple linear form by letting $W_1 = e^{-x}$ and $W_2 = X^{1/2}$. Model C can also be converted to a multiple linear regression form where $W_1 = e^{-X1}$, $W_2 = X_2^2$, and $W_3 = X_3$.

15.4.1 Determining Model Complexity

In polynomial regression, the user must choose the order of the model, the p-number of variables in Equation (15.31). Two few terms, not enough flexibility can lead to a model that does not fit the overall trend well. At the other extreme with too many terms the model will attempt to fit the noise in the data and have absurd shapes. Figure 15.2 provides an example.

The figure was part of an enrollment study of incoming Chemical Engineering (ChE) students (freshmen and transfers). The vertical axis represents the number of new students declaring ChE as their major at Oklahoma State University. The horizontal axis represents the number of chemical engineering students graduating nationally. The suspicion for the enrollment trend was that low graduation rates, nationally, create a seemingly high industrial demand for ChEs. News of this appearance of an immense demand takes time to be

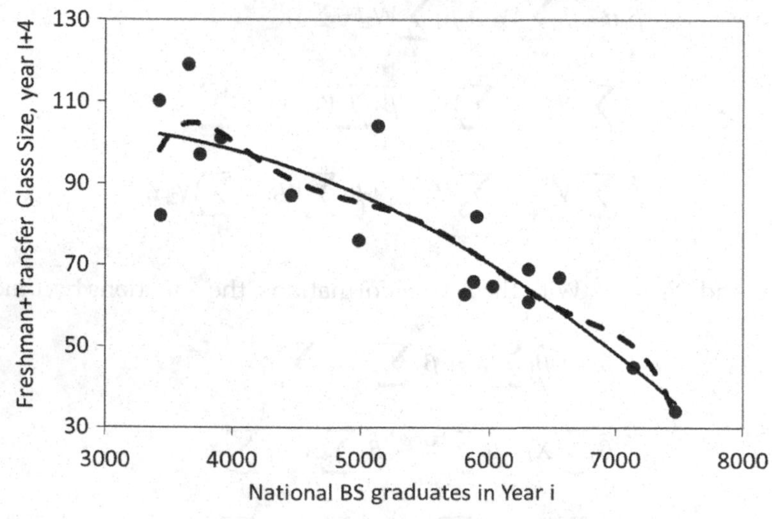

FIGURE 15.2
Matriculation study.

processed and affects matriculation choices about 4 years later. The "lag-4" description means that the graph shows matriculation rates 4 years after the national graduation rate. On the other side, when we graduate 7,500 nationally, with about 6,000 job openings, some graduates cannot get employment. When the news "ChEs are not employable" influences matriculation decisions about 4 years later, students choose another major.

Not knowing what the true functionality is, the data are modeled with a power series regression. The solid line is a three-term quadratic model. The dashed line is a seven-term sixth-order model. The quadratic model seems to reveal the general trend of the data. By contrast the sixth-order model has a lower SS_E, which might suggest that it is a better model. However, the sixth-order seems to suggest conceptually unreasonable trends: on the left side, as national graduation increases, the enrollment rises very strongly. And it suggests that some mysterious alternate mechanism causes an inflection in matriculation decisions in the middle. There are 17 data points in the graph. A 17th-order model could perfectly go through each data point. Imagine what that would look like!

The quadratic model seems better for several reasons. 1) It is simpler than the seven-term model. 2) The lack of inflections indicates it represents one mechanism, not several. Nature seems to have simple behavior (Occam's Razor).

There are guides to automate the choice of the number of functionalities that you might choose for a model. These include Ljung's Final Prediction Error (FPE), Akaike's Information Criterion, Mallow's Cp, and Rhinehart's Equal Concern balance. Generally, they balance goodness-of-fit with model complexity and provide a useful guide.

The simpler of these several approaches is Ljung's FPE, a product of one goodness-of-fit metric, SSD_m, which reduces with model complexity, and a factor that increases with model complexity.

$$FPE_p = \frac{n+p+1}{n-p-1} SSD_{p+1} \tag{15.34}$$

Here $p+1$ is the number of model coefficients (p powers plus an intercept). The complexity factor is effectively quadratic in the number of model coefficients relative to the number of datasets. Starting with low-order models relative to the number of data, an increase in model order will cause a large drop in SSD but only a modest increase in the complexity factor. After the model fits the general trend in the data, increasing model order has a lower and lower improvement in SSD, asymptotically approaching zero, but the complexity factor will markedly rise when p approaches n. In the FPE method, start with a low-order model, perform the regression, and calculate FPE. Increase p by one, and repeat. Initially FPE will drop. Continue increasing p by one, until FPE rises. The method directs that the right p is the one before FPE rises.

Often your modeling routines may return the r-squared value, not the SSD_{p+1}. An equivalent FPE can be used by substituting SSD_m from Equation (15.26) into (15.34).

$$FPE'_p = \frac{FPE_p}{SSD_o} = \frac{n+p+1}{n-p-1}\left(1-r_{p+1}^2\right) \tag{15.35}$$

These methods may not exactly match the "right" balance for every application context. In the data of Figure 15.2, FPE directs the use of a quadratic model, matching our sensibility. But in cases where there are a lot of data, these methods would allow more complex models.

We suggest that you progressively increase the order of your polynomial model, observe the model, and use your judgement to select the right order for your data. This is termed *forward addition* meaning that progressive terms are added one at a time.

15.4.2 Culling Irrelevant Model Functionalities

Consider that nature is generating data with a simple quadratic relation $y = x^2$, with a zero intercept and no linear relation. Since you don't know what the truth of the model is you first use a linear model, and you find that the intercept and slope are both statistically different from zero.

Then you add another term to obtain a three-term $p = 2$ model, such as Equation (15.1). Since the quadratic term alone is all that is needed to match the data, the coefficients on the intercept and linear term in the quadratic model will ideally be zero. However, due to experimental vagaries, they will not be exactly zero. However, if you perform a t-test using the coefficient value and associated standard error, you may find that you cannot reject the hypothesis that the coefficient value is zero. In this case, adding the right functionality makes the intercept and linear terms irrelevant. As you add functionalities to your model, remove the terms that no longer have statistically significant coefficient values, and redo the regression with the remaining terms. This is termed *backward elimination*.

15.4.3 Extrapolation of Polynomial Models

Polynomial models, such as Equation (15.31) probably do not contain exactly the functionality that nature is using to generate the data. For example, in reaction kinetics a first-principles representation of how reaction rate $A + B \rightarrow C$ depends on temperature is the Arrhenius model $r = k_0 e^{-E/RT}[A][B]$. This model does not have the power series functionalities of $r = a + bT + cT^2 + dT^3 \ldots$ but you might be using a power series to best fit the data. However, that is not what nature is using. Although the power series model is a wrong model, you might find that it provides a reasonably good fit within the range of temperatures of your experiment.

As a caution, just because your wrong model appears to give a good fit, that does not mean it is the right model. It may be useful for matching the data, and interpolating, but it may be very poor for extrapolating. To see this, extrapolate the quadratic model in Figure 15.2 to either extreme of zero or 10,000. Does it make sense?

Do not think that your power series model is the truth about nature just because it gives a reasonable fit to a limited range of data. Be very careful in extrapolating to conditions outside of your experimental range.

15.5 Functional Linearization of Models with Nonlinear Coefficients

For regression, from the viewpoint of coefficient functionality, a nonlinear model is any one in which the derivative of Y w.r.t. a model coefficient is not independent of that or other model coefficients. Three common nonlinear models are the exponential, power, and general exponential laws, e.g.,

$$Y = \alpha_0 e^{\alpha_1 X} + \varepsilon' \tag{15.36}$$

$$Y = \alpha_0 X^{\alpha_1} + \varepsilon' \tag{15.37}$$

$$Y = \alpha_0 \alpha_1^X + \varepsilon' \tag{15.38}$$

Unfortunately, the normal equation equivalents to solve for the model parameter values which minimize the sum of squares of deviations are also nonlinear. Consequently, there may be no unique solution. Optimizers that iteratively approach the optimum are usually needed to solve for the model coefficient values that best fit the data. This is in contrast to solving the unique, deterministic, and usually simpler linear algebra equation set when coefficients appear linearly in the model. So, it is common to seek and use functional transformations of models that are linear in the coefficients.

In some cases, a nonlinear model may be linearized by using a functional transformation. In the three cases above, the logarithm will work. In other cases, algebraic rearrangement, reciprocals, an inverse sine, or inverse tangent transformation is applicable.

For example, Equation (15.36) can be linearized by the following six-step procedure:

1. Subtract ε' from both sides.

$$Y - \varepsilon' = \alpha_0 e^{\alpha_1 X} \tag{15.39a}$$

2. Take the natural logarithm of the result.

$$\ln(Y - \varepsilon') = \ln \alpha_0 + \alpha_1 X \tag{15.39b}$$

3. Factor out Y from the left-hand side.

$$\ln\left[Y\left(1 - \frac{\varepsilon'}{Y}\right)\right] = \ln \alpha_0 + \alpha_1 X \tag{15.39c}$$

$$\ln Y + \ln\left(1 - \frac{\varepsilon'}{Y}\right) = \ln \ln \alpha_0 + \alpha_1 X \tag{15.39d}$$

4. Realize that ε' should be much less than Y to make sense out of the data. Therefore $\varepsilon'/Y \ll 1$, and using a Taylor series expansion for the $\ln(1 - \varepsilon'/Y)$ term about 1, we have

$$\ln\left(1 - \frac{\varepsilon'}{Y}\right) = \ln(1) + \frac{1}{1!}\left(-\frac{\varepsilon'}{Y}\right) - \frac{1}{2!}\frac{1}{1^2}\left(-\frac{\varepsilon'}{Y}\right)^2 + \cdots$$

$$= 0 - -\frac{\varepsilon'}{Y} + \text{terms on the order of } \left(\frac{\varepsilon'}{Y}\right)^2 \text{ or smaller} \tag{15.39e}$$

$$\ln\left(1 - \frac{\varepsilon'}{Y}\right) \cong -\frac{\varepsilon'}{Y}$$

5. Substitute Equation (15.39e) into (15.39d) and rearrange.

$$\ln Y = \ln \alpha_0 + \alpha_0 X + \frac{\varepsilon'}{Y} \tag{15.39f}$$

6. Substitute Z for ln Y, β_0 for ln α_0, β_1 for α_1 and ε for ε'/Y into Equation (15.39f).

$$Z = \beta_0 + \beta_1 X + \varepsilon \tag{15.39g}$$

Equation (15.39g) is now a linear functional transformation version of Equation (15.36), and estimates of the values of parameters β_0 and β_1 can be determined by the standard linear regression techniques presented earlier in this chapter. Many nonlinear models can be linearized by a similar procedure.

Note: This is not a local linearization of the function by using a truncated Taylor Series approximation. This functional transformation is exact and global, not a local linearization.

However, you must be aware of two cautions. First, Step 4 assumes that $\varepsilon'/Y \ll 1$ so that terms of order 2 and higher can be neglected in the Taylor series expansion. If experimental error is large relative to Y, Equation (15.39g) may not be valid.

Second, even if ε' represents errors of uniform size throughout the range of Y, ε, the error of the transformed equation, is not uniform. At small values of Y, ε will be larger than at large values of Y. If Y covers a wide range (small to large) then the linear regression assumption that ε is uniform will not be valid.

On the other hand, a large number of data points can have local errors compensate locally and still return a good model. Further, placing more data in the Y-region that would cause errors to be magnified will provide a similar balance.

The transformed errors may behave ideally, but you can't be sure of that. A good way to check the results of a linearizing functional transformation is to perform the regression, test all parameters in the transform domain, reconvert to the original domain, use the parameter estimates to predict the values of the dependent variable, and compare the original and predicted values. Hopefully, the variance in the residuals appears uniform through the x and y range, and the distribution appears normal.

The assumptions of normality and homogeneity of the error distribution may not be justified for the original data, but those properties may be more closely approached as a result of a transformation. A case in point is the log-normal distribution used for particle size analysis in air pollution studies.

In addition to the exponential model of Equation (15.36), other nonlinear models that can be linearized are shown below with their corresponding transforms.

Power or log-log:

$$Y_i = \alpha X_i^{\beta} + \varepsilon' \tag{15.40}$$

$$Z_i = \ln Y_i = \ln \alpha + \beta \ln X_i + \varepsilon \tag{15.41}$$

General exponential:

$$Y_i = \alpha \beta^{X_i} + \varepsilon' \tag{15.42}$$

$$Z_i = \ln Y_i = \ln \alpha + (\ln \beta) X_i + \varepsilon \tag{15.43}$$

Multiplicative:

$$Y_i = \alpha X_{1i}^{\beta} X_{2i}^{\gamma} X_{3i}^{\delta} + \varepsilon' \tag{15.44}$$

$$Z_i = \ln Y_i = \ln\alpha + \beta \ln X_{1i} + \gamma \ln X_{2i} + \delta \ln X_{3i} + \varepsilon \tag{15.45}$$

Exponential or semi-log:

$$Y_i = \exp(\alpha + \beta X_{1i} + \gamma X_{2i}) + \varepsilon' \tag{15.46}$$

$$Z_i = \ln Y_i = \alpha + \beta X_{1i} + \gamma X_{2i} + \varepsilon' \tag{15.47}$$

Reciprocal:

$$Y = \frac{1}{\alpha + \beta X_{1i} + \gamma X_{2i}} + \varepsilon' \tag{15.48}$$

$$Z_i = \frac{1}{Y_i} = \alpha + \beta X_{1i} + \gamma X_{2i} + \varepsilon \tag{15.49}$$

Mixed:

$$Y_i = \frac{1}{1 + \exp(\alpha + \beta X_{1i} + \gamma X_{2i})} + \varepsilon' \tag{15.50}$$

$$Z_i = \ln\left(\frac{1}{Y_i} - 1\right) = \alpha + \beta X_{1i} + \gamma X_{2i} + \varepsilon \tag{15.51}$$

15.6 Takeaway

Linear regression means that the coefficients in the model appear linearly. It does not necessarily mean that the dependent variable is linearly related to the independent variables.

Just because you have the coefficient values that make your model best fit the data, does not mean your model represents the truth that generated the data. Consider, you can use a quadratic model to best fit a range of data from an exponential relationship, but that does not mean the relationship is truly quadratic.

Commonly, regression is accepted as the best fit by minimizing the vertical least squares as the objective. Since models with linear coefficients and the vertical least squares criterion permit extensive analysis, these have become the standard tools, and many seek to find functional transformations that permit such tools.

Functional transformations, however, could lead to distorted results when un-transformed. If you do regression on transformed variables, check the results on the original variables.

Take caution with extrapolating empirical models beyond the data range used to get the model coefficients. If the model is not mechanistically true, it may not extrapolate well.

When choosing model complexity, be sure that the model does not reveal untenable trends.

15.7 Exercises

1. Derive the functional transformations that linearize Equations (15.40), (15.42), (15.44), (15.46), (15.48), and (15.50). Confirm the answers in the equations that follow each of those.

2. The classic Arrhenius chemical reaction kinetic relation for $A+B \rightarrow C+D$ is $r = k_0 e^{-\frac{E}{RT}}[A][B]$. Where r is the reaction rate, the dependent variable. T is the reaction temperature, and $[A]$ and $[B]$ are the concentrations of the reactants, the independent variables. K_0 and E are model coefficients to be determined by regression. Use functional transformations to get a model that is linear in coefficients.
3. The irritant factor Y of polluted air can be determined as a function of the concentrations of SO_2 and NO_2 in the atmosphere. The following data are available where X_1 = parts NO_2 per 10 million parts of air and X_2 = parts SO_2 per 100 million parts of air. Determine the irritant factor as a function of X_1 and X_2.

y	x_1	x_2
65	10	12.5
72	12	15
82	15	18
95	16	22
110	19	26
122	21	29
125	25	33
130	29	37

4. Accelerated life tests were conducted on three different proprietary materials as a function of time. The results are tabulated below and represent percent degradation (Y) and time (t) in hours. As we cannot distinguish subsequent degradation of an already damaged location, the process appears to be first order, i.e., only degradation on previously undamaged sites is observed upon repeated examination of the specimens as time progresses. The physical phenomenon leads to the selection of a negative exponential model:

$$Y_i = \beta_0\left(1-e^{-\beta_1 t_i}\right)+\varepsilon$$

Determine values for β_0 and β_1 for each of the three materials.

Time (hrs)	Material 1	Material 2	Material 3
0	0.000	0.000	0.000
8	0.029	0.017	0.018
16	0.046	0.026	0.032
24	0.060	0.034	0.042
32	0.069	0.041	0.053
40	0.076	0.047	0.062
48	0.082	0.056	0.068
56	0.088	0.063	0.075
64	0.096	0.068	0.083
72	0.100	0.074	0.089
80	0.104	0.078	0.092

This model could be approximately linearized by the natural logarithm transformation (often a more time-efficient approach). A more exact approach requires the use of nonlinear regression, which does not result in a skewed error distribution.

5. The overall heat transfer coefficient U, Btu/hr ft^2 °F, shows a log-log dependence on v_{max}, ft/sec, in heat exchangers with parallel banks of finned tubes. Typical data are below. The U values were replicated, two or three measurements for flow rates. For these data, determine the relation between U and v_{max}.

U	v_{max}
8.4, 7.9	300
9.7, 9.1	500
9.7, 9.9	600
10.4, 10.1, 10.1	800
11.3, 11.6, 11.4	1000
12.9, 14.0, 13.8	1200
14.8, 16.0, 15.1	1500

Also show that any constants you obtain are probably nonzero.

6. A rotameter with a spherical stainless steel float was calibrated for helium service at 20 psig inlet pressure at 74°F. The true flow rate values were obtained by a soap-film bubble meter made from a 100 ml buret. The data are below. As we want a calibration curve, let Y = flow rate and X = scale reading. Because of slight irregularities in the position of the float, it was often necessary to readjust the inlet valve, wait for the float to stabilize at the desired scale reading, and remeasure the flow rate. Some flow rate values were from replicate readings. As the plot of flow rate versus scale reading appeared quadratic in shape, that model was selected for the regression analysis. Was that model a proper choice? If so, what is the mathematical relation between flow rate in ml/min and scale reading in mm?

Scale reading (mm)	Flow rate (ml He/min)
10	9.2, 9.5
16	15.0
20	21.3
25	29.4
30	41.0, 40.5
35	54.0, 55.0
40	68.8, 68.0
45	86.0, 88.1
50	103.2, 104.6
55	124.0, 123.1
60	144.0, 145.3, 145.1

7. Use the Excel Data Analysis Regression program to generate a quadratic model for the data of Example 15.7.

Create a column for the squared values, then include both the x and x^2 columns as the input variables.

y	x	x^2
0	0	0
2	10	100
4	20	400
6.6	30	900
8.3	40	1600
10.3	50	2500
12.3	60	3600
15.7	70	4900
17.7	80	6400
21	90	8100
23	100	10000

Which model terms are significantly different from zero? Remove that (them) and redo the regression. Note that the squared input variable can have a value that is 100 times larger than the x input, and comment on the value of the regression coefficients. Rescale the x^2 column (divide by 100) so that it has a similar range of values as the x column. Re-do the regression model and comment on the displayed coefficient values.

8. Derive Equations (15.6) and (15.7) from (15.5).
9. Derive Equation (15.25) from (15.24).

16

Nonlinear Regression – An Introduction

16.1 Introduction

Regression is a procedure for adjusting coefficient values in a mathematical model to have the model best fit the data. In nonlinear regression the model coefficients are not linear in the model, and there are many categories for nonlinear models, which prevent linear methods (Chapter 15) from being applied. This chapter is an overview of the issues and a redirection to sources of techniques for nonlinear regression.

Dynamic Models: The model and data can represent either steady-state (alternately called static or equilibrium) or a transient (time-dependent) response. This contrasts with the methods of Chapter 15, which was for steady-state models only. Nonlinear dynamic models would probably be solved numerically, in incremental time steps. This time discretization creates striations on the model response surface, and associated discontinuities in the optimization objective (the response to adjustable coefficients), which can confound model coefficient adjustment.

Measures of Goodness-of-Fit: Further, the variables (both model inputs and outputs) and model coefficients do not have to be continuum-valued as presumed in Chapter 15. The variables could be limited to integer or other discretized values, such as ranks, or even have category (nominal, string, text, classification) values.

If the model outputs are categories, then the best fit objective might not be to minimize a least squares measure, but to maximize the number of right classifications with a penalty for very wrong classifications. For instance, optical character recognition (OCR) uses optical patterns and models of the visual patterns to interpret letters. The input variables might be which pixels are dark and light, and the modeled outputs might be the letters "A", "a", "B", "b", etc. If the model claims a "B" is an "8" or a "O" is an "0" the output would be marginally wrong, but if it thinks an "M" is an "e" that could bring a stronger error penalty.

The model output might be integer variables, such as rank. The output values might be ranks 1, 2, 3, 4, etc. If the data is a rank 2 and the model predicts 2, that is perfect. If the model predicts a rank 3, then it is marginally wrong, and if a rank 6, that is very wrong. Nominally, the measure of the model error might be the absolute value of the error but weighting large errors as very wrong the objective might be the square of the error. However, getting a rank 1 wrong is very important, but getting a rank 8 wrong may have little consequence. So, the error squared objective might be scaled by the expected rank: $J = \Sigma \frac{(R_{\text{data}} - R_{\text{model}})^2}{R_{\text{data}}}$. (This is minimizing a chi-squared statistic, not minimizing the vertical sum of squared deviations.)

DOI: 10.1201/9781003222330-16

There are many variants, but the common feature is that the objective function associated with classifications or ranks is discontinuous, not differentiable.

The criterion for regression quality with continuum-valued variables could be the classical vertical sum of squared deviations (SSD) between model and data responses, or it could be any number of other criteria such as maximum likelihood, normal deviation (Akaho's approximation, or total squared deviation), etc.

Constraints: Nonlinear models typically have several types of constraints. If a coefficient is within a logarithm or square root functionality, then the coefficient value is constrained to keep the function argument positive, and the particular value will change with other coefficients or data values. There are many types of constraints, on the several types of variables.

Conditionals: Whether models are linear or nonlinear if there are IF-THEN rules (conditionals) in the calculation, they create sharp ridges (or valleys) or even discontinuities (cliffs) in the surface (model outcome w.r.t. coefficients), which can confound optimizers.

Procedure to Adjust Coefficients: The objective in nonlinear regression continues to be that of maximizing measures of goodness-of-fit by adjusting model coefficient values. This is an optimization statement. In linear regression, the optimization procedure was an analytical method: Take the derivative of the goodness-of-fit criteria w.r.t. model coefficients, set the equations to zero, and solve the N linear equations for the N coefficient values. In contrast, in nonlinear regression, it may be impossible to take the analytical derivative of the function, or even if that is possible, it may be impossible to solve the resulting nonlinear equations. Necessarily, optimization is an iterative numerical search procedure. Many search techniques are commonly used (Levenberg–Marquardt, back propagation, particle swarm, GRG, Hooke–Jeeves, etc.), and depending on your software environment you may have some of these, or alternate optimization procedures. In our experience leapfrogging (LF) is an excellent generic optimization algorithm when considering generality, robustness to constraints and striations, probability of finding the global optimum, code simplicity, and the computational burden.

These techniques are much like what you would do with a trial-and-error manual heuristic search: Adjust a model coefficient a bit. If it improves the model fit to the data, adjust it a bit more in the same direction. If it makes the fit worse, then adjust it in the other direction, perhaps by a smaller amount. If any adjustment of that model coefficient makes the fit worse, then begin adjusting another coefficient, etc. These are iterative techniques. Make a move, see the results, make another move, reevaluate ….

Convergence Criteria: Iterative techniques require a criterion for defining convergence – to identify when the optimizer is close enough to a solution, to determine when it can stop. Often this criterion is based on the incremental changes of the model coefficient values (the decision variables, DV), requiring a user to specify a tolerance (or precision) threshold on the DV-value increments. However, this requires the user to forecast what is a meaningful increment, and often cannot be truly done until after the model is identified, permitting analysis of model sensitivity to coefficient value and model uncertainty relative to data variation. Too small a threshold, and the optimizer takes excessive iterations. Too large, and the model is not good enough. The user choice can have a significant impact.

Recommended here, the convergence criterion is based on the relative improvement of the model w.r.t. data variation. In this procedure: After each iteration, a random subset of about 30 to 50% of the data is selected. The objective function, the goodness-of-fit metric, such as the sum of squared deviations of model to data, is calculated from the randomly selected data subset, and plotted w.r.t. iteration. This is a signal that relaxes from an initial high value to a noisy steady-state when the model is best fitting the data. When the signal is perceived to be at steady-state, when the optimizer is not making any detectable model improvement, relative to data-to-model residuals, then convergence should be claimed.

Global and Local Minima: Nonlinear regression is not guaranteed to have a single global optimum. Any optimizer might get stuck in local minima. Accordingly, one should start an optimizer N times from randomized initializations and take the best of N solutions as the reportable solution. N can be calculated from a user's desire for a confidence, c, that the best of N trials will have found one of the solutions within a best fraction, f, of all possible solutions. $N = \ln(1 - c)/\ln(1 - f)$.

Uncertainty on Model and Coefficient Values: In regression, where models are linear in coefficients, Chapter 15, we have analytical approaches to calculate the standard error of both model coefficients and the model prediction. There is not an analytical procedure for models that are nonlinear. However, a numerical technique, Bootstrapping (see Section 19.3), can reveal model uncertainty.

16.2 Takeaway

Nonlinear regression means that the coefficients in the model do not appear linearly in the regression context, and/or that other nonlinearities cause difficulty in adjusting model coefficients. The scope is too large to be included in this book on traditional statistical approaches. One author has written a book which may be helpful (Rhinehart, R. R., *Nonlinear Regression Modeling for Engineering Applications*: Modeling, Model Validation, and Enabling Design of Experiments, Wiley, New York, NY, 2016. ISBN 9781118597965. 361 pages with companion web site www.r3eda.com). Of course, there are other books with similar titles, and also you'll find much information about all of the issues and procedures on the internet.

16.3 Exercises

1. Accelerated life tests were conducted on three different proprietary materials as a function of time. The results are tabulated below and represent percent degradation (Y) and time (t) in hours. As we cannot distinguish subsequent degradation of an already damaged location, the process appears to be first order, i.e., only degradation on previously undamaged sites is observed upon repeated examination of the specimens as time progresses. The physical phenomenon leads to the selection of a negative exponential model:

$$Y_i = \beta_0 \left(1 - e^{-\beta_1 t_i}\right) + \varepsilon$$

Determine values for β_0 and β_1.

Time (hrs)	Material 1	Material 2	Material 3
0	0.000	0.000	0.000
8	0.029	0.017	0.018
16	0.046	0.026	0.032
24	0.060	0.034	0.042
32	0.069	0.041	0.053
40	0.076	0.047	0.062
48	0.082	0.056	0.068
56	0.088	0.063	0.075
64	0.096	0.068	0.083
72	0.100	0.074	0.089
80	0.104	0.078	0.092

This model could be approximately linearized by the natural logarithm transformation (often a more time-efficient approach outlined in Section 15.5). A more exact approach requires the use of nonlinear regression, which does not result in a skewed error distribution. You might use Excel's Solver, but you'll find that the results are dependent on the initialization values and convergence criterion. Alternately, you might just use a human heuristic search (trial-and-error search) for the β_0 and β_1 values.

17

Experimental Replicate Planning and Testing

17.1 Introduction

This chapter is about the number of trials you should take to test treatments to be able to make a use or not-use decision. For example, you might currently be using a raw material or an ingredient from Supplier A, or perhaps a Procedure A that has been developed and works well, or perhaps a software or hardware Provider A, etc. Then, treatment B is proposed, and the question is, "Should we change from A to B?"

Sometimes the change from A to B is required, such as in the case that Supplier A goes out of business, or the daughter of your company's owner marries into the B option owner's family. But, many times the change is optional and test trials are used to assess the benefit of changing to Treatment B.

There are costs associated with the change. First will be the costs of the trials to qualify B; then subsequently, the costs associated with managing the change (training people, revising documents) and coping with any unforeseen risks of the change. We'll label the trial cost, *TC*, to represent the cost of one trial. It includes the purchasing, monitoring, evaluating, interference with production, and time aspects. The cost of n number of trials is n^*TC. Subsequently, the Management of Change, *MoC*, costs include all costs associated with the change, for instance operator training, inventory changeover, documentation, and the risk associated with unknown outcomes. These will sum to the one-time costs associated with a switch from A to B, totaling $n * TC + MoC$, and may have the value of $k. If the switch from A to B is justified, then the benefit of B over A needs to justify the costs of the change. The benefit associated with B over A might be improved product yield, increased production rate, decreased utility use, longer life, or the like. Often people evaluate benefit with an economic index termed Pay Back Time, *PBT*, which is the time for the benefit of the switch to repay the cost of change. Using B_k as the benefit of the kth choice, evaluated in $k/year, then *PBT* is calculated as

$$PBT = \frac{n * TC + MoC}{B_B - B_A} = \frac{n * TC + MoC}{\Delta} \tag{17.1}$$

Of course, any of several economic indices could be used. Typically, a company specifies a threshold value for the economic index, such as *PBT*, and only invests in the option if the calculated *PBT* is less than the threshold. Rearranging Equation (17.1), the required difference in benefit needs to be

$$\Delta_{\text{threshold}} = B_{B,\,\text{threshold}} - B_A \geq \frac{n * TC + MoC}{PBT_{\text{threshold}}} \tag{17.2}$$

DOI: 10.1201/9781003222330-17

But prior to the trials, one cannot know either the required number of replicate trials to make a confident decision or the actual improvement in benefit of B over A. Further, although *TC* and *MoC* values could be reasonably estimated before the trials and actual changeover, their true values are uncertain until after the event.

In considering a change from A to B, one needs to estimate values for n, *TC*, *MoC*, and $B_B - B_A$ to even postulate that the change might be justified. Your situation/organization may have different considerations of how to categorize the costs or may prefer an alternate profitability index to assess the change. Regardless, there will be an estimate of the unknown values, and a postulate of whether the change should be accepted or rejected.

If the trial outcome could be deterministic, without variability, then $n = 1$, one trial is all that would be needed to determine the value of $B_B - B_A$; and from that, whether the *PBT* exceeded the threshold. However, if the trials and their analysis have variation, then more than one trial will be needed to determine the variance and the possible mean of B_B. For example, the μ_{B_B} value might be \$100k/yr, exceeding \$95k/yr, the \$80k/yr benefit of Treatment A plus the threshold of \$15k/yr associated with the costs to change. However, the variation on trial outcomes may be relatively large, such as σ_B of \$10k/yr, in which case a single trial outcome might result in a measurement $X_B = \mu_B - 1.96\sigma_B$ which has a value of \$100k/yr–1.96 × \$10k/yr = \$80.4k/yr, which would indicate a decision to reject the B replacement, which would be a Type-I (T-I) error.

On the other hand, the μ_{B_B} value might be \$85k/yr, not exceeding the \$95k/yr threshold (\$80k/yr benefit of Treatment A plus the \$15k/yr justification to change). Again, if the variation on trial outcomes is relatively large, such as σ_B of \$10k/yr, then a single trial outcome might result in a measurement $X_B = \mu_B + 1.96\sigma_B$ which has a value of \$90k/yr + 1.96 * \$10k/yr = \$109.6k/yr, which would indicate a decision to accept the B replacement, which would be a Type-II (T-II) error.

Figure 17.1 illustrates these two situations. The horizontal axis represents the benefit in \$k/yr, and the vertical axis the *CDF* values. The right-most curve has its mean at \$100k/yr, which exceeds the threshold of \$95k/yr (indicated by the vertical dashed line) and should be accepted. Notice that there is about a 30% chance that the should-be-accepted Treatment B will give a single trial data value below the threshold, marked by the vertical dashed line. If the single trial value is 90 \$k/yr, Treatment B might get rejected. This is a T-I error, $\alpha \cong 0.3$.

On the other side of the situation, the left-most curve has its mean at 85 \$k/yr. If that represents Treatment B, then it should be rejected, but observe that due to variability in a single trial outcome, the *CDF* value at the accept-reject value of \$95k/yr is about 0.84,

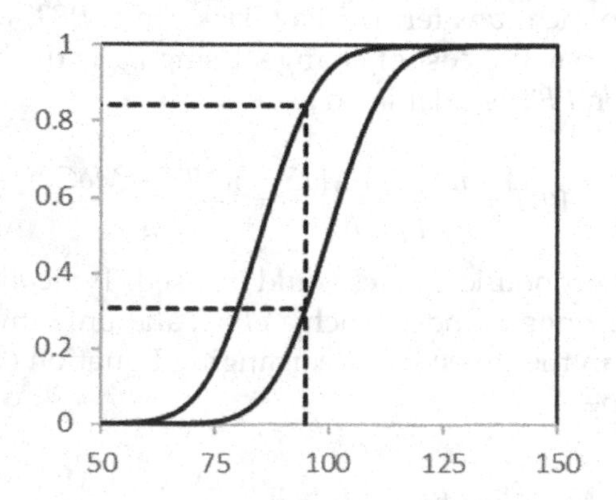

FIGURE 17.1
Illustration of T-I and T-II errors if $n = 1$.

84%. There is about a 100% – 84% = 16% chance that the should-be-rejected B would give a single trial value above the threshold, which might lead to acceptance. This is a T-II error, $\beta \cong 0.16$.

If the number of replicate trials is increased, then the average of n trials will have less variation than the single trial values. The same μ_B situation as before is illustrated in Figure 17.2, except that there are $n = 10$ trials and the curves represent the *CDF* of the averages of $n = 10$ trials, not the individuals as in Figure 17.1. Now the probability of rejecting the option with a mean of \$100k/yr is about 6%, the T-I error is about 6%, $\alpha \cong 0.06$. And now the probability of accepting the option with a mean of \$85k/yr is about 0.1%, the T-I error is about 0.1%, $\beta \cong 0.001$. In either situation, with an increased number of trials, the chance of making a wrong decision is greatly reduced.

But, of course, the increased number of trials to permit greater assurance cost time and money. That means that the threshold, as calculated by Equation (17.2) will increase. Perhaps it increases from \$95k/yr to \$99k/yr, as illustrated in Figure 17.3. Now the T-I error of rejecting B if $\mu_{B_B} = \$100\text{ k/yr}$ is about 17%, $\alpha \cong 0.17$. And the T-II error of accepting B if $\mu_{B_B} = \$85\text{ k/yr}$ is about 0.01%, $\beta \cong 0.0001$.

Note: If the right-most curve has a mean closer to the threshold, then the T-I error, α, will be higher and the number of replicate trials, n, will need to be increased to return to an acceptable T-I error.

Increasing the number of trials has two effects. Beneficially, the variation of the average is reduced, and the average approaches the mean. This reduces both T-I and T-II errors. But, unfortunately, increasing the number of trials increases the threshold to recover the costs of the extra trials, requiring a larger benefit of Treatment B to justify the change. In determining the number of replicate trials, we wish to balance both of these effects.

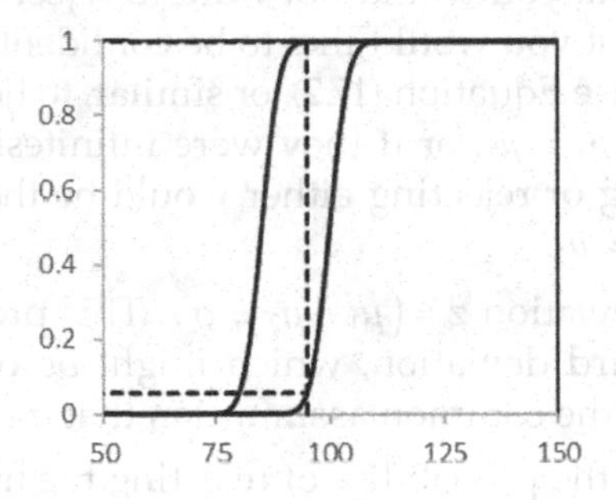

FIGURE 17.2
Reduction in T-I and T-II errors if $n = 10$.

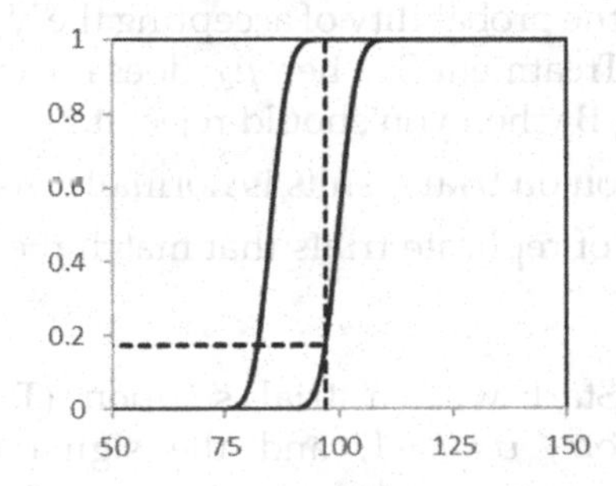

FIGURE 17.3
Illustration of T-I and T-II errors if $n = 10$ and auxiliary costs are included.

But there is more. There are two stages to determining whether to switch or not.

The first stage is the experimental design – the *a priori* planning of the number of replicate experiments needed to get the values for n, σ_{B_B} the variation on B_B, TC the cost of a trial, and an estimate of μ_{B_B} the mean of B_B. In this planning stage, we need to use the *a priori* estimate of the yet unknown economic values. *A priori* means before the evidence, knowledge obtained from concepts.

Then there is the second stage. If n from Stage 1 is estimated to be 20 trials, and after the 7th trial it is obvious that μ_{B_B} exceeds the value needed to make an accept-or-reject decision, then there is no sense in performing the next 13 planned trials. So, the second stage is the *a posteriori* trial-to-trial progressive analysis to make the decision (either to accept or reject, or to continue trials). *A posteriori* means after the evidence, knowledge obtained from tests.

17.2 *A Priori* Estimation of *N*

17.2.1 Classic Estimation of *n*

The classic statistically based approach to determining the number of replicate trials in Stage 1 of the experimental plan, the pre-experimentation planning, does not consider the economics. It uses an estimated value of $\Delta\mu/\sigma$, and user-specified T-I and T-II errors (α and β) to determine the number of trials, n. The procedure goes like this:

1. Determine the mean that you would not want to reject, μ_1, a desirable value, and determine the mean that you would like to be confident in rejecting, μ_2, an undesirable value. Perhaps use Equation (17.2), or similar, to help get those values. They cannot be the same. If $\mu_1 = \mu_2$, or if they were infinitesimally different, then the probability of accepting or rejecting either would be the same. For this analysis, larger is better, and $\mu_1 > \mu_2$.
2. Calculate the scaled deviation $z = (\mu_1 - \mu_2)/\sigma_X$. This presumes an estimate of σ_X, the trial-to-trial standard deviation, which might be obtained from experience with Treatment A, and the common assumption that $\sigma_{B_B} = \sigma_{B_A} = \sigma_X$.
3. Choose α, the T-I error, the probability of rejecting the hypothesis that $\mu_{B_B} \geq \mu_1$, of finding that the test result does not exceed that threshold when it actually does. The probability of rejecting B when you should accept it.
4. Choose β, the T-II error, the probability of accepting the hypothesis that μ_{B_B} exceeds the threshold to accept Treatment B, when μ_{B_B} does not exceed that threshold. The probability of accepting B when you should reject it.
5. Assume that the variation on trial results is normally distributed.
6. Determine the number of replicate trials that match the value of β, given choices for α and z.

Step 6 has several parts. Start with a trial solution (TS) for n, an initial guess. Calculate degrees of freedom, $\upsilon = n-1$, and the sigma on the n-sample average, $\sigma_{\bar{X}} = \frac{\sigma_X}{\sqrt{n}} = \sigma_{B_B} / \sqrt{n}$. Use the inverse of the t-distribution to calculate a t-value associated

TABLE 17.1

Classical Approach to Determining the Number of Replicate Trials for a *t*-Test of a Single Mean (Single-Sided Alpha and Beta). $Z = \frac{\mu_1 - \mu_2}{\sigma_X}$. For a Two-Sided Test Use $\alpha_{table} = \alpha_{decision}/2$

alpha =	**0.01**	**0.01**	**0.01**	**0.01**	**0.025**	**0.025**	**0.025**	**0.025**
beta =	**0.01**	**0.025**	**0.05**	**0.1**	**0.01**	**0.025**	**0.05**	**0.1**
Z								
0.25	350	297	255	211	297	249	211	171
0.3	244	207	178	148	207	174	147	119
0.35	180	153	132	109	153	128	109	88
0.4	139	118	102	84	118	99	84	68
0.45	111	94	81	67	94	79	67	54
0.5	90	77	66	55	77	64	55	45
0.55	75	64	55	46	64	54	46	37
0.6	64	54	47	39	54	46	39	32
0.65	55	47	40	34	47	39	33	27
0.7	48	41	35	30	41	34	29	24
0.75	42	36	31	26	36	30	26	21
0.8	38	32	28	23	32	27	23	19
0.85	34	29	25	21	29	24	21	17
0.9	30	26	23	19	26	22	19	16
0.95	28	24	21	18	24	20	17	14
1	25	22	19	16	22	18	16	13
1.1	22	19	16	14	19	16	13	11
1.2	19	16	14	12	16	14	12	10
1.3	17	14	13	11	14	12	10	9
1.4	15	13	11	10	13	11	9	8
1.5	13	12	10	9	12	10	9	7
1.6	12	11	9	8	11	9	8	7
1.7	11	10	9	8	10	8	7	6
1.8	10	9	8	7	9	8	7	6
1.9	10	8	8	7	8	7	6	6
2	9	8	7	6	8	7	6	5
2.5	7	6	6	5	6	5	5	
3	6	5	5	5	5	5		
alpha =	**0.05**	**0.05**	**0.05**	**0.05**	**0.1**	**0.1**	**0.1**	**0.1**
beta =	**0.01**	**0.025**	**0.05**	**0.1**	**0.01**	**0.025**	**0.05**	**0.1**
Z								
0.25	255	211	176	139	211	171	139	107
0.3	178	147	123	97	148	119	97	75
0.35	132	109	91	72	109	88	72	55
0.4	102	84	70	56	84	68	56	43
0.45	81	67	56	44	67	54	44	34
0.5	66	55	46	36	55	45	36	28
0.55	55	46	38	30	46	37	30	24
0.6	47	39	32	26	39	32	26	20
0.65	40	33	28	22	34	27	22	17

(Continued)

TABLE 17.1 (CONTINUED)

Classical Approach to Determining the Number of Replicate Trials for a *t*-Test of a Single Mean (Single-Sided Alpha and Beta). $Z=\frac{\mu_1-\mu_2}{\sigma_X}$. For a Two-Sided Test Use $\alpha_{\text{table}}=\alpha_{\text{decision}}/2$

alpha =	**0.05**	**0.05**	**0.05**	**0.05**	**0.1**	**0.1**	**0.1**	**0.1**
beta =	**0.01**	**0.025**	**0.05**	**0.1**	**0.01**	**0.025**	**0.05**	**0.1**
Z								
0.7	35	29	24	20	30	24	20	15
0.75	31	26	22	17	26	21	17	14
0.8	28	23	19	16	23	19	16	12
0.85	25	21	17	14	21	17	14	11
0.9	23	19	16	13	19	16	13	10
0.95	21	17	14	12	18	14	12	9
1	19	16	13	11	16	13	11	9
1.1	16	13	11	9	14	11	9	7
1.2	14	12	10	8	12	10	8	7
1.3	13	10	9	7	11	9	7	6
1.4	11	9	8	7	10	8	7	5
1.5	10	9	7	6	9	7	6	5
1.6	9	8	7	6	8	7	6	5
1.7	9	7	6	5	8	6	5	
1.8	8	7	6	5	7	6	5	
1.9	8	6	6	5	7	6	5	
2	7	6	5		6	5		
2.5	6	5			5			
3	5				5			

with α. In Excel, $t_\alpha=T.\text{INV}(\alpha,\upsilon,1)$. The *t*-value will be negative, if $\alpha<0.5$. From this, calculate the deviation from the desirable mean, $\bar{X}=\mu_1+t_\alpha\sigma_{\bar{X}}$. Then calculate the number of deviations that value is from the undesirable mean. $T=\left(\bar{X}-\mu_2\right)/\sigma_{\bar{X}}$. Finally, determine beta associated with the *n* trial solution (TS), $\beta_{TS}=1-T.\text{DIST}(T,\upsilon,1)$. If the trial solution value for *n* is correct, then the calculated β_{TS} will equal the target β from Step 4. Repeat choosing *n*-values until the choice of *n* makes the calculated β_{TS} equal the target β.

Note: The *n* trial solution value must be an integer, and it must be 2 or greater so that $\upsilon\geq 1$. With only integer values of *n*, it is unlikely that the calculated β_{TS} will exactly equal the target β. Here, we choose to find the *n* that minimizes the deviation between β_{TS} and the target β:

$$\min_{\{n\}} J=\left(\beta_{TS}-\beta\right)^2 \tag{17.3}$$

Then, if $\beta_{TS}<\beta$, to use the next largest trial count to ensure that β_{TS} is as close to the target as possible without being larger. This is a constrained integer optimization. We used leapfrogging as the optimization algorithm.

Table 17.1 shows how *n* changes with various choices of *z*, α, and β. The *n*-values match those of other publications, except for occasional differences of a count of 1; which we suspect is due to convergence criterion, or choice of using either the *t*- or *z*-distribution, or the rule of rounding up or down. We only report values of $n\geq 5$.

Note: The table header is for a one-sided alpha. For a two-sided test of a mean to a hypothesized value, the alpha value is twice that shown in the table. In the table, $\hat{\sigma}$ and δ are not independently expressed but are combined into $Z = \delta / \sigma = (\mu - \mu_0) / \hat{\sigma}$.

Note: Classic DoE determination of the number of replicates does not consider the cost of the trials. Only the statistical impact.

Two examples will serve to illustrate the use of these tables.

Example 17.1: A production lot of paper rolls is acceptable for making grocery bags if its mean breaking strength on a standard sample is at least 40 lb_f. Twenty samples were used. They produced a mean breaking strength of 39 lb_f with a standard deviation of 2.4 lb_f. If desired values are $\alpha = 0.05$ and $\beta = 0.05$, are the results acceptable, i.e., was the sample large enough?

For this situation, the null hypothesis is H_0: $\mu \geq \mu_0 = 40$ lb_f. A one-tailed test of single-class data is involved. Assuming that the sample parameter values approach the population values,

$$Z = \frac{\mu_0 - \mu}{\sigma} \approx \frac{\mu_0 - \bar{X}}{s} = \frac{40 - 39}{2.4} = 0.41667$$

From Table 17.1, under α (single-sided test) = 0.05 and $\beta = 0.05$, we find by interpolation that about 65 samples (replicates) should have been used instead of 20.

Example 17.2: One of the design criteria for a machine producing molding powder is that it will produce cylindrical pellets by extrusion through $\frac{1}{8}$ in holes. These pellets are produced by cutting the strings of extruded polymer into pieces $\frac{3}{16}$ in long (0.1875 in). The specification permits a deviation of ± 0.0095 in. To determine whether the unit was performing properly, a sample of 100 pellets was selected and examined. The mean length determined was 0.197 in. The standard deviation was 0.005 in. If desired $\alpha = 0.02$ and $\beta = 0.05$, how many observations would have been needed?

For this problem, the null hypothesis *is* H_0:$\mu = \mu_0$, and we need to use a two-sided test. The data show that

$$\mu_0 = 0.1875 \text{ in.}$$

$$\Delta = 0.0095 \text{ in.}$$

$$s = 0.005 \text{ in.}$$

From these values, $Z = (0.0095)/0.005 = 1.9$. From Table 17.1, for $\alpha = 0.01$ (this example is for a double-sided test, but the entry is for a single sided test) and $\beta = 0.05$, only eight samples are needed for all future pellet-length tests on this machine.

There is a complementary table to Table 17.1, for the case of comparing two means (both experimentally determined) as opposed one mean against a given value. In this case, the number of trials is also predicated on n for each treatment being the same, $n_1 = n_2$. Essentially, this case reveals that $n_{total} = n_1 + n_2$ is nearly twice that from Table 17.1.

There are several objections with this classic approach. First, the number of replicate trials could be very large, and the cost of a large number of trials to determine if Treatment

B is desirable could undermine any potential economic benefit. This approach does not account for the economics.

Second, how does a user decide appropriate values for the T-I and T-II error or have definite knowledge of the value of sigma prior to trials? The exact values for n (or n_1 and n_2) are predicated on much uncertainty, and the table entries should be taken as an ideal guide, not definitive perfection.

Certainly, we don't want to reject a desirable option or accept an undesirable option. But if alpha and beta are both nearly zero, the number of trials to be nearly 100% confident will be excessive. The 95% confidence ($\alpha = 0.05$) is a standard choice, but what makes this right? We would prefer an economic optimization to determine n.

17.2.2 Economic Estimation of *n* – Method 1

As n increases, the economic benefit of B over A must increase to pay for the extra trials. One question is what value of n minimizes the required benefit to be able to choose Treatment B.

For example, consider that the trial cost is \$1k per trial, that the *MoC* cost is \$2k, and that the *PBT* is 1 year. This means that μ_B must be

$$\mu_B \geq \mu_A + \frac{n\,TC + MoC}{PBT_{\text{threshold}}} \tag{17.4}$$

Consider that we are evaluating operation costs, and that the current Treatment A costs \$800k/year to use. The "benefit" with Treatment A is −\$800k/yr. We'll run Treatment B for a trial to see if its operating cost is low enough to justify the change to B. If the trial-to-trial sigma on determining the operating cost is \$20k/yr and we want to be 99% confident ($\alpha = 1-0.99 = 0.01$) in not rejecting a good Treatment B then the average of the trials must be

$$\bar{X}_B \geq \mu_A + \frac{n\,TC + MoC}{PBT_{\text{threshold}}} + t_{v,\alpha}\frac{\sigma_X}{\sqrt{n}} \tag{17.5}$$

In the limit of large n, the third term on the right-hand side diminishes and the required $\bar{X}_B$ increases linearly with n. By contrast, in the small n limit, the trial cost has a minimal impact and the required $\bar{X}_B$ diminishes with increasing n. There is a value of n that minimizes the required $\bar{X}_B$ value.

Figure 17.4 illustrates this particular case using Equation (17.5). Six trials, $n = 6$ minimizes the benefit B must have to be accepted. The trial number remains in the range of 4 to 8 for a range of reasonable choices of T-I errors. You can easily use Equation (17.5), or your organization's equivalent, with economic values that match your situation.

Note: The minimum required $\bar{X}_B$ value indicated in Figure 17.4 is about −\$780/yr. If the expected benefit of Treatment B is not equal or better, then there is no justification to run the trials.

17.2.3 Economic Estimation of *n* – Method 2

In Method 1 you need to specify a desired confidence, or $(1 - \alpha)$, but the right value for the application might not be known. In Method 2, hypothesize a possible μ_B value and seek

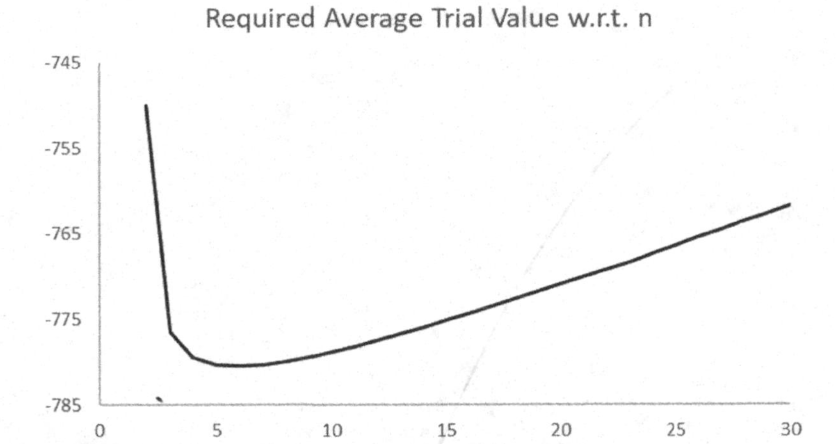

FIGURE 17.4
An illustration of economic estimation of n by method 1.

the number of trials to either maximize confidence or the economic benefit. The possible μ_B value may come from any number of *a priori* sources.

The potential economic benefit is $(\mu_B - \mu_A) - \frac{n\,TC + MoC}{PBT_{\text{threshold}}}$. This decreases with increasing number of trials. If you should accept μ_B, and do accept it, the benefit is above. The probability of accepting μ_B is $1 - \alpha$. However, if you reject Treatment B after n trials, then the associated cost of the trials is ($n\,TC$). The probability of rejecting μ_B is α. The associated t-value is

$$T = \left[(\mu_B - \mu_A) - \frac{n\,TC + MoC}{PBT_{\text{threshold}}}\right]\frac{\sqrt{n}}{\sigma} \tag{17.6}$$

from which, in Excel,

$$\alpha = T.\text{DIST}(T, \upsilon, 1) \tag{17.7}$$

then the probable outcome is

$$P_{\text{benefit}} = (1-\alpha)\left[(\mu_B - \mu_A) - \frac{n\,TC + MoC}{PBT_{\text{threshold}}}\right] + \alpha(-n\,TC) \tag{17.8}$$

Given a minimum of $n = 2$ trials to determine the sigma, the best possible outcome could be

$$P_{\text{benefit}} = (1-0)\left[(\mu_B - \mu_A) - \frac{2\,TC + MoC}{PBT_{\text{threshold}}}\right] + 0(-n\,TC) \tag{17.9}$$

and the ratio of probable to best possible is

$$R = \frac{\left\{(1-\alpha)\left[(\mu_B - \mu_A) - \frac{n\,TC + MoC}{PBT_{\text{threshold}}}\right] + \alpha(-n\,TC)\right\}}{\left[(\mu_B - \mu_A) - \frac{2\,TC + MoC}{PBT_{\text{threshold}}}\right]} \tag{17.10}$$

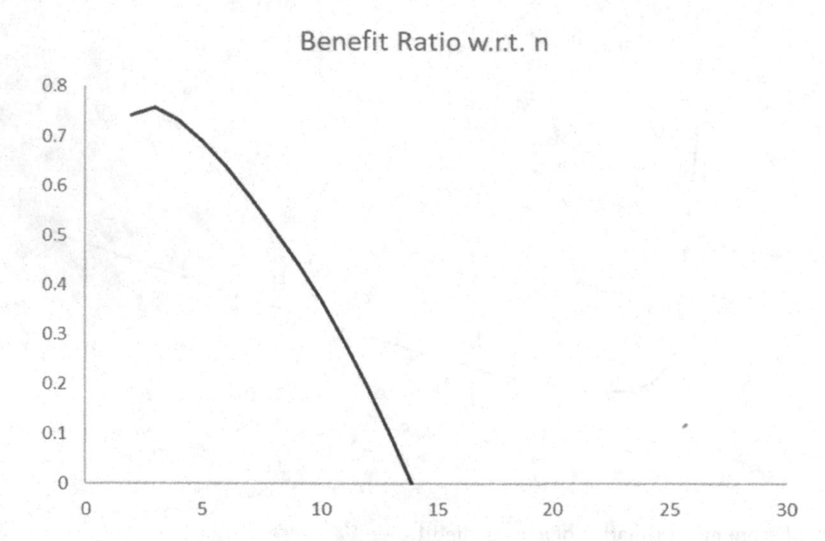

FIGURE 17.5
An illustration of economic estimation of n by method 2.

In this case, Figure 17.5 indicates how the potential benefit ratio changes with number of trials. Here, $\mu_B = -\$780\ \text{k/yr}$, $\mu_A = -\$800\ \text{k/yr}$, $\sigma = \$20\ \text{k/yr}$, $TC = \$1\text{k}$, $\text{MoC} = \$2\text{k}$, and $\text{PBT}_{\text{threshold}} = 1\,\text{yr}$.

With the stated economic values, the Figure 17.5 optimum is $n = 3$ trials, but n ranges from 3 to about 7 with variations in the economic values.

In Figure 17.5 at the optimum n, the ratio of probable to maximum possible benefit is about 78%. The number of trials above 2 and the *MoC* costs diminish the possible best benefit. If the best possible ratio of benefit is a small value, again, one might reject B even prior to running trials.

17.2.4 Economic Estimation of *n* – Method 3

In Method 1, you have to specify a level of confidence, which is often arbitrary. In Method 2 you do not, but Method 2 only seeks to accept (not reject) a hypothesized good, new Treatment B. Method 2 does not seek to reject a bad Treatment B. Method 3 does both – accept if good, reject if not good, without having to specify the α and β values.

In Method 3, also hypothesize the mean of Treatment B, the good, new, treatment to replace Treatment A, current practice. If B is to replace A, then the benefit of B over A, needs to exceed the cost of trials to qualify B and the costs of the management of the change. If there is no value benefit of switching from A to B, then there is no justification to do the trials. The hypothesized value benefit might be indicated by laboratory testing, or even claims by the promoter of B. Alternately, one might accept B if it is equivalent to A, because of a desire to alleviate the concern about supply security that an alternate provider causes. In this case there is no anticipated economic advantage, and for this Method 3 procedure, assign an economic benefit that would be equivalent to the concern driving the evaluation of Treatment B.

In Figure 17.6a, the horizontal axis is the number of standard deviations from the mean of Treatment A, current practice, and the vertical axis is the *CDF*. The right-most curve is the sigma-scaled value of the hypothesized value of Treatment B over that of A,

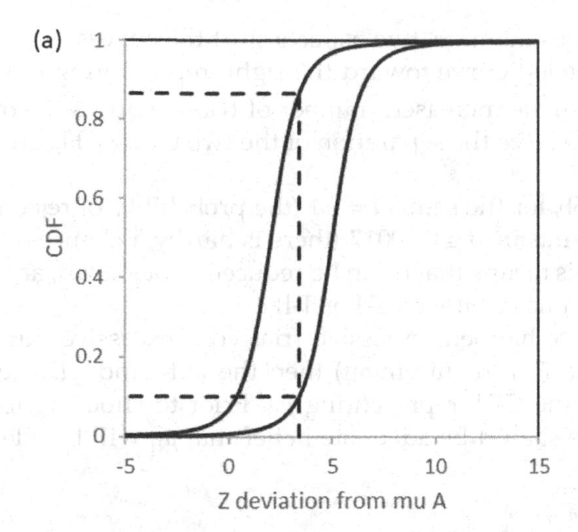

FIGURE 17.6a
Illustration of T-I and T-II errors comparing Treatment B to the minimum required to justify the cost of trials and *MoC*, $n = 4$.

$Z = \left(\bar{X}_H - \mu_A\right)/(\sigma_X / \sqrt{n})$. It represents the distribution of averages that we might get if Treatment B has the hypothesized mean $\mu_{\text{Hypothesized}} = \mu_H$. If we determine that the treatment mean is $\mu_B \geq \mu_H$ we'll accept Treatment B. In this illustration, μ_H has a z-value of about 5 (it is 5 $\sigma_{\bar{X}} = \sigma_X / \sqrt{n}$ increments greater than zero, the location of the *CDF* curve for μ_A.

The left-most curve is the minimum that B must be to justify acceptance of B. The mean of the left-most curve is the known mean of Treatment A, plus the cost of trials and *MoC*.

$$\mu_{\min} = \mu_A + \frac{n\,TC + MoC}{PBT_{\text{threshold}}} \tag{17.11}$$

here μ_A and $\mu_{\min}$ might be in \$/year. In the numerator of the second term, n is the number of trials run, and *TC* is the cost per trial, and *MoC* is the cost associated with implementing the change. The dimensional units of the numerator might be \$. The denominator, $PBT_{\text{threshold}}$ is the minimum *PBT* to justify a change in the units of years, required to make the equation for $\mu_{\min}$ dimensionally consistent. Scaling $\mu_{\min}$ to a consistent Z relative to μ_A, $Z = \left(\frac{n\,TC + MoC}{PBT_{\text{threshold}}}\right)/(\sigma_X/\sqrt{n})$. In this illustration, $\mu_{\min}$ is about 2 increments greater than zero, the location of the *CDF* curve for μ_A.

The lower horizontal line is at a *CDF* of 0.1, and the vertical dashed line represents the z-value where a one-sided α = 0.1 to reject the hypothesis intersects the μ_H curve. The z-value is about 3.5. If an $\bar{X}$ from trials is to the right of the vertical line, then we accept that $\mu_B \geq \mu_H$. The probability of rejecting B when in fact B could represent the desirable μ_H is $\alpha = 0.1 = 10\%$.

The vertical line intersects the $\mu_{\min}$ curve with a *CDF* of about 0.88. Here β = 0.12. If an $\bar{X}$ from trials is to the left of the vertical line, then we accept that $\mu_B \leq \mu_{min}$, and we reject Treatment B. The probability of accepting B when in fact B could be inadequate is $\beta = 0.12 = 12\%$.

If the number of trials increase, two aspects shift the curves. First, the cost of the trials increases, shifting the left curve toward the right, representing the sampling *CDF* from the $\mu_{\min}$ curve. Second, the increased number of trials reduces the $\sigma_{\bar{X}} = \sigma_X / \sqrt{n}$ value for each curve, which increases the separation of the two curves. Figure 17.6b represents this effect.

Note: In Figure 17.6b, for the same $\alpha = 0.1$ (the probability of rejecting Treatment B if in fact $\mu_B \geq \mu_H$), now results in $\beta \cong 0.00017$ (there is hardly a chance of accepting Treatment B if it is $\mu_B \leq \mu_{\min}$). This means that α can be reduced. Not shown, at $\alpha = 0.01$, $\beta \cong 0.0022$. A choice of α nearly eliminates either a T-I or T-II error.

Note: If any of these happen (excessive trial cost, excessive number of trials, excessive *MoC*, too small a *PBT* requirement) then the left-hand *CDF* representing $\mu_{\min}$ will move to the right of the *CDF* representing μ_H. Prior to choosing to run trials to evaluate Treatment B there should be adequate belief that μ_H will be adequately greater than $\mu_{\min} = \mu_A + \dfrac{n\,TC + MoC}{PBT_{\text{threshold}}}$.

There are two conditions, either Treatment B is $\mu_B \geq \mu_H$ and worth the change, or Treatment B is $\mu_B \leq \mu_{\min}$ and not worth the change. There are two conditions, accept or reject, leading to 4 cases (Table 17.2).

The two desirable cases are to Accept B if $\mu_B \geq \mu_H$ and to Reject B if $\mu_B \leq \mu_{\min}$.

If $\mu_B \geq \mu_H$ and we appropriately accept Treatment B, then the economic benefit is at least $\mu_H - \dfrac{n\,TC + MoC}{PBT_{\text{threshold}}} - \mu_A$.

If $\mu_B \geq \mu_H$ and we falsely reject Treatment B, then the economic penalty is $-(n\,TC)/PBT_{\text{threshold}}$, the costs of the trials scaled by $PBT_{\text{threshold}}$ to have units consistent with μ. Note that *MoC* is not included. If you don't accept B then there is no cost of the change.

If $\mu_B \leq \mu_{\min}$ and we appropriately reject Treatment B, then the economic penalty is again $-(n\,TC)/PBT_{\text{threshold}}$.

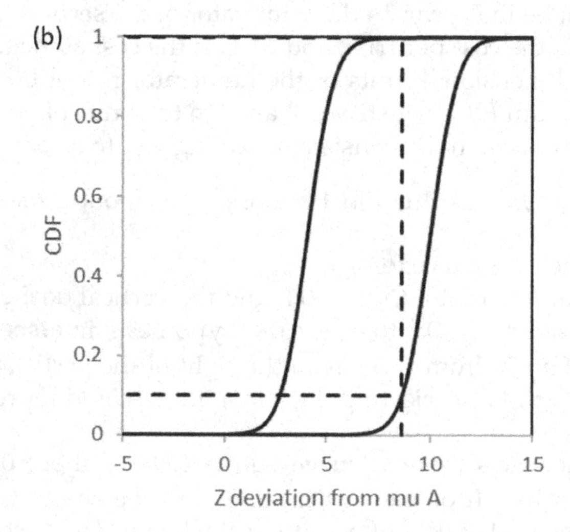

FIGURE 17.6b
Illustration of T-I and T-II errors comparing Treatment B to the minimum required to justify the cost of trials and *MoC*, $n = 16$.

TABLE 17.2
Decision Probability

	Action is to accept B	Action is to reject B
Truth is $\mu_B \geq \mu_H$	Probability = $1-\alpha$	Probability = α = T-I error
Truth is $\mu_B \leq \mu_{min}$	Probability = β = T-II error	Probability = $1-\beta$

TABLE 17.3
Decision Economic Impact

	Action is to accept B	Action is to reject B
Truth is $\mu_B \geq \mu_H$	$\mu_H - \frac{n\,TC + MoC}{PBT_{threshold}} - \mu_A$	$-\frac{n\,TC}{PBT_{threshold}}$
Truth is $\mu_B \leq \mu_{min}$	$-\frac{n\,TC + MoC}{PBT_{threshold}}$	$-\frac{n\,TC}{PBT_{threshold}}$

If $\mu_B \leq \mu_{min}$ and we falsely accept B, then the economic penalty is $\mu_B - \mu_{min} = \mu_B - \frac{n\,TC + MoC}{PBT_{threshold}} - \mu_A$. However, often Treatment A and B are equivalent, so set the economic benefit as $-\frac{n\,TC + MoC}{PBT_{threshold}}$ the cost of the trials and managing the un-justified change to B (Table 17.3).

The probable economic impact is the probability times the impact.

There is one more issue. Two mutually exclusive situations have been described: 1) The truth is $\mu_B \geq \mu_H$, and 2) The truth is $\mu_B \leq \mu_{min}$. If you are running trials, there must be a belief that $\mu_B \geq \mu_H$, otherwise you would not take the time and effort to run the trials. But it is not a certainty that $\mu_B \geq \mu_H$, because you need to run the trials to make that determination. Perhaps you are 80% sure that B is the right choice (because of strong laboratory experience). Or perhaps only 50% sure (because the belief is based on provider claims). Let P be the *a priori* probability that $\mu_B \geq \mu_H$, and $1-P$ be the *a priori* probability that $\mu_B \leq \mu_{min}$.

To determine the number of trial replicates, *a priori*: Hypothesize what a desirable value, or an expected good value is for μ_B, μ_H. Hypothesize the sigma for B, probably equating $\sigma_B = \sigma_A$ from experience with Treatment A. Estimate the anticipated cost of a trial, *TC*, and the *MOC*. Assign a probability (belief) that Treatment B might actually live up to its promise.

Let an optimizer choose n and α. From these, calculate a *T*-value associated with $\nu = n-1$ and α. In Excel $T = T.\text{INV}(\alpha,\nu)$. Calculate $\sigma_{\bar{X}} = \sigma_B / \sqrt{n}$, then the associated lower extreme $\bar{X}$ value if $\sigma_B = \sigma_H$. This is $\bar{X}_{crit} = \mu_H + T\sigma_B / \sqrt{n}$. The "+" sign might be unexpected, but the *T*-value will be negative if $\alpha < 0.5$, which it should be. Determine the *T*-value relative to μ_{min}, $T = \frac{\bar{X}_{crit} - \mu_{min}}{\sigma_B / \sqrt{n}}$, and from this the beta value. In Excel $\beta = 1 - T.\text{DIST}(T,\nu,1)$. Now that alpha and beta are known, calculate the probable impact of each possible outcome from Table 17.4, and sum them. Scale this by the best possible outcome, if $\alpha = \beta = 0$, $P(\mu_B = \mu_H) = 1$, and $n = 2$, the minimum number of trials, per Table 17.5.

The *a priori* number of trials is determined by seeking to maximize the scaled probable economic outcome.

$$\min_{\{n,\alpha\}} J = \frac{\text{Sum of Table 4 elements}}{\text{Sum of Table 5 elements}} \tag{17.12}$$

TABLE 17.4

Probable Economic Impact

	Action is to accept B	Action is to reject B
Truth is $\mu_B \geq \mu_G$	$P(1-\alpha)\left[\mu_H - \frac{n\,TC + MoC}{PBT_{\text{threshold}}} - \mu_A\right]$	$-P\alpha \frac{n\,TC}{PBT_{\text{threshold}}}$
Truth is $\mu_B \leq \mu_{\min}$	$-(1-P)\beta \frac{n\,TC + MoC}{PBT_{\text{threshold}}}$	$-(1-P)(1-\beta)\frac{n\,TC}{PBT_{\text{threshold}}}$

TABLE 17.5

Best Possible Economic Impact ($\mu_B \geq \mu_H$, and We Accept B)

	Action is to accept B	Action is to reject B
Truth is $\mu_B \geq \mu_H$, P=1	$\left[\mu_H - \frac{2\,TC + MoC}{PBT_{\text{threshold}}} - \mu_A\right]$	0
Truth is $\mu_B \leq \mu_{\min}$, P=0	0	0

This is mixed integer-continuum decision variables with constraints. We used either leap-frogging optimization, or an enumeration search over n, and for each, the Excel Solver to search for the optimum alpha. The results are the same for either optimization approach.

Table 17.6 provides some results for a variety of economic values. The header of the table presents the economic parameter values. The current Treatment is A and "mu A" represents the costs of using A, in \$k/yr. This should be well-known from operating experience. The "trial sigma" row represents the uncertainty of a single measurement in a trial, again the value is \$k/yr. The value represents the σ_B, but since trials with B have not been performed, a best estimate might be the σ_A from experience with A. The row labeled "mu H" is the hypothesized cost of using Treatment B, again in \$k/yr. The hypothesized value is not yet known. The trials have not been run. So, this value will be a "guestimate" of what might be the cost of Treatment B, perhaps extrapolated from lab trials or vendor claims. In this table the "mu H" values are 2 to 5 sigma from the "mu A" values. The row "PBT" is the threshold pay-back time, in years, needed to economically justify an investment. The row "MoC" is the anticipated cost of managing the change, in \$k. Finally, the row "P(H)" represents the probability that the hypothesized "mu H" might actually be true. The first column in the header of Table 17.6a reveals the base case of the economic values used in this study.

Below this header is the left-most column "trial cost", with the values in \$k/trial. Note that the first entry is 0.01 meaning a cost of \$10 per trial. This would be extremely rare. The "5" for a trial cost would mean \$5,000 per trial. Note that there are no entries for that row. If a trial cost is \$5,000 per trial, and the *MoC* is \$10,000, and the payback time threshold is 0.5 years, then with $n = 2$ trials the cost of testing and switching to B would be $\frac{2 \times 5{,}000 + 10{,}000}{0.5} = \$40{,}000$ which nearly exceeds the hypothesized benefit of switching to Treatment B.

The first section below the header represents the number of trials that optimizes the economic benefit for the investigation. Note that for reasonable trial costs, the number of trials is in the single digits, which is a significant contrast to the number of trials presented in Table 17.1, derived to meet alpha and beta conditions, without regard for the economics.

TABLE 17.6

A Priori Number of Trials Using Economic Method 3

mu A	−1,000	−1,000	−1,000	−1,000	−1,000	−1,000
trial sigma	20	20	20	20	20	20
PBT	0.5	0.5	0.5	0.5	0.5	1
MoC	10	10	10	10	10	10
mu H	−950	−950	−950	−900	−965	−950
P(H)	0.7	0.5	0.9	0.7	0.7	0.7
Trial Cost			*N*			
0.01	17	17	15	8	36	14
0.02	14	15	13	7	27	12
0.05	11	11	9	6	15	10
0.1	8	9	7	5	8	8
0.2	6	6		4	4	6
0.5	4	3		4		5
1		3		3		
2						
5						
			OF = ratio of Equation (17.12)			
0.01	0.687	0.486	0.888	0.698	0.633	0.696
0.02	0.677	0.476	0.880	0.697	0.588	0.693
0.05	0.654	0.451	0.861	0.693	0.495	0.686
0.1	0.623	0.418	0.839	0.688	0.408	0.677
0.2	0.579	0.368		0.679	0.323	0.661
0.5	0.492	0.269		0.658		0.628
1		0.154		0.633		
2						
5						
			alpha			
0.01	0.00198	0.00306	0.00159	0.00015	0.01422	0.00043
0.02	0.00431	0.00526	0.00271	0.00038	0.02942	0.00093
0.05	0.00986	0.01526	0.00786	0.00096	0.07391	0.00211
0.1	0.02231	0.02737	0.01395	0.00250	0.12716	0.00486
0.2	0.04039	0.06198		0.00685	0.18698	0.01163
0.5	0.07777	0.15279		0.00741		0.01956
1		0.17022		0.02200		
2						
5						
			beta			
0.01	0.00710	0.00462	0.02269	0.00112	0.02675	0.00383
0.02	0.01550	0.00795	0.03886	0.00268	0.05649	0.00830
0.05	0.03546	0.02308	0.11316	0.00651	0.14870	0.01847
0.1	0.07972	0.04139	0.20168	0.01592	0.26722	0.04135
0.2	0.14248	0.09295		0.03936	0.39262	0.09398
0.5	0.26377	0.21649		0.04273		0.15321

(Continued)

TABLE 17.6 (CONTINUED)

A Priori Number of Trials Using Economic Method 3

1		0.24522		0.10680		
2						
5						
mu A	−1,000	−1,000	−1,000	−1,000	−1,000	−1,000
trial sigma	20	20	40	30	10	5
PBT	2	0.5	0.5	0.5	0.5	0.5
MoC	10	5	10	10	10	10
mu H	−950	−950	−950	−950	−950	−950
P(H)	0.7	0.7	0.7	0.7	0.7	0.7
Trial						
Cost				*N*		
0.01	13	12	40	28	9	6
0.02	12	10	31	23	8	5
0.05	10	8	19	16	6	4
0.1	8	6	11	11	5	4
0.2	7	5		6	4	3
0.5	5				3	3
1					3	3
2						3
5						
			OF = ratio of Equation (17.12)			
0.01	0.698	0.693	0.665	0.677	0.694	0.697
0.02	0.697	0.689	0.641	0.660	0.690	0.694
0.05	0.694	0.677	0.589	0.622	0.679	0.687
0.1	0.690	0.661	0.536	0.578	0.663	0.678
0.2	0.683	0.638		0.521	0.638	0.663
0.5	0.667				0.582	0.625
1					0.498	0.552
2						0.361
5						
			alpha			
0.01	0.00019	0.00093	0.00639	0.00378	0.00059	0.00023
0.02	0.00030	0.00209	0.01289	0.00737	0.00116	0.00082
0.05	0.00077	0.00486	0.03205	0.01883	0.00458	0.00309
0.1	0.00204	0.01163	0.05753	0.03665	0.00952	0.00321
0.2	0.00342	0.01904		0.07183	0.02046	0.01291
0.5	0.01005				0.04755	0.01456
1					0.05646	0.01807
2						0.02997
5						
			beta			
0.01	0.00365	0.00830	0.02585	0.01454	0.00188	0.00065
0.02	0.00574	0.01823	0.05394	0.02886	0.00364	0.00226

(Continued)

TABLE 17.6 (CONTINUED)

A Priori Number of Trials Using Economic Method 3

0.05	0.01430	0.04135	0.14516	0.07632	0.01374	0.00805
0.1	0.03608	0.09398	0.28768	0.15431	0.02758	0.00836
0.2	0.05882	0.14816		0.31243	0.05604	0.03030
0.5	0.15700				0.11844	0.03425
1					0.14336	0.04273
2						0.07223
5						

The second section in the table represents the Economic Objective Function value, Equation (17.12). Note that numbers are lower than the probability of the hypothesis being true. If there is a 70% chance of the probability of Treatment B matching the mean from the H distribution (μ_H), then 70% of the time it will be a right choice. But even if the hypothesized μ_H is a true representation of μ_B, with the cost of trials and *MoC*, one cannot achieve the full idealized benefit.

The following two sections of Table 17.6 reveal the alpha (T-I error, the probability of rejecting Treatment B if μ_H is true) and the beta (T-II error, the probability of accepting Treatment B even if the μ_B is just at the minimal threshold).

There are no entries in the table for either of two situations. One is if the cost of a switch cannot be justified by the hypothesized benefit. The other is if the beta value is above 0.5. A large beta means that any average of trials of B will be accepted, the T-II error.

There are several observations from data in Table 17.6:

- If the cost of trials increases, the economic benefit decreases, and the number of trials that can be justified decreases. But also, the value of alpha and beta that support the economic optimum change. Their values should not be independent of the trial cost.
- If the probability of the hypothesized μ_H value for Treatment B can be increased, fewer trials are needed to confidently accept it.
- If either the economic benefit of Treatment B improves, or the trial sigma decreases then fewer trials are needed.
- If either the *PBT* threshold increases, or the *MoC* costs decrease, then fewer trials are required for the economic optimum number.
- For a reasonable range of economic parameter values, the number of trials to test if Treatment B is better, the Table 17.6 data, is in the single digits. This is a contrast to the double-digit estimate of the number of trials using exclusively statistical, noneconomic calculations of Table 17.1.

17.3 *A Posteriori* Estimation of *N*

The previous methods estimated the number of trials needed to determine if we should switch to Treatment B or remain with Treatment A, prior to having data, using hypothesized values for μ_B and σ_B. That forecast number of trials is predicated on assumptions, unknown values. If the promised benefit of B justifies trials, then start the trials. But, after each trial re-assess the situation with progressive data to estimate μ_B and σ_B.

Consider that the *a priori* plan indicates that $n = 7$ replicate trials are needed to have enough information to make a decision. But if $\bar{X}_B > \mu_H$, then after each trial the belief that we should switch to Treatment B is reinforced. Alternately, if $\bar{X}_B \le \mu_{\min}$ then after each trial the belief that we should reject Treatment B is reinforced. It could be that after 4 trials there is much certainty about the decision, and therefore no need to waste the time or expense with continuing to run the $n = 7$ replicate trials from the original plan.

A Bayes Belief method appears to be an appropriate method to make that determination.

One could consider a variety of cases representing the truth of the situation. Here, we'll illustrate either $\mu_B > \mu_{\min}$ (a desirable outcome) or $\mu_B \le \mu_A$ (an undesirable outcome). For either case, because of trial-to-trial variation, the average, $\bar{X}$,might indicate that we should accept Treatment B, or that we should reject Treatment B. Table 17.7 indicates this situation.

The table entries of α_1 and α_2 are the complement to the T-I and T-II error probabilities. If the truth is that $\mu_B > \mu_{\min}$ and the data indicates that Treatment B should be rejected, that is a T-I error. Alternately, if the truth is that $\mu_B \sim \mu_A$ and the data indicate that Treatment B should be accepted, that is a T-II error.

After n number of trials, one can calculate both the average and standard deviation of the benefit of Treatment B, $\bar{X}_B$ and s_B. Assuming that $\bar{X}_B$ represents the value for μ_B, and that s_B represents the value for σ_B, and using $\mu_{\min} = \mu_A + (n\,TC + MoC)/PBT_{\text{threshold}}$. The experimental T-value is the benefit scaled by the standard error of the benefit.

Note: Entries in Table 17.7 are not grounded in *a priori* hypothesized values of μ_B or σ_B.

Note: The values for α_1 and α_2 will change after each trial as the values of v, n, $\bar{X}_B$, s_B, and $\mu_{\min}$ (and maybe even TC) each change with the new data.

Note: The value for s_B can only be calculated after the second trial. Accordingly, start the Bayes Belief calculations after the second trial.

Note: The top row in Table 17.7 represents the desired belief that μ_B of Treatment B is better than μ_{min}. The bottom row represents an alternate belief that μ_B is equivalent to μ_A. There could be any of many alternate beliefs. One could give Treatment B the benefit of the doubt and permit that μ_B is halfway between μ_A and the advertised μ_H, $\mu_B \sim (\mu_H - \mu_A)/2$. This would change the T-value calculated in the lower-right cell to $T = \left(\mu_{\min} - (\mu_H - \mu_A)/2\right)\sqrt{n}/s_B$.

TABLE 17.7

Probabilities of T-I and T-II Error

Truth \ Experimental outcome	$\bar{X}_B > \mu_{\min}$ (Indicates accept B)	$\bar{X}_B \le \mu_{\min}$ (indicates reject B)
$\mu_B > \mu_{\min}$ (should accept B)	$T = (\bar{X}_B - \mu_{\min})\sqrt{n}/s_B$ $\alpha_1 = \text{T.DIST}(T, n-1, 1)$ $P_1(\text{this event}) = \alpha_1$	$P_2(\text{this event}) = 1 - \alpha_1$
$\mu_B \sim \mu_A$ (should reject B)	$P_3(\text{this event}) = 1 - \alpha_2$	$T = (\mu_{\min} - \mu_A)\sqrt{n}/s_B$ $\alpha_2 = \text{T.DIST}(T, n-1, 1)$ $P_4(\text{this event}) = \alpha_2$

We will next consider that the *a priori* belief is $\mu_B > \mu_{min}$ (to accept Treatment B). If there were not such a belief, there would not be justification for doing the trials. Prior to the trials, the initial belief to accept Treatment B, might be 0.7. $B_{k=0} = 0.7$. If the *a priori* belief was 100%, $B_0 = 1.0$, a certainty that we should switch to Treatment B, then again, there would not be a reason to do trials. If the *a priori* belief was very low, such as $B_0 = 0.1$, then again, there is not much incentive to perform the trials. So, likely, the initial belief to accept Treatment B, might be on the order of $B_0 \cong 0.7$.

Belief after the kth trial is calculated from the prior belief and the probability values associated with the Accept or Reject decision in Table 17.7. If $\bar{X}_B > \mu_{min}$, then the evidence would suggest accepting Treatment B. In this case Bayes Belief that we should switch to Treatment B is updated as

$$B_k = \frac{B_{k-1}(P_1)}{B_{k-1}(P_1)+(1-B_{k-1})(P_3)} \tag{17.13}$$

Alternately, if $\bar{X}_B \leq \mu_{min}$ then the evidence would suggest rejecting Treatment B. In this case Bayes Belief to accept Treatment B is updated as

$$B_k = \frac{B_{k-1}(P_2)}{B_{k-1}(P_2)+(1-B_{k-1})(P_4)} \tag{17.14}$$

One might be adequately confident to stop trials and accept Treatment B, if the belief is very high, perhaps $B_k > 0.99$. Alternately, one might be adequately confident to stop trials and reject Treatment B, if the belief is very low, perhaps $B_k < 0.05$. There is no need for these two thresholds to be complements of each other.

There have now been several heuristically chosen values $B_0 = 0.7$, and the two thresholds to accept or reject Treatment B, $B_k > 0.99$, or $B_k < 0.05$. The human managing the choice should be able to choose reasonable values within the context. More extreme thresholds for B_k require more trials to have that certainty. Larger values for the initial belief B_0 also require more trials to reject, but fewer to accept the belief.

One can simulate the experiments to explore the method. Specify values for μ_B and σ_B, then sample from that population. Calculate the $\bar{X}_B$ and s_B values from the past k number of trials, and after each sample use equations in Table 17.7 to calculate the T-values, then the four probabilities. If $\bar{X}_B > \mu_{min}$, use Equation (17.13) to update the belief, otherwise use Equation (17.14).

Over a wide variety of values explored in the authors' simulations, the number of trials to a decision (either accept or reject) is about $n = 5$, being between about $n = 2$ and $n = 8$, matching the analysis of Sections 17.2.2, 17.2.3, and 17.2.4. Again, this is in contrast to the double-digit estimate of the number of trials using statistical, noneconomic calculations of Table 17.1.

Figure 17.7 presents results of 1,000 simulations each of the trials to make the belief hit the threshold for a large variety of critical economic and probability factors.

The horizontal axis is a modified z-statistic, which indicates the scaled magnitude of the hypothesized benefit when the number of trials, $n = 2$, is the minimum for obtaining statistics.

$$z' = \left[\mu_B - \mu_A - (2\,TC + MoC)/PBT_{threshold}\right]/\sigma_B \tag{17.15}$$

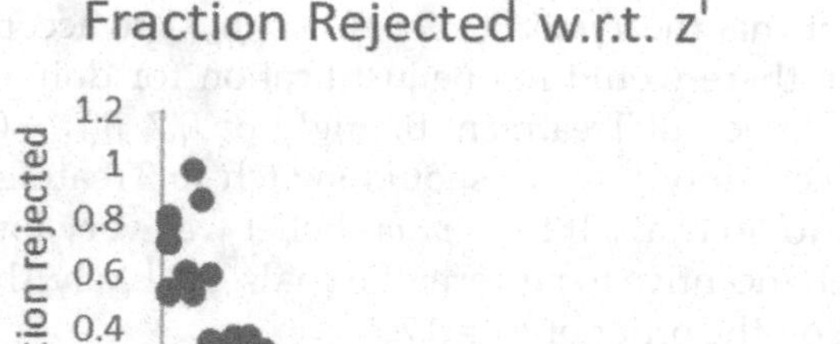

FIGURE 17.7
Results from Bayes belief simulations.

The vertical axis is the fraction of trials that were rejected. Since z' is positive in all cases, $\mu_B > \mu_A + (2\,TC + MoC)/PBT_{\text{threshold}}$, indicating that Treatment B should be accepted. However, when Treatment B is marginally acceptable, the cost of trials and the probability that data will lead to an unfavorable $\bar{X}_B$ and s_B will lead to a reject decision. For example, if $z' \cong 0.5$, about 20% of the tests will reject a good Treatment B.

Figure 17.7 suggests that one should not undertake trials to consider Treatment B unless the after-trials benefit is expected to be greater than about one standard deviation, $z' > 1$.

17.4 Takeaway

The number of replicate trials required to make either a statistical or economic decision depends on the variation of the trial outcome measurement and the user's choice of T-I and T-II errors. The economic impact can help reveal rational context-relevant choices for α and β values.

After each trial, use the *a posteriori* information to reconsider the *a priori* assumptions and trial design.

17.5 Exercises

1. Make up your own scenario for the trial result variation, desired difference in the A and B treatments, and desired T-I and T-II error probabilities. Determine the *a priori* number of trials.
2. Make up your own scenario for the trial result variation, cost of trials, desired difference in the A and B treatments, *PBT*. Determine the *a priori* number of trials, and associated T-I and T-II error probabilities.

3. Make up your own scenario. Use the Bayesian *a posteriori* number of trials method to determining n. Use consistent outcomes, either A is better than B, or B is better than A.
4. The uncertainty on a probability is $\sigma_p = \sqrt{pq/n}$. If one desires that the 99% confidence interval on the experimental probability value is no more than 10% of the probability, then $z_{0.995}\sigma_p \le 0.1p$. A) Defend that formula. B) Derive a generic equation to determine n required to experimentally determine the value of p. C) Describe the cost implications of n trials w.r.t. to the implications of an error on the value of p.

18

Experimental Design for Linear Steady-State Models – Screening Designs

18.1 Introduction

The greatest body of knowledge about the design of experiments is for linear steady-state relations between a response and treatments (or inputs, influences). These are models of the form

$$y = a + bx_1 + cx_2 + dx_3 + ex_4 + \cdots \tag{18.1}$$

The x_i terms may represent independent input variables, but they may also represent independent functionalities of the input variables, such as a difference between input variables 5 and 4 $(x_5 - x_4)$, or a product x_1x_3, or a ratio x_2/x_1, or a squared value x^2, or an exponential factor e^{-x}, or any of many others. The model is linear in coefficients and linear in the functionalities (even though the functionality may represent a nonlinear function of the input variable or treatment value). These models do not necessarily represent the true functionality of the response; however, they are convenient and commonly used in developing an initial, approximate understanding of a response.

They are useful for "screening" purposes. One application is to acquire data for ANOVA analysis (as in Chapter 12) to determine if treatments are significantly different. Another is to create mathematical models of the response. This chapter will focus on design of experiments to develop models, but the issues and considerations for both ANOVA and modeling applications are similar.

Linear models, as in Equation (18.1), are easily created, easily fit to the data, and are computationally rapid. They reveal an approximate shape of the response, which is useful for defining optimum operating conditions, or identifying a region that justifies greater experimental investigation. Additionally, comparison of models to data can identify response regions or the model that are not well fit to the data, suggesting alternate functionalities that should be included in the model.

Rarely should these linear-coefficient models with generic additive functionalities be considered as representing the true functionality of nature. Seek mechanistic models (alternately termed phenomenological, or first-principles models) to represent the functionality of nature. However, models of Equation (18.1) can be defended theoretically as truncated Taylor series representations (which means they remain missing some functionality, and may not express represented functionality truthfully), or as parsimonious (adequate for the purpose and requiring minimal effort).

Mathematical analysis of variance of the linear models is analytically tractable (relatively easy compared to that of models that are nonlinear in coefficients). So, there is much

DOI: 10.1201/9781003222330-18

more, and more sophisticated information on design of experiments (DoE) for models that are linear in coefficients.

In general, screening model DoE objectives are to minimize uncertainty in model coefficient values with minimal experimental effort. An assumption in DoE for screening models is that the model functionality matches that of the process. Other assumptions are that the data are only corrupted by measurement noise (not, for instance, error in exact input values), and that the uncertainty is independent and uniform over the entire experimental range (homoscedastic).

Experimental design is the sequence of steps taken to ensure that the data will be obtained in such a way that the subsequent ANOVA will maximize statistical confidence. In general, screening DoE indicates you should: 1) Place experimental points to include the center of the region of interest and to maximize the distance of other points about the center. 2) Replicate some points to obtain information about experimental variability. 3) Avoid linear correlation of the x_i terms. And 4) have about 3 or more experimental data points per model coefficient. But note, even if you follow best experimental design, that does not mean that the screening model is functionally correct to the process.

Three design principles that will help you obtain high-quality data at minimum cost are replication, randomization, and control. Without replication, no estimate of the magnitude of experimental error is possible. Nor will there be any way to decide whether an effect is significant or not.

Randomization of the design sequence is used to eliminate bias, to ensure that none of the x_i values are correlated to uncontrolled influences that are also changing in time (ambient temperature, atmospheric pressure, equipment warmup, process, or material aging, etc.) or space (position in a factory). Randomization helps to minimize any such correlations so that the statistical analyses can be carried out under the assumption of independence.

18.2 Random Ordering of the Experimental Sequence

Experiments should be run in a random sequence to prevent uncontrolled variables from biasing the results. For example, suppose one wishes to determine the effect of binder, in weight %, on nonwoven polyester fabric strength. The experiment may be conducted as follows. A batch of fiber is blended and continuously fed into a laboratory card (a machine that separates and lays fibers in a batting) that continuously makes a 1 ft wide, 2 oz/yd^2 web, which is taken up in a roll. About 5 linear yards are made. A latex emulsion binder is poured into a 1 ft-square, 3 in-deep pan and diluted to 50% with water. The web is unrolled a little and a 6-in square is cut from it. The fiber web sample is sandwiched between screens, dipped into the emulsion until saturated, roller-squeezed to remove excess emulsion, and oven-dried. The latex is diluted by adding water, and another sample is cut, sandwiched, dipped, squeezed, and dried. Each web sample picks up roughly the same amount of liquid on any dip. Because the latex emulsion was diluted, the second web will have less latex binder. The latex emulsion is again diluted, and another sample is dipped. Ten such samples are prepared. During drying, the evaporating water migrates to fiber crossovers conveying latex particles which bead-up there. After drying, the samples are cured at a high temperature (the binder glues the fibers) and the nonwoven fabrics are tested for strength.

The independent variable is the amount, in wt. %, of binder added to the fabric. In this procedure, several other factors which may also affect fabric strength are correlated to the independent variable in both time and space. One such factor is latex composition. There are finish oils, antistatic agents, and surfactants on the carded fiber. At each dip into the latex emulsion, a portion of these finishes are washed off, which are progressively added to the emulsion. The progressively increasing level of impurity in the emulsion is therefore correlated to decreasing binder add-on wt. % in time.

Another correlated variable is latex chemistry. The preparation of the 10 samples might take an entire morning. During that 4-hour period, the latex in the dip pan is exposed to both light and air. If either oxygen or light affects the latex-emulsion properties, then a change in the latex chemistry is also correlated in time with binder add-on wt. %.

Finally, just prior to carding the polyester web, the card may have been used on a batch of rayon. Consequently, the initial section of polyester web will have residual rayon finishes from the card. Since rayon finishes are different from polyester finishes, the last few web samples prepared (from the front of the roll) will contain foreign finishes. Those chemicals will have a unique effect on bond adhesion and ultimately on web strength. The level of foreign finish in the web is correlated to binder add-on wt. % in space.

When uncontrolled variables are correlated to the independent variable, one cannot distinguish whether the dependent variable is affected by the designated independent variable or by one of the uncontrolled variables. To eliminate such correlated uncontrolled effects, you should randomize the experimental order in both time and space. In this example, web samples should not be sequentially cut from the web. They should be cut in random positions throughout the 5-yard web. Systematic variation in binder add-on and latex chemistry should be prevented by avoiding progressive dilution of the previous emulsion. Eliminate those biasing effects by preparing fresh binder dilutions at random.

To randomize the order of your experimental sequence, first, list all of the experimental conditions. Usually, for convenience of understanding the experiment, the listing is done in some sort of order, but the order is inconsequential. Then generate a random number for each trial. In Excel this is the cell call "=Rand()". Run the experiments in ascending (or descending) order of the assigned random number.

Although statistical considerations strongly encourage random experimental order in both time and space, other considerations may require that experiments be run in a particular order. For instance, if two different reactor manifold designs are each to be evaluated at 10 different operating conditions, statistics would prefer that the trials be sequenced so that the manifolds are randomly chosen at each trial. However, if it takes a day of downtime to change manifolds, experimental time and costs may dictate that all the tests for one manifold design be run and then all the tests for the other design.

So, as a strong preference, randomize the treatment combinations in time and space to the maximum extent permitted by economic and other constraints, but don't demand randomization as a requirement.

18.3 Factorial Experiments

If you need a preliminary evaluation of several variables (a screening test) to determine the nature and ranges of future detailed experiments, any one of several factorial experimental designs may be suitable.

In factorial experiments a sufficient number of levels (it could be just two high and low; or three high and medium and low; etc.) of each factor are examined to obtain an approximate evaluation of its effects and its interactions with other variables. By reducing either the number of the experiments or the number of levels for each variable, the response will be covered at the expense of detail and the ability to estimate the model error. Increasing the number of experiments and number of levels per variable will improve model veracity but at the expense of extra time and effort. This is a characteristic trade-off involved in factorial designs.

Consider a simple linear model that is presumed to represent how y responds to x.

$$y = a + bx \tag{18.2}$$

Only two experimental (x, y) pairs are needed to determine coefficient values, a and b, for the model. But just because you can determine values, does not either 1) affirm that the model has the correct functionality, or 2) indicate the uncertainty associated with the model.

If you placed several experimental points within the x-range of interest, then you might detect curvature indicating that quadratic or reciprocal functionality should be included in the model. A rule of thumb, a heuristic, is that there should be about three times as many data points as there are degrees of freedom (number of adjustable coefficients) in the model.

Further, if the model provides a reasonable fit to the data, and if the y-variance is uniform throughout the x-range, then the residuals (the difference between model and measured y-value) should only be represented by experimental noise (error, uncertainty, variability), and the residuals should have a standard deviation that is the same as sigma that would be obtained from replicate experiments. If the standard deviation of the residuals is significantly greater than the standard deviation from replicate tests, this indicates that the model is not properly fitting the data. You should reevaluate the functionality included in the model. If the replicate sigma at one end if the x-range is significantly different from that of the other end of the x-range, then you should question the homoscedastic (uniform variance) assumption that is implicit in classic linear regression.

The experiments required for replicate testing should not be the extra ones used to reveal the $y(x)$ functionality. If for instance, the model is represented by Equation (18.2) indicating that 6 experiments would be recommended, and three were replicates at the x_1 value, and three at the x_1 value, then 1) there would be no in-between points to validate or contest the linear model assumption, and 2) the variability in the residuals would match that of the replicates pretending that the model was perfect.

The number of replicate trials (at the same conditions) should be in addition to the number of trials throughout the range to determine the coefficient values.

If there are two factors, x_1, and x_2, and the model was cubic in each factor, but quadratic with interactions, then the model would be:

$$y = a + bx_1 + cx_2 + dx_1^2 + ex_2^2 + fx_1^3 + gx_2^3 + hx_1x_2 \tag{18.3}$$

There are now eight coefficients to be evaluated, recommending about 24 trials. If the experimental plan were to have the same number of investigations for each x_i factor, then a 2×2 array of four levels (low and medium low and medium high and high) of each factor would provide $2^4 = 16$ data points, or a 2×2 array of five levels of reach factor would provide $2^5 = 32$ data points. Either choice bounds the recommended 24 data points. But either choice would be OK. And as an alternate, you do not need to do a square design, you could do a 4×5 design allocating the 5 levels to a more critical factor, with a total of

20 experimental data points. If you also wanted 5 replicate values at each of 2 points, this would add 2 × (5 – 1) more experiments to make a total of 28.

If there are 3 factors, x_1, x_2, and x_3, and the model was quadratic (with interactions), then the model would be:

$$y = a + bx_1 + cx_2 + dx_3 + ex_1^2 + fx_2^2 + gx_3^2 + hx_1x_2 + ix_1x_3 + jx_2x_3 \tag{18.4}$$

There are now ten coefficients to be evaluated, recommending 30 trials. If the experimental plan were to have the same number of investigations for each x_i factor, then a cube design with three levels of each factor would provide $3^3 = 27$ data points, close enough to the heuristic 30 trials.

With n number of levels for each factor, l number of levels for each, and m number of model coefficients, the heuristic is

$$n^l = 3m \tag{18.5}$$

Alternately, if n_i is the number of levels for the ith factor, then the heuristic is

$$\prod_{i=1}^{l} n_i = 3m \tag{18.6}$$

If we presume that the variance is uniform throughout the entire x_1, x_2, x_3 space, and choose 10 replicates (nearly enough to confidently assess variance) at a central point, then easily one could have an experimental design requiring about 30–40 trials for relatively simple models. If one did not want to place all replicates at one point, wanting to test if the uniform variance presumption is valid, one could place fewer than ten replicates at a few conditions, and use a pooled variance as a representative value. This, however, would reduce the precision of testing the homoscedastic assumption.

These are termed factorial experiments, ones that explore several levels of each factor. Factorial experiments do not necessarily have to be "square" – they do not have to have the same number of levels for each factor investigated. As described above, these were complete factorial experiments, for any level of one variable each level of the others is explored.

For screening trials, one could choose several options to reduce the experimental effort.

1. Choose fewer than the recommended heuristic number of trials. Perhaps, plan for 2 independent trials per model coefficient. The minimum, of course, to determine coefficient values is 1 trial per coefficient. However, fewer trials means that 1) the model may be fitting experimental noise, not true trends, and 2) you lose the ability to see if the model contains correct functionalities and to assess model uncertainty.
2. Do not include replicate trials. Sacrifice assessment of data variability to focus on the modeled trend.
3. Eliminate the interaction terms in the initial model. If the trial evidence (residual variation is greater than replicate variation) suggests that interactions might be important, then add them to the model, and correspondingly add additional trials for a second round of testing.
4. Lower the postulated highest order. If the trial evidence suggests that higher-order terms might be needed, include them in the model, and add additional trials for a second round of testing.

5. Keep the number of levels of each variable but perform an incomplete factorial design. Retain the most important sets of trial conditions. You could even add levels for each variable for a better look at how the residual changes relative to any x-variable. If the data indicates greater detail is warranted, add new points to a second set of testing.
6. Change the modeling approach to one that requires fewer coefficient values, perhaps use dimensionless groups rather than primitive variables.

18.3.1 Constraints

The discussion of factorial design is predicated on being able to implement combinations of extreme values on the trials. But it could be that the value of one input variable restricts possible values of another. As an example, in a distillation column the boil-up vapor leaving the top of the column is condensed and some is returned as liquid reflux to "wash impurity back down to the bottom". The reflux rate must be less than the boil-up rate. One cannot return more reflux than that which is condensed from the boil-up. Reflux choices cannot be greater than a boil-up choice. An experimental plan that considers all inputs as independent might have infeasible designed conditions.

18.3.2 Missing Data

Events within the enterprise might cut trials after a few have been completed. There are many reasons why a plan could be overridden with newly arising institutional priorities. This leads to an incomplete set of trials. Additionally, after a trial, it may become known that a particular trial was corrupted, and the data needs to be discarded.

Although a completely randomized order through the trial sequence might be preferred to minimize the impact of correlation with uncontrolled conditions, one might somewhat override complete randomization to start with the most important trials relative to how a model could be developed. If it is discovered that one trial needs to be repeated, it can be scheduled. If a change in priorities cuts the plan short, the most important trial conditions have been achieved.

18.3.3 Confounding

Confounding occurs in either 1) incomplete block experiments (where data from one block is missing), 2) in factorial designs when the number of data is lower than the minimum needed to develop all of the model coefficient values, 3) in an experiment with one trial in each block, or 4) when the range of input levels has a smaller effect on the response than the experimental variability. Here, no independent estimate of experimental error is possible. And, if there are fewer independent trial conditions than model coefficients, evaluating the effects of higher-order interactions is usually sacrificed to be able to determine lower-order functionalities.

Consider these cases to develop coefficient values for the Equation (18.2) model:

1. One experiment is performed, which is fewer than the number of model coefficients. The data has one (y_1, x_1).
2. Two experiments are performed, which is equal to the number of model coefficients, but they are at the same x-value. The data has two sets (y_1, x_1) and (y_2, x_1).

3. Two experiments are performed, which is equal to the number of model coefficients, and they are at different x-values. Yeah! The data has two sets (y_1, x_1) and (y_2, x_2). However, the range on the x-values causes a smaller change in the y-value than the experimental variation in y.

18.3.4 Alternate Screening Trial Designs

A full factorial design has n_1 values for one variable and n_2 values for another. Visually graphed, the experimental points for a two-factor experiment would form a rectangle with a regular grid of points. A three-factor experiment would appear as a regular 3-D rectangular prism, again with a grid of internal points. However, there are many other patterns. If the number of experimental points meets the three-per-coefficient heuristic, some internal grid points could be discarded and not significantly reduce model precision.

Alternately, there are many other shapes for the placement of points which seek to cover all of the space: The grid might not have regular Δx intervals. The pattern might be aligned with alternate axes (perhaps principal components). The pattern could be a star shape.

18.4 Takeaway

Factorial designs, screening trials, are easy to understand and visualize. Determining coefficient values from screening models is relatively easy. They are useful for identifying general trends in a region of interest.

Try to get the number of trials to be three times the number of model coefficients. If you want replicates to reveal process variability, count these as extra.

Do not think that a successful set of trials and a statistically significant screening model represents nature.

18.5 Exercises

1. Use Equation (18.5) or (18.6) to design trial conditions for a linear model in each of three x-variables. $y = a + bx_1 + cx_2 + dx_3$.
2. Use Equation (18.5) or (18.6) to design trial conditions for a quadratic model in each of two x-variables. $y = a + bx_1 + cx_1^2 + dx_2 + ex_2^2 + fx_1x_2$.
3. Consider that $y = 3 + 0.2x$ and that $\sigma_y = 2$, alternately that $R_y \cong 5$. But this is not yet known. The screening model is $y = a + bx$ and the x range of interest is $5 \le x \le 6$. How will the impact of the x-range on y-values compare to experimental uncertainty on y? If unacceptable, what can you change?
4. Explain why model-to-data residuals in screening models cannot be used to estimate the experimental variance on the y-measurements.
5. Use a nonlinear first principles model to represent nature, to generate $Y(\underline{x})$ data from a factorial design. Best fit screening model coefficients to the data. Compare the screening model to the nature-surrogate. Explore the impact of the screening model structure and the experimental design pattern.

19

Data-Based Model Validation

19.1 Introduction

We use mathematical models for design, optimization, analysis, and theoretical affirmation of concepts for natural phenomena and man-made products, procedures, and processes. So, it is relatively important that the mathematical model matches the real item. Unfortunately, nature seems always to be one step more complicated than our understanding, so, we cannot expect that our models will be absolutely true. We hope that testing of the models finds the models to be acceptable, or using more appropriate statistical terminology, that we cannot reject the models.

Data-based model validation will compare model outputs to experimental data. Logic-based validation compares the behavior of model variables to expectations. Although logic-based validation is important, the expectations are usually grounded in the same concepts that were used to derive the model. If there were no errors in the derivation from concept to mathematical formula to numerically computing the mathematical models, then one should expect the models to meet the logic-based validations. This chapter will focus on statistical techniques for comparing model to data.

There are several classes of models:

1. Steady-state (or equilibrium) deterministic models predict an end value. It could be a property of an assembled product, or the ending value after a fully relaxed transient. Examples could be the temperature of a mix of hot and cold water as calculated by volume and temperatures of the hot and cold water, how the speed of light transmission through material is affected by electron density, or the breaking strength of a rope as dependent on the number of fibers and the pitch of components. These typically are an average or expected value.
2. Alternately, steady-state stochastic models predict possible outcomes (or the distribution of possible outcomes) after steady-state or at the end. For instance, the distribution of breaking strength of a seatbelt as influenced by strength variation of the fiber elements.
3. Finally, the models may be dynamic (transient, or time-dependent) models that reveal how variables change in time. An example could be how the car speed evolves in time with a change in accelerator pedal position or a change in the road angle. Another example could be how the internal temperature of a cake changes when it is placed in an oven.

DOI: 10.1201/9781003222330-19

Validate and Verify are separate concepts. Both are important relative to model development and unfortunately sometimes the terms are used interchangeably. We distinguish them as:

Verify: To confirm the correctness of the solution, to check that no mathematical errors were made in solving the model or encoding errors in converting the model to executable code. Verify does not mean that the model concepts and mechanisms are correct.

Validate: A conscientious attempt by a critical and knowledgeable person, to be unable to reject the model at some stated confidence level by comparing the model to real process behavior. Validate means that the data could not be used to find fault with the model. It is not that the model is true, it just means that the experimental variance masks the difference.

This chapter is about validation.

All models are imperfect. Assumptions, idealizations, and simplifications are present, even in the most rigorous model. Validation, therefore, will not conclude that the model is correct. Validation will conclude that either "the model is so bad that it must be rejected" or "the model is not bad enough to be rejected". Acceptance means that the model was not rejected; it does not mean that the model was true (correct).

As a final concept, a model that is rejected by the data, a model that is statistically not valid, still may be fully functional. The model may not be perfect and may even have statistically significant deviations from the data, but in a balance of sufficiency with perfection, the imperfect model may still be adequate for its purpose.

19.2 Data-Based Evaluation Criteria and Tests

For steady-state deterministic models we would like the data to reveal:

1. The model outputs, y, should substantially go through the center of the experimental response data, throughout the entire data range, for each output variable. If there is only one model output, then after regression fitting of the model to data, the model likely goes through the center of the data. If not, suspect an error in the regression method. But if there are several output predictions, in an attempt to best fit all outputs with an inadequate model, some may show a consistent bias. Use a Wilcoxon Signed-Rank test on each output variable to determine if there is a bias in residuals. If there is a bias, you have evidence that the model can be rejected.
2. A residual is the difference between model and data. There should not be any trend in the residuals of any output variable w.r.t. any variable (input, output, chronological order, spatial order, or treatment). If the model is true to the data, then residuals should be independently distributed (randomly above and below). If there is an unexpectedly long sequence of residuals of like sign, then this is an indication of a model-data mismatch. Use an Autocorrelation of Lag-1 test, or alternately a Runs test, to see if there is sequential correlation in the data. Test

residuals for each output w.r.t. every other variable (input, output, chronological order, special order, or treatment), because the sorting of the sequence of residuals by one variable may not reveal autocorrelation, but it might when sorted by another variable. If a test reveals autocorrelation in a particular region for a particular variable, this can be a clue as to where the model needs to be improved. It might sound like a lot of work, but it is very simple to do by sorting data in a spreadsheet.

3. The variability expectation from either propagation of uncertainty in the data model or from replicate trials should match the magnitude of the residuals. Use a chi-squared or *F*-test to compare the variance of the residuals to either the estimated variance from propagation of uncertainty in the measurements or replicate experimental values. If you are propagating uncertainty to estimate measurement variance, and the ratio is unexpectedly too small or too large, it suggests that you do not understand the process in estimating variance. If you are comparing residuals to replicate variance and the residuals are unusually small, then likely you have too many degrees of freedom in the model (adjustable coefficients) relative to the number of data and are able to fit the model to noise. If the residuals are unusually large relative to the replicate variability, then likely the model is not a good fit to the data.
4. Use Bootstrapping to determine the 95% (or other) confidence interval on the predictions. The model is likely composed of functional choices of the human (either phenomenologically derived or heuristically assigned), and has coefficients adjusted to best make the model match the data. Bootstrapping randomly selects data with replacement, and develops many models based on the many sets of data samplings. If 95% of the data values (or about) are within the 95% confidence interval of the model prediction, and the outliers are not clustered in one region, then the model cannot be rejected. Use a hypothesis test on a proportion to see if your count of outliers could reject the model. Note if the models are linear, then you can use the analytical techniques from Chapter 15 to estimate the model confidence intervals.

For steady-state stochastic models the *CDF* of the model (whether it is a mathematical probability distribution or a numerical data distribution) should match the empirical *CDF* of the data. The deviations between the two distributions should be small, and the deviations should be independently above and below.

1. Use a Kolmogorov–Smirnov D-test to compare the model and experimental data distribution. If rejected, this suggests features of the model that need to be revisited. Alternately use the chi-squared goodness-of-fit test. If you choose the χ^2 test, we suggest partitioning the data into equal-area (equal count) cells.
2. Use a runs test, or autocorrelation lag-1 to see if the distribution of deviation signs is randomly and independently scattered throughout the range of the independent variables, when data are sorted w.r.t. other variables.

For a transient deterministic model to match the data, there are any of several aspects that could be of concern. These could be the steady-state final value, the rate of change of the model after an input change, the delay in model response, the value of the modeled output after a desired time (before steady-state), or any of several others. For any of these deterministic values that are important, use the four tests described for the steady-state deterministic models:

1. A Wilcoxon signed-rank test to determine if there is a bias. If there is a bias, you have evidence that the model can be rejected.
2. An autocorrelation of lag-1 test, or alternately a runs test, to see if there is sequential correlation in the data. Test residuals for each output w.r.t. any variable (input, output, chronological order, special order, or treatment), because the sorting of the sequence of residuals by one variable may not reveal autocorrelation when sorted by another variable.
3. A chi-squared or *F*-test to compare the variance of the residuals to the estimated variance from propagation of uncertainty in the measurements.
4. Use bootstrapping to determine the 95% (or other) confidence interval on the predictions, then a hypothesis test on proportion within (or without) to see if your count of outliers could reject the model.

In any case, the design of the experiment that generates data for model validation needs to generate data that can critically, legitimately challenge the model. Here are three criteria:

1. There needs to be a sufficient number of independent data values within the input variable range of interest: 20 to 40 for each input variable would be a desirable number. These should not be replicates, but independent values throughout the entire range. Certainly, adding replicates will aid in estimating experimental variance, but replicates do not reveal the trend.

Fewer than 20 input values for each input variable would be less expensive, but too few would begin to lose power to "see" model deviations. Desirably, use at least about 10 independent input values for each input variable. If you are considering a model with three input variables, this might seem to direct $20 \times 20 \times 20 = 8{,}000$ experiments! But there is no need for a complete factorial design. (See Chapter 20.) More than 40 values for each variable would provide a more critical test of the model but acquiring excessive data from a designed experiment might be impractically expensive or time consuming.

2. Further, there needs to be a large enough range on each input variable to be able to ensure that the change in the output is large relative to the experimental variation. If not, the model trend will not be distinguishable from the noise. The range on each y must be much greater than the standard error of the deviations, $R_y \gg \sigma_e / \sqrt{n}$.

Subjectively using "much greater" to represent an order of magnitude,

$$R_y \geq 10 S_e / \sqrt{n} = 10 \frac{\left(\sum e^2 / (n-1)\right)^{1/2}}{\sqrt{n}}, \text{ for each } y \tag{19.1}$$

Since $n \geq 10$, $\sqrt{n-1} \approx \sqrt{n}$, and since the factor of 10 in the inequality is subjective, the condition can be computationally simplified to

$$R_y \geq 10 \frac{\left(\sum e^2\right)^{1/2}}{n}, \text{ for each } y \tag{19.2}$$

If this condition is not met, the scatter will be too great to "see" skewed or off-center data.

3. Finally, there should be no gaps in the data regions as viewed by any variable (input or output). If there are gaps, then there cannot be any claims made about that region of the model.

Example 19.1: Model validation of an *e* vs *u* graph. A process that has two inputs and three outputs has been modeled. Shown below are the $n = 15$ values of one input, u_2, and the residuals of one output, e_3. From the data, one can see the range on u_2 is 2.9. Since the *e* vs *u* tests do not require y_3 and $\hat{y}_3$ data, those data are not shown, although the range on $\hat{y}_3$ is 1.7. Is the modeled u_2 *to* y_3 relationship valid?

u_2	0.6	2.2	1.1	3.2	2.7	1.4	1.7	3.5	3.0
e_3	−0.3	−0.4	−0.5	+ 0.2	−0.4	−0.3	−0.4	−0.3	−0.3
u_2	2.2	1.0	3.3	0.7	2.3	3.2			
e_3	+0.1	−0.2	−0.2	+0.1	−0.1	−0.4			

The e_3 versus u_2 plot is presented in the figure below and shows that the residuals on output variable 3, e_3, are biased toward the negative side throughout the range of u_2. This situation indicates a systematic error in the model

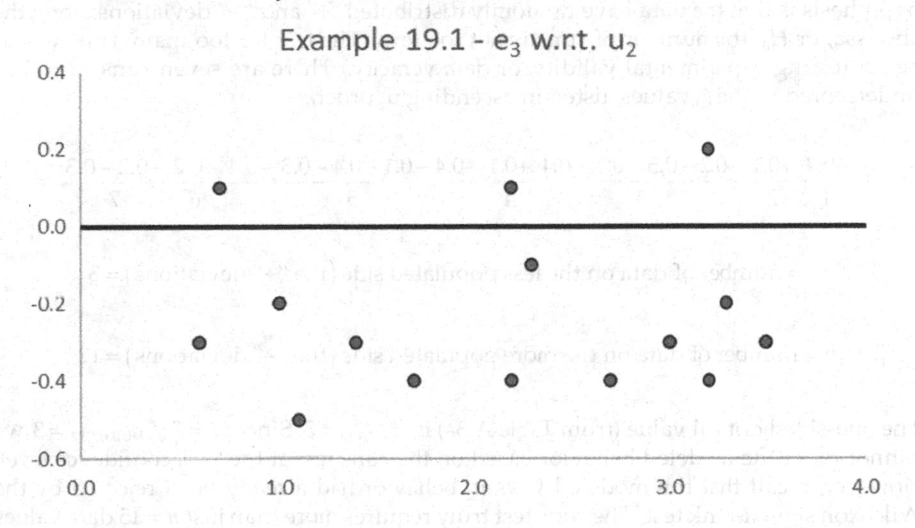

Number criteria: With $n = 15$, there is a minimally adequate number of data points for confidence in a statistical test. Proceed with tests.

No missing region criteria: There are no wide empty spaces on the u_2 axis. Data permits comparison of the model to data throughout the range. Proceed.

Range test to determine if the range-to-scatter (signal-to-noise) ratio is large enough to permit statistical discrimination. From the problem statement, $R_{y_3} = 1.7$. The standard error of the deviations is

$$\frac{S_e}{\sqrt{n}} = 0.08165 \approx \frac{\left(\sum_{k=1}^{n} e_k^2\right)^{1/2}}{n} = 0.07888 \approx 0.08$$

$$R_{y3} = 1.7 \approx 10\frac{S_e}{\sqrt{n}} \approx 0.8$$

The range of 1.7 is more than 10 times bigger than 0.08, the standard error of the data. The "signal" is much larger than the "noise". Consequently, one can see trends, and statistical discrimination is allowable. Proceed.

Wilcoxon Signed Rank: If $n \leq 30$ and the data are supposed to be symmetric, apply the Wilcoxon signed-rank test to determine if the data are centered about the model prediction. If the number of data is greater than 30, the sign test will be easier to apply and nearly as powerful. The Sum of Ranks of positive residuals is 9 (rank values are shared when the absolute values of the residuals are equal). The critical values (two-sided, alpha = 0.05) for the Wilcoxon signed-rank test (Table A.2) are 25 and 95. Since the data value of 9 is beyond one of the critical values, the test rejects the null hypothesis of the residual median being zero. Reject the model on the zero-bias condition, at the 95% confidence level.

Autocorrelation: The sum of the product of the 14 adjacent residuals is 0.44, and the sum of the 15 residuals squared is 1.4. The r-lag-1 value is $0.44/1.4 = 0.31428\ldots$ The two-sided critical values for the r-lag-1 statistic (Equation 13.6) are $\pm 0.43334\ldots$ Since the data value is not beyond the critical value, the test cannot reject the null hypothesis of zero autocorrelation. Accept the model on the zero-autocorrelation condition, at the 95% confidence level. However, recall that the modeled $\hat{y}_3$ vs u_2 behavior had already been rejected by the Wilcoxon signed-rank test.

A runs test is to determine if there is any skew or curvature in the data band about the model (higher-order systematic error in the model). When the data are viewed in ascending order, a run is a contiguous group of data with the same sign. The null hypothesis is that the data have randomly distributed "+" and "−" deviations along the abscissa, or H_0: the number of runs is not too few. (The test for too many runs would be for testing experimental validity, or data veracity.) There are seven runs, $U = 7$, as underscored in the e_3 values, listed in ascending u_2 order.

$$\underset{1}{\underline{-0.3}}\ \underset{2}{\underline{+0.1}}\ \underset{3}{\underline{-0.2-0.5-0.3-0.4}}\ \underset{4}{\underline{+0.1}}\ \underset{5}{\underline{-0.4-0.1-0.4-0.3-0.4}}\ \underset{6}{\underline{+0.2}}\ \underset{7}{\underline{-0.2-0.3}}$$

$$m = \text{number of data on the less populated side} \left(\text{the "+" deviations}\right) = 3$$

$$n = \text{number of data on the more populated side} \left(\text{the "−" deviations}\right) = 12$$

The one-sided critical value (from Table A.3a) is $u_{0.05,3,12} = 3$. Since $U = 7 \nleq u_{0.05,3,12} = 3$ we cannot reject the modeled behavior based on the runs test at the 95% confidence level. However, recall that the modeled $\hat{y}_3$ vs u_2 behavior had already been rejected by the Wilcoxon signed-rank test. The runs test truly requires more than just $n = 15$ data values to critically test the data.

Although more data would be desirable, the Wilcoxon signed-rank test reveals that the model does not match the data.

Example 19.2: Someone has hypothesized that a population is normal, has sampled the population 430 times, calculated $\bar{X} = 15.327$ mm and $S = 0.768$ mm, partitioned the range into 15 cells, and counted the number of observations in each cell. The data are tabulated below. Is the population normal? The partitions were chosen so that if the population is normal, each of the 15 cells should contain 1/15 = 6.666...% of the observations or 28.666... observations.

Cell	Range	Number of observations	Expected number	Dev. sign
1	$0 < x \le 14.18$	31	28.666	–
2	$14.18 < x \le 14.47$	33	28.666	–
3	$14.47 < x \le 14.68$	34	28.666	–
4	$14.68 < x \le 14.85$	34	28.666	–
5	$14.85 < x \le 15.00$	33	28.666	–
6	$15.00 < x \le 15.13$	26	28.666	+
7	$15.13 < x \le 15.26$	27	28.666	+
8	$15.26 < x \le 15.39$	31	28.666	–
9	$15.39 < x \le 15.52$	34	28.666	–
10	$15.52 < x \le 15.66$	29	28.666	–
11	$15.66 < x \le 15.81$	29	28.666	–
12	$15.81 < x \le 15.97$	19	28.666	+
13	$15.97 < x \le 16.18$	24	28.666	+
14	$16.18 < x \le 16.48$	27	28.666	+
15	$16.48 < x \le \infty$	19	28.666	+

Apply the chi-squared goodness-of-fit test. Since there are more than five observations in each category and very many observations, and since a critic will probably not claim experimenter bias in the choice of partitions, we will use the chi-squared goodness-of-fit test for distribution deviations. (In fact, without having been given the individual observation values we cannot apply the Kolmogorov–Smirnov test.) The hypothesis being tested is that the number of observations in each cell represents a normal distribution, or that deviations from the expected number of observations are small (this is a one-sided test).

Choose $\alpha = 0.05$. The degrees of freedom are $v = n - 1 - k = 15 - 1 - 2 = 12$. Calculate χ^2 as:

$$\chi^2 = \frac{(28.666-31)^2}{28.666} + \frac{(28.666-33)^2}{28.666} + \dots + \frac{(28.666-19)^2}{28.666} = 12.395\dots$$

The critical value is $\chi^2_{12,0.95} = 21.0261$. Therefore, $\chi^2 = 12.395\ldots \ngtr \chi^2_{12,0.95} = 21.0261$. We conclude, using the χ^2 goodness-of-fit test on 430 observations partitioned into 15 uniform-area cells, that we cannot reject the hypothesis that the population is normally distributed at the 95% confidence level. We accept that the population could be normally distributed.

Apply the runs test to determine if there is some systematic pattern to the deviations indicating skew in the distribution. If there is skew, the "+" and "–" deviations in the observed and expected frequencies in each cell would be clustered, and there would be too few runs.

Hypothesize that the "+" and "–" deviations each have a $p = 0.5$ probability of occurring and we choose $\alpha = 0.05$. There are $m = 6$ "+" and $n = 9$ "–" deviations and $U = 4$ runs. The critical value is $u_{0.05,6,9} = 4$. Now, $U = 4 \le u_{0.05,6,9} = 4$. Accordingly, based on the runs test on the deviations of observed and expected numbers of observations in 15 equal-area cells containing 430 observations, we reject the hypothesis that the population is normally distributed at the 95% confidence level. There seem to be long trends of residuals above and below the model.

Even though the deviations from the normal model are small, the population does not appear to be normally distributed. It is significantly skewed.

Note: Two tests were performed on the hypothesis that the data were normally distributed. The chi-squared goodness-of-fit data could not reject the hypothesis because the deviations between observed and hypothesized distributions were not too big. However, the runs test determined that the deviations were improbably grouped, indicating a skew in the shape of the population distribution from that hypothesized, and found just cause to reject the hypothesis.

Caution: The standard alternate statement for "cannot reject" is "accept". You may have a tendency to apply one statistical test, find that the hypothesis is "accepted", and then not apply the other required tests. There are several characteristics of a good model. Each nonparametric statistical procedure tests for only one characteristic. For a model to be "good", it must be "accepted" by each of the tests.

19.3 Bootstrapping to Estimate Model Uncertainty

In regression, model coefficient values are selected to minimize the deviations between data and model. It could be a purely empirical model with many coefficients, or one that was developed from a phenomenological (mechanistic, either first-principles or rigorous) approach with only a few adjustable coefficients. Here, one set of experiments provides the data to best fit the model to data. However, a replicate set of experiments will provide similar but different data, due to the experimental vagaries, providing different model coefficient values. The question is, "How does irreproducibility in the data affect the uncertainty on the model?" Bootstrapping is a technique to estimate the uncertainty in model predictions due to uncertainty in the experimental data.

One assumption in bootstrapping is that the experimental data that you have represents the entire population of all data realizations, including all nuances in relative proportion. It is not the entire possible population of infinite experimental runs, but it is a surrogate of the population. A sampling of that experimental data then, represents what might be found in an experiment. Another assumption is that the model cannot be rejected by the data, that the model expresses the underlying phenomena. In bootstrapping:

1. Obtain N datasets experimentally. Include all expected range of values and vagaries. This dataset will be the surrogate for the entire population.
2. Sample from the experimental data (the surrogate population) with replacement (retaining all data in the draw from the original set) to create a new set of data. The new set should have the same number of items in the original, but some items in the new set will likely be duplicates, and some of the original data will be missing. This represents an experimental realization from the surrogate population.
3. Determine the model coefficient values that best fit the dataset realization from Step 2. This represents the model that could have been realized. It does not matter whether the objective function is vertical or total least squares, SSD or rms, maximum likelihood, etc. Use whichever is appropriate.
4. Record the model coefficient values.
5. For independent variable values of interest, determine the modeled response. You might determine the y-value for each experimental input x-set. This would

be simple. Alternately, if the model is needed for a range of independent variable values, you might choose key x-values within the range and calculate the model y for each x-set. This is likely more useful.

6. Record the modeled y-values for each of the desired x-values.
7. Repeat Steps 2–6 many times (perhaps over 1,000, perhaps $n = 100{,}000$).
8. For each desired x-set, create the *CDF* of the n model y-predictions. This will reflect the distribution of model prediction values due to the vagaries in the data sample realizations. The variability of the prediction will indicate the model uncertainty due to the vagaries within the data.
9. Choose a desired confidence interval value. The 95% range is commonly used.
10. Use the cumulative empirical distribution of the n model predictions to estimate the confidence interval on the model prediction. If the 95% interval is desired, then the confidence interval will include 95% of the models, or, 5% of the modeled y-values will be outside of the confidence interval. As with common practice, split the too-high and too-low region of values into equal probabilities of 2.5% each, and use the 0.025 and 0.975 *CDF* values to determine the y-values for the 95% confidence interval. With $n = 1{,}000$ trials, sorted on the y-value, the 25th and 975th values represent the 95% interval. With many other choices for the number of trials, you will have to interpolate between data values.

This bootstrapping approach presumes that the original data have enough samples covering all situations so that it represents all features of the entire possible population of data. Then the new sets (sampled with replacement) represent legitimate realizations of sample populations. Accordingly, the distribution of model prediction values from each re-sampled set represents the distribution that could arise if the true population were independently sampled.

If you are seeking to model transient data, you might have set up one situation then take 100 measurements over time. But this is one realization, not 100. All ensuing 100 samples are related to the same initialization. If the set-up would have mixing, composition, temperature, etc. variation from one set-up to another, then the 100 sequential samples do not represent that true variation. They are simply one realization. You should run many replicate trials of the 100 measurements to include the set-up to set up variability.

Similarly, if you create a large sample of material, or set up an experimental condition, then perform multiple tests on the same material (or set-up) all the tests reflect a single set-up. These replicate part of the source of variability, but not all of it. The data excludes the variation due to set-up, which is in effect a single realization. For example, in rheology we make a mixture, place a sample of it in a device, and measure shear stress for a variety of shear rates (perhaps 20 different shear rates). Then we plot stress w.r.t. shear rate and use regression to match a rheology model to the 20 sets of data. But there is error in the sample preparation, and the 20 sets of data represent one sample preparation realization, not the population. To make the data reflect the population, perform the tests on replicate set-ups (independent material preparations).

Bootstrapping assumes: The limited data represents the entire population of possible data, that the experimental errors are naturally distributed (there are no outliers or mistakes, not necessarily Gaussian-distributed, but the distribution represents random natural influences), and that the functional form of the model matches the process mechanism. Then a random sample from your data would represent a sampling from the population and for each realization the model would be right.

If there are N number of original data, then sample N times with replacement. Since the Central Limit Theorem indicates that variability reduces with the square root of N, using the same number keeps the variability between the bootstrapping samples consistent with the original data. In Step 2, the assumption is that the sample still represents a possible realization of a legitimate experimental test of the same N. If you use a lower number of data in the sample, M, for instance, then you increase the variability on the model coefficient values. You could accept the central limit theorem and rescale the resulting variability by square root of M/N. But the practice is to use the same sample size as the "population", to reflect the population uncertainty on the model.

In Step 7, if only a few re-samplings, then there are too few results to be able to claim the variability range with certainty. As the number of Step 7 re-samplings increases, the Step 10 results will asymptotically approach the representative 95% values. But, the exact value after infinite re-samplings is not the truth, because it simply reflects the features captured in the surrogate population of the original N data, which is not actually the entire population. So, balance effort with precision. Perhaps $n = 100$ re-samplings will provide consistency in the results. On the other hand, it is not unusual to have to run 100,000 trials to have Monte Carlo results converge. After 100 or so re-samplings observe how the *CDF* evolves, and keep adding bootstrapping trials until the *CDF* reasonably settles.

As described, bootstrapping assumes that the data have no systematic bias. If there is a systematic bias, then all the data would be shifted up or down. Here, bootstrapping is an analysis of random fluctuation. However, if you want to determine the joint impact of random and systematic errors on the model prediction, either: 1) Use propagation of variance to estimate the systematic error, then the square root of the sum of variances. Or 2) Include a random realization of possible systematic error in each bootstrap sample as described in Chapter 9. Then the model will reveal the joint impact.

One can estimate the number of re-samplings, n, needed in Step 7 for the results in Step 10 to converge from the statistics of proportions. From a binomial distribution the standard deviation on the proportion, p, is based on the proportion value and the number of data:

$$\sigma_p = \sqrt{p(1-p)/n} \tag{19.3}$$

Desirably, the uncertainty on the proportion will be a fraction of the proportion:

$$\sigma_p = fp \tag{19.4}$$

Where the desired value of f might be 0.1.

Solving for the number of data required

$$n = \left(\frac{1}{p} - 1\right) / f^2 \tag{19.5}$$

If $p = 0.025$ and $f = 0.1$, then $n \approx 4{,}000$.

Although $n = 10{,}000$ trials is not uncommon, and $n = 4{,}000$ was just determined, for most engineering applications 1,000 re-samplings may provide an appropriate balance between computational time and precision. Alternately, you might calculate the 95% confidence limits on the y-values after each re-sampling and stop computing new realizations when there is no meaningful progression in its value, when the confidence limits seem to be approaching a noisy steady-state value.

In Step 10, if you assume that the distribution of the $\tilde{y}$-predictions are normally distributed, then you could calculate the standard deviation of the $\tilde{y}$-values and use 1.96 times the standard deviation on each model prediction to estimate the 95% probable error on the model at that point due to errors in the data. Here, the term "error" does not mean mistake, it means random experimental normal fluctuation. The upper and lower 95% limits for the model would be the model value plus/minus the probable error. This is a parametric approach.

By contrast, searching through the n = 1,000, 4,000, or 100,000 results to determine the upper and lower 97.5% and 2.5% values is a nonparametric approach. The parametric approach has the advantage that it uses values of all results to compute the standard deviation of the $\tilde{y}$-prediction realizations and can get relatively accurate numbers with much fewer samples. Perhaps, $n = 20$. However, the parametric approach presumes that the variability in $\tilde{y}$-predictions is Gaussian (bell-shaped and symmetric). It might not be. The nonparametric approach does not make assumptions about the underlying distribution, but only uses two samples to interpolate each of the $\tilde{y}_{0.025}$ and $\tilde{y}_{0.975}$ values. So, it requires many trials to generate truly representative confidence interval values.

Unfortunately, the model coefficient values are likely to be correlated. This means, if one value needs to be higher to best fit a data sample, then the other will have to compensate (perhaps be lower). If you plot one coefficient w.r.t. another for the n re-samplings and see a trend, then they are correlated. When the variability on input data values is correlated, the classical methods for propagation of uncertainty due to coefficient uncertainty are not valid. They assume no correlation (no co-variance) in the independent variables in the propagation of uncertainty.

Also, Step 10 has the implicit assumption that the model matches the data, that the model cannot be rejected by the data, that the model expresses the underlying phenomena. If the model does not match the data, then bootstrapping still will provide a 95% confidence interval on the model, but you cannot expect that interval to include the 95% of the data. As a caution: If the model does not match the data (if the data rejects the model) then bootstrapping does not indicate the range about your bad model that encompasses the data, the uncertainty of your model predicting the true values.

Figure 19.1 reveals the results of a bootstrapping analysis on a model. The model is of the controller output required to get a desired air flow rate, from one author's pilot-scale experimental units. The model is based on first-principles fluid mechanic concepts. The circles represent data, the inside thin line is the modeled value from the entirety of original $N = 15$ data, and the darker lines indicated the 95% limits of the model based on 100 realizations of the dataset.

In Figure 19.1, two data points out of 15 are beyond the 95% limits. This is cause for suspicion, but not enough violations as expected to confidently reject the model. Three out of 15 would have been. However, as it turned out, the most extreme data point at about $x = 70$ was subsequently rejected by using Chauvenet's criterion (see Section 15.2.4).

If the experimental procedure is valid, the deviations between replicate sets would be small and the model-to-model prediction variability would be small, but it would exist.

If the model matches the underlying phenomena, then ideal natural experimental vagaries (the confluence of many, small, independent, equivalent sources of variation) should result in residuals that have a normal (Gaussian, bell-shaped) distribution. If this is the situation 1) model matches phenomena, 2) residuals are normally distributed, 3) model coefficients are linearly expressed in the model, and 4) experimental variance is uniform over all of the range (homoscedastic), then analytical statistical techniques have been developed to propagate experimental uncertainty to provide estimates of uncertainty on model

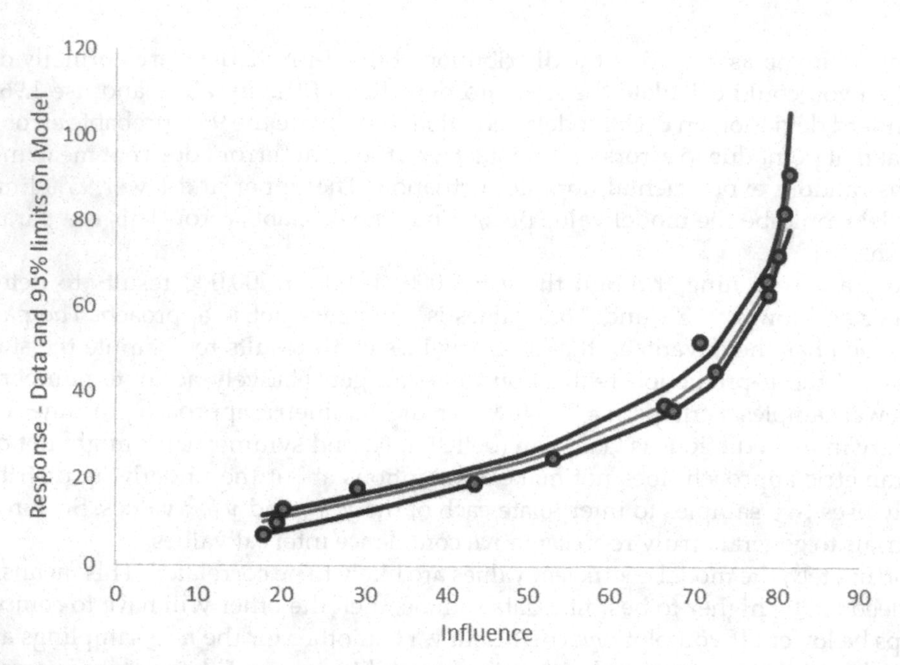

FIGURE 19.1
Bootstrapping estimate of model uncertainty due to data.

coefficient values, and on the model. It gives the 95%, or so, probable range for the model. However, if the variation is not normally distributed, if the model is nonlinear in coefficients, if variance is not homoscedastic, or the model does not exactly match the underlying phenomena, then the analytical techniques are not applicable. In this case numerical techniques are needed to estimate model uncertainty. Bootstrapping is recommended. It seems to be understandable, legitimate, simple, and is widely accepted.

Figure 19.2 reveals a bootstrapping analysis of the response of electrical conductivity to salt concentration in water. The data are generated with [salt] as the independent variable, and conductivity as the response, but as an instrument to use conductivity to report [salt] the x and y axes are switched. Here is a calibration graph for laboratory data analysis. The model is a polynomial of order 3 (a cubic power series with four coefficients).

The conductivity measurement results in a composition error of about +/− 2 mg/dL in the intermediate values and higher at the extremes. The insight is, if +/− 2 mg/dL is an acceptable uncertainty, then the calibration is good in the intermediate ranges. If not, the experimenters need to take more data to use more points to average out variation. Notice that the uncertainty in concentration has about a +/− 7 mg/dL value in the extreme low or high values. So, perhaps the experiments need to be controlled so that concentrations are not in the extreme low or high values where there is high uncertainty.

Figure 19.3 shows an example of laboratory analysis data from calibrating index of refraction with respect to mole fraction of methanol in water. Because the response gives two mole fraction values for one IR value, the data represents the original *x*–*y* order, not the inverse that is normally used to solve for *x* given the measurement. This model has a sine-like functionality, because it gives a better fit than a cubic polynomial.

How to interpret? If the measured index of refraction, gets a value of 1.336, the model (the middle of the five lines) indicates that the high methanol composition is about

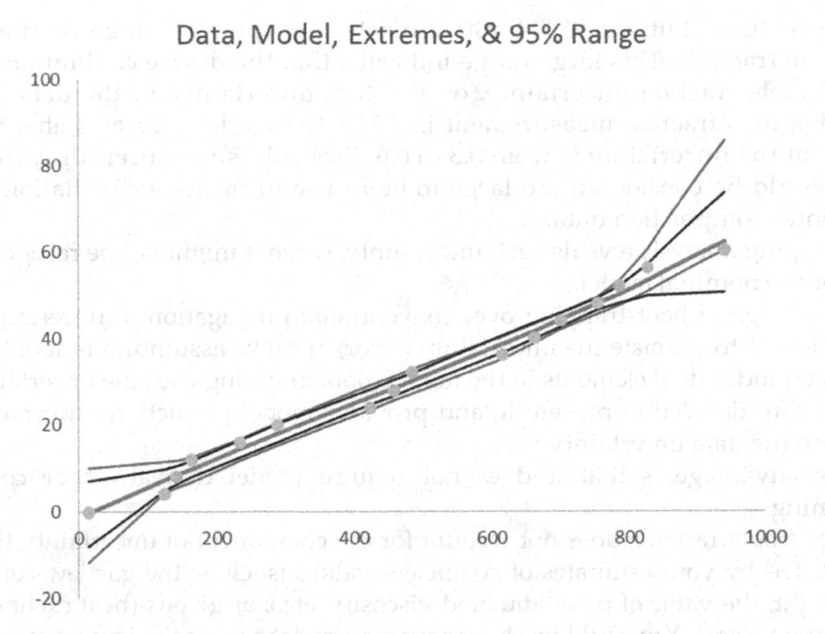

FIGURE 19.2
Bootstrapping on conductivity data. The abscissa is the instrument reading, and the ordinate is composition.

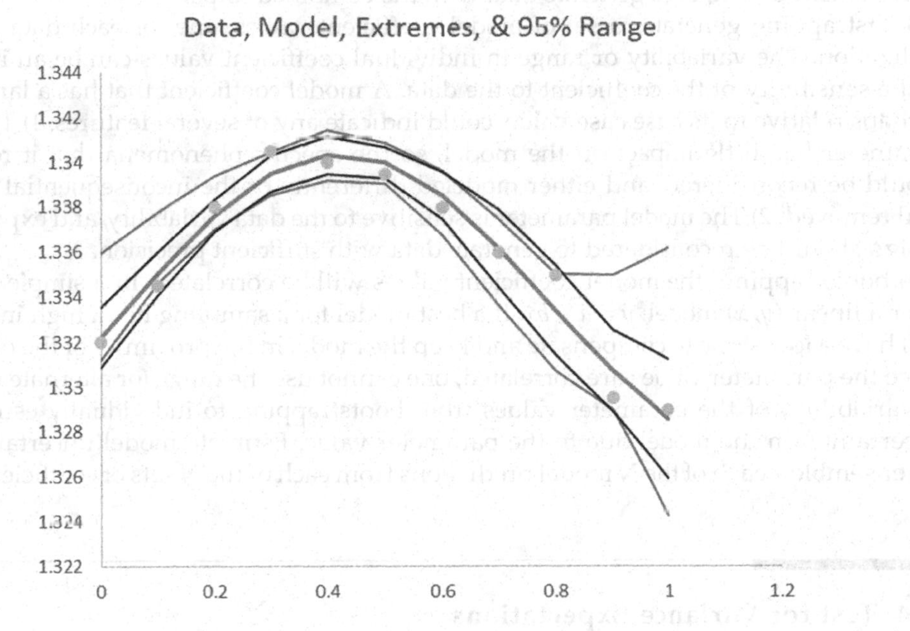

FIGURE 19.3
Bootstrapping on index of refraction (IR) data. IR is on the ordinate, and methanol mole fraction is on the abscissa.

0.72 mole fraction. But, the 95% limits indicate that it might range from about 0.65 to 0.80 mole fraction. This large range indicates that the device calibration produces +/− 0.075 mole fraction uncertainty, or +/− 10% uncertainty in the data. However, if the index of refraction measurement is 1.340, the model indicates about 0.4 mole fraction, but the uncertainty is from 0.3 to 0.6. Probably, such uncertainty in the mole fraction would be considered too large to be of use in fitting a distillation model to experimental composition data.

Bootstrapping analysis reveals such uncertainty, when it might not be recognized from the data or the nominal model.

One advantage of bootstrapping over conventional propagation of uncertainty is that you do not have to estimate the uncertainty (error) or make assumptions about error distributions on individual elements in the model. Bootstrapping uses the uncertainty in the data, as nature decided to present it, and provides model-prediction uncertainty corresponding to the data uncertainty.

Another advantage is that it does not require model derivatives or complicated programming.

However, bootstrapping does not account for the component of uncertainty that would be contributed by your estimates of coefficient values (such as the gas law constant, the speed of light, the value of pi, a tabulated viscosity, etc.), or givens (heat exchanger duty, production rate, etc.). You could use bootstrapping to determine the impact of experimental uncertainty on the model prediction, ε_y. Then use propagation of uncertainty to combine that with a propagation of uncertainty from model parameters and givens to generate an estimate of total model error.

Alternately, as described in Chapter 9, inject realizations of these uncertain values into each bootstrap sample to generate data with the combined impact.

Bootstrapping generates a set of model coefficient values, one for each data sampling realization. The variability or range in individual coefficient values can be an indication of the sensitivity of the coefficient to the data. A model coefficient that has a large range, perhaps relative to its base case value, could indicate any of several features: 1) The model parameter has little impact on the model, so the specific phenomena that it represents should be reconsidered, and either modeled differently or the inconsequential phenomenal removed. 2) The model parameter is sensitive to the data variability, and experimental design should be reconsidered to generate data with sufficient precision.

In bootstrapping, the model coefficient values will be correlated. In a simple case consider a linear (y, x) model, $y = a + bx$. If a best model for a sampling has a high intercept, it will have a low slope to compensate and keep the model in the proximity of the other data. Since the parameter values are correlated, one cannot use the range (or alternate measures of variability) of the parameter values from bootstrapping to individually estimate the uncertainty on the model due to the parameter value. Estimate model uncertainty from the ensemble – each of the N model predictions from each of the N sets of coefficient values.

19.4 Test for Variance Expectations

If the model functionality matches the process, and the experiment to generate data is properly performed and understood, then the residuals (model to data deviations) should have a variance that matches the propagation of variance on the experimental data model (Chapter 8), and the model uncertainty from bootstrapping (Section 19.3).

The variance in the residuals should match that expected from propagation of uncertainty in the data model. One could compare the range of residuals to either replicate trials or the propagation of probable error on the data model (the method used to calculate a data value from measurements). If the residual range is much larger than expected, this could imply that the process model is wrong, or that the experiment was not controlled or understood as expected. If the residual range is much smaller than expected, this could mean that the experimental uncertainty was much smaller than expected. In any case, something was not properly represented.

A statistically proper way to compare variances is with an *F*-statistic. However, for the comparison of residuals to those expected from a propagation of variance, the propagation of variance is just an estimate based on linearization, independence, and human estimates of several of the component uncertainties. It is a reasonable estimate, not the truth. Our experience has usually been that the number of data in question is too low (too few residuals) to lead to a definitive test, and the human choice as to how to partition the data could lead to challenges.

Although use of an *F*-statistic to test variances might be the "should", the test would be infected with human choices. Accordingly, it seems that a human judgment as to whether the residual variance matches the expectation and whether the data is homoscedastic best balances the perfection/sufficiency balance of values.

Probable error from propagation of uncertainty on data model should be equivalent to the standard deviation (or effectively the rms) of the residuals. The *F*-test indicates that a 4:1 ratio might not be unexpected for experiments of $N = 15$ or so data. So, this test will only be revealing if the ratio is larger than about 5:1. If the expected probable error is much smaller than the rms then you may be underestimating your experimental uncertainty, or the model might not be matching the data leading to large deviations. If the expected probable error is much larger, then you may be overestimating the experimental uncertainty, your model may have too many adjustable coefficients and be fitting the noise, or you might not have reached steady-state when you took the data. A look at residual variance and that expected from propagation of variance would provide a check on your understanding of the data-generating process.

19.4.1 Trouble Shooting Variance Indications

You hope to see that:

1. Model-predicted variation is ≅ experimental (within 4:1 or 1:3 ratio). If so:
    1.1. This probably confirms the analysis and affirms understanding of the process.
    1.2. You can use the model to see key contributors to variation and estimate the improvement if effort is invested to reduce the source of variation.
    1.3. You can use the model to see if alternate processing conditions would reduce variation, e.g., shorter samples for tensile testing (weakest link in a chain analysis), longer cure times to ensure approach to equilibrium or completeness, etc.
    1.4. The variation on an influence, a model input, an independent variable, may be due to several factors; if important, use propagation of uncertainty to see what aspect is most important. If no model is possible, run tests to seek to eliminate variation. Perhaps use a single technician to prepare samples from

common chemical batches on a single oven, etc. This may eliminate variability. If the resulting sigma is improved, then seek the course of variation and seek to eliminate that source.

But you might find that:

2. Model-predicted variation is >> experimental (greater than four times experimental). If so:
   2.1. Your estimate of some source of variability is too large.
   2.2. The experimental data is described by the standard error, $s/\sqrt{N}$, not standard deviation, s.
   2.3. Fewer than 10 samples in experimental s calculation, provides a poor estimate.
   2.4. Variation in influences may be correlated (if a technician rounds up, then factors in the numerator and denominator will tend to cancel).
3. Model-predicted variation is << experimental (less than experimental/3).
   3.1. Model does not include an influence.
      3.1.1. Such as T or %RH on load – generate a model, perhaps by regression, then include the new variables in the analysis.
      3.1.2. Variation in experimental is due to multiple machines, multiple dual operators, etc. Isolate data from separate devices or procedures.
      3.1.3. Variation in experimental is due to sample preparation (composition, materials, technique). Run an experiment to have one person make up samples from one batch of chemicals, cooked in one oven, etc. Design a procedure to eliminate what might be sources of variation. See if this reduces variation.
      3.1.4. Variation is due to sloppy technique (sample handling, placement in machine, etc.).
   3.2. Estimate of some source of variability is too small.
   3.3. Experimental data is not limited to replicates but combines disparate data and calculates sigma from things that should not be identical.
   3.4. Less than 10 samples in experimental σ calculation.
   3.5. Variation in influences may be correlated (if a technician rounds up then all dimensions will be high, and area will be doubly high).

19.5 Closing Remarks

The original definition was that validation is the failure of a conscientious attempt by a critical and knowledgeable person to use data to reject the model at some stated confidence level. The validation procedure is as follows:

1. Ensure that a good experimental technique was used and assume that the data are correct. It is imperative that the data accurately reflect the process if any sort of conclusions (not just model validation) are to be drawn about the process.

2. Assume that the model is correct, and that the data are randomly perturbed.
3. Anticipate what the model/experimental data plots should look like if both model and data are correct. This step identifies the hypotheses to be tested. Validation requires a conscientious and critical attempt to find ways to demonstrate that the model is wrong by a knowledgeable person. The most knowledgeable person is the one who developed the model (who likely also wants it to be a good model) and who may suffer from a conflict of interest. Your particular model may require a comparison not suggested in our brief presentation. Be your own devil's advocate. If you are not critical, someone else will be, and the validity of your entire work can be questioned if one point is questioned.
4. Choose a level of confidence. Normally 95% is used, but, where life and safety are at risk, often 99.999% confidence may be appropriate.
5. Determine the number of experimental data required for validation. Methods are based on the level of significance and confidence interval desired. In general, we recommend 20 or more experimental data points. However, experiments can be expensive, and 10 industrial trials could cost more than what assurance of the model is worth. In addition, 10 industrial or pilot trials may require a 2- to 3-month period. Unfortunately, the process will be different from the experiment start date to the finish date. Such data will lack consistency. Use your judgment when choosing the required number of data.
6. Choose your experimental program, or design.
7. Apply the tests that you selected in Step 3. If the model/data mismatch in any of the tests is worse than that expected at your level of confidence, invalidate that portion of the model.
8. Report each of the four items (reject/accept, level of confidence, test used, and H_0), as was done in the examples.

The autocorrelation test presumes both zero bias and uniform variance throughout the data sequence, which may not be true. If the data fails the test for bias, then the model is rejected, and there is no reason to consider autocorrelation. If the model is not rejected for bias, then accept that the bias is zero, and inspect the data (as they are independently arranged on each graph) for uniform variance.

If the variance is one region is high, then the data in that region dominate the numerator and denominator terms in the autocorrelation ratio, making the r-lag-1 value focus on that region. So, only use the autocorrelation test if variance is uniform throughout the range of the y and x variables. Visual inspection may be an adequate test for uniform variance.

Figure 19.4, from Rhinehart's PhD thesis (modeling a coal gasification process), illustrates a case of variance not being uniform. Note the collective nearness to the 1:1 line of the data in the left side (less than about 300 mol/hr) compared to the collective variability of the data to the right. With such a data pattern, when calculating an autocorrelation r-lag-1, the data in the upper right will dominate the data in the center and lower left, and the r-lag-1 value will not be representative of the whole.

If you suspect that the variance is not uniform, the runs test can be used to detect the presence of curvature or skew. See Section 7.5. The runs test is nonparametric, it does not presume distribution features. Although this is an advantage in applicability, it makes it less efficient, requiring a greater number of data points to be able to confidently reject the model based on curvature or skew.

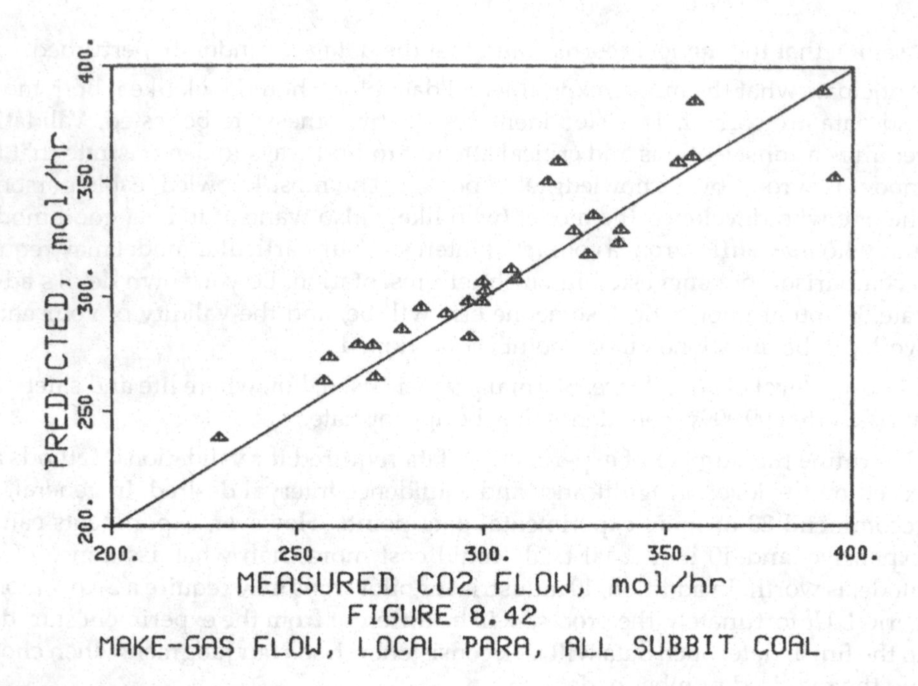

FIGURE 19.4
An example of nonuniform variance.

As a note, some people prefer the Durban–Watson, Spearman Rank, or other similar tests for autocorrelation in data. All are good. The r-lag-1 is simple and effective.

19.6 Takeaway

Your model is not perfect, and it cannot possibly exactly and properly account for every mechanism affecting the process. Validation is a procedure to see if the process/model mismatch is so bad that the model can be rejected by the data.

Use both logic-based and data-based criteria to accept or reject both models and the experimental procedure. In data-based validation, first check for bias in the residuals, then organize residuals by y, and each x, and experimental sequence, and test for autocorrelation (or runs). Although r-squared and ANOVA are appropriate for screening models, neither test address mechanistic validation. The t-test for bias and r_1 (autocorrelation of lag-1) test for skew or curvature are preferred, because these parametric tests are more sensitive (need less data) than the nonparametric tests. But, if there is evidence or suspicion that the residuals are not normally distributed with a uniform variance, then use the Wilcoxon signed-rank test for bias and the runs test for skew or curvature.

You likely will not have the luxury of enough or of properly spaced data to be able to apply the statistical tests. You may have to use visual appearance to make conclusions, but the issues discussed in this chapter should guide your visual inspection.

Even if a model is rejected, because there is sufficient evidence that it does not match the data trends, the model may still be fully adequate for its intended purpose. Balance perfection with sufficiency.

19.7 Exercises

1. Although grounded in ideal concepts, the Ideal Gas Law mismatch to data led to many other attempts to better match nature (Van der Waals, corresponding states, etc.). Yet the Ideal Gas Law is often accepted as adequately functional. Provide another example of a model or procedure that is wrong from a perfection point of view but permissible from a practical, functionally adequacy view.
2. Apply techniques in this chapter to your own data and associated model.
3. Use the data from Example 15.4 and test if the linear model is valid.
4. Use the data from Example 15.7 and test if the model is valid.
5. Use the data from Exercise 15.2 and test if a linear model, $y = a + bx_1 + cx_2$, is valid.
6. Use the data from Exercise 15.4 and compare bootstrapping to the analytical method of determining confidence limits on the model.

20

Experimental Design for Data-Based Model Validation

20.1 Introduction

Validation is the testing of models to determine if they are functionally representative of the process or item of interest. Design of Experiments (DoE) for model validation has a different objective from the DoE to determine coefficient values for screening models. In screening DoE, the purpose was to minimize the number of experiments and to minimize uncertainty in coefficients that are linear in the models. Validation is the comparison of a model to data that is designed to critically test the model – to find its faults, deficiencies, or weaknesses. The purpose of DoE in this chapter is to generate the data that is useful for critically testing the functionality of a model, per procedures in Chapter 19.

The focus will be on steady-state models, but the principles are applicable to stochastic and dynamic models.

20.2 Patterns Desired and Undesired

Some examples of desired and undesired features in the experimental design will be illustrations from the Rhinehart PhD Thesis (Dynamic Modeling and Control of a Pressurized Fluidized Bed Coal Gasification Reactor, 1985, North Carolina State University, Raleigh).

Figure 20.1 is an illustration of a desired data pattern. This is a plot of the predicted value w.r.t. the empirical measured value (a parity plot). Key features are: There are many data points. The data have a large range, relative to the residual (data-to-model mismatch). The data cover all internal regions of the range (there are no void areas). And finally, in this parity presentation, the data surrounds the 1:1 line with no evidence of bias, skew, or curvature. A conclusion is that the model well-matches the experimental functionality.

In contrast there are several data features that would be undesirable. In Figure 20.2 there are too few number of data to be able to critically challenge the model.

In Figure 20.3 there is a large void space – a region of the variable without data. There might be a suspicion that the trend in the data curves about the 1:1 line, but there is no evidence to either indicate that it does or does not. Because of this, one cannot critically test the model.

Figure 20.4 illustrates the problem that the experimental range is not much larger than the variability between model and data. Although the model is true to the data on average,

DOI: 10.1201/9781003222330-20

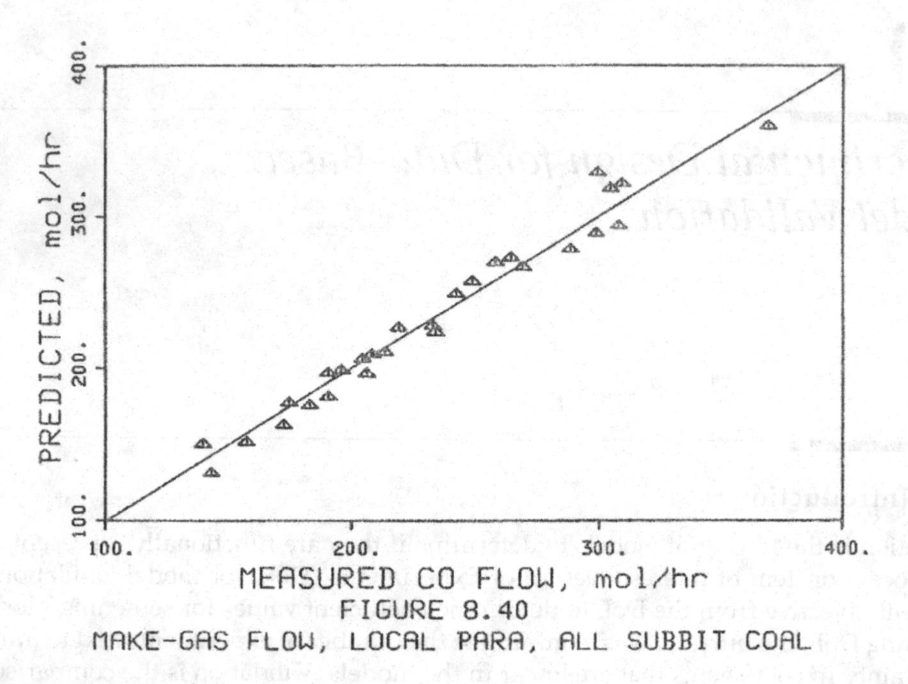

FIGURE 20.1
A data pattern that is desired for validation.

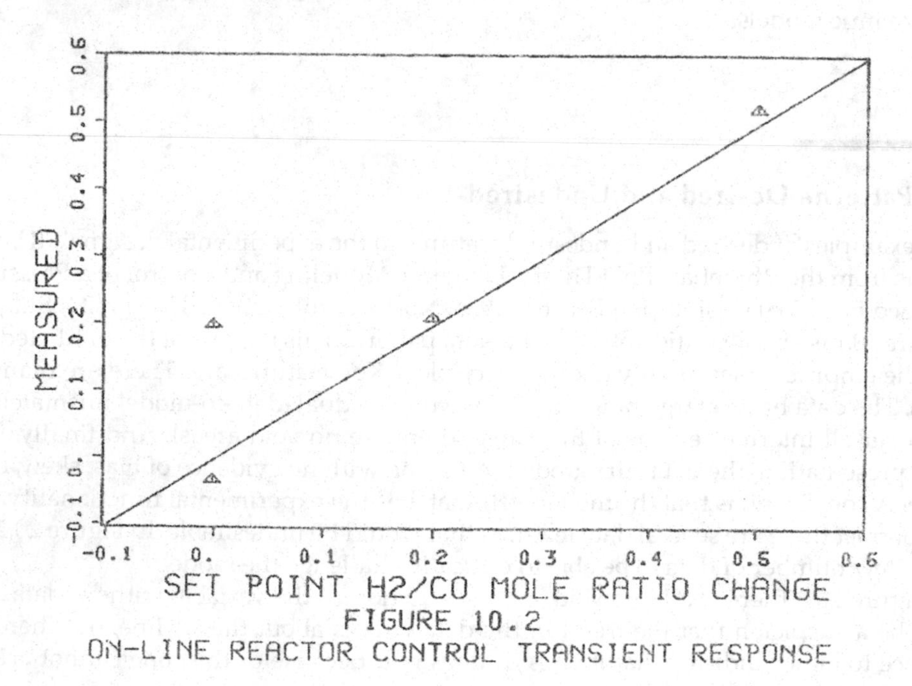

FIGURE 20.2
Illustration of too few number of data.

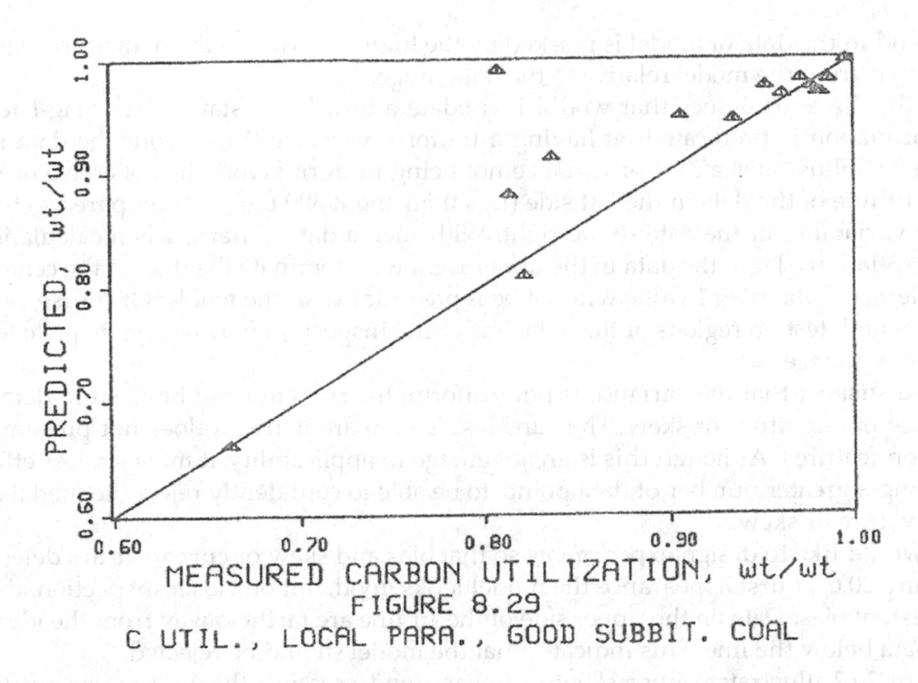

FIGURE 20.3
Data void region.

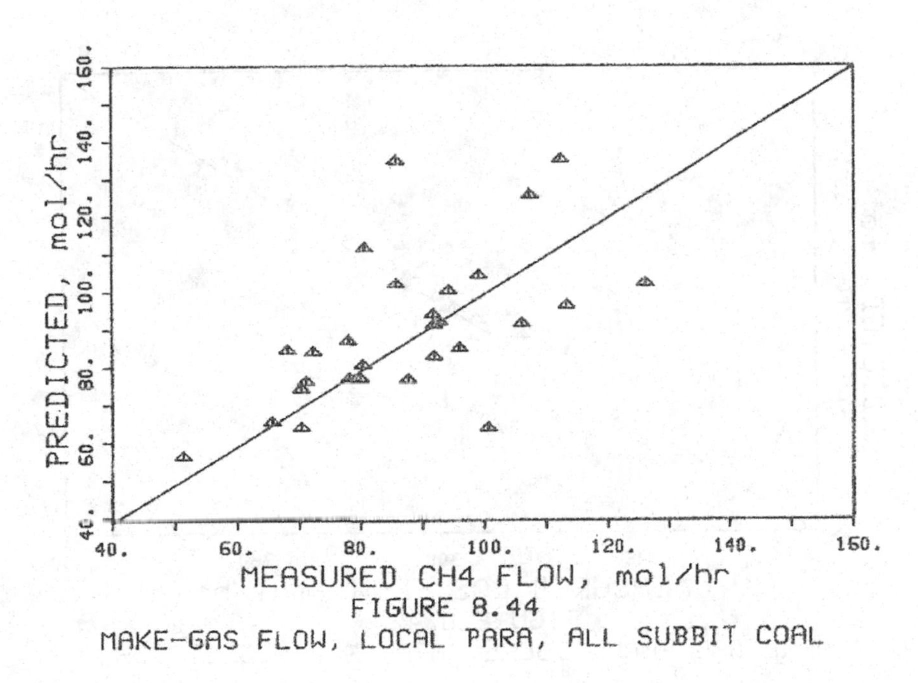

FIGURE 20.4
Inadequate range relative to variability.

any trend in the data or model is masked by the high variation in the data as indicated by deviations from the model relative to the data range.

Finally, there are issues that would invalidate a test. For instance, the r-lag-1 test for autocorrelation is predicated on having a uniform variance throughout the data range. Figure 20.5 illustrates a case of variance not being uniform. Note the collective nearness to the 1:1 line of the data in the left side (less than about 300 mol/hr) compared to the collective variability of the data to the right. With such a data pattern, when calculating an autocorrelation r-lag-1, the data in the upper right will dominate the data in the center and lower left, and the r-lag-1 value will not be representative of the whole. Of course, one can perform an *F*-test on regions of the data, but visual inspection may be an adequate test for uniform variance.

If you suspect that the variance is not uniform, the runs test can be used to detect the presence of curvature or skew. The runs test is nonparametric, it does not presume distribution features. Although this is an advantage in applicability, it makes it less efficient, requiring a greater number of data points to be able to confidently reject the model based on curvature or skew.

We would like to design experiments so that bias and skew or curvature are detectable. In Figure 20.6, at first appearance the model looks good, but on closer inspection, there is a consistent bias. Data on the upper side of the 1:1 line are farther away from the ideal line than data below the line. This indicates that the model should be rejected.

Figure 20.7 illustrates either skew (a linear trend crossing the 1:1 line) or curvature. Which it is, cannot be confidently stated. This indicates that the model should be rejected.

Figure 20.8 indicates a final undesired experimental plan. There are only two levels of the variable. Although the model predicts nearly perfectly, on average, it is not possible

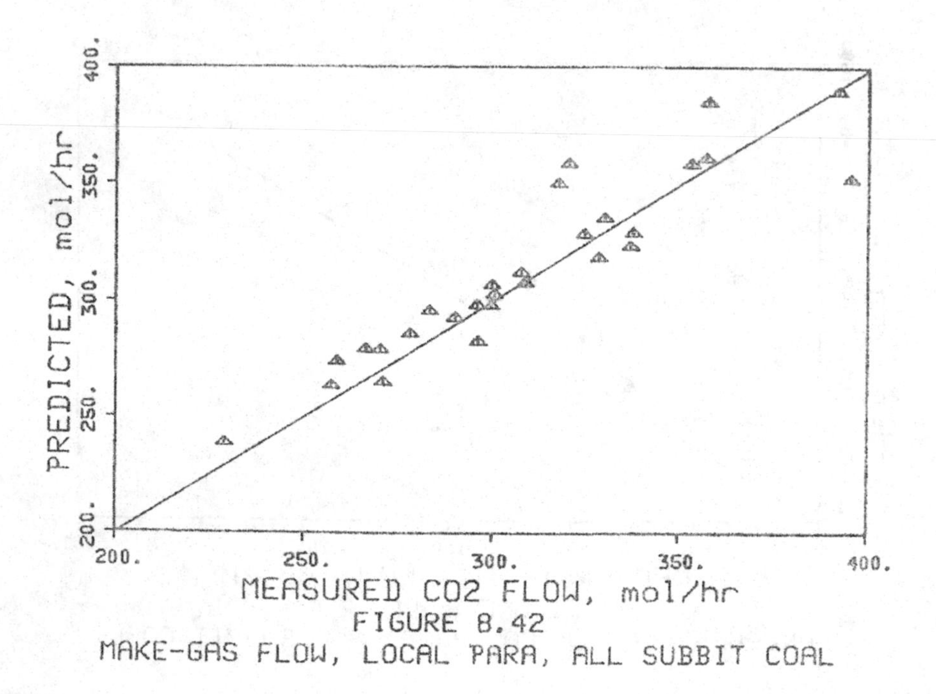

FIGURE 20.5
A case of nonuniform variance.

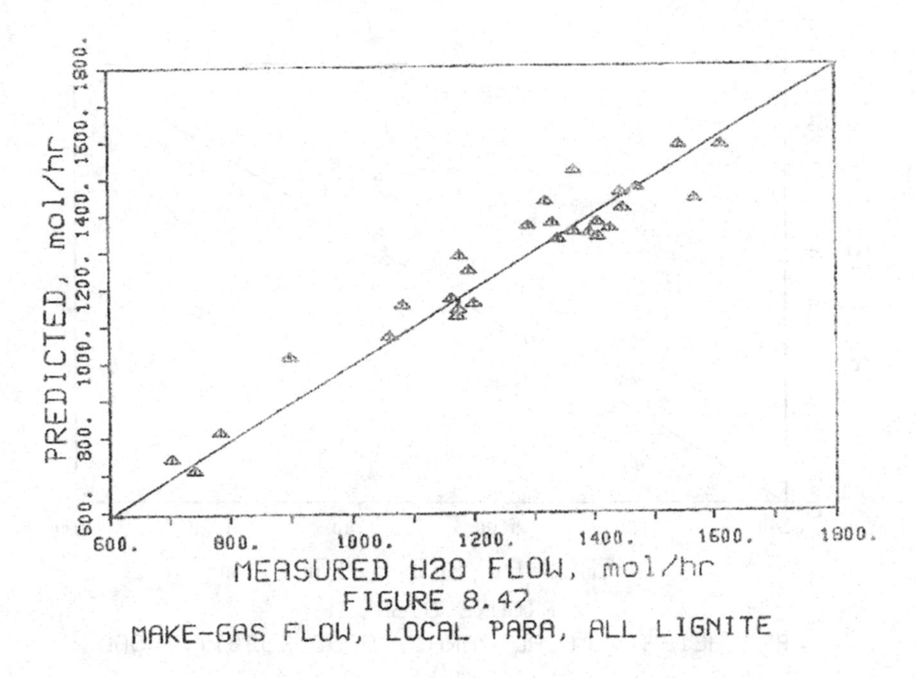

FIGURE 20.6
Illustrating bias.

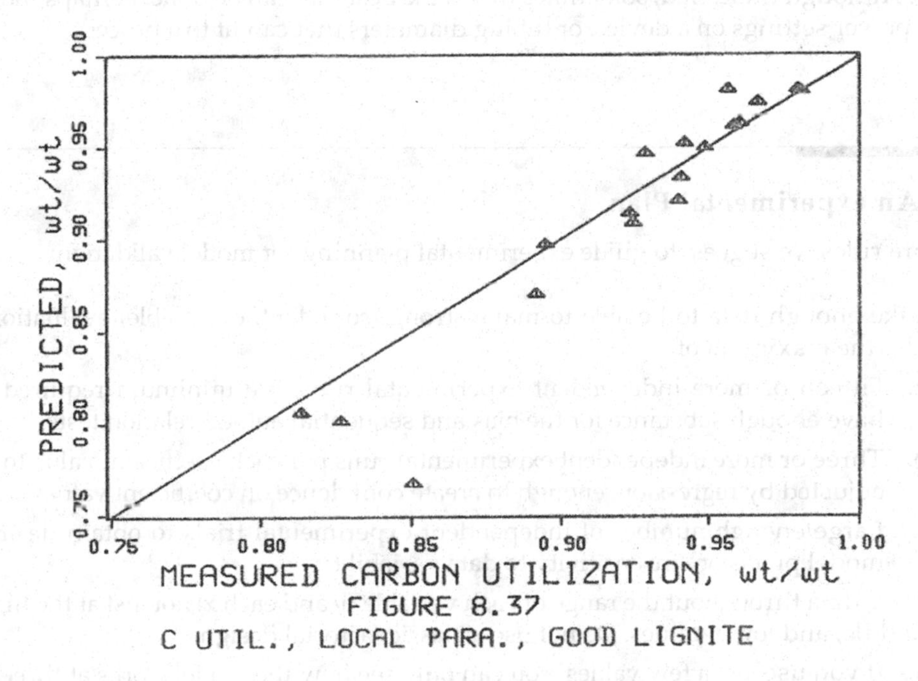

FIGURE 20.7
Illustrating either skew or curvature.

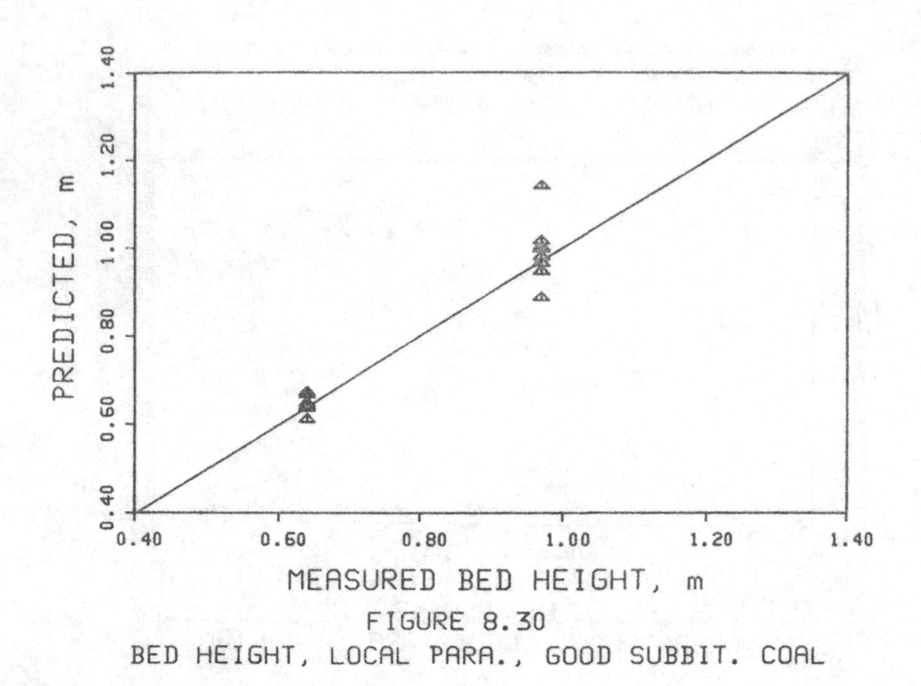

FIGURE 20.8
Illustrating inadequate number of experimental values.

to determine if the model might be good for in between or beyond the two experimental values. Although undesired, sometimes this is the best that can be done. Perhaps there are only 3 power settings on a device, or tubing diameters that can fit to a process.

20.3 An experimental Plan

Here are rules we suggest to guide experimental planning for model validation:

1. Take enough data to be able to make strong (confident, defensible) evaluations. Use the maximum of:
   a. Fifteen or more independent experimental runs, the minimum required to have enough substance for the bias and sequential autocorrelation tests.
   b. Three or more independent experimental runs per each coefficient value to be adjusted by regression, enough to create confidence on coefficient values.
   c. Large enough number of independent experimental trials to obtain desired model precision (insensitivity to data variability).
2. Place data throughout the range of each variable (y and each x) not just at the high, middle, and low x-values. Do not use a classic factorial design.
   a. If you use just a few values, you can only see how the model works at three or so values, not throughout the range.

b. In thinking this you will probably select x-values uniformly within the x-range. It does not have to be a uniform interval. Values could be at randomized intervals.

c. An exhaustive experimental design, a complete factorial, is not required. For example, if there are 3 independent x variables, and it is desired to have 20 levels for each, it is not necessary to have $20^3 = 8{,}000$ experimental runs. Randomly pair the x-values. For example, place the 20 values in Hat #1 for x_1, 20 in Hat #2 for x_2, etc. Then draw one value from each hat, the set representing one experimental condition. This randomly defines 20 experimental conditions (trials or runs) covering 20 values in each variable. Of course, you do not need hats. Random number generators can generate the pairing. Nor must it be 20 values and 20 trials.

3. The random pairing might miss combinations of interest, such as high values of all variables, or it might randomly lead to a correlation between variables, or it might create pairing that violates constraints. So, feel free to shape the randomization with certain selections, or reshape datasets. That was the *a priori* action, before the trial. Even if the x-values are uniformly distributed throughout their ranges, unexpected nonlinearities in the process model and data model may cluster too many y-points in the high (or low) range and leave sparsely populated y-values in the other range or in a region of rapid change. So, as the experimental results evolve, add additional sets or re-shape remaining sets of input conditions that are designed to fill in the empty spaces, questionable spaces, or locations of high rate of change, on the y-variable values.

4. Seek a range on each x that produces a great enough influence on y so that the resulting range on y is much greater than the experimental uncertainty on y, permitting a critical inspection of the x-functionality in the model. Exactly what this range is, may not be known until some of the experimental data reveals the sensitivity of y to each x and uncertainty on y. Equation (19.2) reveals a desire that the range on a dependent variable is about 10 times the standard error. If the response is linear, this can be converted to a desired range on an input variable to achieve the desired range on the response variable. Note that both range on x and the number of trials are DoE choices.

$$R_x \geq 10\left(\hat{\sigma}_y/\sqrt{N}\right)\Big/\left|\frac{\partial \tilde{y}}{\partial x}\right|, \text{ for each } y \tag{20.1}$$

Here, the $\hat{\sigma}_y$ symbol indicates that the value is estimated, perhaps from propagation of uncertainty on the data model, perhaps from replicate experiments. Also, the value of $\frac{\partial \tilde{y}}{\partial x}$ may not be knowable until after initial trials generate data from which to define model coefficient values.

5. Randomly sequence the individual experimental runs. Preferentially, do not let experimental sequence (chronological order, run time) be correlated with any of the variables. Randomization prevents changes in experimental technique, drifts in uncontrolled variables, progressive improvement of technique, etc. to be correlated with input variables, which could make a right model appear to have a skew trend.

a. Of course, this preference may have to be overridden if changing one variable takes significant time, effort, cost, or waste generation. Then cost, speed, and convenience could override the random sequence desire.
b. Select the initial trials to represent data near to the extremes to best develop initial understanding and models for subsequent reshaping of the experimental plan.
c. Select initial trials to do the most important ones first, because something may come up that blocks your ability to perform subsequent trials – the equipment might break, supply material may get interrupted, institutional priorities may divert funding, scheduling, etc.
d. Toward the end of the planned runs, the trial conditions might be redirected to regions that have become of high interest.

6. Calibrate measurement devices that you will use to obtain both the y and the x data.
   a. Calibration is a small cost relative to the cost of the experiments. If the measurements are wrong, then the experimental investment is devalued.
   b. Be sure that the measurement technique minimizes both bias and variability (accuracy and precision) of the experimental error. You may want to use a more precise measurement device or increase the sampling rate or sample size than those devices or procedures used for normal plant monitoring and control. This also would include measurement, y, and input, x, variability.
   c. If there are several instruments, labs, technicians, or historical periods in generating or analyzing data, be sure they are internally consistent. Perhaps test controls and bias one set of data by the difference in results.
7. Choose measurement conditions so that the uncertainty on the measured y-values is inconsequential to an in-use desire.
   a. Run replicate trials to reduce data uncertainty.
   b. Be sure that steady-state (SS) is achieved for SS models.
   c. Run long enough at SS so that averaging of noisy measurements provides desired precision.
   d. Run long enough at SS so that the averaging window includes the expected uncontrolled influence vagaries.
   e. Don't let samples spoil (oxidize, agglomerate, cool, react, evaporate) between sampling and analysis, or between preparation and use.
8. As experimental evidence evolves, adjust the experimental plan:
   a. As constraints are revealed that prohibit particular pairing of input conditions, choose new constraint-free parings. Attempt to retain the desire to have no voids, a large range-to-uncertainty in the y and x data, and little or no correlation in inputs.
   b. As empty spaces are revealed in data plots, add experimental conditions to fill them in. Look at data organized by y and each x. Even if data are uniformly distributed in the x-range, nonlinear relations could leave voids within the y-range. If there are voids in a y-variable, shift subsequent x-values to add data in the y-void region.

c. As regions are revealed in which there is much change in y relative to x, shift the experimental plan or add runs to place conditions preferentially in the regions of higher change.

d. As regions are revealed in which there is little change in y relative to x, shift the experimental plan to add diversity of results or delete runs to lessen effective duplication.

e. If developing results seem to indicate that there might be a transition where y behaves differently before and after (a slope or level discontinuity in the y w.r.t. x plot) then add points near the transition point. This may happen with a flow transition from laminar to turbulent, the onset of flashing/degassing, the onset of crystallization, or another mechanism.

f. As the sensitivity of x on y is revealed, change the range of any x-variable to ensure its influence on y can be definitely observed relative to the noise on y.

g. As data variability (noise, uncertainty) is revealed, change (increase or decrease) the number of experimental runs. Increase the number if variability seems high enough to mask trends. Decrease, if the trends can be clearly seen with fewer data.

h. As developing data patterns in the residual w.r.t. y and x graphs raise suspicions of autocorrelation of the residuals (indicating mismatch of model functionality to data), place additional experiments in that range.

i. As regions appear where the residual values are large (relative to the other model-to data differences), place additional experiments in that range.

j. Use bootstrapping to analyze the post-data model uncertainty. If uncertainty on the model is adequately low, stop the experiments. If not, add new experiments.

9. Don't attempt to get the input conditions exactly at the planned values. If the experimental implementation has approximately the right value, accept it and move on. Experiments take time and consume resources. If you are selecting a dozen x-values from a range of low to high, it does not matter if a slightly different set were chosen. Set the knobs, and if the resulting conditions are reasonably close to the target, use that data. Don't progressively fine-tune the knobs to get the input conditions to exactly match the planned values.
10. However, control the input conditions to a value that remains constant in time. For example, if you are interested in the flow-rate response to a heat transfer coefficient, and set the valve to a fixed position, the upstream pressure might change during a run, which would have flow rate change in the middle of the run. Since flow rate is the process influence, control it, not the signal to the valve, not the valve position, not something indirectly related to the process influence. As another example, consider the water temperature in a water heater. It rises when the heater is on and falls when cold inflow mixes with the in-tank water when the hot water is removed. If the heater thermostat is set to a single value, the water used in your experiment might actually be cycling by several degrees during a run.
11. Be sensitive to the other humans. As you adjust the experimental plan, don't upset the humans in the enterprise by doing so. This includes operators who made preparations for the initial plan, and managers who authorized the initial experimental design. Preview the how and why of the plan revisions.

12. Always consider environmental, health, safety, and loss prevention (EHS&LP) issues. Although this chapter has a focus on experimental design to determine valid models, this purpose cannot come at the expense of EHS&LP.
13. Don't take all the data then start to work up the results. There is a tendency, once trials are started, to continue to focus on the trials and data collection, until that job is finished. Instead, after each data point, completely work-up the data, and evaluate models, data patterns, and appropriately revise the plan.

20.4 Data Sources and Other Modeling Objectives

There are many ways to test theories including retrospective studies, observational studies, and designed experiments. If you are seeking data to defend a hypothesized mechanism, then you need data that can validate a cause-and-effect relationship and reject alternate cause-and-effect relationships.

Retrospective studies (a term common to the medical community) are those that the take data from natural processes or from a historical database, which seek to find variables that correlate to each other. This is also common in the "Big Data" era of seeking patterns in purchasing data that might reveal possible fraud in credit card use. If a correlation is significant, the correlation may be a clue as to a cause-and-effect relation, but then again, it may simply reveal that two variables are caused by another unrecognized cause. Like correlating gas mileage to tire wear, it may not mean that poor gas mileage causes rapid tire wear. The two variables may each be the result of the same phenomena (perhaps aggressive driving, or a heavy car, or towing a trailer). Similarly, the retrospective data may reveal a strong correlation of gray hair and skin wrinkles, a correlation does not reveal the mechanism. Retrospective data was not controlled to validate the mechanistic models.

Observational studies are those that infer a population property from a sample, where the independent variables are not wholly under the control of the investigator. In some designed experiments it might not be possible to impose strict controls. This includes a broad class of human investigations (psychology, social, medical studies, etc.) in which various ethical or legal issues place constraints on the conditions. It could also include physical trials, such as on production lines, in which there is some flexibility in setting conditions, but many other conditions (shift assignments, product scheduling, equipment faults, environmental factors, etc.) are beyond the control of the investigator. Here, the outcomes will have a stronger reveal of the underlying mechanisms than retrospective studies, but there is usually a question because alternate mechanisms cannot be rejected.

Designed experiments on laboratory physical or chemical processes rarely have ethical or legal constraints, and all of the suspected inputs that might have an effect on the outcome can be controlled. Many of these are "screening experiments" which seek to generate data for an empirical model, such as

$$y = a + bx_1 + cx_1^2 + \cdots + dx_2 + ex_2^2 + \cdots + fx_1x_2 + \cdots \tag{20.2}$$

Such models may be useful to identify input conditions that make the best y outcome, scale up a unit operation, forecast the change in one input value if another is changed, etc. However, these models do not reveal the cause-and-effect mechanism. And, the degree of the monomials and whether cross product terms are included in the model are all user choices which are not phenomenologically based.

Typically, as above, the empirical models are linear in coefficients, and the associated experimental design places input values in patterns that ideally (uniform variance, no physical constraints on combinations, accurate measurements, etc.) generate data to minimize uncertainty on the coefficient values. The patterns are termed "factorial", "star", "Latin square", "complete block", and suchlike.

The prior three approaches (retrospective, observational, screening) are in contrast with controlled experiments that seek to reveal mechanistic cause-and-effect relationships.

Designed experiments may also be those designed to validate a theoretical model (an equation) or a conceptual mechanism (If this situation, then that outcome). Again, the input conditions, the variables that might affect an outcome are controlled. Often a mechanism is postulated, and experiments are designed to validate the mechanism, and also to reject alternate mechanism conjectures. If a joker claims a mechanism "multiplication is the same as addition" and does experimental tests to show $2+2=2\cdot 2$, and $1+2+3=1\cdot 2\cdot 3$, this does not prove the claim. One must also design tests that would reject the claim, such as $3+3=6\neq 3\cdot 3=9$.

An entire book (C. H. Kepner and B. B. Tregoe, *The New Rational Manager*, Princeton Research Press, Princeton, NJ, 2013) is devoted to methods to rationally determine mechanisms underlying observations, as are the teachings and practices of Statistical Process Control and Six Sigma. Here Chapters 19 and 20 are devoted to the concepts and procedures for validation of theory, models, and experimental procedures.

20.5 Takeaway

Design a model-validation experiment to generate data that could reject the model, not to accept the model.

Even if a model is rejected, because there is sufficient evidence that it does not match the data trends, the model may still be fully adequate for its intended purpose. Balance perfection with sufficiency.

Work up data as each point is obtained, and let the progressive understanding reshape the remainder of the experimental plan.

If your experimental plan was approved by a cautious hierarchy of your organization, don't go changing the plan on your own. Don't usurp their claim to ownership. If you believe the plan should be changed, justify your reasons, and explain the change to get their approval. And, maybe, don't do this very often!

20.6 Exercises

1. Use your own data and models to identify the desirable and undesirable features as illustrated in Figures 20.1 through 20.8.
2. Use the guide in Section 20.3 to design data collection trials and to adjust the plan as information is revealed.

21

Statistical Process Control

21.1 SPC Concepts

Statistical process control (SPC) is an application of statistics to manufacturing processes. SPC provides evidence that feeds back information to the process manager that something has changed, relative to normal variation. It is an element in six sigma practices. SPC is not automatic feedback control that keeps a process at a set point. Instead, it indicates when something has happened, and triggers an investigation to change something to either prevent the undesired event from happening again, or to ensure a fortuitous event becomes part of normal practice. Progressive prevention of upsetting events leads to improved uniformity, quality, and process yield.

Imagine a process that has noise but has neither measurable changes nor systematic drifts. In such a case, there is no need for either automatic or human control action. A graph of some process variable (PV) versus time may appear like that illustrated in Figure 21.1.

Control action based on normal noise is both unnecessary and unwanted. There are several reasons: Such action increases valve wear, servomechanism wear, etc. and can upset other PVs. Further, if the process is on target, input adjustment causes real change, which requires subsequent counter adjustment, and this "tampering" actually increases the process variability and burdens the organization. (To visualize the concept of tampering, search the Internet for W. Edwards Deming's Funnel Experiment.)

The process of Figure 21.1 is in statistical control because there is no evidence that the PV has significantly changed in level or variation. The trend is characterized as random variation with the same noise characteristics about the same mean. There is no evidence of an *assignable cause*. An assignable cause means that something has happened and created statistical evidence that either the level or the variation change. The term *assignable* means that the cause might not be known, but it could be uncovered with investigation. Managerial control action is warranted only when an assignable cause happens.

Figure 21.2 illustrates some PV response to some systematic change. One example of a systematic change or assignable cause is an orifice flowmeter calibration change due to erosion. The result might be a change in blended composition or *level* of one of the products in a refinery.

Figure 21.3 illustrates some PV response to some other assignable cause that changed the process variability. In this case, perhaps loosened bolts caused increased vibration in a man-made fiber extrusion process, resulting in an increase in fiber thickness variability.

Trends in overall level, sudden shifts in level, cycling in level, and changes in variability all indicate the presence of an assignable cause. The assignable cause need not be known. Nor need it be identified. However, it represents some real change, as it either shifted the mean or changed the variability (or both) and therefore deserves action.

DOI: 10.1201/9781003222330-21

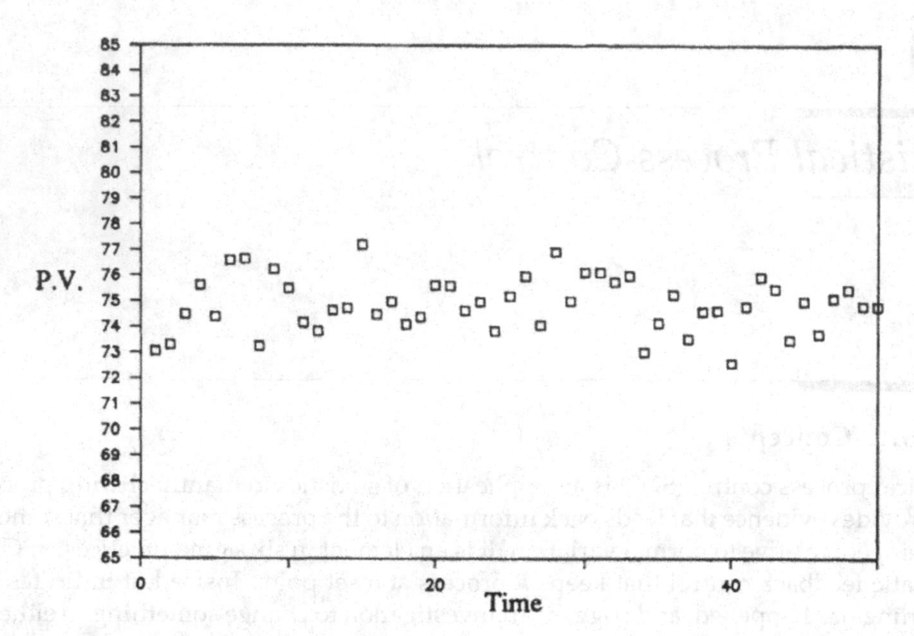

FIGURE 21.1
Process variable versus time (no changes in level or variability).

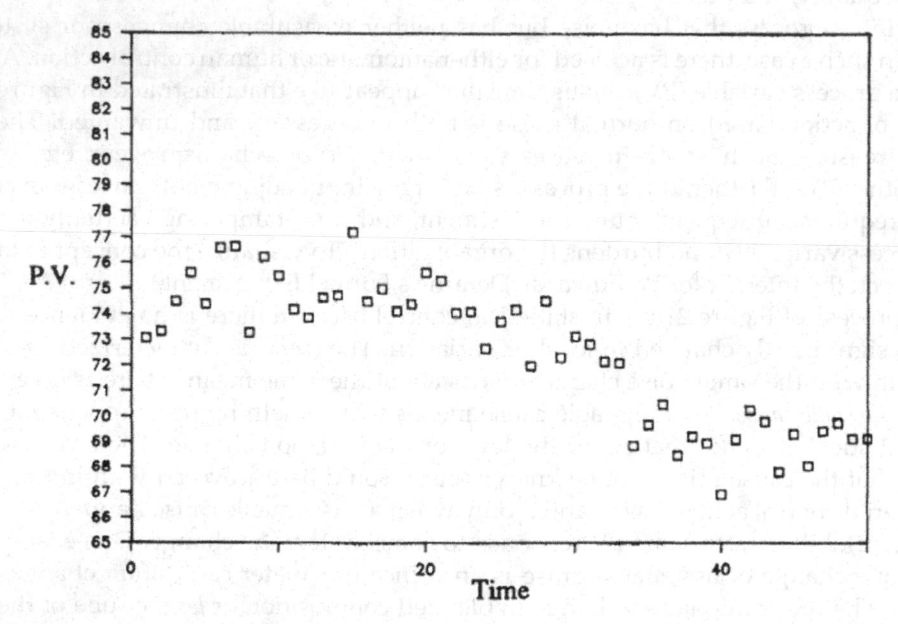

FIGURE 21.2
Process variable versus time (change in level).

One of the objectives of SPC is to determine where the assignable cause occurred so that the operator knows where in the process to take corrective action. This procedure requires sampling appropriate variables at strategic points in the process. Another of the objectives of SPC is to determine when an assignable cause occurred. So, in opposition to classical automatic control that works constantly, SPC is a supervisory strategy that identifies

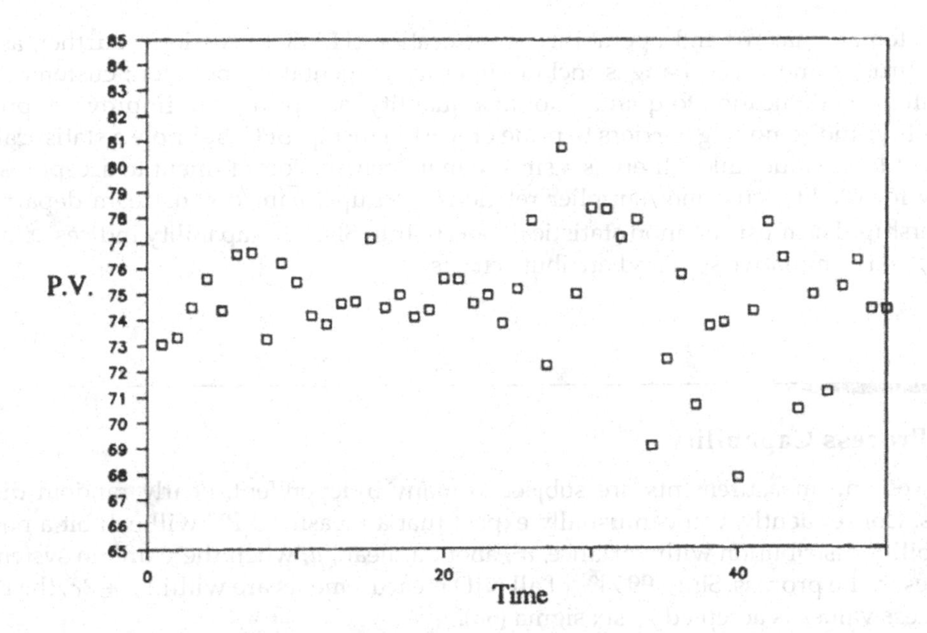

FIGURE 21.3
Process variable versus time (change in variability).

when operator or management intervention is justified. Procedures for rational diagnosis to determine what happened and to determine appropriate ways to fix things are part of the body SPC and of six sigma. This chapter will focus on the statistical procedure for detection.

Statistically, an assignable cause occurs when an observed change in either PV level or variability has about a 0.26% or less chance of occurring.

Why was $\alpha = 0.0026$ chosen as the significance level? It represents the $\pm 3\sigma$ limits and was chosen from experience to balance the occurrence of Type 1 and Type 2 errors, i.e., justified intervention action and not reacting to false alarms. If, for instance, the $\pm 2\sigma$ limits (95.45% CI, $\alpha = 0.0455$) were chosen, then normal process variability would appear to have an assignable cause 4.55% of the time. If 1,000 PVs were being observed in a plant each day, 46 assignable causes would seem to occur, and 46 possibly unjustified control actions would be taken. On the other hand, if the $\pm 4\sigma$ limits (99.994% CI, $\alpha = 0.00006$) were chosen, then unjustified control action would be taken only twice per month. However, at the $\pm 4\sigma$ limits, some assignable causes would go unrecognized until there was enough information to be 99.994% sure that the PV behavior was not simply due to normal variability.

The $\pm 3\sigma$ limit is a subjective choice. There are other conventions. In some regions, the convention is to use the $\pm 3.09\sigma$ limits, which encompass 99.8% of the normal data. Further, where subsampling of $n = 2$ is used, the 96.5% confidence limits are often used. Although the 99.74% confidence interval is one convention, you may choose to modify those limits based on your process experience.

SPC will not replace online primary regulatory process control; however, SPC-triggered action can be automatically implemented within a feedback control loop as an alternate to filtering to temper response to noise.

In the big picture, SPC is simply one set of tools associated with a growing manufacturing philosophy to "do it right". Other tools include Pareto (common problem identification) and Fishbone (cause and effect) charts as well as brainstorming (creative problem solving) and systematic hazard and operability identification (HAZOP) sessions. Further, associated attitudes and style changes include interdepartmental cooperation; customer/supplier alliance; dedication to quality, not just quantity; accepting, but improving, process variability; and removing barriers to pride of workmanship. SPC is simply a statistical tool that is of little value unless there is a sincere management commitment to accept responsibility for quality, customer/supplier relations, and upstream/downstream department partnership. The most common statistical tools within SPC are capability indices, $\bar{X}$ and R charts, and cumulative sum and attribute charts.

21.2 Process Capability

Processes and measurements are subject to many independent, nearly random disturbances. Consequently, you can usually expect that a measured PV will exhibit a normal probability distribution with variance, σ^2, about a mean, μ, when there are no systematic changes in the process. Since 99.74% of all NID measurements are within $\mu \pm 3\sigma$, the range of process values is accepted as six sigma (6σ).

Processes are operated at some set point or target value, x_{sp}, and have an upper specification limit (USL) and a lower specification limit (LSL). If $6\sigma <$ (USL – LSL), the process is said to be capable of meeting the specification limits. The smaller the value of 6σ when compared to (USL – LSL), the more capable is the process. The process capability index, Cp, is defined as

$$\mathrm{Cp} = \frac{\mathrm{USL} - \mathrm{LSL}}{6\sigma} \tag{21.1}$$

and is a measure of the capability of the process to meet specifications. If Cp < 1, the process will make more than 0.26% out-of-specification product and is termed "not-capable" in SPC philosophy. If Cp = 1, 0.26% of the product will be out-of-specification, a situation that is acceptable in SPC philosophy. If Cp ≥ 1, the process is capable of being within statistical control.

Other similar indices are used, but their meaning is the same. For instance, if USL and LSL are not symmetric about the measured process mean $\bar{X}$, which should be equal to the set point, one uses

$$\mathrm{Cp_k} = \min\left\{\frac{\mathrm{USL} - \bar{X}}{3\sigma}, \frac{\bar{X} - \mathrm{LSL}}{3\sigma}\right\} \tag{21.2}$$

as the capability index. When calculations are performed by hand instead of computer, it is easier to calculate an average range, $\bar{R}$, associated with subgroup sampling than to calculate σ. (A *subgroup* is a collection of repeated observations of the PV at the same operating conditions.) Then the relative precision index (RPI) is defined as

$$\mathrm{RPI} = \frac{\mathrm{USL} - \mathrm{LSL}}{\bar{R}} \tag{21.3}$$

For a process to be capable, the RPI must be greater than roughly 2.5, but that value depends on the subgroup size. Finally, if a PV is not normally distributed (e.g., particle size, polymer molecular weight), then the index of Equation (21.4) should be used, or

$$\mathrm{Cp_k} = \min\left\{\frac{\mathrm{USL} - \overline{\mathrm{X}}}{X_{.9987} - \overline{\mathrm{X}}}, \frac{\mathrm{LSL} - \overline{\mathrm{X}}}{X_{.0013} - \overline{\mathrm{X}}}\right\} \tag{21.4}$$

where $X_{.0013}$ and $X_{.9987}$ are the PV values corresponding to areas under the normalized distribution curve corresponding to probabilities of 0.0013 and 0.9987.

21.3 Mean and Range Charts

Introduced by Walter A. Shewhart in the 1920s, the $\overline{X}$ (X-bar, average) and R (range) charts are some of several Shewhart charts devised to detect when an assignable cause occurred in a continuous PV. Although the $\overline{X}$ and S charts are statistically more rigorous, the $\overline{X}$ and R charts are fully functional, were preferred in the pre-computer age, have advantages for small subgroup size sampling, and have become an accepted SPC standard. We will develop the $\overline{X}$ and S charts first with an example to illustrate the principles involved.

Table 21.1 lists measured values of a continuous PV that are based on subgroup sampling size of four. In this example, polyester staple yarn is being measured for denier (a measure of thickness, in decitex or dtex, gm/10,000 m) once per shift. The past 25 shift samplings are listed. The average of each subgroup, $\overline{X}$, and the standard deviation of each subgroup, are also listed. The range value on each subgroup is the difference between the largest and smallest of the four samples. The average of the entire set of 100 individuals, $\overline{\overline{X}}$, is 74.6985 dtex and the standard deviation, σ, of the entire set of individual measurements is 1.27297 dtex. The average of the subgroup standard deviations, $\overline{S}$, is 1.049861 dtex, and the average range $\overline{R}$ is 2.2974 dtex.

Example 21.1: Develop an $\overline{X}$ chart for the data in Table 21.1.

Figure 21.4 shows the $\overline{X}$ chart, on which the past 25 subgroup averages are plotted against the sequential sample number. The upper and lower control limits, UCL and LCL, are:

$$\mathrm{UCL} = \mu + \frac{3\sigma}{\sqrt{n}} \cong \overline{\overline{X}} + \frac{t_{\alpha,\upsilon}\, s}{\sqrt{n}} \cong \overline{\overline{X}} + A_3\overline{S} \tag{21.5}$$

$$\mathrm{UCL} = \mu - \frac{3\sigma}{\sqrt{n}} \cong \overline{\overline{X}} - \frac{t_{\alpha,\upsilon}\, s}{\sqrt{n}} \cong \overline{\overline{X}} - A_3\overline{S} \tag{21.6}$$

where n is the subgroup sample size (of $n = 4$) and $\sigma/\sqrt{n}$ represents the standard deviation of the $\overline{X}$, the subgroup means. The ideal first part of the equations presumes that the standard deviation of the individual measurements is known and that $\overline{\overline{X}} = \mu$, the process mean. If the process is in statistical control (only subject to random independent

TABLE 21.1
Sampled Yarn Deniers

Sample no.	Shift	Date	Sample values (decitex)				Subgroup average	Subgroup std. dev.	Range
1	2	8/19	73.07	73.33	74.49	75.62	74.1275	1.170934	2.55
2	3	8/19	74.38	76.57	76.63	73.26	75.21	1.669073	3.37
3	1	8/20	76.22	75.49	74.17	73.83	74.9275	1.120209	2.39
4	2	8/20	74.63	74.72	77.18	74.47	75.25	1.290814	2.71
5	3	8/20	74.98	74.10	74.37	75.59	74.76	0.6645796	1.49
6	1	8/21	75.58	74.62	74.96	73.84	74.75	0.7252611	1.74
7	2	8/21	75.19	75.96	74.07	76.91	75.5325	1.202285	2.84
8	3	8/21	75.00	76.11	76.10	75.73	75.735	0.5209285	1.11
9	1	8/22	75.96	73.02	74.15	75.25	74.595	1.287234	2.94
10	2	8/22	73.55	74.59	74.62	72.60	73.84	0.9648132	2.02
11	3	8/22	74.77	75.93	75.46	73.49	74.9125	1.06127	2.44
12	1	8/23	74.98	73.72	75.09	75.42	74.8025	0.7454901	1.70
13	2	8/23	74.79	74.79	75.05	73.84	74.6175	0.5326279	1.21
14	3	8/23	73.16	75.97	75.81	73.13	74.5175	1.586221	2.84
15	1	8/24	74.77	75.39	75.63	74.42	75.05251	0.5559604	1.21
16	2	8/24	75.41	73.91	76.02	74.87	75.0525	0.8948886	2.11
17	3	8/24	76.22	74.96	76.19	74.53	75.475	0.8611035	1.69
18	1	8/25	75.09	77.77	74.32	75.74	75.73	1.47867	3.45
19	2	8/25	75.93	75.12	75.35	75.79	75.5475	0.3772181	0.81
20	3	8/25	75.13	74.42	76.23	76.00	75.445	0.8314843	1.81
21	1	8/26	73.43	74.83	73.77	74.51	74.135	0.6465035	1.40
22	2	8/26	74.98	73.56	74.12	72.59	73.8125	1.002642	2.39
23	3	8/26	72.77	71.23	75.16	75.41	73.6425	2.00069	4.18
24	1	8/27	76.48	72.08	71.42	74.22	73.55	2.289948	5.06
25	2	8/27	72.27	72.57	73.39	71.54	72.4425	0.7655659	1.85

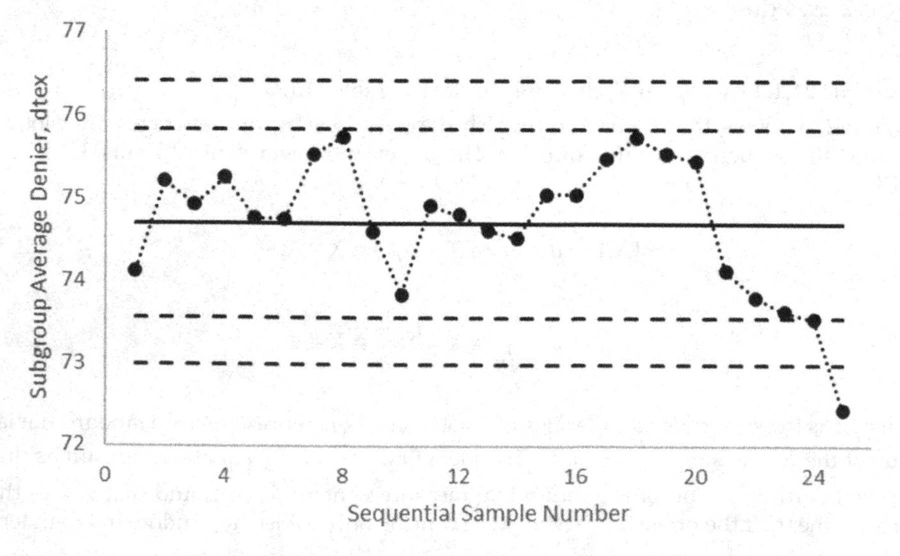

FIGURE 21.4
$\bar{X}$ chart for Example 21.1 (showing UCL, UWL, $\bar{\bar{X}}$, LWL, and LCL values).

perturbations), the average and standard deviation of the many individuals (100 in this case) could be used as estimates and the t-statistic value would replace the z-value of 3. However, if the process is not in statistical control, if an assignable cause has occurred, then the 100 individuals would not represent the same mean or variance, and a safer estimate of the standard deviation would be the average of the subgroup standard deviations. The last part of Equations (21.5) and (21.6) represent how the UCL and LCL values in an X-bar chart are calculated. The value for A_3 comes from Table 21.3.

Note that UCL and LCL are the process $\pm 3\sigma$ capability limits and are *not* the product specification limits. The product specifications are not normally shown on the control chart. The process $\pm 2\sigma$ limits are occasionally illustrated and are termed the upper and lower warning limits (UWL and LWL). The UWL and LWL values are useful when a process capability index is low. In such a case, UWL or LWL violations are used to trigger a second sampling and, if warranted, subsequent action.

The $\bar{X}$ chart is a moving window on the recent history of the data. As each scheduled sampling occurs, all the data points are indexed to the left, the oldest one is discarded, and the new $\bar{X}$ value is added to the right. Also $\bar{\bar{X}}$, UCL, and LCL (and UWL and LWL if used) are recalculated and added. The index is usually sample time, here, it is sample number, with #25 being the most recent.

To use the $\bar{X}$ chart to identify when an assignable cause occurred, one looks for events that have less than a 0.26% chance of occurring during periods in which there are no systematic changes in stationary process behavior. One such event is the violation of a control limit by sample 25 in this example. Another such improbable event is too long a run of data on one side of the $\bar{\bar{X}}$ line. Using Figure 21.4, in the 2nd to 8th sample, seven data points are on one side of the X-bar-bar line. Using the binomial distribution, the probability of 7 contiguous data on one side of the $\bar{\bar{X}}$ line is $\binom{8}{0} 0.5^0\, 0.5^7 = .007812\ldots$ and indicates 99.22% confidence that an assignable cause occurred. Other conventional rules are listed in Table 21.2. Many other rules are sometimes used.

Note, the assignable cause may not have occurred at the same time when the pattern indicates something real has happened. If the most recent several points had not been included, the $\bar{\bar{X}}$ value would be greater, and there would not be 7 in a row on one side of the $\bar{\bar{X}}$ line. When the most recent data shifted the $\bar{\bar{X}}$ line, then the early data points were also able to flag that something has happened.

TABLE 21.2

Common Rules to Identify an Assignable Cause Occurrence

1.	A single violation of the UCL or LCL indicates a change.
2.	A run of 6 or 7 consecutive data on one side of $\bar{\bar{X}}$, $\bar{S}$ or $\bar{R}$ indicates a change.
3.	A run of 6 or 7 consecutive data with a consistent upward or downward trend indicates a trend.
4.	2 out of 3 consecutive data violating a warning ($\pm 2\sigma$) limit indicate a change.
5.	4 or more crossings of $\bar{\bar{X}}$, $\bar{S}$ or $\bar{R}$ out of 15 consecutive data indicate cycling.
6.	4 out of 5 consecutive data outside a $\pm 1\sigma$ line indicate a change.
7.	10 out of 11 consecutive data on one side of $\bar{\bar{X}}$, $\bar{S}$ or $\bar{R}$ indicate a change.
8.	12 out of 14 consecutive data on one side of $\bar{\bar{X}}$, $\bar{S}$ or $\bar{R}$ indicate a change.

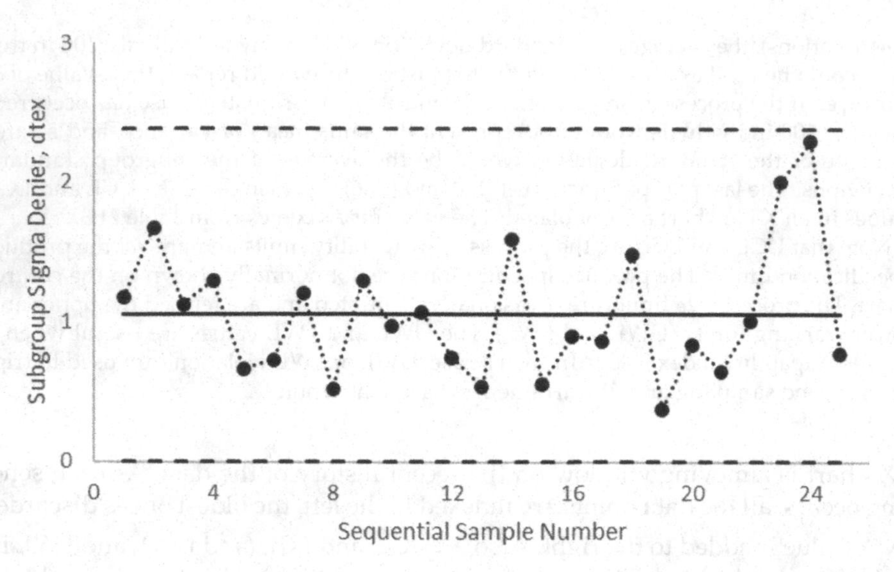

FIGURE 21.5
S chart for Example 21.2.

Even though this approach is grounded in statistical theory, subjectivity is present in both the sampling program and rules for action. Experience is needed to set the sampling frequency and the subgroup size to responsively observe an assignable cause. Similarly, you must use a time "window" or data horizon of sufficient length that long-term process drifts may be observed with confidence. Further, you must balance the cost of increasing both sampling frequency and subgroup size against the economic benefit of improved control. Conventionally, 100 samples are represented on an X-bar chart. In Example 21.1, a subsample size of $n = 4$ and a horizon of 25 samplings, as illustrated in Table 21.1, are used.

The action rules in Table 21.2 generally reflect 99.74% confidence, but some rules reflect a 98% confidence limit based on individual experiences.

Example 21.2: Develop an *S* chart for the data in Table 21.1.
Figure 21.5 illustrates the *S* chart for the yarn denier data of Table 21.1. In this figure, subgroup standard deviations are plotted in chronological order. $\bar{S}$ is the average of the sample standard deviations

$$\bar{S} = \frac{1}{k}\sum_{i=1}^{k} S_i \tag{21.7}$$

which is not the σ of the entire dataset. The number of subgroups is k; in this example, $k = 25$.

There are several ways to set the UCL and LCL. Since S^2 is distributed as chi-squared and has $n - 1 = 3$ degrees of freedom, the 99.8% confidence interval for S can be determined by

$$\sigma\sqrt{\frac{\chi^2_{3,0.001}}{3}} < S < \sigma\sqrt{\frac{\chi^2_{3,0.999}}{3}} \tag{21.8a}$$

TABLE 21.3

Selected Values of Coefficients for Calculation of Control Limits on $\bar{X}$ and S with $\bar{\bar{X}}$ and either σ, $\bar{S}$, or $\bar{R}$ Known

Sample size	A	B_1	b_2	B_3	B_4	A_3	A_2
2	2.121	0.0013	3.291	0	3.267	2.659	1.880
3	1.732	0.032	2.628	0	2.568	1.954	1.023
4	1.500	0.090	2.329	0	2.266	1.628	0.729
5	1.342	0.151	2.150	0	2.089	1.427	0.577
6	1.225	0.205	2.026	0.030	1.970	1.287	0.483
7	1.134	0.252	1.935	0.118	1.882	1.182	0.419
8	1.061	0.292	1.864	0.185	1.815	1.099	0.373
9	1.000	0.327	1.807	0.239	1.761	1.032	0.337
10	0.949	0.358	1.760	0.284	1.716	0.975	0.308

Selected entries from ASTM STP 15-C: Manual on Quality Control of Materials, Copyright 1951, ASTM, used with permission.

This, again, presumes that the true standard deviation could be known. Using values from Table 21.3

$$\text{LCL} = B_3\bar{S} < S < B_4\bar{S} = \text{UCL} \tag{21.8b}$$

Use of the S chart parallels that of the $\bar{X}$ chart. Violations of the UCL indicate a significant increase in variability due to an assignable cause. Violations of the LCL indicate a significant reduction in variability. If beneficial, the cause should be found and intentionally instituted. However, excursions below the LCL may indicate instrument failure, simply an "apparent" process improvement, and the necessity for instrument repair.

In practice, experience indicates that the 99.74% confidence limits should be relaxed to about 97% for small sample subgroup sizes in order to maintain the balance of responsiveness to change while minimizing false alarms. Often, the LCL is near zero and not shown. The US convention is to calculate the UCL and LCL for $\bar{X}$ and S as

$$\text{UCL}_{\bar{X}} = \bar{\bar{X}} + A\sigma \tag{21.9}$$

$$\text{UCL}_{\bar{X}} = \bar{\bar{X}} - A\sigma \tag{21.10}$$

$$\text{UCL}_S = B_2\sigma \simeq B_4\bar{S} \tag{21.11}$$

$$\text{UCL}_S = B_1\sigma \simeq B_3\bar{S} \tag{21.12}$$

Selected values of A, B_1, B_2, B_3, and B_4 are presented in Table 21.3 for samples of $n = 2$ to $n = 10$.

In statistical process control work, larger samples are generally not used. Further, the control limits on $\bar{X}$ may be similarly calculated from either $\bar{S}$ or the average sample range, $\bar{R}$, which are both estimates of σ. The formulas are:

$$\text{UCL}_{\bar{X}} = \bar{\bar{X}} + A_3\bar{S} \tag{21.13}$$

$$UCL_{\bar{X}} = \bar{\bar{X}} - A_3\bar{S} \tag{21.14}$$

$$UCL_{\bar{X}} = \bar{\bar{X}} + A_2\bar{R} \tag{21.15}$$

$$UCL_{\bar{X}} = \bar{\bar{X}} - A_2\bar{R} \tag{21.16}$$

Selected values of A_3 and A_2 are also listed in Table 21.3.

Example 21.3: Develop an *R* chart for the data in Table 21.1.
The principles of construction and the interpretation of a *R* chart are similar to those of the *S* chart. The *R* chart for the data of Table 21.1 is shown as Figure 21.6. The upper and lower control limits are calculated as

$$UCL_R = D_4\bar{R} \tag{21.17}$$

$$UCL_R = D_3\bar{R} \tag{21.18}$$

where selected values for D_3 and D_4 are listed in Table 21.4.

You may wish to compare Figure 21.6 to Figure 21.5. The *R* chart and *S* chart are essentially indistinguishable. The *R* chart is easier for some people to understand and use. Although less statistically rigorous, the *R* chart is an accepted manufacturing standard.

21.4 Modifications to the $\bar{X}$ and *R* Charts

The $\bar{X}$ and *R* charts developed in Section 21.3 assumed that multiple samples of uniform sample size were available. However, it may not be possible to obtain repeated observations

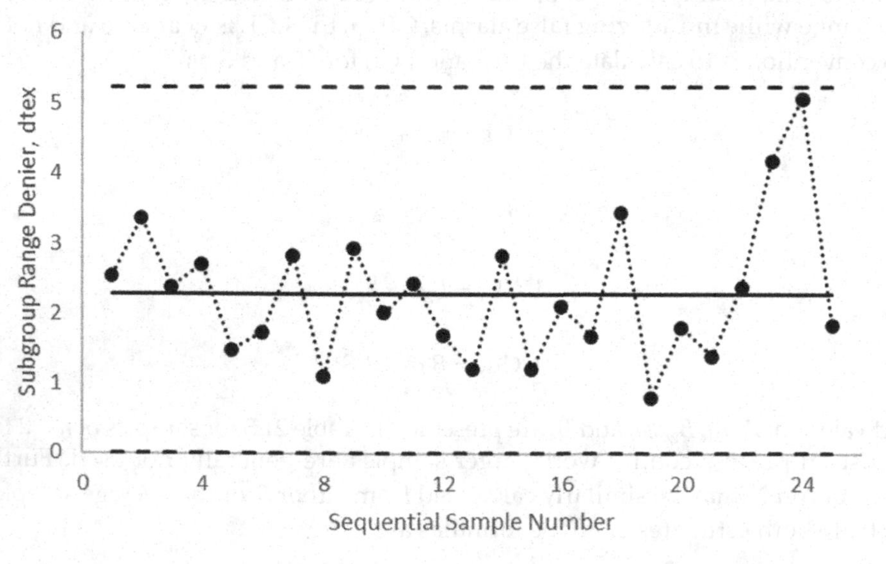

FIGURE 21.6
R chart for Example 21.3.

TABLE 21.4

Selected Values of D_3 and D_4 for Calculation of Control Limits on R with $\bar{R}$ Known

Sample size (n)	D_3	D_4
2	0	3.267
3	0	2.575
4	0	2.282
5	0	2.114
6	0	2.004
7	0.076	1.924
8	0.136	1.864
9	0.184	1.816
10	0.223	1.777

Selected entries from ASTM STP 15-C: Manual on Quality Control of Materials, Copyright 1951, ASTM, used with permission.

within a timespan smaller than that for process changes. For instance, an online gas chromatograph may be sampling the products of several distillation columns. With a 15-minute purge and analysis time per sample, a single sample per column each hour may be the best that can be obtained. It would normally take 4 hours to collect four samples, but the variables describing the thermodynamic state (temperature, pressure, composition) of a distillation column can change considerably in that time. Another difficulty may arise if it is not possible to maintain a uniform subgroup size. For instance, to minimize use of an expensive or toxic intermediate, the sampling device may be designed to extract just enough material for triplicate laboratory determinations. But if a lab technician makes an error on one determination, only two analyses would be reported.

Where multiple samples are not available, the use of a moving or filtered average and the use of variable UCL and LCL are common modifications to the $\bar{X}$ and R charts. In the moving-average method, $\bar{X}$ and R are calculated by using the most recent n data points. Each data point is equally weighted. In the filtered-average method, the data points are weighted, usually to enhance the contribution of the most current point to the average. A geometric filter weights the data points according to a geometric progression such as

$$\left(\frac{1}{2}\right)^n = \left[\frac{1}{2}, \frac{1}{4}, \frac{1}{8}, \frac{1}{16}, \frac{1}{32}, \frac{1}{64}, \ldots\right]$$

For example, using the four latest data points, the geometric filtered average is

$$\frac{\frac{1}{2}X_{25} + \frac{1}{4}X_{24} + \frac{1}{8}X_{23} + \frac{1}{16}X_{22}}{\frac{1}{2} + \frac{1}{4} + \frac{1}{8} + \frac{1}{16}}$$

Although this is illustrated for a $\left(\frac{1}{2}\right)^n$ series, you could choose $\left(\frac{1}{3}\right)^n$ or $\left(\frac{1}{4}\right)^n$ to amplify the weighting of the most recent point. An exponential or first-order filter weights the data according to an exponential decay. This method uses all previous data points, is common in automatic control noise filtering, and is very easily implemented as

$$\overline{X}_{new} = \alpha X_{new} + (1-\alpha)\overline{X}_{old} \quad (21.19)$$

where the choice of α in the range $0 \leq \alpha \leq 1$ defines the weighting of recent points. Although the exponential filter uses all of the previous points, if α is 0.6, the cumulative contribution of all points beyond the five most recent ones is only about 1% of the new average.

Where subgroup size is variable, one adjusts the upper and lower control limits to reflect the subgroup size. Tables 21.3 and 21.4 list values of A, A_1, A_2, B_3, B_4, D_2, and D_4 to be used for some subgroup sample sizes. The $\overline{X}$ chart may appear as in Figure 21.7.

In Figure 21.7 the sample number 8 had fewer replicates, so the UCL and LCL widen. At sample number 15, there were a larger number of replicates, so the UCL and LCL narrow. Notice that sample 15 violates its UCL, even though it is within the nominal UCL for all the other subgroups. This indicates the presence of an assignable cause.

Occasionally, the UCL or LCL may violate a physical constraint. For instance, a UCL on mole fraction may be calculated as a nonsensical 1.032, when the maximum possible value is 1.000. Similarly, the LCL on an impurity may be calculated as −0.37% when the minimum physically possible is 0.00%. Such events are likely when $\overline{\overline{X}}$ occurs near the physical constraint or when sampling variability is large. In such cases, replace the nonsensical UCL or LCL with the constraint value as illustrated in Figure 21.6, and use Rules 2 through 8 of Table 21.2 to detect assignable causes. Do *not* replace the UCL or LCL with the specification limits.

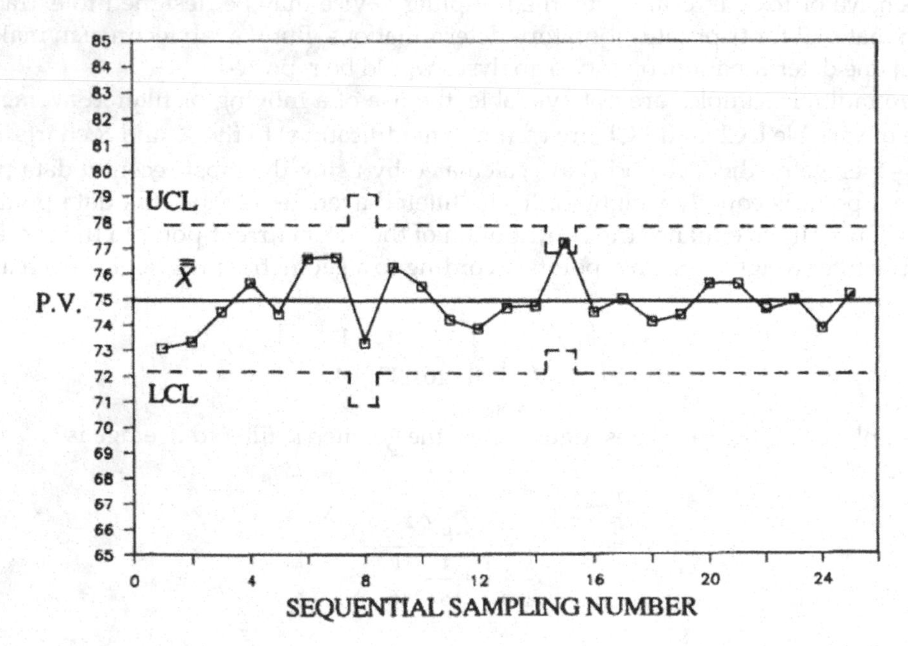

FIGURE 21.7
Illustration of an $\overline{X}$ chart with control limits reflective of changing sample size.

21.5 CUSUM and RUNSUM Charts

A CUSUM chart plots the cumulative sum of $\bar{X}$ deviations from the *target* value, or set-point. The CUSUM chart makes trends more observable than the $\bar{X}$ chart does and also shows deviations from targets, not from $\bar{\bar{X}}$. In its elementary form,

$$\text{CUSUM} = \sum_{i=1}^{k} \left(\bar{X}_i - \text{target}\right) \tag{21.20}$$

one would expect a plot of CUSUM versus sequential sample number to exhibit a random walk about a value of zero. If, however, $\bar{\bar{X}}$ is not "on target", CUSUM will progressively move away from zero. The rate at which CUSUM moves away from zero is related to the local deviation of $\bar{\bar{X}}$ from the target value.

There are other cumulative sum methods, but the principle is the same. One is the running sum of the number of standard deviations by which the process differs from the target value,

$$\text{RUNSUM} = \sum_{i=1}^{k} \text{INT}\left[\frac{\bar{X}_i - \text{target}}{S_i / \sqrt{n_i}}\right] \tag{21.21}$$

In Equation (21.21), S_i represents the standard deviation of the individual samples and n_i is the number of samples in the ith subgroup. When RUNSUM accumulates a value of ± 3 or greater, action should be taken to bring the process back to the control target.

21.6 Attribute Charts: Nonconforming

Many processes use product attributes (good–bad, full–short, pass–fail) to measure whether a product conforms or does not conform to specifications. Examples include computer chips in which one nonfunctional component makes the entire chip worthless, or food packaging in which not attaining a vacuum seal makes the jar or can of product not suitable for sale. *Conforming* products meet specifications. *Nonconforming* products do not meet specifications. The number of nonconforming products, n_{nc}, is of interest. Here, n_{nc} is the product of the total number of products times the portion of them that are nonconforming. $n_{nc} = n \cdot p$.

Where production or lot size is variable, the fraction of nonconforming units, p, is used to normalize the data. The distributions of the attribute variables n_{nc} and p are described by the binomial distribution. As a clarification, the number of nonconforming items is different from the number of defects. The number of defects per unit follows the Poisson distribution and is treated in Section 21.7. Furthermore, as variables, both are different from a continuous PV, which follows a normal distribution and is treated with $\bar{X}$, S, and R charts.

The principle behind the n_{nc} and p chart is exactly the same as that behind the $\bar{X}$ chart. The product is sampled at regular intervals, perhaps once per shift or batch, the samples are tested, and the number of nonconforming units is recorded as n_{nc_i}. If the sample size n_i is variable, p_i is calculated as

$$p_i = \frac{n_{nc_i}}{n_i} \tag{21.22}$$

and recorded. After about 25 scheduled samplings, either n_{nc} or p is plotted versus sequential sample number. The average number and average proportion of nonconforming items are calculated by the following equations:

$$\overline{n_{nc}} = \frac{1}{\text{k}} \sum_{i=1}^{k} n_{nc_i} \tag{21.23}$$

and

$$\overline{p} = \frac{\sum_{i=1}^{k} n_{nc_i}}{\sum_{i=1}^{k} n_i} \tag{21.24}$$

Since n_{nc} and p follow a binomial distribution,

$$\sigma_{n_{nc}} = \sqrt{\frac{\overline{n_{nc}}\left(n - \overline{n_{nc}}\right)}{n}} \tag{21.25}$$

And

$$\sigma_p = \sqrt{\frac{\overline{p}\left(1 - \overline{p}\right)}{n}} \tag{21.26}$$

for samples of size n.

The UCL and LCL are calculated as

$$\text{UCL}_{n_{nc}} = \overline{n_{nc}} + 3\sigma_{n_{nc}} \tag{21.27}$$

$$\text{UCL}_{n_{nc}} = \overline{n_{nc}} - 3\sigma_{n_{nc}} \tag{21.28}$$

$$\text{UCL}_p = \overline{p} + 3\sigma_p \tag{21.29}$$

$$\text{UCL}_p = \overline{p} - 3\sigma_p \tag{21.30}$$

Then either n_{nc} or p from each sampling is plotted versus sequential sample number with either $\overline{n_{nc}}$ or $\overline{p}$ and with the UCL and LCL indicated. The SPC use of the n_{nc} or p chart is identical to that of the $\overline{\text{X}}$ chart.

As an additional caution, the binomial distribution is not symmetric and, ideally, the UCL and LCL are not at the $\pm 3\sigma$ limits. However, if the sample size is large enough so that the number of nonconforming sample units statistically represents the entire production segment, and p is not near to the 0 or 1 limits, then the n_{nc} and p distribution will approach symmetry. The $\pm 3\sigma$ limits therefore have become the accepted standard.

21.7 Attribute Charts: Defects

Product attributes can also be used to measure the number of faults in a unit – for instance, 3 missed welds/frame, 5 missed stitches/square yard, or 1 scratch/table. And, in such cases, the product may be acceptable, albeit imperfect. In such instances, the number of defects, c, or the number of defects per unit, u, is sequentially plotted. Here, however, both c and u follow the Poisson distribution; even so, the principles behind the construction and use of the c and u charts are exactly the same as those for the $\bar{X}$, np, and p charts. The standard deviations and confidence limits are calculated from the following equations:

$$\sigma_c = \sqrt{\bar{c}} \tag{21.31}$$

$$\sigma_u = \sqrt{\frac{\bar{u}}{n}} \tag{21.32}$$

$$\text{UCL}_c = \bar{c} + 3\sigma_c \tag{21.33}$$

$$\text{UCL}_c = \bar{c} - 3\sigma_c \tag{21.34}$$

$$\text{UCL}_u = \bar{u} + 3\sigma_u \tag{21.35}$$

$$\text{UCL}_u = \bar{u} - 3\sigma_u \tag{21.36}$$

Because the Poisson distribution is not symmetric, the sample size must be large enough so that c and u are relatively normally distributed.

21.8 Takeaway

This chapter presented the fundamental statistical analysis of SPC (or six sigma, or quality control methods). After an assignable cause is recognized, the important tasks of trouble shooting, rational diagnosis, and decision making need to be performed.

21.9 Exercises

1. If the product specification limits are between 6.8 and 7.3 wt.%, and the process σ is 0.04 wt.%, what could be the target (or set point) wt.% to make the process capable?

2. Why not calculate σ from all 100 PV sample values in Table 21.1 instead of using $\overline{s}$ to determine the UCL and LCL values?
3. Suppose the sample average values in Table 21.1 were not normally distributed. How would you calculate UCL and LCL?
4. Use Figures 21.4 and 21.5 to see if they reveal any patterns listed in Table 21.2.
5. Derive Rule 2 in Table 21.2 from the binomial distribution.
6. If there were too many zero crossings on an X-bar chart, 1) what might it mean, and 2) what criterion could be used to flag it?
7. What would be the impact if the X-bar chart contained just a few samples (perhaps n = 10), or very many samples (perhaps n = 1,000)?
8. Derive values for the entries in Columns A, B_1, and B_2 in Table 21.3. As a hint, think $3/\sqrt{n}$ and $\sqrt{\chi^2/(n-1)}$.
9. Generate a simulator to produce data such as that in Table 21.1, and to display an X-bar chart. As data is sequentially generated have the X-bar chart reveal the moving window. Display the X-bar-bar and UCL and LCL values. Have the simulator make a change in level of the PV and see when the chart recognizes the change.

22

Reliability

22.1 Introduction

"The reliability of an item is the probability that it will adequately perform its specified purpose for a specified period of time under specified environmental conditions" (L. M. Leemis, *Reliability*, Prentice-Hall, Englewood Cliffs, NJ, 1995). Note 1) Reliability is a probability, not a certainty. 2) The criteria is that it will adequately perform, not that it is best in class or equivalent to a more expensive product. 3) Reliability is based on the specified purpose of the item, not on other measures of quality, such as it remains looking like new. 4) The specified period of time does not mean forever. 5) The specified environmental conditions do not mean that the product can be abused by leaving it outside in the rain, or otherwise stressed beyond normal expected use.

Reliability should not be confused with quality, performance, cosmetic appeal, etc.

Reliability is an indication of the consistent functionality of a thing to meet some chosen criteria. Because there is a wide variety of criteria, and things, and many measures of consistency, there are many measures of reliability. For instance, the thing may be an item, a procedure, or a process. The criterion may be that the item is always available, that it always works, or that it has a certain size or color. Alternatively, the criterion may be that a procedure always gives an answer, or that it discriminates between several possible causes. The criterion may be that the process always makes product of a desired quality. Consistent functionality may refer to short-term uniformity, on-stream time, or lack of change over an extended period.

In the following examples, the item is used intermittently, and the expectation is that it works on demand: When rolled, the die shows a 6. When kicked, the ball scores a goal. When switched to ON, the light works. These are end-of-batch, final outcomes, end-of-process outcomes, and the binomial distribution will govern probabilities.

However, in many situations the device, process, or procedure is in continual use over time, and then reliability relates to the probability of an event during a time interval. Here, measures of reliability could be on-stream-time (the fraction of time that the item works as desired), or the average number of particular events within a time period or accumulated operating time, or the probability of 0, 1, 2, etc. events within a time duration. Here, the Poisson distribution will be the appropriate probability model.

Finally, the events could be over a space interval, such as the number of rust spots on an automobile, or defective shingles on a house roof. Here also, the Poisson distribution will be the appropriate probability model.

However, regardless of the user's definitions of consistency or criteria or of the thing being considered, probability distributions and composite probability concepts are involved.

DOI: 10.1201/9781003222330-22

22.2 Probability Distributions

Often, reliability is concerned with time-dependent behavior. For instance, if the useful life of a car is to be the measure of reliability, and if car A lasts for 5 years while car B lasts for 7 years, then car B seems to be more reliable. Normally, though, one cannot know the life of any one particular car in a model series, because each car is subject to independent treatment and has its own confluence of manufactured variations. However, given several hundred thousand cars of a model series sold in one year, you could graph the number of remaining cars on the road for each successive year. Graphs for car types A and B may appear as shown in Figure 22.1. The actual data (counts) are represented by the histogram, and the general trend is illustrated by the continuous curve. You can use Figure 22.1 to obtain a measure of the car life. Starting with 500,000 cars of each type, half (250,000) of the type A cars are still on the road after 5 years. Such a half-life is one measure of the model reliability. Another might be the weighted average life.

Figure 22.1 illustrates an exponential distribution in which the probability of failure, or disappearance, is proportional to the number of items remaining. Classical examples of the exponential distributions are radiation attenuation through shielding, chemical reaction dependence on species concentration, life of light bulbs, and particle age in a continuous-flow mixer. For this car example, a failure mechanism that causes the failure rate to be proportional to the remaining items might be crashes. If car A is a red sports car, it may appeal to drivers with a risk-taking style and have a higher probability of accident involvement than car B, a family minivan. Formulas for the exponential distribution (and many others) are presented in Chapter 3.

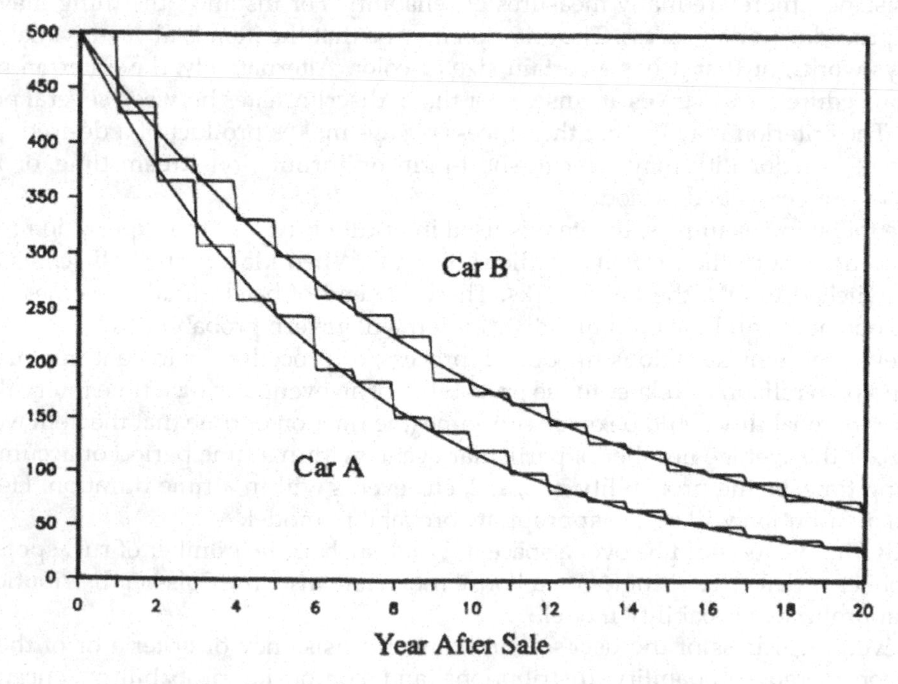

FIGURE 22.1
Histograms of the number of cars on the road after the year of their sale.

Many failure mechanisms are not simply related to the number of items remaining. For instance, manufacturing defects usually cause failure early in a product life, and those items that make it through the "burn-in" or warranty period will likely function as designed, except for occasional catastrophic events. Catastrophic events might include random in-use stress exceeding design strength or mistreatment such as being dropped. Toward the end of a product life, effects such as mechanical wear, chemical corrosion, and oxidative or mechanical embrittlement can cause parts to fail in their normal function. As the product ages, such aging mechanisms increase the failure rate. These three failure mechanisms – manufacturer's defects, catastrophic events, and aging – combine to make a life histogram similar to that shown in Figure 22.2. The vertical axis is the number of items remaining functional, and the horizontal axis is time in use.

The failure rate is the negative derivative of the smoothed curve in Figure 22.2 and represents the rate (number of) items failing in any particular time interval. Such curves are often referred to as "bathtub" curves. Figure 22.3 shows the "bathtub" curve corresponding to the smoothed histogram of Figure 22.2. The vertical axis of the bathtub curve is the rate at which items fail (number per time interval).

Although it may no longer be politically correct, the early-life high rate of product failure has been termed "infant mortality". It represents manufacturing defects (or weaknesses due to manufactured product variation). As these weak items are eliminated by use or by manufacturer "burn-in" the failure rate drops to a normal failure rate due to normal in-use events. It is termed the "intrinsic rate" and is often modeled as an extended flat period. In Figure 22.3 this could be considered as the interval between 4 and 8 years. The right-hand side represents "wear-out" where the product fails because of extended use and other time-related ageing factors (oxidation, actinic, or biological degradation, etc.).

Note: The $f(t)$ curve represented in Figure 22.3 cannot rise unbounded as time progresses. The normalized area under the curve must go to unity as time goes to infinity.

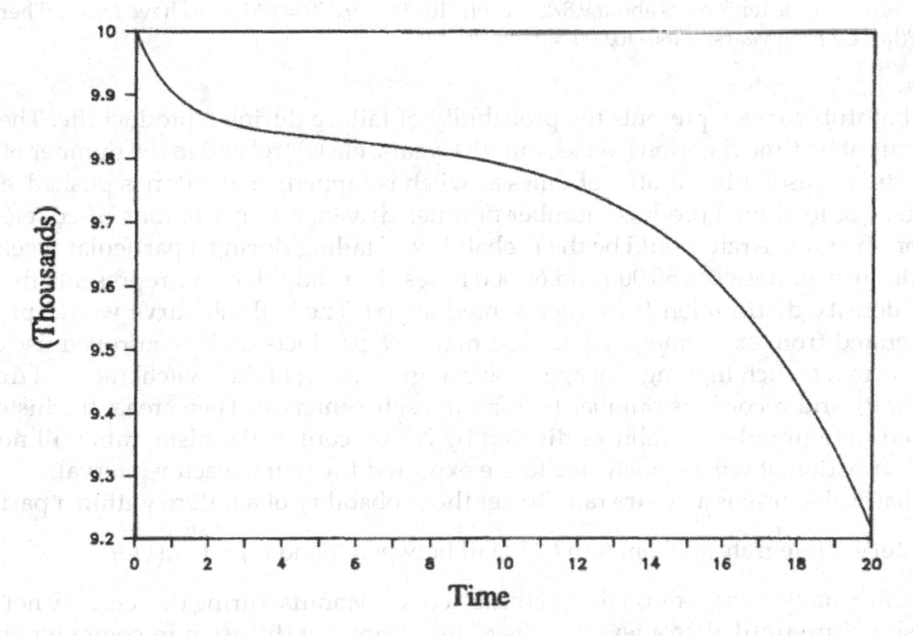

FIGURE 22.2
A representative histogram of number of items not failed versus time.

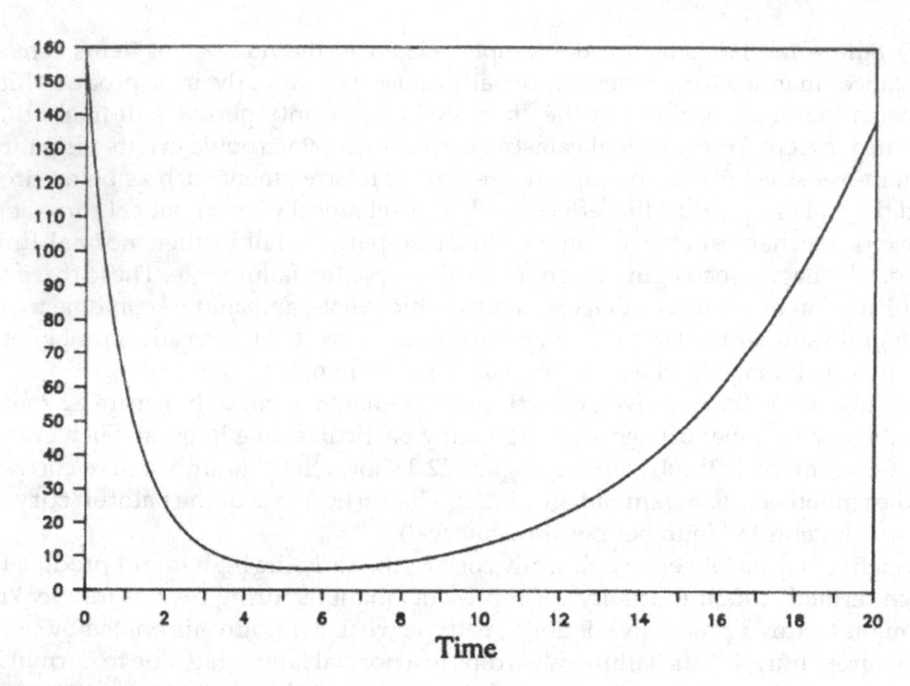

FIGURE 22.3
A representative "bathtub" curve of failure rate versus time.

Example 22.1: Use Figure 22.2 to determine the probability of failure after in-use for times of 2 and 5 years.

The original number of items is 10,000. After 2 years about 9,870 are left. 10,000 – 9,850 = 130 items have failed. Then $P(\text{fail by } t = 2 \text{ years}) = 130/10{,}000 = 0.013$.

Similarly, after 5 years about 9,872 are left. 10,000 – 9,820 = 180 items have failed. Then $P(\text{fail by } t = 5 \text{ years}) = 180/10{,}000 = 0.018$.

The bathtub curve represents the probability of failure during a product life. The units of life might be time duration (weeks, months, years, etc.) or related to the number of times the product is used (the number of times a switch is flipped, or a button is pushed, etc.), or the extent of load on a product (number of miles driven, number of tons lifted, etc.). The units on the failure rate would be the probability of failing during a particular week, or at the 200th use, or between 50,000 and 60,000 miles. The bathtub curve represents the probability density distribution (*pdf*), represented as $f(x)$. The bathtub curve would probably be generated from experimental data. Test many, N, products under controlled use conditions (normal switch flipping, not aggressive angry hitting of the switch) (normal driving, not racing), and record the number that fail in each x-interval. Then create the histogram of scaled data (number of failures divided by N). Of course, the histogram will not be a smooth function, it will be noisy due to the expected fraction in each x-interval.

The bathtub curve is a failure rate. To get the probability of a failure within a particular time interval integrate $f(t)$ from t_1 to t_2. $P(\text{fail between } t_1 \text{ and } t_2) = \int_{t_1}^{t_2} f(t)dt$.

There are many variations on the "bathtub" curve. Manufacturing defects may not begin to cause failure until after a few months of use. Impurity diffusion in computer chips is such an example. In such a case, initial failure rate may be low, then rise as defects become important, then fall, as the robust components continue full performance. The "bathtub" curve may have a lip in the early-life stage. The "wear-out" period might also begin to flatten at very long times, perhaps because items subject to light use, care and maintenance have a longer life than others.

The Weibull distribution and the gamma distribution both have two adjustable parameters that make them flexible. Either can be used to approximate experimentally observed or postulated failure rates. As a result, both are used in reliability studies. The Weibull and gamma distribution formulas are presented in Chapter 3.

Alternately, "burn-in" or other finishing touches or "cull-out" testing could eliminate much of the early-stage failures when the product is in use. The shape of the bathtub curve depends on what the product is, and for any product the materials used, quality control, and many features. The failure rate could be based on the original number of products, or the remaining items after failed items are removed. And, of course, the failure rates could be significantly affected by the severity of use of the item.

The exponential, gamma, and Weibull distributions are continuous functions and are valid whenever the number of items is very large. However, you may often encounter an item composed of a limited number of parts, and if any individual part fails, the item fails. For example, a bridge may have four bents, a paratrooper may have one parachute, a seam may have two welds, or a reactor may have three feed controllers. The binomial and Poisson distributions (Chapter 3) describe such discrete events.

Be sure that the probability of failure is defined per time interval, and per space interval or per item for continual use, or per instance of use for end-of-batch or on-demand uses.

These represent common elementary distribution situations. But you may need additional considerations. For instance, light bulbs, or some items that you purchase, may have already failed, and will not work when first tried. In those situations, the cumulative probability distribution has a positive intercept at time equal to zero as shown in Figure 22.4a. Alternatively, a cumulative probability distribution may never reach unity because the part may be scrapped for a nonfailure reason such as obsolescence. This situation is represented by Figure 22.4b. Finally, there may be two distinct modes of failure that are additive and occur at different product ages as shown in Figure 22.4c.

Example 22.2: Use Figure 22.4a to determine the probability of an item failing by time 5.

$F(t)$ represents the *CDF* of failed items over time. Note the $t = 0$ value if $F(t)$ is about 0.3. This means that 30% of the products were defective when first opened or put into use.

At $t = 5$, $F(t) \approx 0.68$.

However, since about 30% were defective prior to use, the portion of those that initially worked failing by time of 5 is $p = \dfrac{0.68 - 0.30}{1.00 - 0.30} \approx 0.54$.

Depending on how the failure probability is interpreted, it could be either 68% or 54%.

22.3 Calculation of Composite Probabilities

An *individual event* or a primary event is defined by the user and has a known probability of occurring. The probability could be theoretically derived, but more likely it would be based on prior experience with similar situations, and intuitively adjusted to match the new situation. Such an event may be as simple as whether a set of tossed coins lands on "Heads" or "Tails"; however, the event may be as complicated as a car accident that is the

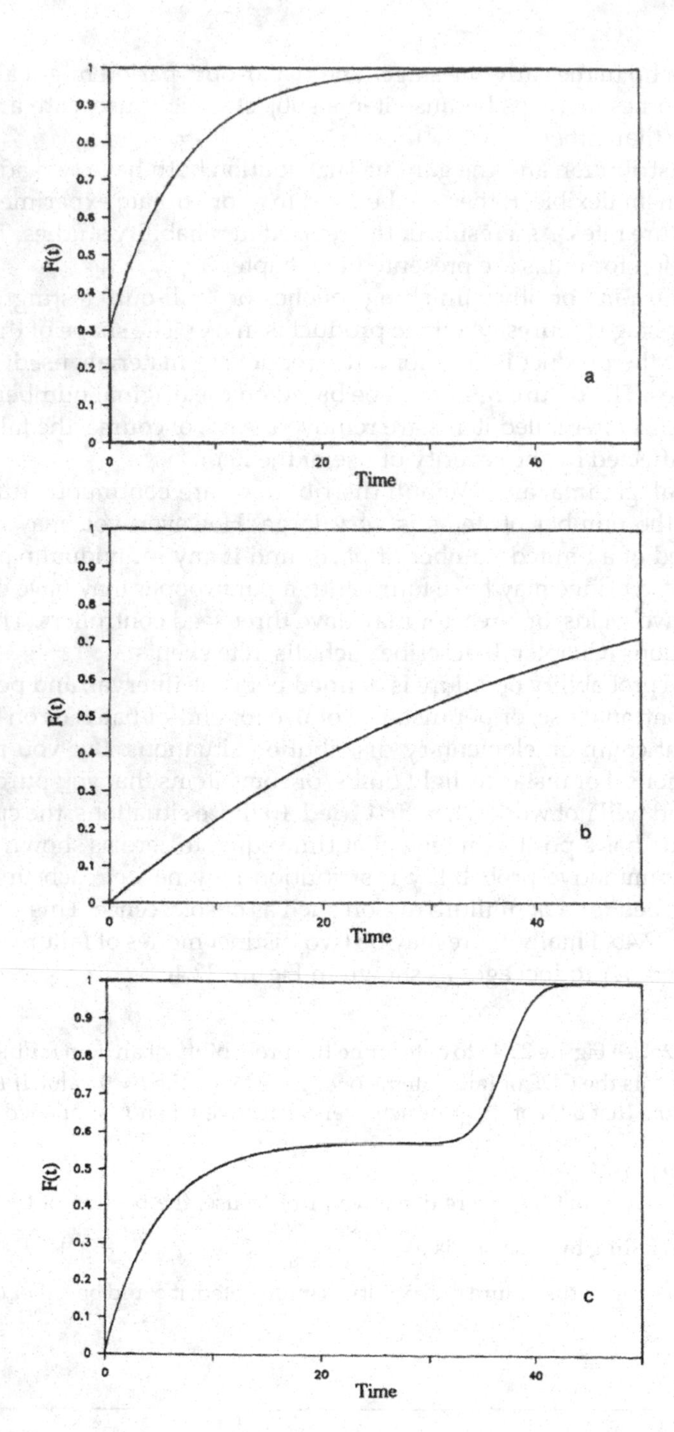

FIGURE 22.4
Cumulative probability functions: (a) Items failed before use. (b) Some items become scrapped before failing. (c) Dual failure modes.

result of several more basic events. A *composite event* is the result of the conjunction of several individual events. The composite event probability of rolling a 12 in a two-dice game, for example, depends on the individual event probabilities of rolling a 6 on each die.

The probability may represent on-demand outcomes (will the back-up generator work on demand), or within-time outcomes (will the generator fail during a month of steady operation).

System reliability is a composite event. Both on-stream time and mean-time-between-failures depend on the conjunction of individual event probabilities. For many cases and many measures of reliability, the composite event probability can be calculated from simple rules. A few are reviewed here.

Throughout the elementary analysis of this section we will assume that individual events are independent of each other and that there are no "common" or "shared" causes. For example, the probability that your wireless router and desktop computer fail simultaneously from independent events is rare. However, if lightning strikes and causes one to fail, it is likely that the other will fail due to that *common cause*. Common causes are not unusual. If changes in maintenance personnel cause an improper pump seal adjustment procedure, that change will eventually affect all pumps. If a new batch of raw material contains a catalyst poison, all reactors will be affected. If one microprocessor on a space probe fails due to radiation exposure, it is likely that other microprocessors will also fail. If one gymnast on a team is taught to perform the compulsory exercises incorrectly and scores low in competition, it is likely that her teammates' performances will also score low.

This introduction to composite probabilities also assumes that there are two states of an item, complete functionality or complete failure. The population is dichotomous. In reality, there are many failure states. For example, a truck could function as designed, or it could be totally useless. However, it could perform well, but because its headlights are broken, it would be limited to day use, partially functional. In another situation, the truck could be missing several of its 18 tires and be restricted to lower weight service, again, partially functional.

For on-demand processes, probabilities are the fractions of instances that an event could occur when demanded. It is not necessarily the fraction of time that an item is not working. For example, a portable oxygen analyzer could be used once per shift for routine inspection of the inert blanket gas in a volatile organic liquid storage tank. The analyzer may work on the average of 995 times out of 1,000 or have a failure probability of 0.005. When the analyzer fails, it may take 10 days for repair. Since it is used about 1,000 times per year, it will fail about 5 times and will be out of order for about 50 days per year. It will be out of service 50/365 or approximately 14% of the time. On-stream time is dependent on failure probability but also depends on time-to-repair. A second oxygen analyzer may be needed so that the unit blanket can be checked once per shift. As the performance of one unit does not affect the other, the probability of simultaneous failure is 0.005×0.005, or 25×10^{-6}. But more importantly is the probability of the back-up failing during the time that the first is in repair.

Finally, for this analysis, we assume that all significant contributing events have been incorporated into the calculated probability of the composite event. If we forgot to include the fact that a dead battery is a possible cause for the analyzer not working, then we would have overlooked and excluded a significant mechanism. We will almost certainly calculate an incorrect analyzer functioning reliability.

22.3.1 "And" Events

There are two basic types of composite events. One is the "and" or "multiplicative" event, in which several individual events must occur simultaneously for a composite event to happen. If $E_1, E_2, \ldots, E_n$ are independent individual events with a probability of occurring of $P(E_1), P(E_2), \ldots, P(E_n)$, respectively, then the probability of E_1 *and* E_2 ... *and* E_n occurring is

$$P(E_1 \text{ and } E_2 \ldots \text{and } E_n) = P(E_1) \cdot P(E_2) \cdot \ldots \cdot P(E_n)$$

$$P(E_1 \text{ and } E_2 \ldots \text{and } E_n) = \prod_{i=1}^{n} P(E_i) \tag{22.1}$$

The "and" conjunction and "multiplicative" nomenclature is obvious from Equation (22.1). Such a composite event could describe a transportation system with several possible paths from point A to point B. If any one path is open, material can flow. Only if all the paths fail simultaneously does material flow stop.

Example 22.3: There are three bolts holding a seam together. The seam will function if any one bolt remains functional. Bolts are either failed or fully functional and are inspected regularly. Independent events between inspections contribute a 0.001 probability (one chance out of a thousand) that any bolt has failed. What is the probability that the seam fails?

Applying Equation (22.1),

$$P(\text{seam failure}) \left| \begin{array}{l} = P(\text{bolt one failed and bolt two failed and bolt three failed}) \\ = P(\text{bolt \#one failed}) \cdot P(\text{bolt \#two failed}) \cdot P(\text{bold \#three failed}) \\ = (0.001)(0.001)(0.001) \end{array} \right.$$

$P(\text{seam failure}) = 10^{-9}$.

22.2.2 "Or" Events

The other basic type of composite event is the "or" conjunction, or "additive," event, in which a composite event occurs if any one or more of several individual events happen. Again, using E_i to represent an individual event and $P(E_i)$ as its probability, then the probability of E_1 *or* E_2 ... *or* E_n occurring can be calculated. There are several ways that the composite event could happen. A few ways include:

$$\begin{gathered} E_1 \text{ only} \\ E_2 \text{ only} \\ \vdots \\ E_1 \text{ and } E_2 \text{ but not } E_3, \ldots, E_n \\ \vdots \\ E_3 \text{ and } E_5 \text{ and } E_{10} \text{ but not} \ldots \\ \vdots \end{gathered}$$

Consequently, rather than enumerate every possible combination of individual events that could cause the composite event, it is easier to use the complementary events.

For a dichotomous population,

$$P(\text{not event}) = 1 - P(\text{event}) \tag{22.2}$$

Since the composite event occurs if any one or more of the individual events occur, the composite event will not occur only if each of the individual events does not occur.

$$P(\text{not composite}) = P(\text{not } E_1 \text{ and not } E_2 \text{ ... and not } E_n)$$

$$1 - P(\text{composite}) = [1 - P(E_1)][1 - P(E_2)]...[1 - P(E_n)]$$

$$P(E_1 \text{ or } E_2 \text{ ... or } E_n \text{ or any combination}) = 1 - \prod_{i=1}^{n}[1 - P(E_i)] \tag{22.3}$$

For the special $n = 2$ and $n = 3$ event cases, Equation (22.3) becomes

$$P(E_1 \text{ or } E_2 \text{ or both}) = P(E_1) + P(E_2) - P(E_1)P(E_2) \tag{22.4}$$

$$P(E_1 \text{ or } E_2 \text{ or } E_3 \text{ or any combination}) = P(E_1) + P(E_2) + P(E_3) +$$

$$-P(E_1)P(E_2) - P(E_1)P(E_3) - P(E_2)P(E_3) + P(E_1)P(E_2)P(E_3) \tag{22.5}$$

It would be easier to use Equation (22.3) than the expanded form of Equation (22.5). However, often, the probabilities of the individual events are small. In such a case, the products in Equations (22.4) and (22.5) are several orders of magnitude smaller than the individual event probabilities and are therefore usually neglected. In addition, if the events are mutually exclusive (for instance, if a process is shut down because of a high-temperature limit, the high-pressure limit cannot be reached), then the composite probabilities become

$$P(E_1 \text{ or only } E_2 \ldots \text{or only } E_n) = \sum_{i=1}^{n} P(E_i) \tag{22.6}$$

Such a composite event could describe the failure of a doorbell to ring.

Example 22.4: What is the probability that a doorbell fails to ring if the independent individual events are the button fails, $p = 0.001$; the wiring fails, $p = 0.0001$; and the ringer fails, $p = 0.0008$.

The events are not mutually exclusive. If the wiring fails, one would still try the doorbell. The bell won't ring if either the button, the wire, the bell, or any combination fails. Therefore, Equation (22.3), or its three-event equivalent, Equation (22.5), describes the composite event.

Using Equation (22.3):

$$P(\text{no ring}) = 1-(1-0.001)(1-0.0001)(1-0.0008)$$
$$= 1-0.999\cdot 0.9999\cdot 0.9992 = 0.0018990200\ldots$$

Using Equation (22.5):

$$P(\text{no ring}) = (0.001)+(0.0001)+(0.0008)-(0.001)(0.0001)$$
$$-(0.001)(0.0008)-(0.0001)(0.0008)+(0.001)(0.0001)(0.0008)$$
$$= 0.0018990200\ldots$$

Using the small p approximation of Equation (22.6):

$$P(\text{no ring}) = (0.001)+(0.0001)+(0.0008) = 0.0019$$

Note that the products of terms are each several orders of magnitude less than the individual terms and ignoring them gives an answer of 0.0019, which is in error of only 0.05%, which is likely smaller than the impact of uncertainty on any of the p-term values.

Example 22.5: If a college student either forgets to set his wake-up alarm, or if his car won't start, or if he and his buddy decide to go fishing, the student will miss class. The events are independent and have probabilities of 0.1, 0.02, and 0.05, respectively. What is the probability of the student missing a class?

The events are independent but not mutually exclusive. On any particular day, he could forget to set his alarm and his car may not start. Accordingly, Equation (22.3) or its three-event equivalent Equation (22.5) describes the composite event probability:

$$P(\text{missed class}) = (0.1)+(0.02)+(0.05)-(0.1)(0.02)-(0.1)(0.05)-(0.02)(0.05)$$
$$+ (0.1)(0.02)(0.05)$$

P(missed class) = 0.1621

Note: If the product terms had been ignored, the answer would have been 0.17, an error of about 5%, which is probably of a smaller impact than the impact of uncertainty on any of the individual p-values. Reconsider the outcome if 0.11 was used for the probability of forgetting to set the alarm.

Example 22.6: Proper operation of a sump pump requires successful operation of both the liquid-level sensor and the pump motor. If the probabilities of the sensor and of the motor working when required (on demand) are 90% and 95%, respectively, what is the composite probability of the sump failing to empty?

Intentionally, in this example, the probability of failing is mixed with the probability of success. You must carefully define events and be consistent throughout composite calculations. The probabilities of items having failed on demand are 1 – 0.90 = 0.10 and 1 – 0.95 = 0.05. The events are independent: the sensor or motor may fail regardless of which state the other is in. Since failure of either item means failure of the sump, Equation (22.3), or the two-item equivalent, Equation (22.4), describes the composite probability.

$$P(\text{sump failing on demand}) = 0.10+0.05-0.005 = 0.145$$

22.3.3 Combinations of Events

System reliability is often improved by having several spare items in parallel and ready to replace a working item that fails. Usually, several working items share the load so that spare item size and hence capital investment is reduced. An example is an exhaust system that circulates solvent-laden room air in a manufacturing area to a solvent-recovery system. A fan failure might necessitate a production slowdown to reduce solvent evaporation into the room. Specifically, three fans may be operating with one spare on standby. If one operating fan fails, the spare can take over. However, if two or more of the fans fail, production must decrease.

For this within-operation example, the probability would need to be defined within a time interval, such as the probability of failure within a week, or a month, etc. Additionally, the time interval needs to be short enough so that the event of two failures for an item can be ignored.

The binominal distribution (Chapter 3) describes the composite probability of k independent events occurring out of n identical events when each event has a probability of p.

$$P(k|n) = \frac{n!}{k!(n-k)!} p^k (1-p)^{n-k} \tag{22.7}$$

The variables n and k are integers with $0 \le k \le n$. Normally, if k or more events out of n occur, the system is in a failed state.

$$P(i \ge k|n) = \sum_{i=k}^{n} \frac{n!}{i!(n-i)!} p^i (1-p)^{n-i} \tag{22.8}$$

Example 22.7: Five spot welds are equally stressed in a loadbearing section of a seam. The load is 200 lb_f, and the breaking strength of each weld is 75 lb_f. Due to independent events, the probability that any weld has failed is 0.001. What is the probability that the seam will fail?

The number of welds required is 200 lb_f / 75 lb_f/weld = 2.6666, indicating that three welds are required. If zero, one, or two welds fail, the seam will still function. If any three or more welds fail, the seam will fail. Using Equation (22.8),

$$P(i \ge 3|5) = \frac{5!}{3!2!}(0.001)^3(0.999)^2 + \frac{5!}{4!1!}(0.001)^4(0.999)^1 + \frac{5!}{5!0!}(0.001)^5(0.999)^0$$

$$= 9.98001 \times 10^{-9} + 4.995 \times 10^{-12} + 1 \times 10^{-15}$$

$$P(i \ge 3|5) \cong 1 \cdot 10^{-8}$$

the probability of a seam failing on demand is one in about 100 million.

One additional case of combination failures must be described. Standby process equipment has routinely scheduled downtime for preventive maintenance. During that time, working equipment can fail. However, since one spare is not available for immediate use, the probability of system failure increases.

Example 22.8: There are four parallel roads through the mountains and any two can adequately handle the traffic. However, if only one passage is open, resulting traffic jams hinder emergency vehicles and the highway system has failed. The frequency of each road closing due to accidents or avalanches or rock-slides is one in about 4 years. What are the failure probabilities if one and if two roads are simultaneously closed for repair for a week?

As long as one or fewer roads are open, the system has failed. The frequency is about

$$\lambda \cong \frac{1\text{ event}}{4\text{ years}} \frac{7\frac{\text{days}}{\text{week}}}{365.25\frac{\text{days}}{\text{year}}} = 0.00479124 \text{ events per week}.$$

From the Poisson distribution $P(0\text{ events, no unscheduled closing}) = 0.99522$. Then, in a given week $P(\text{an unscheduled closing}) = 0.00478$.

If no roads were closed for repair, the probability of one or fewer roads open out of four roads is

$$P(i \le 1|4) = \frac{4!}{0!4!}(0.99522)^0(0.00478)^4 + \frac{4!}{1!3!}(0.99522)^1(0.00478)^3$$

$$P(i \le 1\,|\,4) = 4.35235 \times 10^{-7} \approx 0.00004\%$$

If one road is closed for repair,

$$P(i \le 1\,|\,3) = \frac{3!}{0!3!}(0.99522)^0(0.00478)^3 + \frac{3!}{1!2!}(0.99522)^1(0.00478)^2$$

$$P(i \le 1\,|\,3) = 6.832 \times 10^{-5} \approx 0.0075\%$$

If two roads are closed for repair,

$$P(i \le 1\,|\,2) = \frac{2!}{0!2!}(0.99522)^0(0.00478)^2 + \frac{4!}{1!1!}(0.99522)^1(0.00478)^1$$

$$P(i \le 1\,|\,2) = 9.5367 \times 10^{-3} \approx 1\%$$

A 1% chance of system failure may be excessive! Do not close two roads simultaneously. Repair one at a time.

22.3.4 Conditional Events

The probability of some events can depend on whether previous events have occurred. For example, if you draw a card from a full deck, there is a 26 out of 52 or 0.5 probability that the card is red. Without replacement, at a second drawing, the probability of drawing a red card is 26 out of 51 (or 0.5098...) if the first card was black, or 25 out of 51 (or 0.4901...) if the first card was red. Conditional probability, or the without-replacement situation, is common in quality assurance sampling aspects of reliability. For such a situation,

$$P(\text{conditional event}) = P(E_1) \cdot P(E_2\,|\,E_1) \tag{22.9}$$

where $P(E_2|E_1)$ is the probability of event 2 occurring given that event 1 has already occurred.

Example 22.9: A product specification states that less than 5% of your product can have a breaking strength below some minimum value. Your company quality assurance procedure is to randomly sample (and to destructively test) 10 out of every 110 items and, if each of the 10 exceeds the minimum strength, to ship the remaining 100. What is the probability that shipped lots containing 5% defective parts will pass the quality inspection test?

Let p be the fraction of defective parts shipped and $q = 1 - p$ be the fraction of acceptable parts. Then in any lot of 100 items, $100q$ items meet the quality control requirements. In a passable lot of 110 items, $(10 + 100q)$ items meet the specification and $100p$ items are defective. The probability that the first sample is acceptable is $(10 + 100q)/110$. The probability that the second sample is acceptable, given that the first sample was acceptable, is $(9 + 100q)/109$. The probability of a third consecutive acceptable sample, given the first two, is $(8 + 100q)/(108)$, etc. The probability that all 10 samples conform to the product specification is obtained by extending Equation (22.9) as

$$P(\text{all 10 samples conforming with } p \text{ out of 100 defective})$$

$$= \frac{10+100q}{110} \cdot \frac{9+100q}{109} \cdot \frac{8+100q}{108} \cdot \ldots \cdot \frac{1+100q}{101}$$

with $p = 0.05$, or $q = 0.95$,

P(shipping a lot with 5% bad) = 0.615 ...

A better quality assurance technique seems warranted!

22.3.5 Weakest Link

A chain is as strong as its weakest link. If the strength of any link is a randomly distributed variable, then a chain of N links will fail based on the lowest strength of a sampling of N links. The distribution of chain strengths will be the distribution of the minimum value of a random sampling of N independent items. In a chain the items are in series. But there are other minimum-value distributions of engineering relevance that have a parallel configuration. Examples include the most thermally sensitive component in a microprocessor (when one part fails the computer chip is dysfunctional) and the strength of stitches in a sewn seam (when one stitch is stressed to its failure limit and breaks, it loosens the adjacent four stitches, which become essentially non-load bearing, and the increased load on the other $N - 5$ stitches can be catastrophic).

If x is a continuous random variable with a probability density function of $f(x)$, then the probability that $x \geq a$ is

$$\mathrm{P}(x \geq a) = \int_a^{\infty} f(x)\,dx \tag{22.10}$$

The probability that x_1 and x_2 and ... and x_n are greater than a is, from Equation (22.1),

$$P(x_1, \ldots, x_n \geq a) = \left[\int_a^{\infty} f(x)\,dx\right]^n \tag{22.11}$$

The probability that one or more values is not greater than a is

$$P(\text{one or more} < a) = 1 - \left[\int_a^{\infty} f(x)\,dx\right]^n = 1 - \left[1 - \int_{-\infty}^{a} f(x)\,dx\right]^n \tag{22.12}$$

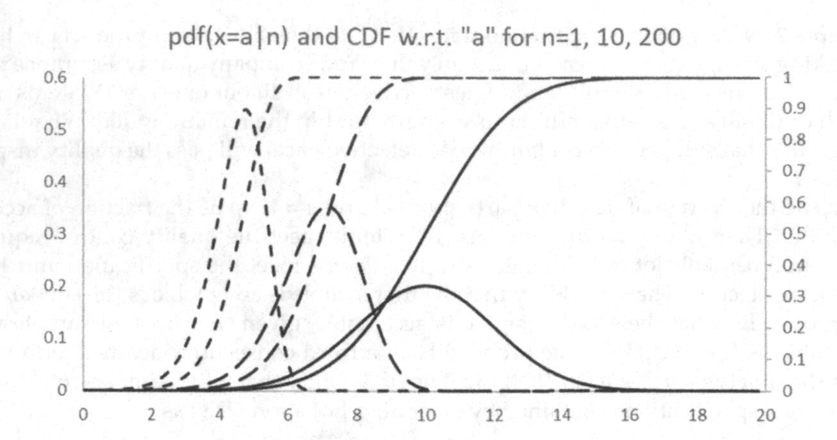

FIGURE 22.5
Examples of CDF and pdf of x for the weakest link in a chain.

The probability density function of the threshold minimum value is

$$f(x_{\min} = a \mid n) = \frac{dP(\text{one or more} < a)}{dx}$$

$$f(x_{\min} = a \mid n) = nf(a)\left[1 - \int_{-\infty}^{a} f(x)\,dx\right]^{n-1} \tag{22.13}$$

Figure 22.5 illustrates Equations (22.12) and (22.13) for a chain. The value of each link is normally distributed with a mean of 10 and a sigma of 2. The solid line represents a one-link "chain", the long-dashed line represents a 10-link chain, and the left-most curves a 100-link chain. The *pdf* value is on the left vertical axis, and the *CDF* value is on the right vertical axis. The abscissa (horizontal axis) is the value of "*a*" the threshold minimum value. For a one-link chain the distribution is Gaussian with a 50th percentile value of 10, the mean of a single link. When there are 10 links in the chain, the 50th percentile value of the chain drops to about a strength of 7, and for a 100-link chain to a value of about 4.5. If the desired critical strength of the chain is 5, then from the *CDF* curves, 99% of the single link "chains" will exceed that value, about 90% of the 10-link chains will exceed that critical value, and only about 30% of the 100-link chains will exceed the critical value.

Example 22.10: If *f(x)* is a uniform probability density function with values from 4 to 6, and if $n = 3$, what is the average value for the "weak link" product?

Since *f(x)* is uniform, its value is 1 / (6 – 4) = 0.5. Consequently, the probability density function for the weak link (hence the product) is given by Equation (22.13), with $f(x) = 0.5$.

$$f(x_{\min} = \alpha \mid n = 3) = 3(0.5)\left[1 - \int_{4}^{\alpha} 0.5\,dx\right]^2$$

$$= 3(0.5)\left[1 - \left(\frac{1}{2}\alpha - \frac{1}{2}4\right)\right]^2$$

$$= 1.5\left(3 - \frac{1}{2}\alpha\right)^2, \quad 4 \le x \le 6$$

The average x_{min} is then

$$\bar{x}_{min} = \int_4^6 \alpha f(x_{min} = \alpha \mid n = 3)\, d\alpha$$

$$= \int_4^6 \alpha\, [(1.5)(3 - 0.5\alpha)^2]\, d\alpha = 4.5$$

Note that, for presentation simplicity, we specified that *f(x)* was uniform so that the integrations were analytically easy. You will most likely encounter a normal *f(x)* for which numerical integration is more convenient.

22.4 Measures of Reliability

The probability of occurrence of an event is one of the common measures of reliability. The rules for calculation of probabilities were given in the preceding section. However, other measures of reliability may be more appropriate, depending on your choice of the event, the criteria, and the definition of consistent functionality. Reconsider the car A and car B example at the beginning of Section 22.2. The half-lives of cars A and B were 5 and 7 years, respectively, which translates to nearly a 14%/yr mortality rate for car A and a 10%/yr mortality rate for car B. Although car B appears to be a superior car based on mortality probability, it could be that drivers who choose the sports car A are less cautious and are involved in more accidents than the drivers who choose the minivan B. If so, failure probability would be more an indication of abuse than of product reliability. Perhaps non-accident repair costs (if cost is important to you) and non-accident in-shop time (if public transportation is not available) are more appropriate measures of car reliability for a mature driver. Several common reliability measures are described next.

22.4.1 Average Life or Mean Time to Failure

Let *f(t)* be the probability density function that describes the failure rate of an item. If the cumulative probability function, *F(t)*, is known, then *f(t)* can be obtained by differentiating *F(t)* with respect to time. With the condition that $f(t) \to 0$ as $t \to \infty$, the average item life then is

$$\bar{t} = \int_0^\infty t\, f(t)\, dt \tag{22.14}$$

The quantity represented by Equation (22.14) is often termed the mean time to failure (MTTF).

Example 22.11: A particular product fails at a rate of 10% per year of the remaining products. What is the average product life?

Let *N* be the number of products in any one year; then 0.1 *N* is the number that fail in that year or

$$\frac{dN}{dt} = -0.1N, \quad N(t = 0) = N_0$$

where N_0 is the original number of new products and t is in years. Integrating, one obtains the exponential distribution

$$N(t) = N_0 e^{-0.1t}$$

for the number of functioning items remaining. Letting M represent the number that failed,

$$M(t) = N_0 - N(t) = N_0\left(1 - e^{-0.1t}\right)$$

Dividing by N_0, one obtains the cumulative probability distribution

$$F(t) = 1 - e^{-0.1t}$$

Differentiating with respect to t gives the probability density function

$$f(t) = 0.1e^{-0.1t}$$

Using Equation (22.14),

$$\bar{t} = \int_0^{t=\infty} t\left(0.1e^{-0.1t}\right)dt$$

$$\bar{t} = 0.1\frac{(-0.1t-1)}{-0.1^2}e^{-0.1t}\Big|_0^\infty = 0 + 10$$

$$\text{MTTF} = \bar{t} = 10 \text{ years}$$

Note that for the special case of the exponential distribution, MTTF is the reciprocal of the failure rate, and equal to the distribution μ.

Example 22.12: A product is hypothesized to have a uniform failure rate. In 104 weeks 10% will have failed. What are values for $f(t)$, maximum life, and MTTF?

$$f(t) = \frac{0.1}{104} = 0.0009615\ldots(\text{per week})$$

$$\text{Max Life} = 1/f(t) = 1040(\text{weeks})$$

$$\text{MTTF} = \bar{t} = \int_0^{1040} t\left(\frac{0.1}{104}\right)dt + \int_{1040}^{\infty} t(0)dt = \frac{0.1}{104}\frac{1}{2}t^2\Big|_0^{1040} = 520(\text{weeks})$$

Example 22.13: A product is hypothesized to have an exponential failure rate, $f(t) = \alpha e^{-\alpha t}$. It is desired for it to have an average useful life of μ. What value of α is required?

$$\bar{t} = \mu = \int_0^{t=\infty} t\left(\alpha e^{-\alpha t}\right)dt = \alpha\frac{(-\alpha t-1)}{-\alpha^2}e^{-\alpha t}\Big|_0^\infty = 0 + \frac{1}{\alpha} = \frac{1}{\alpha}$$

Compare to Equation (3.44).

Example 22.14: A product is a composite of many parts. Consider a car, or a TV, or a vacuum cleaner. If any one of N critical parts fails, the product fails. The desire is that only p portion of the products fail after t_0 time (hours of use). The infant mortality of the components is hypothesized to be zero, and the intrinsic period (the middle of the bathtub curve) for each component is hypothesized to have a uniform failure rate, $(t)=\frac{1}{b}$, each with the same value of b. What value of b is required for each of the N critical components? How does N affect the required reliability of each component?

If any component fails, the product fails. If A fails, or B fails, or C fails, ... the product fails. It is simpler to use the AND conjunction than to consider all the possible combinations of the OR conjunction. So, consider the not-fail complementary condition.

$$P(\text{product failing})=1-P(\text{product not failing})$$

$$P(\text{product failing})=1-P(\text{A not failing AND B not failing AND C not}\ldots)$$

$$P(\text{product failing})=1-P(A\text{ not failing})\cdot P(B\text{ not failing})\ldots$$

Since there are N components with common $F(t)=\int_0^{t_0}\frac{1}{b}dt=\frac{t_0}{b}=P(i\text{ not failing})$

$$P(\text{product failing})=1-\left[1-\frac{t_0}{b}\right]^N$$

desiring that $P(\text{product failing})\leq p$ and solving for b

$$b\geq\frac{t_0}{1-(1-p)^{1/N}}$$

For the second question, consider $f(t)=\frac{1}{b}$ the failure rate. If there were only $N=1$ components, then $b\geq\frac{t_0}{p}$ but with N components the failure rate for each must be reduced by the ratio

$$\frac{b_{N=1}}{b_N}=\frac{1-(1-p)^{1/N}}{p}$$

For illustration, if the desired value for p is 0.05 and there are $N=10$ critical components in the product, the failure rate for each component must be effectively one tenth of the product.

For a uniform distribution, MTTF is

$$\text{MTTF}=\int_0^b t\left(\frac{1}{\alpha}\right)dt+\int_b^\infty t(0)dt=\frac{1}{b}\frac{1}{2}t^2\Big|_0^b=\frac{b}{2}$$

Similarly, if the desired value for p is 0.05 and there are $N=10$ critical components in the product, the MTTF for each component must be effectively ten times greater than that of the product.

Example 22.15: A product contains 10 components. If any one component fails, the product fails. The product and the components each have a uniform distribution of failure rate. The manufacturer desires that 90% of the products last at least 2 years in use. What must be the lifetime of each component?

The *pdf* of the failure rate of the component is uniform. If 90% must last over 2 years, then the area under the *pdf* up to 2 years is 1 – 0.9 = 0.1. This means that the *pdf* value of the product is $f(t) = 0.1/2 = 0.05/\text{yr}$, which means that the maximum product life is 20 years, or the half-life is 10 years.

The probability of a product surviving 2 years, is the probability of all the components surviving 2 years. Since all components have the same *pdf*

$$P(\text{product surving 2 years}) = \prod_{i=1}^{10} P(\text{component } i \text{ surviving 2 years})$$

$$0.9 = \left[P(\text{component } i \text{ surviving 2 years})\right]^{10}$$

$$P(\text{component } i \text{ surviving 2 years}) = \sqrt[10]{0.9} = 0.989519\ldots$$

With a uniform failure rate distribution, and $(1-0.989519)$ area from 0 to 2 years, the $f(t)$ for the component is $\dfrac{1-0.989519}{2} = 0.00524037\ldots$, which means that the maximum life of a component must be 190.826… years or the half-life of a component must be 95.4… years.

The half-life of a component must be much greater than that of the product.

22.4.2 On-Stream Time

On-stream time (OST) is the fraction of time available and relates to the productivity that can be expected. OST is dependent on both mean time to failure (MTTF) and out-of-service time for repair. Since the item, on average, runs for a time (= MTTF) and is down for an average repair time of MTFR (mean time for repair), the fraction of productive time is

$$\text{OST} = \frac{\text{MTTF}}{\text{MTTF} + \text{MTFR}} \tag{22.15}$$

Example 22.16: A particular chemical reactor is presumed to have a uniform probability density function of operating continuously. The longest time any one reactor has ever run without needing repair is 2 years. What is the OST if the mean time for repair is 3 days?

Since the *pdf* is uniform with a 2-year period, the *pdf* value is $\frac{1}{2}\text{year}^{-1}$. Using Equation (22.14),

$$\text{MTTF} = \int_0^2 t\frac{1}{2}\,dt = 1\,\text{year}$$

Using Equation (22.15)

$$\text{OST} = \frac{1\,\text{year}}{1\,\text{year} + 3\,\text{days}} = \frac{365\frac{1}{4}}{365\frac{1}{4} + 3} = 99.18\ldots\%$$

22.4.3 Monte Carlo Techniques

Normally, there are several failure modes in a production unit and each mode has a distinct *pdf*, production consequences, and the distribution of repair duration. Attempts at analytically calculating OST for a process or a weighted annual production rate quickly become mathematically intractable. Even in the examples above with one aspect, we used the uniform *pdf* for simplicity. For complex situations, a Monte Carlo technique, a stochastic simulator, is used. Monte Carlo simulations were originally developed to study long-term gambling percentages, hence the name. However, they are commonly used in stochastic process simulations.

In a Monte Carlo technique, a computer "looks" at a process at frequent intervals and "determines" the failure/ operating/on hold/PM state of the process by "rolling dice". It is much like the fate determination of a battling hero in an adventure role-playing game. Actually, instead of rolling dice, the computer generates a pseudo-random number between 0 and 1 that represents the cumulative probability of a "stress". The inverse of the cumulative probability distribution is then used to generate the value of the stress. If the value is above or below some threshold, then an event happened. Perhaps a seal on one of three parallel pumps fails. The computer records the state of the pump as "down for repair," records the event time, and reduces the production rate by $\frac{1}{3}$. It may take 5 hours, on average, with a $\frac{1}{4}$ hour standard deviation, to repair and recommission the pump. The computer may "look" at the process every 10 minutes and, at each time, "roll the dice" to see whether another pump has failed and whether the first pump has been fixed. See Chapter 9.

One major advantage of Monte Carlo simulations is that none of the restrictions on the composite probability calculations of Section 22.3 apply. One can have common causes, and they can be interrelated in many ways. For example, if one pump seal fails with a probability of 0.001 because of improper maintenance, then the probability of a second pump seal failing for the same reason can be increased to 0.6. Several mechanisms for failure of one item can be incorporated. The population need not be dichotomous. Partial failures can also be incorporated.

At each "look" at the process, the computer can record any characteristic of interest, for example, on-stream time or production rate. At the end of a long-term simulation, the computer can then present a histogram, or average, or any desired statistic to describe the characteristic of interest.

22.5 Reliability in Process Design Choices

One important objective in process design and manufacturing is the optimization of process economics, through choices of equipment type, sizing, and redundancy, while designing a process or production line to meet a desired annual capacity or reliability criteria. Common objectives are the minimization of capital and/or operating expenses. In a more comprehensive analysis, the objective would be to maximize the discounted cash flow rate of return or the long-term return on investment, which balances both capital and total expense. The reliability measure, OST as calculated from Equation (22.15), is a key factor in guiding design choices. We will present two applications, sizing equipment in series and deciding redundancy.

22.5.1 Sizing Equipment in Series

We will introduce the issues using a simple chemical process example in an idealized treatment. Often such calculations are sufficient for design; however, as we will indicate, stochastic simulations are required for greater rigor.

In an elementary consideration, imagine a chemical reactor with a 70% OST producing a mixture of products. The reactor effluent is separated into desired product and recyclable material in a downstream unit. To meet desired production at a 70% OST, the reactor must instantaneously operate at 1/0.70 = 143% of the annualized rate. Consequently, the reactor, the separations unit, and all ancillary equipment must be sized 43% larger than the annualized rate to meet the instantaneous production rate. Roughly then, using the 6/10 power law of capital scale-up with capacity, the entire processing section would cost $(1.43)^{0.6} = 1.24$ or 24% more than a nominal design.

One complication in this example is that the separations unit will be subject to its own preventive maintenance and failures. It will have its own OST, say 90%. Again, in an elementary way, assume that the OSTs of 70% and 90% are the probabilities of the reactor and separations units working properly. Then the probabilities of reactor and separations unit failures are 0.3 and 0.1, respectively. The process will fail if either the reactor or the separations unit, or both, fail. If these failures are independent events, then the probability of process failure is determined by Equation (22.4).

$$P(\text{process failure}) = 0.3 + 0.1 - 0.3(0.1) = 0.37$$

Consequently, the OST is roughly 63%. Then the processes must be sized for a 59% larger (1/0.63 = 1.587...) instantaneous rate than the annualized rate. Using the 6/10 "rule", the capital cost of the process would be roughly 32% greater than a nominal design for 100% OST.

Usually, such a simplified analysis is sufficient for design purposes and indicates design changes that will improve the process. For example, inventory tanks between the reactor and separations unit decouple the units, prevent failure of one from shutting down the other, and reduce the size requirement of each and the capital investment. The inventory tanks need to be large enough to 1) hold a reserve to feed the downstream process while the reactor is off-line, and also to accumulate inventory from the reactor when the downstream process is off-line. However, the use of individual unit OSTs as probabilities to calculate a process OST is based on several assumptions that are not necessarily true.

First, OST does not exclude upset periods in which nonconforming product is made. For the technique just described, OST must only include quality operating periods. Second, OST is the fraction of time that a unit is operating; although it involves a probability of failure, OST is not a probability. The propagation of OSTs as if they were probabilities is defensible only when OST is high. Third, normal downtime of the process units includes scheduled preventive maintenance, scheduled turnarounds, etc. Consequently, the "failure" probabilities are not independent. It is likely that separations unit preventive maintenance will be scheduled during reactor outage periods so that the independent separations unit downtime will be negligible, and the process OST might more nearly match that of the worst unit.

The more rigorous options for developing a composite OST require a greater amount of work and are often unwarranted when one considers the accuracy of OST data. When required, stochastic simulation techniques are usually employed.

22.5.2 Selecting Redundancy

Equipment both fails and requires preventive maintenance and scheduled downtime. Unless a spare item is installed in parallel, the entire process stops. For many processes, including those that support primary production, interruption is unacceptable. The redundant unit must be able to provide the full capacity of the operating equipment. Consequently, the redundant-unit cost will be twice the single-unit cost. Another choice is to use three half-capacity units with two of them always scheduled for operation. The cost is similar. Using the 6/10 power law as representative of capital dependence on capacity, the cost of three half-capacity units is $3 \cdot \left(\frac{1}{2}\right)^{0.6} = 1.97\ldots$ times the single-unit cost, an inconsequential difference over the two full-sized unit cost. (Note that the cost-to-size exponent, 0.6, can vary from 0.4 to 1 which will give different cost results.)

There are, however, advantages to the three half-capacity-unit design. One is that the transfer upset is reduced. When one unit goes down and is replaced by the spare, there is a brief upset to the process. If the units are half-sized, the upset is also half-sized. However, the primary benefit is in the reduction in probability that the process stops completely. When both of the full-sized items fail, the entire process stops. When two of the three half-sized items fail, the process can continue at 50% capacity. Stoppage is usually intolerable because of the shutdown and startup consequences to subsequent process equipment. A reduction in throughput is much more desirable.

Again, allowing that OST is an appropriate measure of probability, one can relatively easily determine annualized production. For the two full-sized units, the process runs at full capacity if either or both items are operable. The process stops if both are inoperable. Letting $p = \text{OST}/100$ represent the probability of being operable, one can construct Tables 22.1 and 22.2. Regardless of the value of p, with independent failure events, the chance of the three half-sized units causing a total stoppage is always less than the chance of the two full-sized units failing simultaneously or $(1-p)^3 < (1-p)^2$.

This technique is approximately valid when the OST is high and when only independent failure mechanisms are involved. For identical parallel processing units or systems, a common failure mode may be likely. Stochastic simulations are a more rigorous approach. However, for many engineering purposes, the quality of OST data does not warrant the more sophisticated stochastic analysis.

Tables 22.1 and 22.2 can also be used to size the equipment based on the weighted production. If failure events are independent, each of the 2 full sized units must be $1/\left(2p - p^2\right)$-fold larger than nominal to provide an instantaneous capacity that will meet the annualized production requirement. Each of the 3 half-sized units must be $2/\left(3p - p^3\right)$-fold larger than nominal. Since a single unit must be sized $1/p$-fold larger than nominal, the cost

TABLE 22.1

Description of Two Full-Sized Units

Process state	Units state	Probability from Equation (22.7)	Weighted process state
Full capacity	*2 of 2 working*	p^2	$1 \cdot p^2$
Full capacity	*1 of 2 working*	$2p(1-p)$	$1 \cdot 2p(1-p)$
Zero capacity	*0 of 2 working*	$(1-p)^2$	$0 \cdot (1-p)^2$
Annualized total		1	$2p - p^2$

TABLE 22.2

Description of Three Half-Sized Units

Process state	Units state	Probability from Equation (22.7)	Weighted process state
Full capacity	*3 of 3 working*	p^3	$1 \cdot p^3$
Full capacity	*2 of 3 working*	$3p^2(1-p)$	$1 \cdot 3p^2(1-p)$
Half capacity	*1 of 3 working*	$3p(1-p)^2$	$\frac{1}{2} \cdot (3p-p)^2$
Zero capacity	*0 of 3 working*	$(1-p)^3$	$0 \cdot (1-p)^3$
Annualized total		1	$\frac{1}{2} \cdot (3p-p^3)$

benefit of 2 full-sized units over a single unit is $\left[p/\left(2p-p^2\right)\right]^{0.6}$, and the cost benefit of 3 half-sized units over a single unit is $\left[2p/\left(3p-p^3\right)\right]^{0.6}$.

Note: This analysis presumes failures are independent, that there is no common cause. This also only considers the capital cost – there is no special consideration for an extra penalty for the time and cost of startup when all units have failed, or the time duration of lost production. The weighting used in Tables 22.1 and 22.2 likely should not be simply the production rate.

Very often, a loss of capacity due to unit downtime can cause a critical situation. For instance, exhaust fans may be removing solvent-laden air from a production facility and sending the air to the solvent recovery section. Loss of fans may mean that solvent-in-air levels will exceed the lower explosive limit. A design criterion may be to design sufficient fan redundancy so that the chance of fan failures causing such a hazard will be less than 0.00001. Conceptually, this design problem can also be solved using OST data to represent probability of adequate fan availability. However, the critical nature of this problem warrants a more rigorous, stochastic simulator analysis in which the cost penalty for degrees of lost production, excessive solvent while a replacement fan is being brought on-line, and delays and costs of process shutdown and startup can be included.

There is no need to limit the number of spares to one. It could be that 8 1/5-sized units (5 operating with 3 on standby) minimize the cost while providing the desired reliability. So, choose the number of spares and their size, and calculate the probability of failures that will cause the undesired composite event. There are probably many combinations of spares and sizes that will meet the design reliability criteria. Let your intuition and progressive insight lead you toward the economically best configuration.

22.5.3 Selecting Reliability of Component Parts

Consider a product that has N critical components (or a procedure that has N critical stages) and if any one critical component (or stage) fails then the product (or procedure) fails.

The product might be used continuously, like electricity in your home. Critical components are high voltage transmission lines, substation, local transformers, local transmission lines. If any one of those critical components fails, then electricity fails. Although we would like the reliability to be 100% (no loss of power ever), a more reasonable target for reliability might be 99.9% availability in a year. The cost of a system that is 100% reliable in spite of its age, airplane crashes, ice storms, sabotage, squirrels, etc. would be excessive.

The product might be used intermittently, on demand, like a flashlight. Perhaps we desire that it function 90% of the instances. Critical components of the flashlight are the battery, lightbulb, and switch. If any one of those critical components fails, then the flashlight fails.

Finally, a procedure might be baking a cake. Critical stages are the oven working, adding baking soda, and taking it out of the oven on time. If any one of those critical components fails, then the outcome of the procedure fails.

The task here is to determine the reliability of each component (or stage) to ensure the desired product (or procedure) reliability. Define $P(P)$ as the probability that the product fails. It could be the fraction of on-demand instances that it fails, the fraction of products that fail in the first year of use, the fraction of time that the product does not work in a year. Each critical component (or stage) must have a higher reliability, because if any one critical component fails the product fails. Using $P(C_i)$ as representing the failure probability of a critical component:

$$P(P)=1-\prod_{i=1}^{N}\left[1-P(C_i)\right] \tag{22.16}$$

If each component has the same failure probability, then

$$P(C_i)=1-\sqrt[N]{1-P(P)} \tag{22.17}$$

However, there is no need to have identical component failure probabilities. If for instance, there are two critical components and a desired $P(P)$ = 0.05 then if $P(C_1)=0.03$, then a value of 0.02061… for $P(C_2)$ will result in the desired product failure rate.

The higher the reliability of a component, the more expensive will be the component. A 100% reliability (a $P(C_i)=0$ probability of failure rate) might be infinitely expensive. As an elementary model, consider the cost of a component to be inversely proportional to its failure rate.

$$C(C_i)=c_i\,/\,P(C_i) \tag{22.18}$$

So, the cost of the product would depend on the sum of the critical component costs (as well as the other contributions, C_0)

$$C(P)=C_0+\sum_{i=1}^{N}c_i\,/\,P(C_i) \tag{22.19}$$

A product or procedure developer might choose $P(C_i)$ values to minimize $C(P)$ subject to the constraint that the $P(C_i)$ values satisfy Equation (22.16).

$$\min_{\{P(C_i)\}} J=C(P)=C_0+\sum_{i=1}^{N}c_i/P(C_i) \tag{22.20}$$

$$\text{S.T. } P(P)=1-\prod_{i=1}^{N}\left[1-P(C_i)\right]$$

Example 22.17: A product has two critical components, and the product is desired to have a probability of failure of less than or equal to 0.05 (95% reliability) over a specified duration. The cost of the components is inversely proportional to their failure rate, with $c_1 = 3$ and $c_2 = 1$. What reliability values of the two components minimize product cost?

One can choose a value for $P(C_1)$ then calculate the $P(C_2)$-value from the constraint on Equation (22.20). Then progressively adjust the $P(C_1)$-value to minimize costs. Although an analytical procedure is possible, the algebra is tedious! From a numerical optimization approach, leapfrogging, the answer is

$$P(C_1) = 0.0319952\ldots$$
$$P(C_2) = 0.0185998\ldots$$

Alternately the reliability of C_1 is about 96.8% and that of C_2 is 98.1%.

Note: The design values for component reliability are not the same, and the product that costs more to make it reliable has a lower reliability (higher probability of failing).

Note: Each component must be more reliable than the desired value of the product.

22.6 Takeaway

Take care to discern end-of-trial (batch, on-demand) probabilities from in-process (during a time interval) probabilities.

Consider uncertainty in the "givens", the basic probabilities in a calculation, on the range of possible outcomes.

22.7 Exercises

1. Graph $pdf\left(\bar{x}_{\min} = \alpha \mid n = 3\right)$ in Example 22.10. Show how this changes with n. Show how $x_{\min}$ changes with n.
2. Reconsider Example 22.8 with reasonable alternate values for the frequency of road closures.
3. Does the graph of Figure 22.1 indicate that Car B is more reliable than Car A. Discuss the considerations and logic of your answer.
4. In Example 22.5 is the $p = 0.1$ value an on-demand or an in-service value?
5. Reconsider Example 22.7 with 4 spot welds per seam.
6. In Example 22.9, how many samples need to be tested to ensure that there is less than a 0.1% chance of shipping a lot of 100 that has 5% or more defective items.
7. Relate OST to λ the average number of failures per time per unit.
8. A product has 10 components, and if any one component fails then the product fails. The manufacturer would like 90% of the products to last two years in normal use. If all components and the product can be considered to have uniform $f(t)$, and all components have the same $f(t)$, then what must be the average life of a component?

9. Develop a table, such as Tables 22.1 and 22.2, to represent the annualized total if there are 5 1/3-sized units.
10. In a Poisson model, λ is the frequency of events per time per interval. If the consideration of the operating lifetime increases from 1 to 10 years, and the interval increases from one unit to 3, show that $\lambda_{10,3} = 30\ \lambda_{1,1}$.
11. If event probabilities could be perfectly known, then Equations (22.1) and (22.3) would provide definitive values. Given uncertainty on event probabilities, however, the composite event will also be uncertain. Propagate uncertainty on either Equation (22.1) or (22.3) to reveal the 99% probable error on the calculated value.
12. Use Equation (22.3) to solve Example 22.14.
13. In Example 22.10, $pdf\left(x_{\min} = \alpha \mid n\right)$, is dependent on the a and b limits of the uniform distribution. Propagate uncertainty to determine how uncertainty on a and b affect $pdf\left(x_{\min}\right)$.
14. Explore the impact of component costs in Example 22.16 on the required reliability of the two components.
15. Determine the 99% probable uncertainty on OST in Example 22.15 due to the implied uncertainty of the failure *pdf* and mean time to repair.
16. A device has 100 independent components, all with identical probability of failure. If any one component fails, the device fails. It is desired that no more than 10% of the devices fail after a defined time-of-use. What must be the failure rate for the individual components?
17. Repeat Example 22.15 but use the exponential model for the *pdf* of the failure rate.

Section 4

Case Studies

Case Studies

Case Study 1 – DJIA and Political Party

The Dow-Jones Industrial Average (DJIA) is a normalized index of the stock market prices in the US, which is one of many indices representing economic growth. The annualized % increase is the compounded annual rate of increase over the term of each US president since 1901, also normalizing the change in the DJIA for the president's number of years in office.

Annualized % Increase in the Dow Jones Industrial Average 1901–2021

Republican president		Democrat president	
Name	**%**	**Name**	**%**
Roosevelt	2.7	Wilson	−0.9
Taft	−0.3	FDR	9.3
Harding	6.9	Truman	8.0
Coolidge	25.5	JFK	4.1
Hoover	−35.6	Johnson	5.3
Eisenhower	10.4	Carter	−0.2
Nixon	−3.2	Clinton	15.9
Ford	8.9	Obama	12.1
Regan	11.3		
Bush I	9.7		
Bush II	−3.5		
Trump	9.4		

Data in the table are reproduced with kind permission of *Bespoke Investment Group @ bespokepremium.com.*

We'll seek to answer the question, "Does the stock market fare better under a president representing the Republican party or the Democrat party?" If the answer can be claimed "yes" or "no" then one party can provide evidence that they are better for the country, a strong one-upmanship claim within the high stakes game of politics. As you do the exercises, consider what choices you need to make, to make the statistical claim support your bias (favoring one party over the other).

Of course, the objective is not to train you on how to distort outcomes, but to help you be aware of what might be done unintentionally, or even intentionally, by others. Enjoy!

The average of the two data columns indicates that over the past 120 years when a Democrat (D) has been president, the US economy has risen faster than when a Republican (R) has been in that office. The correlation is that the US is more prosperous with a D president. But correlation is not causation. So, what might be a mechanism to get that result?

DOI: 10.1201/9781003222330-23

It could be that oncoming wave of prosperity under R control alleviates concern over the economy and gets voters to be more open to other issues, who then elect a D to represent that. Then the D president would be the result of a rising prosperity due to other reasons, and rides the wave started earlier. There are many mechanisms that we could postulate. The correlation might indicate something, but does not reveal why, or prove any proposed explanation. But, we will have fun seeking how to statistically analyze this data with techniques from throughout this book, and even more fun seeking choices that imply there are definitive claims that can be made.

Exercises

1. What is the average annualized % growth under each party? How could this support a claim that one party is better? Which is better?
2. What is the probable error on the average?
3. Are there any outliers in the data that should be removed? How does this affect the average? What level of confidence will you choose for the preprocessing?
4. Should data from Taft, Wilson, and FDR be removed because it represents the aberrational economics of the WWI and WWII periods and can be claimed inconsistent with the party's normal influence? Could you come up with a reason to discard any data in the table?
5. Use a *t*-test to see if the averages can statistically reject a hypothesis. Which of the three cases will you use? What level of confidence? Which of the three hypotheses (R = D, R > D, D > R)?
6. Use a Wilcoxon signed-rank test to see if the averages can statistically reject a hypothesis. Which of the three cases will you use? What level of confidence? Which of the three hypotheses (R = D, R > D, D > R)?
7. Is the stability of the economic growth, as measured by DJIA, worse for one party than another? Use an *F*-test on variance of the data. What level of confidence?
8. Which party has more negative values in annualized growth? Is the number significantly different? Use a chi-squared contingency test. What level of confidence?
9. Can you match the distribution of the annualized % growth to a theoretical model How well does the model match the data? Can the model be rejected? Can the model reject a data point as an outlier? What level of confidence will you choose?
10. Are the distributions of annualized % growth statistically different?
11. Use ANOVA considering that the two parties are two treatments. Does the outcome provide evidence that one is better? What level of confidence?
12. Create any other claim that one would want to make to supporting their preference for one party over another, defend a hypothesis, state a procedure to analyze the data, and do it.
13. Discuss possible limitations to the data to be the basis of a claim.
14. Other metrics used to assess the national economy are inflation, national debt, and loan interest. Can you name some others? If the supposition is that the economy is getting better, would the DJIA metric be the only one needed to test the hypothesis?

15. Discuss how a person's choices of the test type, hypothesis, level of confidence, and pre- or post-processing of the data can shape the statistical outcome, hence claim.

Case Study 2 – PBT Justification for a Change

The situation is: The company is currently using a treatment (raw material, reactor, protocol, recipe, etc.) and is considering a change to a new treatment. So, they start experiments to explore the benefit, *B*, of the new treatment (perhaps \$k/yr). After the experiment, all relevant issues are assessed (product quality, ease of operation, product yield, production cost, etc.) and their impact combined to quantify that trial, B_i. The experiments could be batches, or continuum time running, or separate locations. Here *N* represents the number of tests or duration periods or long enough periods to collect independent samples The new trials have a cost associated with the increased attention, training, duration, sampling, delay of action, and the cost per trial is *c* (perhaps \$k/trial). If the company switches, there will be the Management of Change cost, *M*, associated with changing company documentation, wide-scale training, loss of friendships associated with the old treatment, errors due to those who forgot, etc. So, if there is a change the total cost will be $M + Nc$. Payback Time, *PBT*, is the time it takes the benefit of the new treatment to pay back the associated costs. $PBT = (M + Nc) / \Delta_B$. Where $\Delta_B = B_{\text{proposed treatment}} - B_{\text{current treatment}}$. For simplicity, $\Delta = \Delta_B$.

The company will be using some form of a profitability index to evaluate options (projects, purchases, changes) and *PBT* could be the index in this case. The lower the *PBT* value the better. And typically, there will be a threshold on the profitability index, $PBT_{\text{threshold}}$. A value might be 2 years. Treatment changes with $PBT \le PBT_{\text{threshold}}$ will be given the go-ahead. This puts a limit on the benefit of any change.

$$\Delta_{\text{threshold}} = \frac{M + Nc}{PBT_{\text{threshold}}} \tag{CS2.1}$$

If $\Delta \ge \Delta_{\text{threshold}}$, the project is a "go". The value of Δ will be the average over *N* trials, $\bar{\Delta} = \frac{1}{N} \Sigma \Delta_i$. So, the decision to accept the new treatment will be if

$$\bar{\Delta} \ge \Delta_{\text{threshold}} = \frac{M + Nc}{PBT_{\text{threshold}}} \tag{CS2.2}$$

Notice that the $\Delta_{\text{threshold}}$ value increases with the number of trials. Desirably, one wants a low number of trials, to minimize the investigation time and cost. However, with a low number of trials, the confidence that the experimental $\bar{\Delta}$ value represents the true benefit of the treatment is low.

Assume that σ_{B_i} is known from past operation of the process, and that it will be the same for the new treatment. Then the one-sided 95% confidence limit on $\bar{\Delta}$ is

$$\varepsilon_{.95} = 1.644 \frac{\sigma_{B_i}}{\sqrt{N}} \tag{CS2.3}$$

A value of $\bar{\Delta}$ could be above the threshold, justifying a change, but because of experimental variation, it might be as low as $\bar{\Delta}-\varepsilon_{.95}$, and not justify a change. So, to be 95% confident that a change is justified, change if $\bar{\Delta}-\varepsilon_{.95} \geq \Delta_{\text{threshold}}$. Change if there is a 95% confidence.

$$\bar{\Delta} \geq \frac{M+Nc}{PBT_{\text{threshold}}}+1.644\frac{\sigma_{B_i}}{\sqrt{N}} \quad \text{(CS2.4)}$$

The last term in the equation increases the nominal $\Delta_{\text{threshold}}$ value to justify change. With only a few trials the $1.644\frac{\sigma_{B_i}}{\sqrt{N}}$ term will be high, but as the number of trials increases it will asymptotically decrease. On the other hand, the cost of trials increases with N.

Alternately, if early trial values of $\bar{\Delta}$ are very low, one might tend toward rejecting the treatment change. If $\bar{\Delta} < \frac{M+Nc}{PBT_{\text{threshold}}}$, reject the treatment and stop wasting $ on trials. But a low experimental $\bar{\Delta}$ value, even a negative value, might be simply the result of experimental vagaries, and the true Δ value might be higher. The 95% possible best Δ value will be $\bar{\Delta}+\varepsilon_{.95}$. Perhaps, one should not reject the new treatment until one is 95% confident that a possible value is not large enough. Reject if there is a 95% confidence.

$$\bar{\Delta} < \frac{M+Nc}{PBT_{\text{threshold}}}-1.644\frac{\sigma_{B_i}}{\sqrt{N}} \quad \text{(CS2.5)}$$

With added trials, the true Δ value will be revealed. But added trials costs. If one chose to be 99.9% confident that there is no hope that the new treatment is better, then the coefficient would be 3.08, requiring very many trials to be sure that the new treatment should be rejected. If there is suspicion that the new treatment is inferior, one would not run many costly trials to be sure. So, perhaps an adequate confidence is 75% with a coefficient of 0.655. Reject if there is a 75% confidence.

$$\bar{\Delta} < \frac{M+Nc}{PBT_{\text{threshold}}}-0.655\frac{\sigma_{B_i}}{\sqrt{N}} \quad \text{(CS2.6)}$$

Use

$M=10$	\$k
$c=1$	\$/trial
$\sigma_{B_i}=10$	\$k/yr
$PBT_{\text{threshold}}=2$	yr

Exercises

1. Relate this situation to a personal choice. Move from an abstract concept and make the equations relate to something you personally experience. For instance, you might want to explore a new response to the greeting question, "How are you?", or a new outgoing voice message, or a new work-out, a different set of wheels, or a humorous way to ask a person on a date or trying to walk in stilts. What values would be appropriate for M, c, σ_{B_i}, and $PBT_{\text{threshold}}$?

2. Plot Equations (CS2.4) and (CS2.5) and describe how the decision on $\bar{\Delta}$ changes with N.
3. Accept the new treatment and stop trials, if $\bar{\Delta}$ is greater than Equation (CS2.4). Reject the new treatment and stop trials, if $\bar{\Delta}$ is less than Equation (CS2.5). What action would be taken if the value of $\bar{\Delta}$ fell between the lines? Explain.
4. Explore how values of M, c, σ_{B_i}, and $PBT_{\text{threshold}}$ change the decision values associated with $\bar{\Delta}$. Use values that would be appropriate to your own case.
5. Consider how you would choose the confidence interval required to make an accept or reject decision. Use the equations to explore it.
6. Create a simulator to generate sequential Δ_i values. Choose a true mean and use a Gaussian perturbation to model, $NID(0,\sigma_{B_i})$, to see the decision evolve with N. Run many realizations. Does this suggest that the confidence limits should be changed?
7. Suppose that one cannot presume that σ_{B_i} for the new treatment is the same as σ_{B_i} historically, and that σ_{B_i} will be estimated from the standard deviation of the sequence of Δ_i values. $s_B = \sqrt{\frac{1}{N-1}\Sigma(\Delta_i - \bar{\Delta})^2}$. Now the t-values should be used instead of the z-values in Equations (CS2.3) to (CS2.6). No decision can be made until there are two trials. And the t-critical value will depend on N. Create a model for how t-critical changes with N for your choice of confidence values and use it in Equations (CS2.4) and (CS2.5) in place of the z-values. Comment on the impact.
8. Use the Bayes Belief technique to update belief to accept (or to reject) the proposed treatment after each simulated trial outcome. Choose a true (but unknowable) value for Δ and have the simulator generated normally distributed trial outcomes.

Case Study 3 – Central Limit Phenomena and μ and σ

Here are three claims:

1. Theoretically, they say that as the number of observations increases, the standard deviation of the average is reduced from the standard deviation of the samples by the square root of N.

$$\sigma_{\bar{X}} = \frac{1}{\sqrt{N}}\sigma_X \tag{CS3.1}$$

2. They also say that regardless of the distribution providing the observations, as N increases $\bar{X}_j$ tends to become normally distributed. Here, $\bar{X}$ is the average of N observations, and $\bar{X}_j$ represents the average of the jth set of N observations.
3. Finally, Chapter 3 also reports the theoretical μ and σ for a variety of distributions.

Show that these claims are true, supported by data from simulations.

Exercises

1. Create a simulator per guidance in Chapter 9, that generates data for:
   a. Binomial distributed data.
   b. Poisson distributed data.
   c. Uniform continuum distributed data.
   d. Gaussian distributed data.
   e. Exponentially distributed data.
2. Choose your own parameter values, and generate N = 100, 1,000, and 10,000 observations.
3. Use the K–S test to test if the simulated data matches the theoretically expected distribution. When the number of observations exceeds that in the K–S critical table, use the chi-squared contingency test. Defend your level of alpha.
4. Show that as N increases in Exercise 2, the average and standard deviation of the N-dataset matches the theoretical values of Chapter 3. You might want to use N = 100, 200, 500, 1,000, 2,000, 5,000, 10,000, 20,000, 50,000, 100,000, etc. and graph the average and standard deviation w.r.t. the log of N.
5. Show that as N increases, Equation (CS3.1) is true. Use the chi-squared variance test. Defend your level of alpha.
6. Generate J number of datasets of N observations each. Use $N > 10$. Show that as J increases the distribution of $\bar{X}_j$ approaches normal regardless of your choice for the distribution of individual x-values.

Case Study 4 – A Corkboard

The Australian winery, 19 Crimes, places a description of one of 19 crimes on the corks for their product. According to company marketing, the 19 crimes represent those that, when committed in England during the colonial period, caused the felon to be deported to Australia. The labeled corks seem to be randomly chosen to seal a bottle. So, when you buy the product, the cork will surprise you with one of the 19. The descriptions seem somewhat whimsical, today. One is "Impersonating an Egyptian". Another is "Stealing a Shroud out of a Cave". One side of the cork has the company label, the other side has the crime.

One author (Rhinehart) likes woodworking and makes corkboards (bulletin boards) from wine corks. His wife suggested he make one completely of 19 Crimes' corks and make it look like a jail window. There are 140 corks in the window. Desirably, at least three of each of the 19 interesting crimes show, and about 10 more display the company logo. At a minimum then $3 \times 19 + 10 = 67$ of the 19 Crimes' corks are needed. The rest of the 140 can be plain corks.

So, he and many of their friends began collecting corks for the project. After 107 corks, the first #10 finally came in, and everyone was joyous. Here is the tally after collecting 112 corks:

Cork category number	Count in each category
1	4
2	4
3	4
4	6
5	4
6	6
7	3
8	9
9	5
10	1
11	9
12	6
13	8
14	7
15	5
16	12
17	6
18	8
19	5

The question is whether the probability of getting any cork is equal. Or alternately, does this indicate a marketing strategy that holds back on some corks to get purchasers to keep buying to get a complete set? The supposition is that the game is fair, that there is an equal probability of getting any cork category in any bottle.

As a caution, the dual use of numbers could be confusing. You might want to label the cork categories 1–19 with the letters A–S.

Also, there are two distributions in this analysis, one for the expected number of corks between each category, and another for the number of corks within any particular category.

Exercises

1. What distribution should model the expected number of corks per category, between categories? Defend.
2. What distribution should model the expected number of corks within any category? Defend.
3. With those distributions what is the expected mean and sigma for each cork category? What are the 2 sigma and 3 sigma limits on the expected count? If you get a negative number, you used the wrong distribution!

4. Does your model in Exercise 3 match the actual distribution? Can either the chi-squared contingency test or the K–S test reject the model (or suppositions about the data). Defend your choices.
5. Create a Monte Carlo simulator to generate the theoretical distribution. One realization is to sample 112 times and randomly assign the cork to a category. Do this many times, perhaps 10,000, and get an average of the count for any frequency. This is a bit confusing. The category is a number. The count in any one category is a number. The number of times the count is a particular value is a different number. Keep these separate.
6. There are 19 separate cork categories. If the chance of getting any one is equal and random, what is the chance of getting 0, 1, 2 etc. of any one of the 19, after a bunch of samples?
7. After 106 corks, there was no #10. If equal chance and random, you'd expect about 5 or 6 of each cork category. Can zero in one of 19 categories out of 106 be used to reject the equal and random chance?
8. After 112 corks, there were 12 of #16. If equal chance and random, you'd expect nearly 6 of each cork category. Can 12 in one of 19 categories out of 112 be used to reject the equal and random chance? Does this indicate a mechanism of distributing corks that is not random?

Note: After enjoying the above investigations, Rhinehart accepted that the distribution of corks collected was random, and from an equal probability of getting any particular category. He accepted that the game was fair. And shortly after, more #10s came in!

Appendix

Critical Value Tables

Appendix: Tables of Critical Values

Table A.1 Critical Values of r in the Sign Test

N = total number of equally probable dichotomous events.

An event is one of two possible dichotomous outcomes. It could be a sign (+/−), but it could also be a category (H/T, M/F, Pass/Fail, etc.).

R = the smaller of the number of events of either kind.

F = cumulative probability = $1 - \alpha$.

If $R \leq r_C$ there are improbably too few events of one kind at the $C = 100\ F$ % confidence level. Table entries are r_C.

N	r_{90}	r_{95}	r_{99}	$r_{99.5}$	$r_{99.9}$
8	1				
10	1	1			
12	2	2	1		
14	3	2	1	1	
16	4	3	2	2	1
18	5	4	3	2	1
20	5	5	3	3	2
22	6	5	4	4	3
25	7	7	5	5	4
30	10	9	7	6	5
35	12	11	9	8	7
40	14	13	11	10	9
45	16	15	13	12	11
50	18	17	15	14	13
55	20	19	17	16	14
60	23	21	19	18	16
70	27	26	23	22	20
80	32	30	28	27	24
90	36	35	32	31	29
100	41	39	36	35	33

Table entries are calculated from the binomial distribution inverse, with N, $p = 0.5$, and $\alpha = (1 - C/100)/2$. The test is two-sided since either of the two data categories could have a count that is two few in number; or alternately, that one category could have either two few or two many. Entry values represent the highest R value outside of the alpha limit.

Table A.2 Critical Values of *s* in the Wilcoxon Matched-Pairs Signed-Rank Test

N = sample size.

S = sum of ranks of positive paired differences in experimental data.

F = 1 − α.

If either $S \leq s_{\alpha/2}$ or $S \geq s_{1-\alpha/2}$ reject the hypothesis that the treatments are equivalent at the 100 · F% level of confidence. Table entries are paired values of $s_{\alpha/2}$ and $s_{1-\alpha/2}$

α (two-sided)	*N* = 5	*N* = 6	*N* = 7	*N* = 8	*N* = 9	*N* = 10
.10	1, 14	2, 19	4, 24	6, 30	8, 37	11, 44
.05		1, 20	2, 26	4, 32	6, 39	8, 47
.02			0, 28	2, 34	3, 42	5, 50
.01				0, 36	2, 43	3, 52
	***N* = 11**	***N* = 12**	***N* = 13**	***N* = 14**	***N* = 15**	***N* = 16**
.10	14, 52	17, 61	21, 70	26, 79	30, 90	36, 100
.05	11, 55	14, 64	17, 74	21, 84	25, 95	30, 106
.02	7, 59	10, 68	13, 78	16, 89	20, 100	24, 112
.01	5, 61	7, 71	10, 81	13, 92	16, 104	19, 117
	***N* = 17**	***N* = 18**	***N* = 19**	***N* = 20**	***N* = 21**	***N* = 22**
.10	41, 112	47, 124	54, 136	60, 150	68, 163	75, 178
.05	35, 118	40, 131	46, 144	52, 158	59, 172	66, 187
.02	28, 125	33, 138	38, 152	43, 167	49, 182	56, 197
.01	23, 130	28, 143	32, 158	37, 173	43, 188	49, 204
	***N* = 23**	***N* = 24**	***N* = 25**	***N* = 26**	***N* = 27**	***N* = 28**
.10	83, 193	92, 208	101, 224	110, 241	120, 258	130, 276
.05	73, 203	81, 219	90, 235	98, 253	107, 271	117, 289
.02	62, 214	69, 231	77, 248	85, 266	93, 285	102, 304
.01	55, 221	61, 239	68, 257	76, 275	84, 294	92, 314
	***N* = 29**	***N* = 30**	***N* = 31**	***N* = 32**	***N* = 33**	***N* = 34**
.10	141, 294	152, 313	163, 333	175, 353	188, 373	201, 394
.05	127, 308	137, 328	148, 348	159, 369	171, 390	183, 412
.02	111, 324	120, 345	130, 366	141, 387	151, 410	162, 433
.01	100, 335	109, 356	118, 378	128, 400	138, 423	149, 446
	***N* = 35**	***N* = 36**	***N* = 37**	***N* = 38**	***N* = 39**	***N* = 40**
.10	214, 416	228, 438	242, 461	256, 485	271, 509	287, 533
.05	195, 435	208, 458	222, 481	235, 506	250, 530	264, 556
.02	174, 456	186, 480	198, 505	211, 530	224, 556	238, 582
.01	160, 470	171, 495	183, 520	195, 546	208, 572	221, 599
	N = 41	*N* = 42	*N* = 43	*N* = 44	*N* = 45	*N* = 46
.10	303, 558	319, 584	336, 610	353, 637	371, 664	389, 692
.05	279, 582	295, 608	311, 635	327, 663	344, 691	361, 720
.02	252, 609	267, 636	281, 665	297, 693	313, 722	329, 752
.01	234, 627	248, 655	262, 684	277, 713	292, 743	307, 774
	***N* = 47**	***N* = 48**	***N* = 49**	***N* = 50**		
.10	408, 720	427, 749	446, 779	466, 809		
.05	379, 749	397, 779	415, 810	434, 841		
.02	345, 783	362, 814	380, 845	398, 877		
.01	323, 805	339, 837	356, 869	373, 902		

Table A.3a Critical Values of u in the Runs Test for small $N = n + m$

m = the smaller of the number of equally probable dichotomous events.

n = the larger.

U = the number of runs in the experimental data, when data is ordered by some variable.

$F = \alpha$ for a lower one-sided test of too few runs.

$F = 1 - \alpha$ for an upper one-sided test of too many runs.

$F = \alpha/2$ and $F = 1 - \alpha/2$ for a two-sided test.

For example: Choose $\alpha = 0.05$. If a two-sided test then $F_{lower} = 0.025$ and $F_{upper} = 0.975$. If $m = 6$, and $n = 10$, then $u_{critical\ lower} = 4$ and $u_{critical\ upper} = 12$. Reject the hypothesis of independent distributions if either $U \leq 4$ or $U \geq 12$.

If $U \leq u_{m,n,F}$ there are improbably too few runs at the $100 \cdot F\%$ level of confidence. If $U \geq u_{m,n,F}$ there are improbably too many runs at the $100 \cdot F\%$ level of confidence. Table entries are $u_{m,n,F}$.

$m = 2$

n	$u_{.005}$	$u_{.01}$	$u_{.025}$	$u_{.05}$	$u_{.95}$	$u_{.975}$	$u_{.99}$	$u_{.995}$
2					4	4	4	4
3					5	5	5	5
4					5	5	5	5
5					5	5	5	5
6					5	5	5	5
7					5	5	5	5
8				2	5	5	5	5
9				2	5	5	5	5
10				2	5	5	5	5
11				2	5	5	5	5
12			2	2	5	5	5	5
13			2	2	5	5	5	5
14			2	2	5	5	5	5
15			2	2	5	5	5	5
16			2	2	5	5	5	5
17			2	2	5	5	5	5
18			2	2	5	5	5	5
19		2	2	2	5	5	5	5
20		2	2	2	5	5	5	5

$m = 3$

n	$u_{.005}$	$u_{.01}$	$u_{.025}$	$u_{.05}$	$u_{.95}$	$u_{.975}$	$u_{.99}$	$u_{.995}$
3					6	6	6	6
4					6	7	7	7

(*Continued*)

$m = 3$								
n	$u_{.005}$	$u_{.01}$	$u_{.025}$	$u_{.05}$	$u_{.95}$	$u_{.975}$	$u_{.99}$	$u_{.995}$
5				2	7	7	7	7
6			2	2	7	7	7	7
7			2	2	7	7	7	7
8			2	2	7	7	7	7
9		2	2	2	7	7	7	7
10		2	2	3	7	7	7	7
11		2	2	3	7	7	7	7
12	2	2	2	3	7	7	7	7
13	2	2	2	3	7	7	7	7
14	2	2	2	3	7	7	7	7
15	2	2	3	3	7	7	7	7
16	2	2	3	3	7	7	7	7
17	2	2	3	3	7	7	7	7
18	2	2	3	3	7	7	7	7
19	2	2	3	3	7	7	7	7
20	2	2	3	3	7	7	7	7

$m = 4$								
n	$u_{.005}$	$u_{.01}$	$u_{.025}$	$u_{.05}$	$u_{.95}$	$u_{.975}$	$u_{.99}$	$u_{.995}$
4				2	7	8	8	8
5			2	2	8	8	8	9
6		2	2	3	8	8	9	9
7		2	2	3	8	9	9	9
8	2	2	3	3	9	9	9	9
9	2	2	3	3	9	9	9	9
10	2	2	3	3	9	9	9	9
11	2	2	3	3	9	9	9	9
12	2	3	3	4	9	9	9	9
13	2	3	3	4	9	9	9	9
14	2	3	3	4	9	9	9	9
15	3	3	3	4	9	9	9	9
16	3	3	4	4	9	9	9	9
17	3	3	4	4	9	9	9	9
18	3	3	4	4	9	9	9	9
19	3	3	4	4	9	9	9	9
20	3	3	4	4	9	9	9	9

$m = 5$								
n	$u_{.005}$	$u_{.01}$	$u_{.025}$	$u_{.05}$	$u_{.95}$	$u_{.975}$	$u_{.99}$	$u_{.995}$
5		2	2	3	8	9	9	10
6	2	2	3	3	9	9	10	10
7	2	2	3	3	9	10	10	11
8	2	2	3	3	10	10	11	11
9	2	3	3	4	10	11	11	11
10	3	3	3	4	10	11	11	11
11	3	3	4	4	11	11	11	11
12	3	3	4	4	11	11	11	11
13	3	3	4	4	11	11	11	11

(Continued)

$m = 5$								
n	$u_{.005}$	$u_{.01}$	$u_{.025}$	$u_{.05}$	$u_{.95}$	$u_{.975}$	$u_{.99}$	$u_{.995}$
14	3	3	4	5	11	11	11	11
15	3	4	4	5	11	11	11	11
16	3	4	4	5	11	11	11	11
17	3	4	4	5	11	11	11	11
18	4	4	5	5	11	11	11	11
19	4	4	5	5	11	11	11	11
20	4	4	5	5	11	11	11	11

$m = 6$								
n	$u_{.005}$	$u_{.01}$	$u_{.025}$	$u_{.05}$	$u_{.95}$	$u_{.975}$	$u_{.99}$	$u_{.995}$
6	2	2	3	3	10	10	11	11
7	2	3	3	4	10	11	11	12
8	3	3	3	4	11	11	12	12
9	3	3	4	4	11	12	12	13
10	3	3	4	5	11	12	13	13
11	3	4	4	5	12	12	13	13
12	3	4	4	5	12	12	13	13
13	3	4	5	5	12	13	13	13
14	4	4	5	5	12	13	13	13
15	4	4	5	6	13	13	13	13
16	4	4	5	6	13	13	13	13
17	4	5	5	6	13	13	13	13
18	4	5	5	6	13	13	13	13
19	4	5	6	6	13	13	13	13
20	4	5	6	6	13	13	13	13

$m = 7$								
n	$u_{.005}$	$u_{.01}$	$u_{.025}$	$u_{.05}$	$u_{.095}$	$u_{.975}$	$u_{.99}$	$u_{.995}$
7	3	3	3	4	11	12	12	12
8	3	3	4	4	12	12	13	13
9	3	4	4	5	12	13	13	14
10	3	4	5	5	12	13	14	14
11	4	4	5	5	13	13	14	14
12	4	4	5	6	13	13	14	15
13	4	5	5	6	13	14	15	15
14	4	5	5	6	13	14	15	15
15	4	5	6	6	14	14	15	15
16	5	5	6	6	14	15	15	15
17	5	5	6	7	14	15	15	15
18	5	5	6	7	14	15	15	15
19	5	6	6	7	14	15	15	15
20	5	6	6	7	14	15	15	15

$m = 8$								
n	$u_{.005}$	$u_{.01}$	$u_{.025}$	$u_{.05}$	$u_{.95}$	$u_{.975}$	$u_{.99}$	$u_{.995}$
8	3	4	4	5	12	13	13	14
9	3	4	5	5	13	13	14	14
10	4	4	5	6	13	14	14	15

(*Continued*)

$m = 8$								
n	$u_{.005}$	$u_{.01}$	$u_{.025}$	$u_{.05}$	$u_{.95}$	$u_{.975}$	$u_{.99}$	$u_{.995}$
11	4	5	5	6	14	14	15	15
12	4	5	6	6	14	15	15	16
13	5	5	6	6	14	15	16	16
14	5	5	6	7	15	15	16	16
15	5	5	6	7	15	15	16	17
16	5	6	6	7	15	16	16	17
17	5	6	7	7	15	16	17	17
18	6	6	7	8	15	16	17	17
19	6	6	7	8	15	16	17	17
20	6	6	7	8	16	16	17	17

$m = 9$								
n	$u_{.005}$	$u_{.01}$	$u_{.025}$	$u_{.05}$	$u_{.95}$	$u_{.975}$	$u_{.99}$	$u_{.995}$
9	4	4	5	6	13	14	15	15
10	4	5	5	6	14	15	15	16
11	5	5	6	6	14	15	16	16
12	5	5	6	7	15	15	16	17
13	5	6	6	7	15	16	17	17
14	5	6	7	7	16	16	17	17
15	6	6	7	8	16	17	17	18
16	6	6	7	8	16	17	17	18
17	6	7	7	8	16	17	18	18
18	6	7	8	8	17	17	18	19
19	6	7	8	8	17	17	18	19
20	7	7	8	9	17	17	18	19

$m = 10$								
n	$u_{.005}$	$u_{.01}$	$u_{.025}$	$u_{.05}$	$u_{.95}$	$u_{.975}$	$u_{.99}$	$u_{.995}$
10	5	5	6	6	15	15	16	16
11	5	5	6	7	15	16	17	17
12	5	6	7	7	16	16	17	18
13	5	6	7	8	16	17	18	18
14	6	6	7	8	16	17	18	18
15	6	7	7	8	17	17	18	19
16	6	7	8	8	17	18	19	19
17	7	7	8	9	17	18	19	19
18	7	7	8	9	18	18	19	20
19	7	8	8	9	18	19	19	20
20	7	8	9	9	18	19	19	20

$m = 11$								
n	$u_{.005}$	$u_{.01}$	$u_{.025}$	$u_{.05}$	$u_{.95}$	$u_{.975}$	$u_{.99}$	$u_{.995}$
11	5	6	7	7	16	16	17	18
12	6	6	7	8	16	17	18	18
13	6	6	7	8	17	18	18	19
14	6	7	8	8	17	18	19	19
15	7	7	8	9	18	18	19	20

(Continued)

$m = 11$								
n	$u_{.005}$	$u_{.01}$	$u_{.025}$	$u_{.05}$	$u_{.95}$	$u_{.975}$	$u_{.99}$	$u_{.995}$
16	7	7	8	9	18	19	20	20
17	7	8	9	9	18	19	20	21
18	7	8	9	10	19	19	20	21
19	8	8	9	10	19	20	21	21
20	8	8	9	10	19	20	21	21

$m = 12$								
n	$u_{.005}$	$u_{.01}$	$u_{.025}$	$u_{.05}$	$u_{.95}$	$u_{.975}$	$u_{.99}$	$u_{.995}$
12	6	7	7	8	17	18	18	19
13	6	7	8	9	17	18	19	20
14	7	7	8	9	18	19	20	20
15	7	8	8	9	18	19	20	21
16	7	8	9	10	19	20	21	21
17	8	8	9	10	19	20	21	21
18	8	8	9	10	20	20	21	22
19	8	9	10	10	20	21	22	22
20	8	9	10	11	20	21	22	22

$m = 13$								
n	$u_{.005}$	$u_{.01}$	$u_{.25}$	$u_{.05}$	$u_{.95}$	$u_{.975}$	$u_{.99}$	$u_{.995}$
13	7	7	8	9	18	19	20	20
14	7	8	9	9	19	19	20	21
15	7	8	9	10	19	20	21	21
16	8	8	9	10	20	20	21	22
17	8	9	10	10	20	21	22	22
18	8	9	10	11	20	21	22	23
19	9	9	10	11	21	22	23	23
20	9	10	10	11	21	22	23	23

$m = 14$								
n	$u_{.005}$	$u_{.01}$	$u_{.025}$	$u_{.05}$	$u_{.95}$	$u_{.975}$	$u_{.99}$	$u_{.995}$
14	7	8	9	10	19	20	21	22
15	8	8	9	10	20	21	22	22
16	8	9	10	11	20	21	22	23
17	8	9	10	11	21	22	23	23
18	9	9	10	11	21	22	23	24
19	9	10	11	12	22	22	23	24
20	9	10	11	12	22	23	24	24

$m = 15$								
n	$u_{.005}$	$u_{.01}$	$u_{.025}$	$u_{.05}$	$u_{.95}$	$u_{.975}$	$u_{.99}$	$u_{.995}$
15	8	9	10	11	20	21	22	23
16	9	9	10	11	21	22	23	23
17	9	10	11	11	21	22	23	24
18	9	10	11	12	22	23	24	24
19	10	10	11	12	22	23	24	25
20	10	11	12	12	23	24	25	25

(Continued)

$m = 16$								
n	$u_{.005}$	$u_{.01}$	$u_{.025}$	$u_{.05}$	$u_{.95}$	$u_{.975}$	$u_{.99}$	$u_{.995}$
16	9	10	11	11	22	22	23	24
17	9	10	11	12	22	23	24	25
18	10	10	11	12	23	24	25	25
19	10	11	12	13	23	24	25	26
20	10	11	12	13	24	24	25	26

$m = 17$								
n	$u_{.005}$	$u_{.01}$	$u_{.025}$	$u_{.05}$	$u_{.95}$	$u_{.975}$	$u_{.99}$	$u_{.995}$
17	10	10	11	12	23	24	25	25
18	10	11	12	13	23	24	25	26
19	10	11	12	13	24	25	26	26
20	11	11	13	13	24	25	26	27

$m = 18$								
n	$u_{.005}$	$u_{.01}$	$u_{.025}$	$u_{.05}$	$u_{.95}$	$u_{.975}$	$u_{.99}$	$u_{.995}$
18	11	11	12	13	24	25	26	26
19	11	12	13	14	24	25	26	27
20	11	12	13	14	25	26	27	28

$m = 19$								
n	$u_{.005}$	$u_{.01}$	$u_{.025}$	$u_{.05}$	$u_{.95}$	$u_{.975}$	$u_{.99}$	$u_{.995}$
19	11	12	13	14	25	26	27	28
20	12	12	13	14	26	26	28	28

$m = 20$								
n	$u_{.005}$	$u_{.01}$	$u_{.025}$	$u_{.05}$	$u_{.95}$	$u_{.975}$	$u_{.99}$	$u_{.995}$
20	12	13	14	15	26	27	28	29

Reproduced with permission from Tables for Testing Randomness of Grouping in a Sequence of Alternatives by F. S. Swed and C. Eisenhart, *The Annals of Mathematical Statistics*, Vol. XIV, No. 1, March 1943. Copyright The Institute of Mathematical Statistics, Hayward, CA.

Table A.3b Critical Values of the u in the Runs Test for Large N

With equally probable dichotomous data classification, the expectation is that $u = \frac{1}{2}N$, the number of runs is half the number of data values in an ordered sequence.

U = the number of runs in the experimental data.

$F = \alpha$ for a lower one-sided test of too few runs.

$F = 1 - \alpha$ for an upper one-sided test of too many runs.

$F = \alpha/2$ and $F = 1 - \alpha/2$ for a two-sided test.

For example: Choose $\alpha = 0.05$. If a two-sided test, then $F_{\text{lower}} = 0.025$ and $F_{\text{upper}} = 0.975$. If $N = 30$ then $u_{\text{critical lower}} = 9$ and $u_{\text{critical upper}} = 21$. Reject the hypothesis of independent distributions if either $U \leq 9$ or $U \geq 21$.

	α (one sided)					
N	**0.005**	**0.01**	**0.025**	**0.05**	**0.1**	**0.25**
10	1	1	2	2	3	3
11	1	1	2	2	3	4
12	1	2	2	3	3	4
13	2	2	3	3	4	5
14	2	2	3	4	4	5
15	2	3	3	4	5	6
16	3	3	4	4	5	6
17	3	3	4	5	5	7
18	3	4	5	5	6	7
19	4	4	5	6	6	8
20	4	5	5	6	7	8
21	4	5	6	6	7	8
22	5	5	6	7	8	9
23	5	6	6	7	8	9
24	5	6	7	8	8	10
25	6	6	7	8	9	10
26	6	7	8	8	9	11
27	7	7	8	9	10	11
28	7	8	8	9	10	12
29	7	8	9	10	11	12
30	8	8	9	10	11	13
31	8	9	10	11	11	13
32	8	9	10	11	12	14
33	9	9	11	11	12	14
34	9	10	11	12	13	15
35	10	10	11	12	13	15
36	10	11	12	13	14	16
37	10	11	12	13	14	16
38	11	12	13	14	15	16
39	11	12	13	14	15	17
40	12	12	13	14	15	17
41	12	13	14	15	16	18
42	12	13	14	15	16	18
43	13	14	15	16	17	19
44	13	14	15	16	17	19
45	14	14	16	17	18	20
46	14	15	16	17	18	20
47	14	15	16	17	19	21
48	15	16	17	18	19	21
49	15	16	17	18	20	22
50	16	16	18	19	20	22
51	16	17	18	19	20	23
52	16	17	19	20	21	23
53	17	18	19	20	21	24
54	17	18	19	21	22	24

(Continued)

	α (one sided)					
N	**0.005**	**0.01**	**0.025**	**0.05**	**0.1**	**0.25**
55	18	19	20	21	22	25
56	18	19	20	21	23	25
57	18	19	21	22	23	25
58	19	20	21	22	24	26
59	19	20	22	23	24	26
60	20	21	22	23	25	27
61	20	21	22	24	25	27
62	21	21	23	24	26	28
63	21	22	23	25	26	28
64	21	22	24	25	26	29
65	22	23	24	25	27	29
66	22	23	25	26	27	30
67	23	24	25	26	28	30
68	23	24	26	27	28	31
69	23	24	26	27	29	31
70	24	25	26	28	29	32
71	24	25	27	28	30	32
72	25	26	27	29	30	33
73	25	26	28	29	31	33
74	26	27	28	30	31	34
75	26	27	29	30	31	34
76	26	27	29	30	32	35
77	27	28	29	31	32	35
78	27	28	30	31	33	36
79	28	29	30	32	33	36
80	28	29	31	32	34	36
81	29	30	31	33	34	37
82	29	30	32	33	35	37
83	29	31	32	34	35	38
84	30	31	33	34	36	38
85	30	31	33	34	36	39
86	31	32	33	35	37	39
87	31	32	34	35	37	40
88	32	33	34	36	38	40
89	32	33	35	36	38	41
90	32	34	35	37	38	41
91	33	34	36	37	39	42
92	33	34	36	38	39	42
93	34	35	37	38	40	43
94	34	35	37	39	40	43
95	35	36	38	39	41	44
96	35	36	38	39	41	44
97	35	37	38	40	42	45
98	36	37	39	40	42	45
99	36	38	39	41	43	46

(*Continued*)

	α (one sided)					
N	0.005	0.01	0.025	0.05	0.1	0.25
100	37	38	40	41	43	46
101	37	38	40	42	44	47
102	38	39	41	42	44	47
103	38	39	41	43	45	48
104	38	40	42	43	45	48
105	39	40	42	44	45	49
106	39	41	42	44	46	49
107	40	41	43	45	46	50
108	40	42	43	45	47	50
109	41	42	44	45	47	50
110	41	42	44	46	48	51
111	42	43	45	46	48	51
112	42	43	45	47	49	52
113	42	44	46	47	49	52
114	43	44	46	48	50	53
115	43	45	47	48	50	53
116	44	45	47	49	51	54
117	44	46	47	49	51	54
118	45	46	48	50	52	55
119	45	46	48	50	52	55
120	46	47	49	51	53	56
121	46	47	49	51	53	56
122	46	48	50	51	53	57
123	47	48	50	52	54	57
124	47	49	51	52	54	58
125	48	49	51	53	55	58
126	48	50	52	53	55	59
127	49	50	52	54	56	59
128	49	50	52	54	56	60
129	49	51	53	55	57	60
130	50	51	53	55	57	61
131	50	52	54	56	58	61
132	51	52	54	56	58	62
133	51	53	55	57	59	62
134	52	53	55	57	59	63
135	52	54	56	57	60	63
136	53	54	56	58	60	64
137	53	54	57	58	61	64
138	53	55	57	59	61	65
139	54	55	58	59	61	65
140	54	56	58	60	62	66
141	55	56	58	60	62	66
142	55	57	59	61	63	66
143	56	57	59	61	63	67
144	56	58	60	62	64	67

(*Continued*)

	α (one sided)					
N	0.005	0.01	0.025	0.05	0.1	0.25
145	57	58	60	62	64	68
146	57	59	61	63	65	68
147	58	59	61	63	65	69
148	58	59	62	64	66	69
149	58	60	62	64	66	70
150	59	60	63	64	67	70
151	59	61	63	65	67	71
152	60	61	63	65	68	71
153	60	62	64	66	68	72
154	61	62	64	66	69	72
155	61	63	65	67	69	73
156	61	63	65	67	70	73
157	62	63	66	68	70	74
158	62	64	66	68	70	74
159	63	64	67	69	71	75
160	63	65	67	69	71	75
161	64	65	68	70	72	76
162	64	66	68	70	72	76
163	65	66	69	71	73	77
164	65	67	69	71	73	77
165	66	67	69	71	74	78
166	66	68	70	72	74	78
167	66	68	70	72	75	79
168	67	68	71	73	75	79
169	67	69	71	73	76	80
170	68	69	72	74	76	80
171	68	70	72	74	77	81
172	69	70	73	75	77	81
173	69	71	73	75	78	82
174	70	71	74	76	78	82
175	70	72	74	76	79	83
176	71	72	75	77	79	83
177	71	73	75	77	79	84
178	71	73	75	78	80	84
179	72	74	76	78	80	84
180	72	74	76	79	81	85
181	73	74	77	79	81	85
182	73	75	77	79	82	86
183	74	75	78	80	82	86
184	74	76	78	80	83	87
185	75	76	79	81	83	87
186	75	77	79	81	84	88
187	75	77	80	82	84	88
188	76	78	80	82	85	89
189	76	78	81	83	85	89

(*Continued*)

N	α (one sided)					
	0.005	0.01	0.025	0.05	0.1	0.25
190	77	79	81	83	86	90
191	77	79	81	84	86	90
192	78	79	82	84	87	91
193	78	80	82	85	87	91
194	79	80	83	85	88	92
195	79	81	83	86	88	92
196	80	81	84	86	89	93
197	80	82	84	87	89	93
198	81	82	85	87	90	94
199	81	83	85	87	90	94
200	81	83	86	88	90	95
201	82	84	86	88	91	95
202	82	84	87	89	91	96
203	83	85	87	89	92	96
204	83	85	88	90	92	97
205	84	85	88	90	93	97
206	84	86	88	91	93	98
207	84	86	89	91	94	98
208	85	87	89	92	94	99
209	86	87	90	92	95	99
210	86	88	90	93	95	100
211	86	88	91	93	96	100
212	87	89	91	94	96	101
213	87	89	92	94	97	101
214	88	89	92	94	97	102
215	88	90	93	95	98	102
216	89	91	93	95	98	103
217	89	91	94	96	99	103
218	90	91	94	96	99	104
219	90	92	95	97	100	104
220	90	92	95	97	100	104
221	91	93	95	98	100	105
222	91	93	96	98	101	105
223	92	94	96	99	101	106
224	92	94	97	99	102	106
225	93	95	97	100	102	107
226	93	95	98	100	103	107
227	94	96	98	101	103	108
228	94	96	99	101	104	108
229	95	96	99	102	104	109
230	95	97	100	102	105	109
231	96	97	100	103	105	110
232	96	98	101	103	106	110
233	96	98	101	103	106	111
234	97	99	102	104	107	111

(Continued)

N	α (one sided)					
	0.005	0.01	0.025	0.05	0.1	0.25
235	97	99	102	104	107	112
236	98	100	102	105	108	112
237	98	100	103	105	108	113
238	99	101	103	106	109	113
239	99	101	104	106	109	114
240	100	102	104	107	110	114
241	100	102	105	107	110	115
242	101	103	105	108	111	115
243	101	103	106	108	111	116
244	101	103	106	109	112	116
245	102	104	107	109	112	117
246	102	104	107	110	112	117
247	103	105	108	110	113	118
248	103	105	108	111	113	118
249	104	106	109	111	114	119
250	104	106	109	112	114	119

For upper values of $F = 1 - \alpha$, reject if U is greater than the table entries for $u_{N,1-\alpha}$. There are unexpectedly too many runs.

N	$F = 1 - \alpha$ (one sided)					
	0.75	0.9	0.95	0.975	0.99	0.995
10	7	7	8	8	9	9
11	7	8	9	9	10	10
12	8	9	9	10	10	11
13	8	9	10	10	11	11
14	9	10	10	11	12	12
15	9	10	11	12	12	13
16	10	11	12	12	13	13
17	10	12	12	13	14	14
18	11	12	13	13	14	15
19	11	13	13	14	15	15
20	12	13	14	15	15	16
21	13	14	15	15	16	17
22	13	14	15	16	17	17
23	14	15	16	17	17	18
24	14	16	16	17	18	19
25	15	16	17	18	19	19
26	15	17	18	18	19	20
27	16	17	18	19	20	20
28	16	18	19	20	20	21
29	17	18	19	20	21	22
30	17	19	20	21	22	22

(Continued)

	$F = 1 - \alpha$ (one sided)					
N	0.75	0.9	0.95	0.975	0.99	0.995
31	18	20	20	21	22	23
32	18	20	21	22	23	24
33	19	21	22	22	23	24
34	19	21	22	23	24	25
35	20	22	23	24	25	25
36	21	22	23	24	25	26
37	21	23	24	25	26	27
38	22	23	24	25	27	27
39	22	24	25	26	27	28
40	23	24	26	27	28	28
41	23	25	26	27	28	29
42	24	26	27	28	29	30
43	24	26	27	28	29	30
44	25	27	28	29	30	31
45	25	27	28	29	31	31
46	26	28	29	30	31	32
47	26	28	30	31	32	33
48	27	29	30	31	32	33
49	27	29	31	32	33	34
50	28	30	31	32	34	34
51	28	31	32	33	34	35
52	29	31	32	33	35	36
53	29	32	33	34	35	36
54	30	32	33	35	36	37
55	30	33	34	35	36	37
56	31	33	35	36	37	38
57	32	34	35	36	38	39
58	32	34	36	37	38	39
59	33	35	36	37	39	40
60	33	35	37	38	39	40
61	34	36	37	39	40	41
62	34	37	38	39	41	42
63	35	37	38	40	41	42
64	35	38	39	40	42	43
65	36	38	40	41	42	43
66	36	39	40	41	43	44
67	37	39	41	42	43	44
68	37	40	41	42	44	45
69	38	40	42	43	45	46
70	38	41	42	44	45	46
71	39	41	43	44	46	47
72	39	42	43	45	46	47
73	40	42	44	45	47	48
74	40	43	45	46	47	48
75	41	44	45	46	48	49

(Continued)

	$F = 1 - \alpha$ (one sided)					
N	**0.75**	**0.9**	**0.95**	**0.975**	**0.99**	**0.995**
76	41	44	46	47	49	50
77	42	45	46	48	49	50
78	42	45	47	48	50	51
79	43	46	47	49	50	51
80	43	46	48	49	51	52
81	44	47	48	50	51	52
82	45	47	49	50	52	53
83	45	48	49	51	52	54
84	46	48	50	51	53	54
85	46	49	51	52	54	55
86	47	49	51	53	54	55
87	47	50	52	53	55	56
88	48	50	52	54	55	56
89	48	51	53	54	56	57
90	49	52	53	55	56	58
91	49	52	54	55	57	58
92	50	53	54	56	58	59
93	50	53	55	56	58	59
94	51	54	55	57	59	60
95	51	54	56	57	59	60
96	52	55	56	58	60	61
97	52	55	57	59	60	62
98	53	56	58	59	61	62
99	53	56	58	60	61	63
100	54	57	59	60	62	63
101	54	57	59	61	63	64
102	55	58	60	61	63	64
103	55	58	60	62	64	65
104	56	59	61	62	64	65
105	56	60	61	63	65	66
106	57	60	62	64	65	67
107	57	61	62	64	66	67
108	58	61	63	65	66	68
109	59	62	64	65	67	68
110	59	62	64	66	68	69
111	60	63	65	66	68	69
112	60	63	65	67	69	70
113	61	64	66	67	69	71
114	61	64	66	68	70	71
115	62	65	67	68	70	72
116	62	65	67	69	71	72
117	63	66	68	70	71	73
118	63	66	68	70	72	73
119	64	67	69	71	73	74
120	64	67	69	71	73	74

(*Continued*)

	$F = 1 - \alpha$ (one sided)					
N	**0.75**	**0.9**	**0.95**	**0.975**	**0.99**	**0.995**
121	65	68	70	72	74	75
122	65	69	71	72	74	76
123	66	69	71	73	75	76
124	66	70	72	73	75	77
125	67	70	72	74	76	77
126	67	71	73	74	76	78
127	68	71	73	75	77	78
128	68	72	74	76	78	79
129	69	72	74	76	78	80
130	69	73	75	77	79	80
131	70	73	75	77	79	81
132	70	74	76	78	80	81
133	71	74	76	78	80	82
134	71	75	77	79	81	82
135	72	75	78	79	81	83
136	72	76	78	80	82	83
137	73	76	79	80	83	84
138	73	77	79	81	83	85
139	74	78	80	82	84	85
140	74	78	80	82	84	86
141	75	79	81	83	85	86
142	76	79	81	83	85	87
143	76	80	82	84	86	87
144	77	80	82	84	86	88
145	77	81	83	85	87	88
146	78	81	83	85	87	89
147	78	82	84	86	88	89
148	79	82	84	86	89	90
149	79	83	85	87	89	91
150	80	83	86	87	90	91
151	80	84	86	88	90	92
152	81	84	87	89	91	92
153	81	85	87	89	91	93
154	82	85	88	90	92	93
155	82	86	88	90	92	94
156	83	86	89	91	93	94
157	83	87	89	91	93	95
158	84	88	90	92	94	96
159	84	88	90	92	95	96
160	85	89	91	93	95	97
161	85	89	91	93	96	97
162	86	90	92	94	96	98
163	86	90	92	94	97	98
164	87	91	93	95	97	99
165	87	91	94	96	98	99

(*Continued*)

	$F = 1 - \alpha$ (one sided)					
N	**0.75**	**0.9**	**0.95**	**0.975**	**0.99**	**0.995**
166	88	92	94	96	98	100
167	88	92	95	97	99	101
168	89	93	95	97	100	101
169	89	93	96	98	100	102
170	90	94	96	98	101	102
171	90	94	97	99	101	103
172	91	95	97	99	102	103
173	91	95	98	100	102	104
174	92	96	98	100	103	104
175	92	96	99	101	103	105
176	93	97	99	101	104	105
177	93	98	100	102	104	106
178	94	98	100	103	105	107
179	94	99	101	103	105	107
180	95	99	101	104	106	108
181	96	100	102	104	107	108
182	96	100	103	105	107	109
183	97	101	103	105	108	109
184	97	101	104	106	108	110
185	98	102	104	106	109	111
186	98	102	105	107	109	111
187	99	103	105	107	110	112
188	99	103	106	108	110	112
189	100	104	106	108	111	113
190	100	104	107	109	111	113
191	101	105	107	109	112	114
192	101	105	108	110	113	114
193	102	106	108	111	113	115
194	102	106	109	111	114	115
195	103	107	109	112	114	116
196	103	107	110	112	115	117
197	104	108	110	113	115	117
198	104	108	111	113	116	118
199	105	109	112	114	116	118
200	105	110	112	114	117	119
201	106	110	113	115	117	119
202	106	111	113	115	118	120
203	107	111	114	116	118	120
204	107	112	114	116	119	121
205	108	112	115	117	120	121
206	108	113	115	118	120	122
207	109	113	116	118	121	122
208	109	114	116	119	121	123
209	110	114	117	119	122	124

(*Continued*)

	$F = 1 - \alpha$ (one sided)					
N	0.75	0.9	0.95	0.975	0.99	0.995
210	110	115	117	120	122	124
211	111	115	118	120	123	125
212	111	116	118	121	123	125
213	112	116	119	121	124	126
214	112	117	119	122	124	126
215	113	117	120	122	125	127
216	113	118	121	123	125	127
217	114	118	121	123	126	128
218	114	119	122	124	127	128
219	115	119	122	124	127	129
220	115	120	123	125	128	129
221	116	120	123	126	128	130
222	117	121	124	126	129	131
223	117	122	124	127	129	131
224	118	122	125	127	130	132
225	118	123	125	128	130	132
226	119	123	126	128	131	133
227	119	124	126	129	131	133
228	120	124	127	129	132	134
229	120	125	127	130	133	134
230	121	125	128	130	133	135
231	121	126	128	131	134	135
232	122	126	129	131	134	136
233	122	127	130	132	135	137
234	123	127	130	132	135	137
235	123	128	131	133	136	138
236	124	128	131	134	136	138
237	124	129	132	134	137	139
238	125	129	132	135	137	139
239	125	130	133	135	138	140
240	126	130	133	136	139	140
241	126	131	134	136	139	141
242	127	131	134	137	140	141
243	127	132	135	137	140	142
244	128	132	135	138	141	143
245	128	133	136	138	141	143
246	129	134	136	139	142	144
247	129	134	137	139	142	144
248	130	135	137	140	143	145
249	130	135	138	140	143	145
250	131	136	138	141	144	146

A Monte Carlo approach was used to determine the cumulative distribution of the number of runs.

1. Choose a number of data, N.
2. For 1,000,000 trials, for each of N of data, randomly and independently generate a data dichotomous attribute. I choose +1 and −1.
1. Count the number of runs in that trial and increment a counter for that number of runs.
4. After the 1,000,000 trials interpolate the run number representing the α and $1-\alpha$ limits.
5. The critical run value for the lower limits is rounded down (truncated), and that for the upper limits is rounded up to the next higher run value.

The reasoning behind Step 5: You can only have an integer count of number of runs in your N dataset. If the 95% limits from the Monte Carlo analysis gave a value of 32.416 runs, if you had 32 runs you could not report that it was equal to or greater than the 95% limit. However, if you had 33 runs it would exceed the 95% limit and you could report the 95% confidence.

The results here occasionally differ by a count of ±1 run from the Swed and Eisenhart 1943 results (Tables for Testing Randomness of Grouping in a Sequence of Alternatives by F. S. Swed and C. Eisenhart, The Annals of Mathematical Statistics, Vol. XIV, No. 1, March 1943), and also from the binomial distribution as an approximation, count = BINOMIAL.INV(n,p,α). However, the binomial inverse does permit a run of zero, which is not possible experimentally. And, anyone generating data for a similar critical table would use their own models, number of realizations, and truncation rules. So, an occasional ±1 count difference between tables could be expected.

Table A.4 Critical Values of *d* in the Kolmogorov–Smirnov Goodness-of-Fit Test

N = number of observations.
$F = 1-\alpha$.
D = maximum of the absolute value of the CDF differences between the experimental and hypothetical cumulative distributions.
This is a one-sided test.
If $D \geq d_F$ the difference between the experimental and hypothetical cumulative distributions is improbably large at the $100 \cdot F$% confidence level. Table entries are d_F.

N	$d_{.90}$	$d_{.95}$	$d_{.99}$
5	.54	.58	.657
6	.48	.52	.610
7	.43	.48	.575
8	.41	.46	.538
9	.40	.43	.510
10	.374	.408	.488
12	.338	.373	.447
14	.312	.350	.417
16	.297	.326	.391
18	.278	.308	.370
20	.264	.295	.352
25	.238	.263	.316
30	.217	.241	.290
35	.202	.225	.2686
40	.189	.210	.2520
45	.179	.199	.2377
50	.168	.188	.2259
55	.162	.1797	.2157
60	.1549	.1721	.2066
65	.1490	.1655	.1987
70	.1438	.1597	.1915
80	.1346	.1496	.1794
90	.1270	.1411	.167*
100	.1206	.1339	.161*

*Extrapolated.

Data from the F versus C and N table of Z. W. Birnbaum, Numerical Tabulation of the Distribution of Kolmogorov's Statistic for Finite Sample Size, *American Statistical Association Journal*, Vol. 47, No. 259, 1952, with permission of the American Statistical Association, Alexandria, VA. As indicated by the asterisk, some entries are interpolated.

Some researchers have used linear or reciprocal N interpolation; however, we find that C vs $-\log(1-F)$ is nearly linear in the $F > .8$ range and interpolated with that relationship. The last reported digit in each entry is the digit that changed when $d = C/N$ was interpolated from several points.

Index

R